Practical Rendering and Computation with Direct3D 11

Practical Rendering and Computation with Direct3D 11

Jason Zink
Matt Pettineo
Jack Hoxley

CRC Press
Taylor & Francis Group
Boca Raton London New York

CRC Press is an imprint of the
Taylor & Francis Group, an **informa** business

AN A K PETERS BOOK

CRC Press
Taylor & Francis Group
6000 Broken Sound Parkway NW, Suite 300
Boca Raton, FL 33487-2742

First issued in paperback 2020

© 2012 by Taylor & Francis Group, LLC
CRC Press is an imprint of Taylor & Francis Group, an Informa business

No claim to original U.S. Government works

ISBN-13: 978-1-56881-720-0 (hbk)
ISBN-13: 978-0-367-65925-7 (pbk)

Library of Congress Cataloging-in-Publication Data

Zink, Jason.
  Practical rendering and computation with Direct3D 11 / Jason Zink, Matt Pettineo, Jack Hoxley.
    p. cm.
  "An A.K. Peters book."
  Includes bibliographical references and index.
  ISBN 978-1-56881-720-0 (hardback)
    1. Rendering (Computer graphics) 2. Three-dimensional display systems. 3. Direct3D. I. Pettineo, Matt. II. Hoxley, Jack. III. Title.

  T385.Z567 2011
  006.6'836--dc22   2011010362

Visit the Taylor & Francis Web site at
http://www.taylorandfrancis.com

and the CRC Press Web site at
http://www.crcpress.com

I dedicate this book to my wife, Janice, and to my two daughters, Jayla and Juliet. Thank you for your unending love and support—this book would not have been possible without it!

—Jason Zink

My writing is the culmination of many years of experience and support gained from many people in offline and online communities. I thank you all.

—Jack Hoxley

For my parents.

—Matt Pettineo

Contents

Preface

The visual aspects of your computer experience have been improving at a rapid pace for many years now. In a few short years the range of devices where complex graphics are possible has vastly increased, and hardware typically reserved purely for graphics is now so powerful it is being considered for general-purpose computing.

For the end-user this is great progress, but for the developer it presents difficult challenges. Not only are there a wide range of features but performance is equally wide-ranging—from lightweight laptops through to high-end gaming desktops.

What is needed is a technology platform to allow developers to capture this vast set of possibilities—enter Microsoft's Direct3D 11. Targeting the many features and scaling across the vast performance spectrum in a consistent and intuitive way becomes possible.

As the latest iteration in a long history, Direct3D 11 provides a significant expansion in capabilities over Direct3D 10. New multithreading features allow software developers to scale across current multi-core and be prepared for future many-core CPU architectures. Tessellation allows for adaptive scaling of content across both image resolution and GPU performance, thus targeting processing power where it is most beneficial and simplifying content creation. General purpose computation allows for new approaches and algorithms to fully utilize available processing power and further reduce bottlenecks.

With such a wealth of capabilities, the potential uses for the API have also broadened significantly. D3D11 can be used for many more purposes than its predecessors. At the same time, such a large increase in functionality can at times lead to difficulty in figuring out the best way to accomplish a given task. As is always the case in software development, there can be several ways to implement an algorithm, and each implementation technique provides a different set of tradeoffs that need to be considered. The documentation for Direct3D 11 is useful for low-level details about how each API function should be used, but there is often a significant difference between being able to use an API function at a technical level and being able to fully leverage its potential capabilities. There is typically a lack of information about mapping the higher level concepts of rendering and computation to the actual API and hardware pipelines.

The authors have a long history with Direct3D, and have seen it evolve over its many iterations. They are very active in the Direct3D communities and have been helping newcomers adapt to using this fantastic and powerful technology with each new version of the library. Together, they have many years of practical experience using Direct3D, and understand its concepts from the very lowest levels while still comprehending the overall organization of the technology and how it can be used. This book seeks to provide this missing information in a format that was designed to provide a deep understanding of both the high and low level concepts related to using Direct3D 11.

The authors would like to thank Dan Green of Radioactive Software for providing many of the models used in the sample images throughout the book. They wish Radioactive Software great success with its upcoming game, *Gettysburg—Armored Warfare*.

What To Expect From This Book

This book project was started out of a desire to provide a practical, all around guide for rendering with Direct3D 11 (which is why the title was chosen). The primary focus of the book is to give you an understanding of what Direct3D 11 can be used for at the algorithmic level. We have found that there is significantly more value in understanding what a particular feature can be used for, rather than a precise description of the tool itself. As such, there is less focus on individual low level API functions, and more focus on designing and implementing of algorithms. After reading this book, you will have a clear understanding of what Direct3D 11 is comprised of and have experience in applying it to real-world problems.

Even with this in mind, it is still important to be able to use the API directly. We try to strike a balance between these two desires—to inform the reader about the high level concepts used to design algorithms as well as the nuts and bolts concepts of how to implement them. It is our hope that as you read through this book, that these two competing concepts can be clearly understood as different aspects of working with the API.

Our motivation in writing this book is to provide the material that we, as veterans in the field, would want to read, and we hope that you find it both informative and interesting at the same time.

As a further aid to understanding how to perform the low level Direct3D 11 operations, all of the source code from the book is provided for your reference and use in an open source rendering framework. The samples are built on a rendering framework named

Hieroglyph 3. This library has been created to provide examples of how to use the low level API, as well as to provide a framework from which to build the sample programs. It is actively maintained and updated, and both the sample applications and the framework itself can be downloaded at http://hieroglyph3.codeplex.com, so please be sure to check back in from time to time for the latest version. Details about how to get and use the source code as well as some information about how to extend the library to your own applications are provided in Appendix A.

How This Book Is Organized

This book is divided into two halves, each with its own style and content. We recommend you read them in order, but it is possible to focus on specific algorithms from the second half and refer to the first half for the supporting theory.

The first half of the book provides an overview of Direct3D 11, followed by a detailed look at each of the major components of the library. This portion of the book deals with the conceptual details of the Direct3D 11 rendering pipeline and how it maps to the underlying hardware, and also covers the actual programming API used for interfacing with the pipeline.

The second half of the book is intended to provide detailed examples of how to use Direct3D 11 in common rendering scenarios. The knowledge gained from the first portion of the book is used to design and implement various rendering algorithms in the second half. Each topic is presented within its own chapter, where several related algorithms are discussed in detail.

Our hope with these split styles of content is bridge the two aspects of learning to use Direct3D 11—learning the design of the API, and then learning how to use it in a practical setting. We think this is a good approach to covering such an expansive and complex topic in a clear and informative manner.

Part I: Direct3D 11 Foundations

As mentioned above, the first half of the book is devoted to introducing and familiarizing you with Direct3D 11. This content is spread over seven chapters. The following list provides a brief description of what you can expect from each of these chapters.

Chapter 1: Overview of Direct3D 11
Our first chapter provides an introduction to Direct3D 11 and describes the overall structure of the library. This includes a look at the major functional portions of the API, as well as a first look at how an application interacts with it.

Chapter 2: Direct3D 11 Resources

Chapter 2 introduces the many variations of memory-based resources in Direct3D 11. Resources are a critically important concept to understand before learning about the various pipeline stages since they are used to provide both input and output to many operations. Details about each type of resource, how to create them, along with where and when to use them are discussed.

Chapter 3: The Rendering Pipeline

Next we cover the primary concept of Direct3D 11—the rendering pipeline. This chapter goes into great detail about how the rendering pipeline, including discussions about its general structure and a detailed description of each pipeline stage. We cover the mechanics of each stage, their general uses, and how they fit into the overall rendering paradigm.

Chapter 4: The Tessellation Pipeline

Direct3D 11 introduces hardware tessellation as a standard feature of the API for the first time in Direct3D's history. Here we spend a complete chapter examining the details of this new technology. While chapter 3 describes the individual stages and how they work together, this chapter digs deeper into how the tessellation system functions and what the developer can expect from it.

Chapter 5: The Computation Pipeline

In a similar format, we also devote a complete chapter to the new general purpose computation stage—the Compute Shader stage. This new stage is intended to allow the developer to harness the massively parallel nature of the Graphics Processing Unit (GPU) for tasks other than rendering (or more correctly, in addition to rendering). We introduce this interesting new processing concept and detail how it's threading architecture works in addition to describing the various memory systems that are available for use.

Chapter 6: High Level Shading Language

Next we dive into the Direct3D-specific language used to write the programs that are executed in the programmable stages of the pipeline. The majority of rendering and computation with Direct3D is expressed in HLSL such that a solid understanding is of particular importance. This chapter describes HLSL, and looks at the specific syntax, objects, and functions that are available to the developer for efficiently leverage the underlying hardware.

Chapter 7: Multithreaded Rendering

The final major functional addition to Direct3D 11 is its ability to natively support multithreaded use. This can lead to significant performance improvements when executed on multicore CPUs. We provide a complete description of the functionality, and also inspect several scenarios that can take advantage of these new multithreaded features.

Part II: Using Direct3D 11

In the second half of the book, you will find that several related topics are grouped together by a common theme into its own chapter. Each chapter topic builds on its own content, with the intent to provide a better understanding of both the overall concept and how D3D11 can be used in various real world situations. In this manner, a wide variety of topics are covered which demonstrate the versatility of the rendering and computation pipelines. Each of these sample-based chapters begins with a general description of the algorithm itself, which is followed by an implementation section, where the algorithm is mapped onto the D3D11 feature set. The chapter is concluded with a discussion of the main characteristics of the presented implementations as they apply to D3D11. The following list provides a brief description of each of the chapters in the second half of the book.

Chapter 8: Mesh Rendering

Our first sample chapter provides a good starting point to understand how the rendering pipeline is used to render triangle meshes. We cover the basic mathematic concepts needed for rendering before considering three different techniques for representing an object. Each of these techniques provide an increasingly complex representation of an object, and include static meshes, vertex skinning, and then extending vertex skinning with displacement mapping.

Chapter 9: Dynamic Tessellation

Chapter 4 introduced tessellation in a general form and seeded the imagination for the exciting new possibilities it offers. This chapter applies this theory to two practical examples, firstly for an algorithmic approach (Terrain Rendering) and secondly in a mathematical style (Curved/Smooth Surfaces). This deep-dive into tessellation offers implementations that can be readily used in a wide number of graphical applications with minimal changes as well as demonstrating how existing algorithms can be moved from conventional CPU-based processing to more modern GPU architectures.

Chapter 10: Image Processing

The general purpose processing capabilities of the Compute Shader are utilized to implement two common image filters—the Gaussian filter and the bilateral filter. Several implementations of each algorithm are described for each filter, which use the various capabilities of the Compute Shader to achieve different performance characteristics.

Chapter 11: Deferred Rendering

This chapter focuses on deferred rendering, which is an increasingly-popular technique for rendering scenes with large amounts of dynamic light sources. It begins by explaining the basic premise and implementation of two common approaches to deferred rendering, which are "Classic" Deferred Rendering and Light Pre-Pass Deferred Rendering respectively. The following sections explain how the power and flexibility of Direct3D 11 can be used to optimize the performance of both techniques, and also work around some of the common pitfalls and limitations.

Chapter 12: Simulations

Chapter 12 describes how to utilize the Compute Shader to perform a physical simulation, which is subsequently rendered with the rendering pipeline. This combination is demonstrated with a fluid simulation technique, as well as a GPU based particle system. Together these two algorithms provide a general concept of how to process other simulations with Direct3D 11.

Chapter 13: Multithreaded Paraboloid Rendering

Our final chapter utilizes a special form of environment mapping, called dual paraboloid environment mapping, to demonstrate the benefits of multithreaded rendering. The dual paraboloid technique is used to render reflective objects. When multiple reflectors are taken together, there are several levels of reflection displayed at the surface of each object. The generation and display of these reflections are used to stress the CPU processing capabilities, and shows a clear improvement when multithreaded rendering is utilized.

1

Overview of Direct3D 11

The Direct3D 11 rendering API has extended Direct3D in several key areas, while retaining much of the API structure that was introduced with Direct3D 10. This chapter will provide an introduction to Direct3D 11 and will discuss some of these changes to the API. It begins with a brief description of how Direct3D is organized and how it interacts with a GPU installed in your computer. This will provide an idea of what the various software components involved in using Direct3D must do to perform some rendering.

Next is a discussion of the rendering pipeline, with a brief overview of each of the pipeline stages. Then we look at the new additions to the API that allow the GPU to be used for general computation, in addition to its normal rendering duties. Once we have constructed a clear overview of how the pipeline is structured, we will look at all of the different types of resources that can be bound to the pipeline, and the different places where they can be bound. After we understand the pipeline and how it interacts with resources, we will look at the two primary interfaces that are used to work with Direct3D 11—namely, the Device and the Device Context. These two interfaces are the bread and butter of the API and hence warrant a discussion of what they are used for. The discussion of these main interfaces then leads into a walk-through of a basic rendering application that uses Direct3D 11.

This chapter provides an overview of the topics mentioned above. It is intended to introduce a conceptual overview of Direct3D 11, and then to provide a working knowledge of how an application can use it. The first half of this chapter uses few or no actual code samples, while the second half provides a more hands-on introduction to the API. After reading this chapter, you should have a basic understanding of the architecture of Direct3D 11, and of what an application needs to do to manipulate the contents of a window. Each of the concepts discussed in the overview chapter is explored in great detail in the following chapters.

1.1 Direct3D Framework

Direct3D 11 is a native application programming interface (API) which is used to communicate with and control the video hardware of a computer to create a rendered image. The term *native* indicates that the API is designed to be accessed through C/C++, as opposed to by a managed language. So, the most direct way to use the API is from within a C/C++ application. Although there are a number of different native windowing frameworks, we will be working with the Win32 API, simply to reduce the number of additional software technologies needed to get Direct3D 11 up and running.

Even though they are transparent to the application, there are actually a number of different software layers that reside below Direct3D, which together make up the overall graphics architecture used in the Windows client environment. In this section we will discuss how the application and Direct3D fit into this overall architecture and will explore what the other components are used for, to provide a well-rounded idea of what we are actually doing when working with the API.

1.1.1 Graphics Architecture

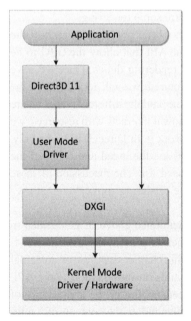

Figure 1.1. The various components of the graphics subsystems used with Direct3D

The general graphics architecture used in Windows is shown in Figure 1.1. In this diagram, we can see that the application sits at the highest level, and primarily interacts with Direct3D. In the next layer down, the Direct3D runtime interacts with the user-mode driver of the video hardware to execute the commands generated by the application. The driver then interacts with the DXGI framework, which is responsible for performing the low level communication with the kernel-mode driver, and for managing the available hardware resources. With such an array of different layers and interfaces, it can be somewhat complex to understand what each of them is responsible for. However, we will briefly walk through each of the layers to gain an insight into what operations each of them perform..

The application is ultimately in control of the content that will end up being presented to the user in her application window. Any two-dimensional or three-dimensional objects, images, text, animations, or other high level content are provided by the application; they are usually specialized to the particular needs of each application.

Direct3D provides a set of API functions that can be used to convert the high-level content of the application into a format that can be interpreted by the user-mode driver. The content required by the application must be formatted according to the rules of Direct3D. This includes the actual data formatting, as well as the required sequencing of function calls. As long as the application follows the required Direct3D usage, it can be sure that it is operating within the graphics architecture.

Once Direct3D has received a series of function calls from the application, it interacts with the user-mode driver to produce a list of commands that are executed by the GPU. The results of these actions are relayed to the DXGI interface to manipulate the contents of hardware resources. DXGI handles the lower-level hardware interfaces, such as providing an "adapter" interface for each video card that is installed in a computer. It is DXGI that also provides access to what are called *swap chains*, which encapsulate the actual buffers that can be used to present the rendered contents to a window.

Although this is not shown in Figure 1.1, it is quite likely that multiple applications will simultaneously be using the Direct3D runtime. This is one of the benefits of the Windows display driver model (WDDM), that the video resources are virtualized and shared among all applications. Applications do not have to worry about gaining access to the GPU, since it is available to all applications simultaneously. Of course, when a user is performing a very graphically intensive operation, like playing a high end game, there are likely to be fewer applications using the GPU, and hence more of its computational power can be directed to the game. This is a departure from the Windows XP driver model (XPDM) which did not explicitly virtualize access to the GPU.

1.1.2 Benefits of Abstraction

Since an application interfaces with the Direct3D API instead of directly to a driver, it only needs to learn the rules of interfacing with one API to produce the desired output instead of trying to adapt to all of the various devices that a user might have installed. This shifts the burden of creating a uniform rendering system with which to interact to the video card manufacturers—they must adhere to the Direct3D standard, as opposed to the application trying to implement its own "standard" rendering system.

The video system of modern computers typically contains specialized hardware that is designed to perform graphics operations. However, it is possible that an entry-level computer doesn't have Direct3D-capable hardware installed, or that it only has a subset of the required functionality for a particular application. Fortunately, Direct3D 11 provides a software driver implementation that can be used in the absence of Direct3D-capable hardware (up to the Direct3D 10.1 feature level). Whether such hardware is present or not is conveniently abstracted from the developer by Direct3D.

Both of these situations greatly benefit the developer. As long as an application can properly interact with Direct3D 11, it will have access to a standardized rendering system, as well as support for whatever devices are present in the system, regardless of if they are hardware- or software-based devices.

1.2 Pipeline

Arguably, the most important concept required for understanding Direct3D 11 is a clear understanding of its pipeline configuration. The Direct3D 11 specification defines a conceptual processing pipeline to perform rendering computations, and ultimately, to produce a rendered image. The term *pipeline* is used since data "flows" into the pipeline, is processed in some way at each stage throughout the pipeline, and then "flows" out of the pipeline. Over the past few iterations of the API, several changes have been made to the pipeline to both simplify it and add additional functionality to it. The pipeline concept itself is composed of a number of stages, which each perform distinctly different operations. Data passed into each stage is processed in some fashion before being passed on to the next stage. Each individual stage provides a unique configurable set of parameters that can be controlled by the developer to customize what processing is performed. Some of the stages are referred to as *fixed-function* stages, which offer a limited customization capability. These stages typically expose a state object, which can be bound with the desired new configuration as needed.

The other stages are referred to as *programmable* stages, which provide a significantly broader configuration capacity by allowing custom programs to be executed within them. These programs are typically referred to as *shader* programs, and the stages are referred to as *programmable shader stages*. Instead of performing a fixed set of operations on the data that is passed into it, these programmable shader stages can implement a wide range of different functions.

It is these shader programs that provide the most flexibility for implementing custom rendering and computational algorithms. Each programmable shader stage provides a common feature set, called the *common shader core*, to provide some level of uniformity in its abilities. These features can be thought of as a toolbox that is available to all of the programmable shader stages. Operations like texture sampling, arithmetic instruction sets, and so on are all common among all of the programmable stages. However, each stage also provides its own unique functionality, either in its instruction set or in its input/output semantics, to allow distinct types of operations to be performed.

From a high level, there are currently two general pipeline paradigms—one which is used for rendering, and one which is used for computation. Strictly speaking, the distinction between these two configurations is somewhat loose, since they can both be used for the other purpose. However, the distinction exists, and there is a clear difference in the two pipelines' capabilities. We will discuss both pipelines, beginning with the rendering pipeline.

1.2.1 Rendering Pipeline

The *rendering pipeline* is the origin from which the modern GPU has grown. The initial graphics accelerators provided hardware vertex transformation to speed up 3D applications. From there, each new generation of hardware provided additional capabilities to perform ever more complex rendering. Today, we have quite a complex pipeline, with significant flexibility to perform nearly any algorithm in hardware. Books are published regularly describing new techniques for using the latest hardware such as (Engel, 2009) and(Nguyen, 2008). The level of advances in real-time rendering has really been quite staggering over the last 5-10 years.

The rendering pipeline is intended to take a set of 3D object descriptions and convert it into an image format that is suitable for presentation in the output window of an application. Before diving into the details of each of the individual stages, let's take a closer look at the complete Direct3D 11 rendering pipeline. Figure 1.2 shows a block diagram of the pipeline, which we will see many times throughout the remainder of the book, in several different formats.

Here, the two different types of pipeline stages, fixed-function and programmable, are depicted with different colors. The fixed-function stages are shown with a green background, and the programmable stages with a blue background. Each stage defines its own required input data format and also defines the output data format that it will produce when executed. We will step through the pipeline and briefly discuss each stage, and its intended purpose.

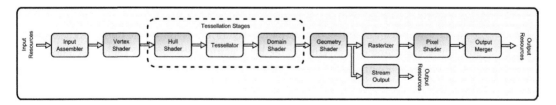

Figure 1.2. The complete Direct3D 11 rendering pipeline.

The entry point to the rendering pipeline is the input assembler stage. This stage is responsible for reading input data from resources and then "assembling" vertices for use later in the pipeline. This allows for multiple vertex buffers to be used and also provides the ability to use *instanced rendering* (described in more detail in Chapter 3). In addition, the connectivity of the vertices is also determined, based on the input resources and the desired rendering configuration. The assembled vertices and primitive connectivity information are then passed down the pipeline.

The vertex shader stage reads the assembled vertex data from the input assembler stage and processes a single vertex at a time. This is the first programmable stage in the pipeline, and it applies the current vertex shader program to each input vertex. It cannot create or destroy vertices; it can only process the vertices that are given to it. In addition, every vertex is processed in isolation—the information from one vertex shader invocation is never accessible in another invocation. The primary responsibility of the vertex shader used to be to project the vertex positions into clip space, but the addition of the tessellation stages (discussed next) has somewhat changed this. In general, any operation that must be performed on every vertex of the input model should be performed in the vertex shader.

The next three stages are recent additions to the pipeline to accommodate hardware tessellation, and they must all be used together. The hull shader stage receives primitives from the vertex shader and is responsible for two different actions. First, the hull shader provides a function that is only run once per primitive to determine a set of tessellation factors. These factors are used by the tessellator stage to know how finely to tessellate or split up the current primitive. The second action that the hull shader stage must perform is executed once for each control point in the desired output control patch configuration. In essence, it must create the control points that will be later be used by the domain shader stage to create the actual tessellated vertices that will eventually be used in rendering.

When the tessellation stage receives its data from the hull shader, it uses one of several algorithms to determine an appropriate sampling pattern for the current primitive type. Depending on the tessellation factors (from the hull shader), as well as its own configuration (which is actually specified in the hull shader as well), it will determine which points in the current primitive need to be sampled in order to tessellate the input primitive into smaller parts. The output of the tessellation stage is actually a set of barycentric coordinates that are passed along to the domain shader.

The domain shader takes these barycentric coordinates, in addition to the control points produced by the hull shader, and uses them to create new vertices. It can use the complete list of control points generated for the current primitive, textures, procedural algorithms, or anything else, to convert the barycentric "location" for each tessellated point into the output geometry that is passed on to the next stage in the pipeline. This flexibility in deciding how to generate the resulting geometry from the tessellator stages' amplified output provides significant freedom to implement many different types of tessellation algorithms.

The geometry shader stage resides in the next pipeline location. As its name implies, the geometry shader operates on a geometric level. In practice, this means that it operates on complete geometric primitives, and also produces geometric primitives. This stage can both add and remove data elements from the pipeline, which allows for some interesting, non-traditional uses. In addition, it can take one type of geometry as input and generate a different type of geometry as output. This allows the conversion of single vertices into complete triangles, or even multiple triangles. The geometry shader is also the conduit through which processed geometry can be streamed out of the pipeline into a buffer resource. This is accomplished in the stream output stage.

After the geometry is sent out of the geometry shader, it has completed the portion of the pipeline that operates at the geometric level. From this point on, the geometry is rasterized and dealt with at the fragment level. A *fragment* is essentially a group of data that corresponds to a pixel in a render target, and that can potentially be used to update the current value of that pixel if it makes its way through the rest of the pipeline. The generation of fragment-level data begins with the fixed-function rasterizer stage. The rasterizer produces fragments from the geometric data passed to it, by determining which pixels of a render target are covered by the geometry. Each fragment receives interpolated versions of all of the per-vertex attributes, to provide the information needed for further processing later in the pipeline. In addition, the rasterizer produces a depth value for each fragment, which will later be used for visibility testing in the output merger stage.

Once a fragment has been generated, the pixel shader is invoked to process it. The pixel shader is required to generate a color value output for each of the render target outputs bound to the pipeline. To accomplish this, it may sample textures or perform computations on the incoming fragment attributes. In addition, it can also override the depth value produced by the rasterizer stage, to allow for specialized algorithms to be implemented in the pixel shader stage.

After the pixel shader has finished its work, its output is passed to the output merger stage. It must correctly "merge" the pixel shader output with the bound depth/stencil and render target resources. This includes performing depth and stencil tests, performing the blending function, and finally, performing the actual writing of the output to the appropriate resources.

This has been a brief high-level overview of each of the pipeline stages, but there are many more details to be considered when using each of them. We will dive much deeper into all of the available configurations and capabilities of each stage in the Chapter 3, "The Rendering Pipeline." Until we reach that point, keep this overview of the pipeline in mind as we continue our overview of the API.

1.2.2 Computation Pipeline

With all of the other new pipeline stages that came along with Direct3D 11, there is one additional stage that has not yet been discussed—the compute shader stage. This stage is intended to perform computation outside of the traditional rendering paradigm, and therefore is considered to execute separately from the traditional rendering pipeline. In this way, it implements a single-stage pipeline devoted to general purpose computation. The general trend of using the GPU for purposes other than rendering has been incorporated directly into the Direct3D 11 API, and is manifested in the compute shader.

Several features have been provided to the compute shader that facilitate a flexible processing environment for implementing more general algorithms. The first new functionality is the addition of a structured threading model that gives the developer significant freedom in using the available parallelism of the GPU to implement highly parallel algorithms. Previously, the invocation of a shader program was restricted to how an input was processed by a particular stage (for example, the pixel shader program was invoked once for each fragment generated), and the developer didn't have direct control over how the threads were used. This is no longer the case with the compute shader stage.

The second new functionality that makes the computer shader stage more flexible is the ability to declare and use a "Group Shared Memory" block. This memory block is then accessible to all of the threads within a thread group. This allows communication between threads during execution, which has typically not been possible with the rendering pipeline. With communication possible, there is a potential for significant efficiency improvements by sharing loaded data or intermediate calculations.

Finally, the concept of random read *and* write access to resources has been introduced (which we will discuss further in the "Resources" section of this chapter). This represents a significant change from the shader stages we have already discussed, which only allow a resource to be bound as either an input or as an output. The ability to read and write to a complete resource provides another avenue for communication between threads.

These three new concepts, when taken together, introduce a very flexible general-purpose processing stage that can be used for both rendering calculations, as well as for general purpose computations. We will take a much closer look at the compute shader in Chapter 5, "The Computation Pipeline," and there are several sample algorithms throughout the second half of the book that make extensive use of the compute shader.

1.3 Resources

The rendering and computation pipelines that we discussed in the previous section are two of the major concepts that one must understand when learning about the Direct3D 11 API. The next major concept that we need to understand is the resources that are connected at various locations throughout the pipeline. If you consider the rendering pipeline to be an automobile assembly line, the resources would be all of the components that come into the line that are put together to create the automobile. When used without resources, the pipeline can't do anything, and vice versa.

There is a wide diversity of different resources at our disposal when considering how to structure an algorithm, and each type has different use cases. In general, resources are split into two groups—textures and buffers. *Textures* are roughly split by their dimension, and can be created to have one-dimensional, two-dimensional, and three-dimensional forms. *Buffers* are somewhat more uniform and are always considered one-dimensional (although in some cases they are actually 0-dimensional, such as a single data point). Even so, there is a good variety of different buffer types, including vertex and index buffers, constant buffers, structured buffers, append and consume buffers, and byte address buffers.

Each of these resource types (both textures and buffers) provides a different usage pattern, and we will explore all of their variants in more detail in Chapter 3, "The Rendering Pipeline." However, there are some common overall concepts regarding the use of resources with the pipeline which we can discuss now. In general, there are two different domains for using resources—one is from C/C++, and the other is from HLSL. C/C++ is primarily concerned with creating resources and binding/unbinding them to the desired locations. HLSL is more concerned with actually manipulating and using the contents of the resources. We will see this distinction throughout the book.

A resource must be bound to the pipeline in order to be used. When a resource is bound to the pipeline, it can be used as a data input, an output, or both. This primarily depends on where and how it is bound to the pipeline. For example, a vertex buffer bound as an input to the input assembler is clearly an input, and a 2D texture bound to the output merger stage as a render target is clearly an output. However, what if we wanted to use the contents of the output render target in a subsequent rendering pass? It could also be bound to a pixel shader as a texture resource to be sampled during the next rendering operation.

To help the runtime (and the developer!) determine the intended use of a resource, the API provides the concept of *resource views*. When a particular resource can be bound to the pipeline in several different types of locations in the pipeline, it must be bound with a resource view that specifies how it will be used. There are four different types of resource views: a *render target view*, a *depth stencil view*, a *shader resource view*, and an *unordered access view*. The first two are used for binding render and depth targets to the pipeline, respectively. The shader resource view is somewhat different, and allows for resources to be bound for reading in any of the programmable shader stages. The unordered access view is again somewhat different and allows a resource to be bound for simultaneous random read and write access, but is only available for use in the compute shader and pixel shader stages. There are still other resource binding types that don't require a resource view for binding to the pipeline. These are typically for resources that are created for a single purpose, such as vertex buffers, index buffers, or constant buffers, and the resource usage is not ambiguous.

With so many options and configuration possibilities, it can seem somewhat daunting when you are first learning about what each resource type can do and what it can be used for. However, as you will see later in the book, this configurability of the resources provides significant freedom and power for designing and implementing new algorithms.

1.4 Interfacing with Direct3D

With a high level overview of Direct3D 11 complete, we now turn our attention to how an application can interface with and use the API. This begins with a look into the two primary interfaces used by the application—the device and the device context. After this, we will walk through a simple application to see what it must do in order to use Direct3D 11.

1.4.1 Devices

The primary interfaces that must be understood when using Direct3D 11 are the *device* and the *device context*. These two interfaces split the overall responsibility for managing the functionality available in the Direct3D 11 API. When the resources mentioned above are being created and interfaced with, the device interface is used. When the pipeline or a resource is being manipulated, the device context is used.

The *ID3D11Device* interface provides many methods for creating shader program objects, resources, state objects, and query objects (among others). It also provides methods for checking the availability of some hardware features, along with many diagnostic- and debugging-related methods. In general, the device can be thought of as the provider of the various resources that will be used in an application. We will see later in this chapter how to initialize and configure a device before using it. Due to the size of this interface and the frequency with which it is used, we will discuss the device methods when their particular subject area is relevant. For a direct list of the available methods, the DXSDK documentation provides a complete linked list for every method.

In addition to serving as the resource provider for Direct3D 11, the device also encapsulates the concept of *feature levels*. The use of feature levels is a technique for allowing an application to use the Direct3D 11 API and runtime on hardware that implements an older set of functionality, such as the Direct3D 10 GPUs. This allows applications to utilize a single rendering API but still target several generations of hardware. We will see later in this chapter how to use this great feature of the API.

1.4.2 Contexts

While the device is used to *create* the various resources used by the pipeline, to actually *use* those resources and manipulate the pipeline itself, we will use a *device context*. Device contexts are used to bind the created resources, shader objects, and state objects to the pipeline. They are also used to control the execution of rendering and computation pipeline invocations. In addition, they provide methods for manipulating the resources created by the device. In general, the device context can be thought of as the consumer of the resources produced by the device, which serves as the interface for working with the pipeline.

The device context is implemented in the *ID3D11DeviceContext* interface. To help in the introduction of multithreaded rendering support, two different flavors of contexts are provided. The first is referred to as an *immediate context*, and the second is called a *deferred context*. While both of these types implement the same device context interface, their usage has very different semantics. We will discuss the concepts behind these two contexts below.

Immediate Contexts

The immediate context is more or less a direct link to the pipeline. When a method is called from this context, it is submitted immediately by the Direct3D 11 runtime for execution in the driver. Only a single immediate context is allowed, and it is created at the same time that the device is created. This context can be seen as the interface for directly interacting with all of the components of the pipeline. This context must be used as the main rendering thread, and serves as the primary interface to the GPU.

Deferred Contexts

The deferred context is a secondary type of context that provides a thread-safe mechanism for recording a series of commands from secondary threads other than the main rendering thread. This is used to produce a command list object which can be "played" back on the immediate context at a later time. Allowing the command list to be generated from other threads provides some potential that a performance improvement could be found on multi-core CPU systems. In addition, Direct3D 11 also allows asynchronous resource creation to be carried out on multiple threads, which provides the ability to simplify multithreaded loading situations. These are important concepts when considering the current and future PC hardware environment, in which we can expect more and more CPU cores to be available in a given PC.

The topic of multithreaded rendering is discussed in more detail in Chapter 7, "Multithreaded Rendering." The topic is itself worthy of a complete book, but in the interests of providing a well-rounded discussion of Direct3D 11, we have attempted to provide design-level information about how and when to use multithreading, rather than trying to explain all of the details of multithreaded programming in general.

1.5 Getting Started

This portion of the chapter is devoted to giving the reader a crash course in how to get a basic Direct3D 11 application running. The description provided here introduces a number of basic concepts used to interact with Direct3D 11 and introduces the basic program flow used in a generic application. This is intended to be a basic introduction with a focus on what actions are needed for using Direct3D 11, and hence it will not explain all of the details of win32 programming. At the end of this section, we will discuss the application framework that will be used for the sample applications provided with this book, which abstracts most of the direct win32 interactions away from the developer.

1.5.1 Interacting with Direct3D 11

Before getting to the actual application details, we first need to cover a few basic Direct3D 11 concepts. This includes the COM based framework in which Direct3D resides, as well as understanding how to call, and evaluate the success of the various API functions that an application must use.

COM-Based Interfaces

Direct3D 11 is implemented as a series of component object model (COM) interfaces in the same way that the previous versions of Direct3D have been. If you are not familiar with COM, it may seem like an unnecessary complication to the API. However, once you understand some of the design implications of using COM, it becomes clear why it has been chosen for implementing Direct3D 11. We won't dig too deeply into COM, but we will cover a few of the most visible details that an application will likely use. COM provides an object model that defines a basic set of interface methods that all COM objects must implement. This includes methods for performing reference counting of the COM objects, as well as methods that allow an object to be inspected at runtime to discover which interfaces it implements.

Reference counting. An application will primarily be interested in the reference counting behavior of the COM objects, since it is used to manage their lifetime. When a Direct3D-based COM object is created with by a function, a reference to that object is returned, typically in a pointer provided by the application, as an argument to the function call. Instead of directly creating an instance of a class with the "new" operator, the objects are always created through a function or an interface method and returned in this way.

When the object reference is returned, it will usually have a reference count of 1. When the application is finished using the object, it will call the object's Release method, which indicates that the object's reference count should be decremented. This is performed instead of directly freeing the object with the "delete" operator. After calling the Release method, the application should erase the pointer reference and treat it as if it were directly deleted. As long as the application accurately releases all of its references to the objects that it creates, the COM framework will manage the creation and destruction of the objects.

This usage allows both the application and the Direct3D 11 runtime to share objects with multiple references. When a resource is bound to the pipeline as a render target, it is referenced by the runtime to ensure that the object is not deleted while it is being used. Similarly, if both the application and the runtime have a reference to an object, and a method call eliminates the runtime reference, the application can be sure that the object will not be destroyed until it releases its reference, as well. This reference-counting object management holds true for any Direct3D 11 object that inherits from the IUnknown interface, and it must be followed to ensure that no objects are left unreleased after an application is terminated.

Interface querying. As mentioned above, the IUnknown interface also provides some methods for querying an object to find additional interfaces that it implements. This is performed with the *IUnknown::QueryInterface()* method, which takes a universally unique identifier (UUID) that identifies the desired interface, and then a pointer to a pointer, which will be filled with an appropriate object reference if the object implements the requested interface. If such a reference is successfully returned in the method call, the reference count of the object is increased and should be managed in the same way as any other object.

In Direct3D, this mechanism is used primarily in conjunction with the ID3D11Device interface. There are several different situations in which additional device interfaces are requested, such as acquiring the debug interface of the device or acquiring the DXGI device interface from the Direct3D device. We will occasionally see this method in action in this text, and the process will be noted as it arises.

Some other properties of COM objects are not used as frequently with Direct3D 11, such as the language independence of the binary interface, or the fact that objects can be instantiated and used from a remote computer. These uses are beyond the scope of this book, but are still interesting to know about. Further information about COM can be found on the MSDN website, and there are also many additional resources available online, due to the longevity of the technology.

Interpreting API Call Results

Direct3D 11 also implements a standard way of handling return codes from functions and methods. In general, the success or failure of a method is indicated by a return code, which is provided to the application as the returned value of the function call. This return code is implemented as an HRESULT value, which is really just a long integer variable.

The list of possible return values can be found in the DXSDK documentation, and additional return codes may added between releases. In some situations, these return codes do not clearly indicate the success or failure of a method, and often the interpretation of the codes depends on the context in which it is returned. To simplify testing of these return codes, a simple macro can be used to test if a method has failed or not. Listing 1.1 shows how the FAILED macro is used.

```
HRESULT hr = m_pDevice->CreateBlendState( &Config, &pState );

if ( FAILED( hr ) )
{
    Log::Get().Write( L"Failed to create blend state!" );
    return( -1 );
}
```

Listing 1.1. A sample usage of the FAILED macro.

In general, any case in which this macro returns a true value should be treated as an error and handled accordingly. If the macro returns a false value, the method has succeeded and should be considered not to have produced an error. This makes testing the result of a method invocation much simpler and can be handled in a relatively simple way.

1.5.2 Application Requirements

With the basics of COM and HRESULT values behind us, we can move on to examine what exactly a win32 application must do to use Direct3D 11. We will begin with an overview of the complete responsibilities of the application, which is then followed by a more detailed look at each of the processes that it must implement.

Program Flow Overview

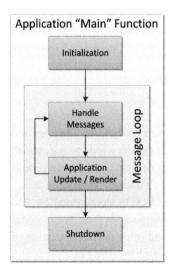

Figure 1.3. The standard operations performed in a win32 application.

The Direct3D 11 application lifecycle follows quite a similar sequence to most other win32 applications. The block diagram for this standard sequence is shown in Figure 1.3.

From Figure 1.3, we can see that the application begins with a one-time initialization action. This is used to obtain references to the device and device context, as well as to create any resources that will be used during the runtime of the application. The resources will vary widely from application to application, while the device and device context creation are normally the same across multiple applications.

After the application has been initialized, it enters the *message loop*. This loop consists of two phases. First, all pending Windows messages are handled with the application's message handler. After all available messages have been processed, the actual application workload can be performed. For our rendering application, this consists of clearing the render targets, then performing the desired rendering operations, and finally presenting the results to the window's client area. Each time these two processes are repeated represents one frame rendering. This will continue until the user requests the termination of the application.

After the application exits the message loop, it performs a one-time shutdown routine. In this routine, all of the resources that have been created are released. After they have been released, the device context, and the device itself, are released as well. At this point, as long as all of the reference counts of the Direct3D 11 objects have been properly managed, the application will have completely released all of its references to all objects. After the shutdown process has completed, the application will end by exiting its main function.

Now that we have seen each of the steps that are performed, we can take a closer look at how each of these steps involves Direct3D 11 and see what the actual code looks like for using the API.

Application Initialization

The application initialization procedure begins with the application creating the win32 window that will present the results of our rendering operations. Basic win32 code for creating a window has been presented many, many times in numerous books and tutorials and won't be repeated here. However, the sample program rendering framework also includes this type of code, and a quick reference to this code within the source distribution can be found in the class files for the Win32RenderWindow class. For the purposes of this discussion, we will assume that the window has been properly registered and created, and that a window handle is available for it.

Device and context creation. After the window has been created, the next step is to obtain a reference to the device. This process is performed with either the **D3D11CreateDevice()** or the **D3D11CreateDeviceAndSwapChain()** functions. For this example, we will use the latter function simply because it creates a swap chain from within the same function, in addition to creating the device and device context.

The prototype for the function is shown in Listing 1.2. We will step through the parameters to this function to gain an understanding of what is needed to create a device.

```
HRESULT  D3D11CreateDeviceAndSwapChain(
        IDXGIAdapter *pAdapter,
        D3D_DRIVER_TYPE DriverType,
        HMODULE Software,
        UINT Flags,
        const D3D_FEATURE_LEVEL *pFeatureLevels,
        UINT FeatureLevels,
        UINT SDKVersion,
```

```
           const DXGI_SWAP_CHAIN_DESC *pSwapChainDesc,
           IDXGISwapChain **ppSwapChain,
           ID3D11Device  **ppDevice,
           D3D_FEATURE_LEVEL *pFeatureLevel,
           ID3D11DeviceContext **ppImmediateContext
     );
```

Listing 1.2. The D3D11CreateDeviceAndSwapChain function prototype.

The first parameter, pAdapter, is a pointer to the graphics adapter that the device should be created for. This parameter can be passed as NULL to use the default adapter. The second parameter is vitally important to properly configure. The DriverType parameter specifies which type of driver will be used for the device created with this function call. The available options are shown in Listing 1.3.


```
  enum D3D_DRIVER_TYPE {
      D3D_DRIVER_TYPE_UNKNOWN,
      D3D_DRIVER_TYPE_HARDWARE,
      D3D_DRIVER_TYPE_REFERENCE,
      D3D_DRIVER_TYPE_NULL,
      D3D_DRIVER_TYPE_SOFTWARE,
      D3D_DRIVER_TYPE_WARP
  }
```

Listing 1.3. The D3D_DRIVER_TYPE enumeration, showing the types of drivers that can be created.

As you can see, it is possible to use hardware based drivers that interact with actual hardware devices. This is the configuration that will be used in the majority of cases, but there are also several other driver types that must be considered. The next driver type is the *reference driver*. This is a software driver that fully implements the complete Direct3D 11 specification, but since it runs in software, it is not very fast. This driver type is typically used to test if a particular set of rendering commands produces the same results with both the hardware driver and the reference drivers. The null driver is only used in testing and does not implement any rendering operations. The software driver type allows for a custom software driver to be used. This is not a common option, since it requires a complete third-party software driver to be provided on the target machine. The final option is perhaps one of the most interesting. The *WARP device* is a software renderer that is optimized for rendering speed and that will use any special features of a CPU, such as multiple cores or SIMD instructions. This allows an application to have a moderately fast software driver available for use on all target machines, making the deployable market significantly larger for many application types. Unfortunately, this device type only supports the functionality provided up to the Direct3D 10.1 feature level,[1] which means that it can't be used to implement most of the techniques discussed in this book!

After the desired device type is chosen, the next parameter allows the application to provide a handle to the software driver DLL in situations when a software driver type is selected. If a non-software device is chosen, this parameter can be set to NULL, which will usually be the case. Next up is the `Flags` parameter, which allows for a number of special features to be enabled when the device is created. The available flags are shown in Listing 1.4.

```
enum D3D11_CREATE_DEVICE_FLAG {
    D3D11_CREATE_DEVICE_SINGLETHREADED,
    D3D11_CREATE_DEVICE_DEBUG,
    D3D11_CREATE_DEVICE_SWITCH_TO_REF,
    D3D11_CREATE_DEVICE_PREVENT_INTERNAL_THREADING_OPTIMIZATIONS,
    D3D11_CREATE_DEVICE_BGRA_SUPPORT
}
```

Listing 1.4. The D3D11_CREATE_DEVICE_FLAG enumeration.

The first flag indicates that the device should be created for single-threaded use. If this flag is not present, the default behavior is to allow multithreaded use of the device. The second flag indicates if the debug layer of the device should be created. If this flag is set, the device is created such that it also implements the `ID3D11Debug` interface. This interface is used for various debugging operations and is retrieved using the COM query interface techniques. It also causes complete error/warning messages to be output at runtime and enables memory leak detection as well. The third flag is unsupported for use in Direct3D 11, so we won't discuss it in any great detail. The fourth flag is used to disable multithreaded *optimizations* within the device while still allowing multithreaded use. This would likely decrease performance, but it could allow for simpler debugging and/or profiling characteristics. The final flag is used to create a device that can interoperate with the Direct2D API. We don't cover Direct2D in this book, but it is still important to know that this feature is available.

The following two parameters also offer a very interesting set of capabilities for creating a device. The first is a pointer to an array of `D3D_FEATURE_LEVEL` values. The concept of a *feature level* is the replacement for the old CAPS bits from older versions of Direct3D. CAPS bits were a name for the myriad number of options that a GPU manufacturer could choose to support or not, and the application was responsible for checking if a particular feature was available or not before trying to use it.

---

[1] The concept of feature levels is discussed in the following pages.

With each new generation of hardware, the number of available CAPS bits was becoming increasingly unmanageable. Instead of requiring the application to parse through all of these options, the Direct3D 11 API groups implementations into categories called *feature levels*. These feature levels provide a more coarse representation of the available features for a given GPU and driver, but using them provides a significantly simpler process to determine what capabilities the GPU supports. The available feature levels are shown in Listing 1.5.

```
enum D3D_FEATURE_LEVEL {
    D3D_FEATURE_LEVEL_9_1,
    D3D_FEATURE_LEVEL_9_2,
    D3D_FEATURE_LEVEL_9_3,
    D3D_FEATURE_LEVEL_10_0,
    D3D_FEATURE_LEVEL_10_1,
    D3D_FEATURE_LEVEL_11_0
}
```

Listing 1.5. The D3D_FEATURE_LEVEL enumeration.

As you can see, there is one feature level for each minor revision that was released for Direct3D since version 9. Each feature level is a strict superset of the previous feature level, meaning that a higher-level GPU will support the lower feature levels if an application requests them. For example, if a particular application only requires features from the 10.0 feature level, it will be able to run on a wider range of hardware than an application that requires the 11.0 feature level. The application passes an array of feature levels that it would like to use in the pFeatureLevels parameter, with the array elements arranged in the order of preference. The number of elements in the array is passed in the FeatureLevels parameter to ensure that the function doesn't exceed the array size while processing the input array. The device creation method will try to create a device with each feature level, beginning with the first element in the array. If that feature level is supported for the requested driver type, the device is created. If not, then the process continues to the next feature level. If no supported feature levels are present in the array, the method returns a failure code.

We have two input parameters left in the device creation function. The SDKVersion parameter is simply supplied with the D3D11_SDK_VERSION macro. This is a defined value that changes with each new release of the DirectX SDK. However, since the value is automatically updated with a new SDK installation, the developer does not need to make any changes when converting to the new SDK.

The final input parameter is a pointer to a swap chain description. As described earlier in this chapter, a *swap chain* is an object created by DXGI that is used to manage the contents of a window's client area. The swap chain object defines all of the lower-level options that

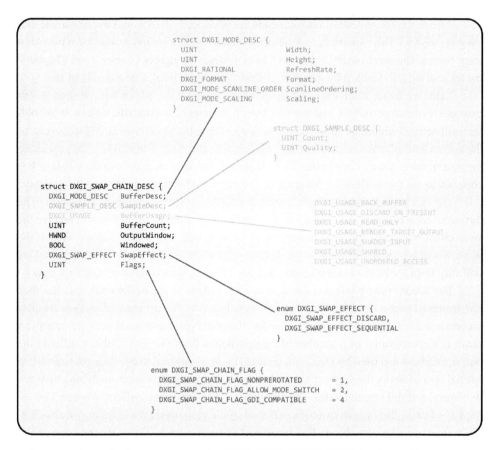

Figure 1.4. A graphical representation of the **DXGI_SWAP_CHAIN_DESC** structure and its members.

are needed to have DXGI properly manage a window's contents. We specify out desired swap chain options in the pSwapChainDesc parameter by filling in a DXGI_SWAP_CHAIN_DESC structure. This structure is shown in Figure 1.4, with the available options for its members.

These structure members each control various aspects of the swap chain's behavior. The BufferDesc parameter provides details about the buffer that will hold the windows contents. Its Width and Height parameters give the size of the buffer in pixels. The RefreshRate parameter is a structure that provides a numerator and denominator to define the refresh rate of the contents of the window. The Format parameter is a member of the DXGI_FORMAT enumeration, which provides all of the available formats for a texture resource to use. The ScanlineOrdering parameter defines the raster techniques for image generation. And finally, the Scaling parameter specifies how the buffer will be applied to the window's client area, allowing either scaled or centered presentation. Many of these options are chosen once and then simply reused as a default behavior.

After the buffer mode description structure, we next specify the options for multisample anti-aliasing (MSAA). This requires that we select the number of samples and the quality of their sampling pattern. This functionality is discussed in more detail throughout Chapter 2 and Chapter 3, but for now we will disable it by specifying a single sample per pixel, with a quality of zero.

Next, we indicate the intended usage of the swap chain, which will be used in our case as a render target output. The BufferCount parameter is used to indicate how many buffers will be used within the swap chain. Having multiple buffers allows DXGI to use one of the buffers to display the rendered image, while at the same time, Direct3D can be generating the next frame in a secondary buffer. This allows for smoother animation in the window. It is common to use two buffers for this purpose, but additional buffers can be used if necessary.

The next two parameters configure the connection to the application window. The OutputWindow parameter is simply the window handle of the window that will receive the swap chain's contents. The next parameter, Windowed, indicates what it sounds like—if the window will appear as a window on the desktop, or if it will occupy the complete screen. This will vary from application to application, but we will stick with windowed mode for now.

The SwapEffect parameter configures what is done to the buffer contents after they are presented to the window. This flag is normally set to discard the buffer contents after presentation, but it can be changed as needed. The final parameter needed to create a swap chain is a combination of a number of miscellaneous flags that can further configure the buffer, and how it is used by DXGI. In general, these are special usage flags that can allow special operations by the application, such as handling monitor rotation, switching between windowed and full screen modes, and allowing GDI access to the buffers. We won't be using any of these flags in our example. After all of the parameters are filled in, we also supply pointers to all of the objects that we expect to receive back. These are the swap chain, the device, the created device's feature level, and the immediate device context.

It is important to always check the return value from API calls, to ensure that the function has succeeded. All of the returned items except for the feature level are COM objects and thus must have their reference counts managed accordingly. After we have the swap chain interface, we must obtain the texture resource interface from it, for use in the rendering pipeline. This is done with the IDXGISwapChain::GetBuffer() method, which works similarly to the COM query interface method described above. The difference is that the index of the buffer we are trying to acquire is passed as the first parameter. The acquisition of the texture interface is shown in Listing 1.6.

```
ID3D11Texture2D* pSwapChainBuffer = 0;
hr = pSwapChain->GetBuffer( 0,
                            __uuidof( ID3D11Texture2D ),
                            (void **)&pSwapChainBuffer );
```

Listing 1.6. Acquiring the texture interface from a swap chain.

In addition, if we created the device with the debug layer installed, we need to acquire the debug interface from the device itself in the same manner. This is shown in Listing 1.7.

```
m_pDebugger = 0;
hr = pDevice->QueryInterface( __uuidof( ID3D11Debug ), (void **)&pDebugger );
```

Listing 1.7. Acquiring the debug interface from the device using the query interface method.

After these interfaces have been acquired, we can consider Direct3D 11 to be initialized and can move on to the next application phase.

Resource creation. After the application has initialized and acquired the device and device context and also has a swap chain to render into, it should then create any resources, shader objects, and state objects that will be used during the rendering portion of the application. It is the best practice to acquire as many of these objects at startup time as possible, since acquiring them during rendering operations can introduce momentary delays in rendering. These glitches can be mostly alleviated with multithreaded rendering, but acquiring the resource during initialization eliminates the problem completely.

Since we haven't yet examined the available resources, or the components of the rendering pipeline, it would not be useful to describe the creation of these objects here. Both of these topics have an entire chapter devoted to them (Chapters 2 and 3), so we will not attempt to provide a glossed-over description here. However, in order to use the texture interface that we acquired from the swap chain, we will need an object called a *resource view* to bind the texture as a render target for the pipeline. The particular type of resource view that is needed is called a *render target view*. As is the case for most objects in Direct3D 11, it is created with the device interface and then used with the device context to perform some action. Listing 1.8 shows a code listing for creating a render target view when the texture resource is already available.

```
ID3D11RenderTargetView* pView = 0;
HRESULT hr = m_pDevice->CreateRenderTargetView( pSwapChainBuffer,
                                               0, &pView );
```

Listing 1.8. Creating a render target view to allow binding the swap chain texture to the pipeline for rendering.

In this case, we pass the texture resource as the first parameter, and instead of a view description, we simply pass NULL to use a default view description. Resource views are reference counted as well, so the application must be sure to properly release the reference when it is finished with it. With the resource view acquired, we are now ready to use our swap chain texture resource.

In addition to the render target, we also need to create a *depth stencil buffer* and a corresponding *depth stencil view* for binding it to the pipeline. The code required to do this is shown in Listing 1.9. Once again, this creation process is covered in detail in Chapter 2.

```
D3D11_TEXTURE2D_DESC desc;
desc.Width = width;
desc.Height = height;
desc.MipLevels = 1;
desc.ArraySize = 1;
desc.Format = DXGI_FORMAT_D32_FLOAT;
desc.SampleDesc.Count = 1;
desc.SampleDesc.Quality = 0;
desc.Usage = D3D11_USAGE_DEFAULT;
desc.BindFlags = D3D11_BIND_DEPTH_STENCIL;
desc.CPUAccessFlags = 0;
desc.MiscFlags = 0;

ID3D11Texture2D* pDepthStencilBuffer = 0;
HRESULT hr = m_pDevice->CreateTexture2D( &desc,
                             0, &pDepthStencilBuffer );

ID3D11DepthStencilView* pDepthView = 0;
HRESULT hr = m_pDevice->CreateDepthStencilView( pDepthStencilBuffer,
                                   pDesc, &pDepthView );
```

Listing 1.9. The process of creating a depth stencil buffer, and a depth stencil view for using it.

Even though we won't be creating additional resources or objects in this sample, we can still consider the types of items that should be created in this initialization routine. Essentially, all Direct3D resources should be created in the initialization phase, unless a particular use case absolutely requires that a resource be created and destroyed during runtime. For example, if a particular texture is procedurally generated, and its size is dependent on some runtime parameter, it would not be easy to create the texture during startup. However, it may be worth considering changing the algorithm to allow early creation. In our example, this could include just declaring one large resource that is shared by all objects that need such a procedural texture. It is also acceptable to create a resource at startup and then modify its contents during runtime, especially if the multithreading capabilities of Direct3D 11 are put to use.

In addition, all pipeline configurations should be pre-created. This includes any state objects that would be used by fixed function stages, as well as the shader program objects that would be used for the programmable shader stages. Since these items are typically planned for well in advance, there shouldn't be any reason that they couldn't be created at startup, instead of by using a dynamic loading scheme. Unless an application uses an enormous number of these objects, they should not cause any issues with memory consumption. In particular, the shader programs would need to read from the hard disk during runtime if these objects are dynamically loaded—which would almost surely cause a brief pause in the application.

Application Loop

After initialization is complete, the application can move on to the looping portion of its execution. This phase consists of three individual sections, which will be described in the following sections.

Handle pending messages. After an application enters its application loop, it must respond to and handle the Windows messages that it receives from the operating system. This is performed by a callback function that is specified when the application window is registered and created. The handling of messages is not directly relevant to Direct3D programming, so we will skip it for now. However, all of the sample programs from this book implement message handling, so the fundamentals can be seen there. The message handler can be found in the **main.cpp** file in the source code distribution.

Update. Our simple test application doesn't perform any simulations or implement special physics calculations, but it does perform a simple animation to change the color of the render target. This is performed in the Update phase of the application loop. This is a typical place to perform such updates, so that any changed state can be reflected in the current frame's rendering, which is performed in the next step of the loop. For our sample application, we simply calculate a sinusoidally varying value that will be used to determine the background color of the window. This is shown in Listing 1.10.

```
float fBlue = sinf( m_pTimer->Runtime() * m_pTimer->Runtime() ) * 0.25f + 0.5f;
```

Listing 1.10. Producing a time-varying parameter for animation.

Here we see a timer class being used to acquire the amount of time that an application has been running. This class is taken from the sample framework of this book, and more detail on its inner workings can be found in the source code distribution in the Timer class files. For now, we will simply assume that it gives us the floating point number of seconds that an application has been running.

Rendering. After updating the state of our application, we next render a representation of our current scene for the user to see. This process begins by binding both the render target view and the depth stencil view to the pipeline, to receive the results of any rendering operations. Since this is a pipeline manipulation, it is performed with the device context. Listing 1.11 shows how this is done.

```
ID3D11RenderTargetView* RenderTargetViews[1] = { pView };
ID3D11DepthStencilView* DepthTargetView = pDepthView;

pContext->OMSetRenderTargets( 1, RenderTargetViews, DepthTargetView );
```

Listing 1.11. Binding the render and depth targets to the pipeline for rendering.

Both the render target and depth stencil target are bound to the pipeline with the same method call. This method for setting render targets takes an array of render target views, so even if you are only using a single render target, it is a best practice to use an array to pass its reference into the function. After the render and depth targets have been bound, we can clear them to prepare for the coming rendering pass. The body of a method that performs this process is shown in Listing 1.12. As you can see, the clearing process is also performed with the device context.

```
ID3D11RenderTargetView* pRenderTargetViews[D3D11_SIMULTANEOUS_RENDER_TARGET_
COUNT] = {NULL};
ID3D11DepthStencilView* pDepthStencilView = 0;

m_pContext->OMGetRenderTargets( D3D11_SIMULTANEOUS_RENDER_TARGET_COUNT,
                                pRenderTargetViews, &pDepthStencilView );

for ( UINT i = 0; i < D3D11_SIMULTANEOUS_RENDER_TARGET_COUNT; ++i )
{
    if ( pRenderTargetViews[i] != NULL )
    {
        float clearColours[] = { color.x, color.y, color.z, color.w }; // RGBA
        m_pContext->ClearRenderTargetView( pRenderTargetViews[i],
                                           clearColours );
        SAFE_RELEASE( pRenderTargetViews[i] );
    }
}

if ( pDepthStencilView )
{
    m_pContext->ClearDepthStencilView( pDepthStencilView,
                                       D3D11_CLEAR_DEPTH, depth, stencil );
}

// Release the depth stencil view
SAFE_RELEASE( pDepthStencilView );
```

Listing 1.12. Clearing the contents of the bound render and depth targets.

Once the render target and the depth stencil target have been cleared, the application can perform whatever rendering operations it needs to. This includes rendering the current scene, any two-dimensional screen elements such as user interfaces, and any text that might be needed for a given frame. Once again, our sample application won't perform any actual rendering. Instead, it simply uses the method shown in Listing 1.12 to clear the render target to the color specified in our update phase.

Presenting results to the window. After all of the rendering for a given frame has been completed, its contents can be presented to the window's client area. This is done quite simply, using a single method call from the swap chain interface. This is shown in Listing 1.13.

```
pSwapChain->Present( 0, 0 );
```

Listing 1.13. Presenting the contents of the swap buffer resource to its window.

When this method is called, the contents of the render target are presented to the client area of the window. The application doesn't need to perform any manual render target manipulation if multiple buffers are used within the swap chain. Instead, this is managed by the swap chain itself, which simplifies the application's responsibilities. After this method has completed, the user will be able to see the render target contents in the window.

After the rendered output has been displayed, the application loop starts over and checks for any pending Windows messages. This looping sequence repeats until the user decides to terminate the application. This is normally done by pressing the escape key or selecting a "quit" option from an application user interface.

Application Shutdown

Before an application can terminate, it needs to clean up all of the open references that it has acquired. With respect to Direct3D 11 objects, this typically means calling the Release method for each reference that the application has used. One point to also consider is that the pipeline itself will keep references to some of the objects that have been bound to it. To ensure that these pipeline references are released, the application can call the ClearState method of the device context. This will clear any references from the pipeline and put it into the default state. The shutdown code for our sample application is shown in Listing 1.14.

```
pContext->ClearState();

SAFE_RELEASE( pView );
SAFE_RELEASE( pDepthView );
SAFE_RELEASE( pDepthStencilBuffer );

SAFE_RELEASE( pSwapChainBuffer );
SAFE_RELEASE( pSwapChain );
SAFE_RELEASE( pContext );
```

```
    SAFE_RELEASE( pDebugger );
    SAFE_RELEASE( pDevice );
```

Listing 1.14. Releasing the various interfaces that have been used in the simple example application.

If there are still some references that have not been properly released, a debug message will be printed to the debug console, indicating which interface object types still have outstanding references that haven't been released. If all references were properly released, the application will exit, just like any other application.

1.5.3 Further Application Considerations

While the example steps that we walked through in the second half of this chapter are indeed important to understand, they are not the most exciting operations to perform for every application that you create. Many of these tasks, such as creating a window or initializing the device, can be handled with library functions that can be reused for all of the applications that you create. In the real-time rendering context, this type of a library is often called an *engine*, a term first coined by John Carmack of id Software.

This is the path that we have chosen to follow for the sample programs in this book. All of the samples are built upon an open source engine developed by the authors, called Hieroglyph 3. The engine and the samples are both freely downloadable from the Hieroglyph 3 project page, which can be found at http://hieroglyph3.codeplex.com. In addition to the samples from this book, many other example applications are included in the source code repository for learning the basics of the library.

By using such an open source library, we can devote more of this book to the concepts and use of Direct3D 11, rather than spending time explaining basic win32 application code or giving repetitive code samples. The library takes care of the basics, so that we can focus on the more interesting portions of the subject matter and ultimately provide a better book. The Hieroglyph 3 library is provided with the MIT license, which includes a liberal set of usage guidelines, and that you can use as the basis of your own projects.

2

Direct3D 11 Resources

In many cases, discussions of modern real-time rendering techniques tend to revolve around the programmable pipeline and what can be done with it. The pipeline is indeed a critically important piece in the real-time rendering puzzle, especially since one of the primary tasks of the graphics developer is writing shader programs for use in the pipeline. However, the importance of the memory resources connected to the pipeline is often overlooked in these discussions. It is just as important to understand what resources are required for a particular algorithm, where they will be used, and how they will be used.

This chapter will inspect the topic of Direct3D 11 Memory Resources in depth. We begin by identifying how resources are organized and managed by the Direct3D 11 API and then consider the two major types of resources—buffers and textures. We will discuss in detail each subtype of resource, how it is created, how it is used, and how it is released when no longer needed. The discussion of resource use will also introduce the concept of a resource view, a type of adapter used to connect a resource to a particular place in the pipeline.

Any discussion about resources must also look at both aspects of their usage. The application is concerned with creating, populating, and connecting resources to the pipeline. However, we must also consider how resources are used inside of the pipeline as well. How they are declared and used within a programmable shader is just as important of a consideration as properly creating the resources. This topic is discussed in detail to provide a sound understanding of this usage within shader programs, which will be built upon in Chapters 3–7.

As we will see later in this chapter, it is very important to have a clear understanding of how a resource will be used before creating it, since its usage domain is predetermined by its creation input parameters. If incorrectly configured, a resource can have several negative effects on your applications, ranging from poor performance all the way to a complete inability to execute the pipeline due to errors. This makes resources a very important topic indeed!

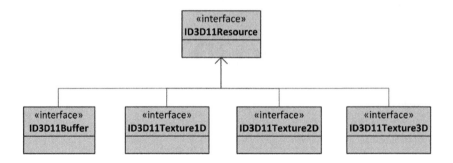

Figure 2.1. The Direct3D 11 Resource Interface Hierarchy.

2.1 Resources Overview

There are two basic categories of resources in Direct3D 11: *buffers* and *textures*. Each of these has several subtypes, and each subtype has different configuration options. The sheer number of available options for creating resources can be somewhat overwhelming at first, so our discussion of the topic will begin with how they are organized within the Direct3D 11 API. By starting at a high level, we can begin to understand the differences in the various resources and then build on this knowledge with additional details about the way that these differences determine how a resource can be used.

The Direct3D 11 resource class structure is organized as shown in Figure 2.1. Here, you can see that there are only four resource classes available for the entire API—buffers, and 1D, 2D, and 3D textures.

The diagram also shows that each of these resource classes is derived from a single common base class, named *ID3D11Resource*. This makes sense and provides an indication of a common theme that we will be revisiting throughout this chapter. That is, the resources are only blocks of memory that can be attached to the pipeline and used for input or output (or in some cases, both input and output). Restated another way, resources are blocks of memory that are made available for the GPU to use and manipulate. Even though they have different names and support different concepts, the only real differences between them are the semantics with which they are bound to the pipeline and the rules about their formatting, access, and use.

2.1.1 Resource Creation

As described in Chapter 1, "Overview of Direct3D 11," the ID3D11Device interface is responsible for creating all memory resources. The created resources can then be attached to the pipeline either directly or with a resource view, where they are then used during a pipeline execution event. The resource creation process uses a different ID3D11Device method for each type of resource, but they all follow the same general pattern.

The creation methods all take three parameters. The first parameter is a structure that specifies all of the various options that a resource can be created with. It is referred to as a *resource description*. Each resource type uses its own description structure, since they each have a different set of available properties, but the structures all serve the same purpose—to define the desired characteristics of the created resource. These structures will be investigated in detail in this chapter, and their proper configuration comprises the bulk of the work required to make a resource do what you want it to. The second parameter in the resource creation methods is a pointer to a D3D11_SUBRESOURCE_DATA structure, which is used to provide the initial data to be loaded into a resource. For example, if a buffer resource will hold static vertex data, this structure would be used to pass a model's vertex data into the buffer. This eliminates the need to manually manipulate the buffer after it is created, if its contents will not be changing. This parameter can also just be set to null if no initialization is required. The final parameter is a pointer to a pointer to the appropriate resource interface, which is where the created resource pointer is stored after a successful resource creation event.

In each of these methods, the real configuration occurs in the resource description structure. As mentioned above, each resource type has its own structure used to define its properties. However, there are some common elements that are shared across all of the structures. These include the usage flags, bind flags, CPU access flags, and miscellaneous flags. Since these parameters are common across all the various resource types, we will discuss them here in detail. The individual parameters that are unique to certain types of resources will be discussed in their respective resource type sections later in this chapter.

Resource Usage Flags

The first resource parameter that we will inspect is the *usage specification*. Stated simply, this parameter is used to specify what the application intends to do with the resource. In general, we can consider a resource to be a block of memory somewhere within the computer. Both video memory and system memory can hold resources, and resources can be copied to and from each type of memory by the Direct3D 11 runtime. To optimize where a resource resides and how it is handled internally by the runtime/driver, the application must specify its intentions with this usage parameter. This differs from previous versions of Direct3D, which allowed the user to specify the memory pool that the resource would reside in. However, by indicating the usage intent of the resource, a similar level of control is provided. The list of available values is encapsulated in the D3D11_USAGE enumeration, which is shown in Listing 2.1.

```
enum D3D11_USAGE {
    D3D11_USAGE_DEFAULT,
    D3D11_USAGE_IMMUTABLE,
    D3D11_USAGE_DYNAMIC,
    D3D11_USAGE_STAGING
}
```

Listing 2.1. The D3D11_USAGE enumeration.

Each of these flags indicates one usage pattern, which identifies the required read/write capabilities of the resource for both the CPU and the GPU. For example, a resource that is only read and written by the GPU and is never accessed by the CPU can be safely placed in video memory. It will never be transferred back to system memory, since the CPU is not able to access it. This provides an optimization by keeping the resource closer to the GPU. Table 2.1 shows the read/write permissions for each of these flags.

Resource Usage	Default	Dynamic	Immutable	Staging
GPU-Read	yes	yes	yes	yes
GPU-Write	yes			yes
CPU-Read				yes
CPU-Write		yes		yes

Table 2.1. The accessibility defined for each usage flag.

Immutable usage. The simplest usage pattern is the *immutable* usage. A resource created with this usage pattern can only be read by the GPU and is not accessible by the CPU. Since the resource can't be written to by the GPU or CPU, it can't be modified after it is created. This requires that the resource be initialized with its data at creation time. Since the resource can't be modified, the usage type's name is appropriately immutable. Common examples of this resource type are static constant, vertex, and index buffers that are created with data that won't change over the lifetime of the application. In general, these resources are used to provide data to the GPU.

Default usage. As can be seen in the table above, the *default* usage provides read and write access to the GPU, but provides no access to the CPU. This is the most optimal usage setting for any resources that will not only be used by the GPU, but that will also be modified by the GPU at some point. Some common examples of this usage pattern are render textures (which are rendered into the GPU and subsequently read from it), and stream output vertex buffers (which have data streamed into them by the GPU and are later read by the

GPU for rendering). Since resources of this type can remain in video memory, the GPU will have the fastest possible access to them, and the application's overall performance will benefit.

Dynamic usage. The *dynamic* usage is the first usage type that allows the CPU to write to a resource that can then be read by the GPU. In this usage case, the contents of the resource are provided by the CPU and are then consumed by the GPU. The most common example of this is a constant buffer, which provides rendering data to the programmable shader stages which changes in every rendered frame. This type of data could be transformation matrices, for example. An important consideration is that the CPU is not able to read back the contents of the resource—the data can only flow in one direction from the CPU to the GPU.

Staging usage. The *staging* usage is the final usage flag, and it provides a special type of usage pattern. The usage patterns that we have discussed up to this point have covered the typical resource usage scenarios when performing rendering. However, there are many situations in which we want to manipulate data with the GPU and then read it back to the CPU for storage or examination. With the introduction of the GPGPU facilities with the DirectCompute technology, along with the ability to stream vertex information out with the stream output stage, it is possible to use the GPU to process information and then read it back to the CPU. Instead of forcing the other usage patterns to allow read access to the CPU, a resource with staging usage can be used as an intermediate instead.

The basic concept is that a desired resource can be manipulated by the GPU, then copied onto a staging resource, and then subsequently read back to the CPU. This requires creating an additional staging resource to retrieve the data, but it lets the other resources used by the GPU remain as close as possible to the GPU and not be concerned with CPU access. We will see additional detail on manipulating resource contents later in this chapter.

CPU Access Flags

After the usage flag of a resource has been determined, the needed CPU access flag must be chosen. The CPU access flag exposes similar information as we have seen in the usage property, but it is restricted to defining the possible CPU access to resources. There are only two possible flags to use for the CPU access flags. These are shown in Listing 2.2.

```
enum D3D11_CPU_ACCESS_FLAG {
    D3D11_CPU_ACCESS_WRITE,
    D3D11_CPU_ACCESS_READ
}
```

Listing 2.2. The D3D11_CPU_ACCESS_FLAG enumerations.

These flags can be combined with a bitwise OR operation to indicate if the resource will be read, written, or read and written by the CPU. These settings must also take into account the usage flags that are specified for this resource as well. As shown in the table above, a staging resource can be both read from and written to by the CPU, allowing both of these flags to be set. In contrast, a default usage is not accessible to the CPU at all, and hence its CPU access flags must be set to 0.

Bind Flags

Our next common resource flag is the *bind flag*. This property indicates where on the pipeline a resource can be bound and includes various flags that represent each available binding location. These flags can also be combined with a bitwise OR to allow for multiple bind locations to be defined for the same resource. If a resource is created without the appropriate flag, and the application tries to bind the resource to the pipeline, then it will cause an error. All of the available binding locations are shown in Listing 2.3 in the `D3D11_BIND_FLAG` enumeration.

```
enum D3D11_BIND_FLAG {
    D3D11_BIND_VERTEX_BUFFER,
    D3D11_BIND_INDEX_BUFFER,
    D3D11_BIND_CONSTANT_BUFFER,
    D3D11_BIND_SHADER_RESOURCE,
    D3D11_BIND_STREAM_OUTPUT,
    D3D11_BIND_RENDER_TARGET,
    D3D11_BIND_DEPTH_STENCIL,
    D3D11_BIND_UNORDERED_ACCESS
}
```

Listing 2.3. The D3D11_BIND_FLAG enumeration.

As you can see from this list, there are only eight types of locations where resources can be bound to the pipeline. The vertex and index buffer flags are used to declare resources that will be attached to the input assembler stage to feed the pipeline with geometry, while the render target and depth stencil flags allow resources to be connected to the output merger stage to receive the rendered output from the pipeline. The *stream output* flag also allows output from the pipeline, although it receives geometry instead of rasterized image data. These flags all represent a single binding point on the pipeline.

In contrast, the `D3D11_BIND_CONSTANT_BUFFER`, `D3D11_BIND_SHADER_RESOURCE`, and `D3D11_BIND_UNORDERED_ACCESS` access flags all indicate that the resource can be

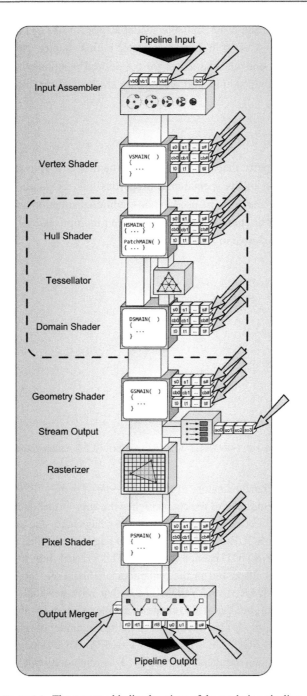

Figure 2.2. The resource binding locations of the rendering pipeline.

bound to some or all of the programmable shader stages for use in the shader programs. These usages will be discussed in more detail later in this chapter, but in general it must be ensured that the proper bind flag is set for the desired pipeline connection points. An overview of these binding points, and where they are located on the pipeline, is shown Figure 2.2.

Miscellaneous Flags

The final common property that we will look at is the miscellaneous flags. This group of flags encapsulates most of the special-case properties that a resource can use. Some of these flags allow for resources to be used in non-traditional situations, such as interoperating with GDI drawing or sharing a resource between multiple ID3D11Device instances. The available flags are shown in Listing 2.4.

```
enum D3D11_RESOURCE_MISC_FLAG {
    D3D11_RESOURCE_MISC_GENERATE_MIPS,
    D3D11_RESOURCE_MISC_SHARED,
    D3D11_RESOURCE_MISC_TEXTURECUBE,
    D3D11_RESOURCE_MISC_DRAWINDIRECT_ARGS,
    D3D11_RESOURCE_MISC_BUFFER_ALLOW_RAW_VIEWS,
    D3D11_RESOURCE_MISC_BUFFER_STRUCTURED,
    D3D11_RESOURCE_MISC_RESOURCE_CLAMP,
    D3D11_RESOURCE_MISC_SHARED_KEYEDMUTEX,
    D3D11_RESOURCE_MISC_GDI_COMPATIBLE
}
```

Listing 2.4. The D3D11_RESOUR_MISC_FLAG enumeration.

Due to the miscellaneous nature of the flags, some of them will be discussed in more detail throughout the remainder of the chapter when the appropriate content has been introduced. However, we will briefly discuss some of these flags here, due to their more general nature. The D3D11_RESOURCE_MISC_SHARED flag indicates that the resource can be shared among multiple ID3D11Device instances. This is used only in advanced applications and is not further discussed in the remainder of this book. Similarly, the D3D11_RESOURCE_MISC_SHARED_KEYEDMUTEX flag indicates that a resource will be used by multiple devices, but that it also supports a mutex system for sharing control of the resource. This functionality is also not discussed in the remainder of this book. And finally, we have the D3D11_RESOURCE_MISC_GDI_COMPATIBLE flag, which indicates that a resource can be used interchangeably with GDI. This functionality is used in some of the demo programs to implement text rendering, but it is not further described in this book.

Releasing Resources

Throughout this chapter, we will explore the various types of memory resources that are available to the developer. All of these resource types implement COM interfaces, and they ultimately have the IUnkown interface in their inheritance chain. This means that they are reference counted, and that they must be released after the application is finished using them. The application must be careful to track the resource references that it retains and release them properly, or else it will result in a memory leak.

2.1.2 Resource Views

Resources also share a common technique for binding them to the pipeline. As we saw above, eight different bind flags are available which identify the locations that a resource is allowed to be bound at. Of these eight binding locations, half of them allow resources to be bound directly to the pipeline. These include the vertex and index buffers, constant buffers, and stream output buffers (the meaning of these buffer types will be further explained in the "Buffer Resources"section of this chapter). However, the other four binding locations all require a type of adapter called a *resource view* to be used when binding a resource to the pipeline. A resource view is an adapter object used to connect a resource to various points in the pipeline. It conceptually provides a particular "view" of a resource. Like the configuration of resources, resource views also provide some configurability. By allowing a resource view to provide some interpretations of the resource, it is possible to use multiple resource views on a single resource for use at different locations on the pipeline. In fact, as we will see later in this section, resource views can also be used to represent subsets of a resource.

Resource View Types

There are four different types of resource views, which each correspond to a particular location that a resource can be bound to on pipeline. These views determine what operations can and can't be done with the resources that they are bound to. These four types of resource views are listed below:

- Render target views (ID3D11RenderTargetView)

- Depth stencil views (ID3D11DepthStencilView)

- Shader resource views (ID3D11ShaderResourceView)

- Unordered access view (ID3D11UnorderedAccessView)

Each of these resource views implies the particular semantics that its resource can use. In addition, each type of resource view provides a number of different configuration options that allow further specification of how a resource can be used. We will briefly discuss the semantics for each of the types of resource views, and their precise usage will be made much clearer, in Chapter 3, "The Rendering Pipeline."

Render target views. The *render target view* (*RTV*) is used to attach a resource to receive the output of the rendering pipeline. This means that the resource that is connected will be written to by the pipeline, and that it will also be read from in some cases to perform blending operations. Traditionally, a render target is a two-dimensional texture, but it is also possible to bind other types of resources. The various configuration options for a render target view depend on the type of resource that is being bound, but include the DXGI format of the resource, as well as various methods to select subportions of a resource to expose to the pipeline.

Depth stencil views. A *depth stencil view* (*DSV*) is similar to a render target view in that it is attached for receiving output from the rendering pipeline. However, it differs from the render target view in that it represents the depth stencil buffer, instead of a color render target. The depth stencil view resource is actively used to perform the depth and stencil tests. This makes the attached depth stencil resource very important for the efficiency of the pipeline. Thus, to further improve performance, the depth stencil view also exposes an additional flag to determine if the attached resource will be written to or not. This allows a resource to be attached as the depth stencil buffer and used for the depth stencil tests, but at the same time to be used in a shader program with a shader resource view (the next resource view that we will inspect). In this configuration, both views are read-only, meaning that the resource will not be modified by either one. With such a usage, the same resource can be bound to the pipeline in multiple places without read/write hazards, which allows for flexible use of the resource. If a standard use of the depth stencil buffer is needed, a second resource view can replace the read-only one to allow writing to take place.

Shader resource views. A *shader resource view* (*SRV*) provides read access to a resource to the programmable shader stages of the pipeline. This view corresponds to the traditional role that a texture would play in a pixel shader—data that is read from and used in the shader program, but it is not written to. A shader resource view can be used with all of the different programmable shader stages.

Unordered access views. The *unordered access view* (*UAV*) provides some of the most interesting new uses for resources in Direct3D 11. Like the shader resource view, a UAV can be used to read information from a resource. However, it also allows the resource to be simultaneously written to from within the same shader program. Further, the output location is not predefined, which allows the shader program to perform scattered writes to any location within the resource. This provides a completely new class of resource access, since it allows the resource to be used in a much more flexible manner. However, it is important to note that this type of view is not available for all programmable shader stages. These resource views can only be used with the pixel shader and compute shader stages.

Resource View Creation

The process of creating a resource view is somewhat similar to the process that we have seen for creating a resource. The ID3D11Device is also responsible for creating resource views, and it provides one creation method for each type. All creation methods follow the same pattern, with three different parameters provided by the application. As an example, Listing 2.5 shows the creation of a shader resource view.

```
ID3D11ShaderResourceView* CreateShaderResourceView( ID3D11Resource* pResource,
                                D3D11_SHADER_RESOURCE_VIEW_DESC* pDesc )
{
    ID3D11ShaderResourceView* pView = 0;
    HRESULT hr = m_pDevice->CreateShaderResourceView( pResource,
                                            pDesc, &pView );

    return( pView );
}
```

Listing 2.5. An example technique for creating a shader resource view.

The first parameter for all of these creation methods is a pointer to the resource that the resource view will represent. The second parameter is a pointer to a description structure, with all of the available options for that particular type of resource view. The third and final parameter is a pointer to a pointer to the corresponding resource view type, which is where the newly created resource view will be stored if the creation call succeeds. The second parameter is the most interesting with respect to configuring a resource view in the way we want it to behave. Each of the four resource view types uses a unique description structure, exposing each of their various unique properties. We will take a closer look at each of these structures to gain an understanding of what options are available to the developer.

Render target view options. We begin by looking closer into the available options for creating a render target view. Figure 2.3 shows the structure that must be filled in and passed to the creation method, **ID3D11Device::CreateRenderTargetView().**

As seen in Figure 2.3, this structure uses several normal member variables, followed by a union of possible structures. This allows all different resource types to use the same description structure, while still providing unique properties for each one. The format parameter specifies the data format that the resource will be cast to when it is read. In some

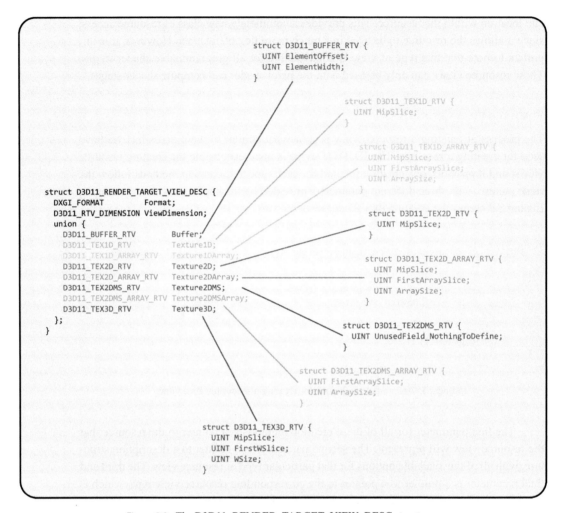

Figure 2.3. The D3D11_RENDER_TARGET_VIEW_DESC structure.

circumstances, this allows the format of the resource to be determined at runtime by the format supplied in the resource view being used. This can be helpful for certain applications, such as video format conversion. This requires that the resource be created with a format that is compatible with the desired view format specified in this structure.

The second variable, ViewDimension, indicates which type of resource will be bound with this resource view. The selection of this variable determines which of the unioned structures will be used to create the resource view. This is how the device knows which version of the unioned structures to interpret, which allows the same creation function and input data structure to be used for all of the various resource types, while still keeping the

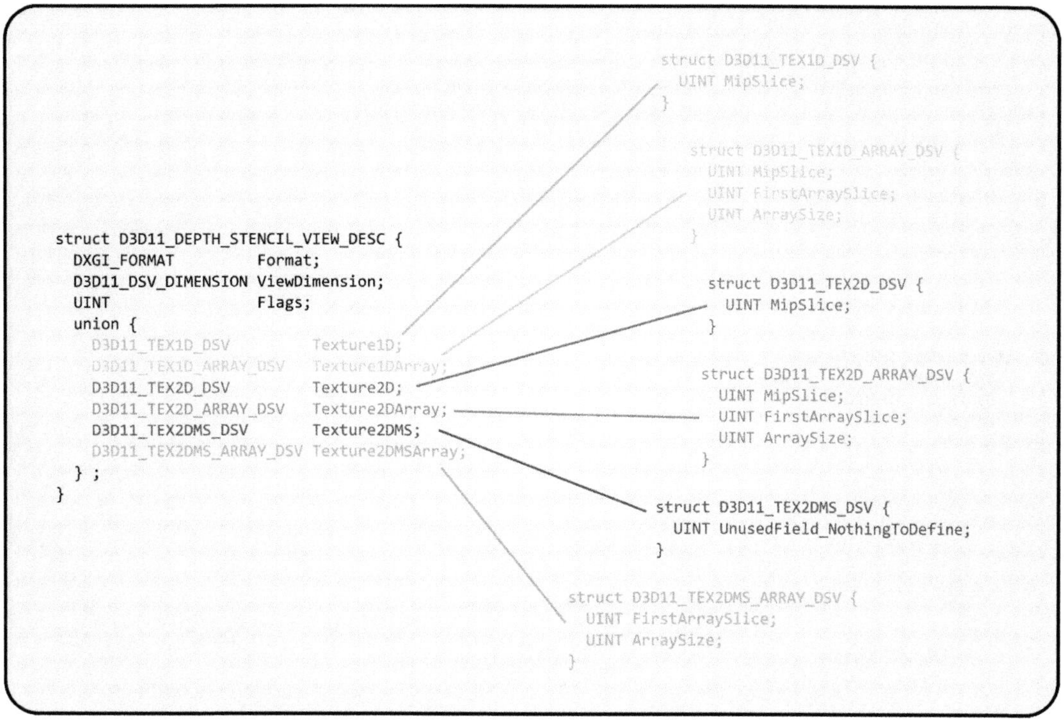

Figure 2.4. The D3D11_DEPTH_STENCIL_VIEW_DESC structure.

size of the structure to a minimum. All of the "union-ed" structures are used to specify what portions of the resource to make available in the resource view. Of course, since each type of resource uses a different memory layout and different options, it makes sense to have different structures for each resource type. Since we haven't covered the various resource types (or their various configurations) in detail, we must defer a detailed look into these properties. When we investigate the various resource types and their layout concepts later in this chapter, we will revisit these structures to see how they provide appropriate subresource selections.

Depth stencil view options. The depth stencil view option structure has a description structure that is similar to the one for the render target view. Its individual components are shown in Figure 2.4.

As can be seen above, this resource view type also requires a DXGI data format and a view dimension property to be specified. One point of interest is that the number of different resource types that can be used for a depth stencil resource view is somewhat smaller than can be used for a render target view. This is due to the specific nature of the operations that are performed on a depth stencil resource.

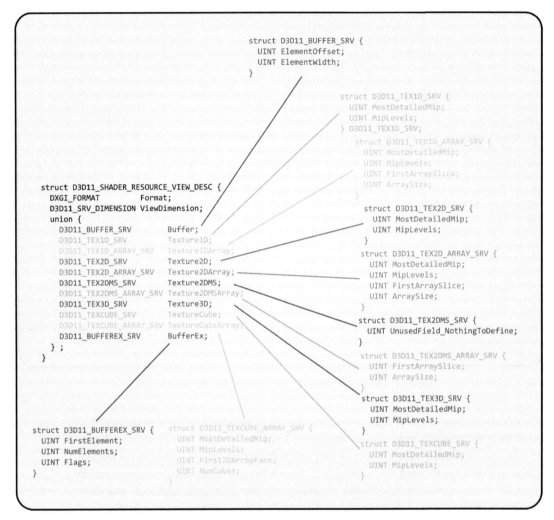

Figure 2.5. The D3D11_SHADER_RESOURCE_VIEW_DESC structure.

Another notable feature of this structure is the addition of a Flags parameter. This allows a bitwise OR combination of flags that specifies if the depth or stencil portion of the resource will be read only to the pipeline with this resource view. This allows multiple views of a depth stencil resource to be bound simultaneously to the pipeline for algorithms that are required to inspect its contents.

Shader resource view options. The *shader resource view* also follows the same basic formula that we have already seen in previous description structures. The format and view dimension operate in the same manner as the resource views we have already inspected.

However, you will notice three new resource types that are available in the unioned structure. Figure 2.5 shows this structure.

The new types of resources listed here are `TextureCube`, `TextureCubeArray`, and `BufferEx`. A texture cube allows reinterpretation of a `Texture2D` Array resource to a cube texture, which allows HLSL programs to use specialized intrinsic functions for sampling the texture. This also holds true for a `TextureCubeArray`, which is basically an array of the same type of cube resource interpretations. We will explore what this means in more detail later on in the "Texture Resources" section, but it provides a good example of the ability of a resource view to take a resource's data and provide a different "view" of it for a particular purpose.

The final new type of resource available for shader resource views is the `BufferEx` type. This is essentially a structure that allows a buffer to be interpreted as a raw buffer. This gives HLSL programs the freedom for structure interpretation within the shader program itself. This will also be discussed in more detail in the "Buffer Resources" section of this chapter.

Unordered access view options. The final resource view type is the *unordered access view*. Its description structure also follows the now standard format, as shown in Figure 2.6.

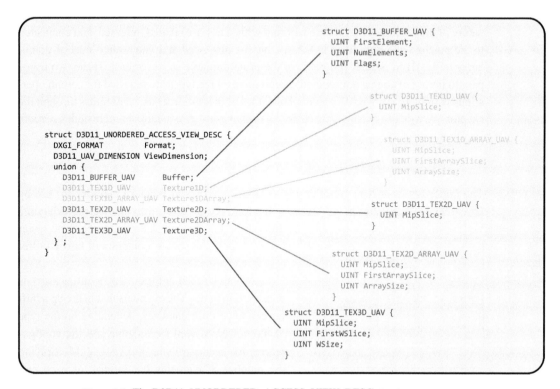

Figure 2.6. The D3D11_UNORDERED_ACCESS_VIEW_DESC structure.

The unordered access view provides a lesser number of available resource types when compared to the shader resource view. However, it does provide several new options for configuring buffer resources. As we will see later in the "Buffer Resources" section, there are some unique types of buffers that can be created for special uses. There are additional flags in this description structure for using buffers as append and consume buffers, as well as another flag to indicate that a built-in counter will be available in the buffer object.

2.2 Resources in Detail

Now that we have learned some of the basic principles of resources and resource views, it is time to begin a more detailed examination of the various types of resources to see what makes them different from one another, and what they can do. This section will introduce each type of resource and discuss each of the available subtypes they can be used to create. We begin with the buffer resource, exploring each of the individual types of buffers that can be used, and their various properties. This is followed by a general discussion about texture resources, followed by a similarly detailed discussion about the various configurations and options available for each texture type. It is also important to understand the dual nature of the resources that we are working with. They are created, released, and connected to the pipeline at a high level in C/C++, but are also used at a much finer level of granularity in the HLSL shader programs of the programmable shader stages. These two usage paradigms are of course intertwined, as certain resource types are required for certain operations within HLSL. We will explore both sides of the resources as we progress through the chapter. We will also explore both sides of a resources usage equation throughout this section.

2.2.1 Buffer Resources

The *buffer resource* type provides a one-dimensional linear block of memory for use by Direct3D 11. A number of different configurations can be used to change the behaviors of a buffer, but they all share this same basic linear layout. Figure 2.7 shows a diagram of how a buffer resource is organized.

As can be seen in Figure 2.7, a buffer's size is measured in bytes. The elements that make up a buffer can have different sizes, depending on the type of the buffer, as well as on some specific configurations for each particular buffer type. Multiplying the number of elements by the size of each element gives the total size of the buffer. This simple

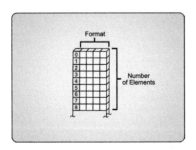

Figure 2.7. The layout of a buffer resource.

array-like layout structure provides a surprisingly wide variety of available buffer types. Some buffer types will be used primarily in the C++ side of an application, while others will be used primarily by HLSL shader programs after being attached to the pipeline. This section will explore each of the different types of buffers, describe what functionality they provide, and demonstrate how to create them. In addition, we will discuss common uses for buffers and provide the basic HLSL syntax to declare and use these resources in shader programs.

Vertex Buffers

The first buffer type that we will examine is the *vertex buffer*. The purpose of a vertex buffer is to house all of the data that will eventually be assembled into vertices and sent through the rendering pipeline. The simplest vertex buffer configuration is an array of vertex structures, where each vertex contains elements such as position, normal vector, and texture coordinates. These vertex elements must conform to the available format and type specifications. However, whatever generic information that is desired within the vertex can also be packed into a vertex structure to allow customized input data for a particular rendering algorithm.

In addition to the simple array-style vertex buffers described above, these buffers also allow for some other, more complex configurations. For instance, it is possible to use more than one vertex buffer at the same time. This allows vertex data to be split among multiple buffers. For example, vertex positions could be stored in one buffer, and vertex normal vectors in another buffer. This allows the application to selectively add in vertex data as it is needed, instead of using one large overall buffer for all rendering scenarios. This technique could be used to reduce the amount of bandwidth required for rendering operations.

It is also possible to perform *instanced rendering*, where one or more vertex buffers provide a model's *per-vertex* data, and an additional vertex buffer provides *per-instance* data instead of per-vertex data. The model defined in the first buffer is then rendered as a series of models, with the per-instance data from the second buffer applied to each instance. Figure 2.8 depicts these different vertex submission configurations. The per-instance data could include

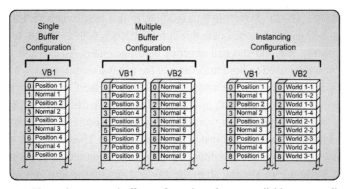

Figure 2.8. The various vertex buffer configurations that are available to an application.

a world transformation, color variations, or whatever else is used to differentiate between the various instances of the model. This setup allows for many objects to be rendered with a single *draw* call, which reduces the overall CPU overhead for rendering operations.

You can see in the diagram that each type of vertex buffer follows our general buffer layout. Each buffer is a one-dimensional array of same-sized elements. The sizes of the individual data elements are always the same as those of the other elements in the same buffer, although if multiple buffers are used, they can have different element sizes. We will see how instanced rendering is performed in more detail in Chapter 3.

Vertex buffer uses. As mentioned above, a vertex buffer's primary purpose is to provide per-vertex information to the pipeline, either directly, in multiple buffer configurations, or through instancing. To this end, the primary place to bind a vertex buffer to the pipeline is in the input assembler stage, which serves as the entry point into the pipeline. In addition to the input assembler stage, vertex buffers can also be attached to the stream output stage to allow the rendering pipeline to stream vertex data into the buffer, so that data can be used in subsequent rendering passes. Both of these stages of the pipeline will be discussed in more detail in Chapter 3, "The Rendering Pipeline," but these connections points are highlighted on the pipeline schematic in Figure 2.9 for easy reference in the future.

Creating vertex buffers. As described in the "Resource Creation" section of this chapter, for each type of location in the pipeline that a resource can be bound to, there is a corresponding bind flag that must be set at resource creation time in order to allow the resource to be bound there. With this in mind, a vertex buffer must always have the `D3D11_BIND_VERTEX_BUFFER` bind flag set. It can also optionally include the `D3D11_BIND_STREAM_OUTPUT` bind flag if it will be used for streaming vertex data out of the pipeline.

In addition to the choice of bind flags, the other major consideration for vertex buffers revolves around the usage scenario that the buffer will experience. Depending on whether or not the vertex buffer's contents will be changing frequently or not, and on if those changes will be coming from the CPU or GPU, different usage flags will be needed. For example, if the data that will be loaded into the buffer will be static, the buffer resource should be created with the `D3D11_USAGE_IMMUTABLE` usage flag. In this case, when the buffer is created it will be initialized with the vertex data through the `D3D11_SUBRESOURCE_DATA` parameter passed to the creation method and will never be modified again. An example of this type of vertex buffer would be used to hold the contents for a static terrain mesh.

However, if the buffer will be updated frequently by the CPU, the buffer should be created with `D3D11_USAGE_DYNAMIC`, and also with a CPU write flag for the `CPU access` flag parameter. An example of this type of vertex buffer usage is when vertex transformations are performed on the CPU instead of on the GPU. These updates would then be copied into the buffer resource every frame. This is a common technique used to condense many *draw* calls into one by putting all of the model data into the same frame of reference, which is typically either world space or view space. Still a third usage type would be to

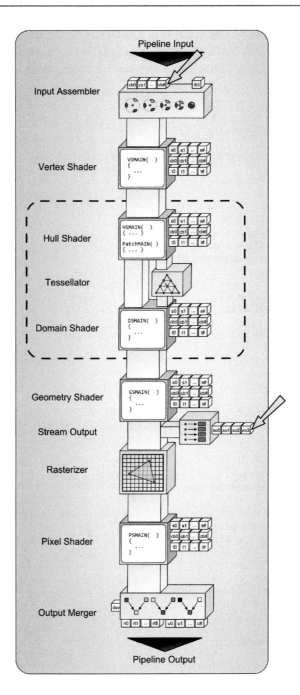

Figure 2.9. The available pipeline binding locations for vertex buffers.

create a buffer that will be updated by the GPU with the stream output functionality. In this case, the `D3D11_USAGE_DEFAULT` usage flag would be used. A sample buffer creation process is provided in Listing 2.6 to demonstrate these options.

```
// Initialize the device using it...
ID3D11Device* g_pDevice = 0;

ID3D11Buffer* CreateVertexBuffer( UINT size,
                                  bool dynamic,
                                  bool streamout,
                                  D3D11_SUBRESOURCE_DATA* pData )
{
    D3D11_BUFFER_DESC desc;
    desc.ByteWidth = size;
    desc.MiscFlags = 0;
    desc.StructureByteStride = 0;

    // Select the appropriate binding locations based on the passed in flags
    if ( streamout )
        desc.BindFlags = D3D11_BIND_VERTEX_BUFFER | D3D11_BIND_STREAM_OUTPUT;
    else
        desc.BindFlags = D3D11_BIND_VERTEX_BUFFER;

    // Select the appropriate usage and CPU access flags based on the passed
    // in flags
    if ( dynamic )
    {
        desc.Usage = D3D11_USAGE_DYNAMIC;
        desc.CPUAccessFlags = D3D11_CPU_ACCESS_WRITE;
    }
    else
    {
        desc.Usage = D3D11_USAGE_IMMUTABLE;
        desc.CPUAccessFlags = 0;
    }
    // Create the buffer with the specified configuration
    ID3D11Buffer* pBuffer = 0;
    HRESULT hr = g_pDevice->CreateBuffer( &desc, pData, &pBuffer );

    if ( FAILED( hr ) )
    {
        // Handle the error here...
        return( 0 );
    }

    return( pBuffer );
}
```

Listing 2.6. A method for creating vertex buffers with various usage considerations.

From this listing, we can see how the buffer description is configured differently for different usage scenarios. The size of the buffer elements and the overall size of the buffer remain the same in both scenarios, while the bind flags, usage flags, and CPU access flags all vary, depending on the intended use of the buffer. We will see a similar pattern throughout each of the buffer description specifications for the other buffer types.

Resource view requirements. Vertex buffers are bound directly to either the input assembler stage or the stream output stage. Because of this, there is no need to create a resource view when using it.

Index Buffers

The second buffer type that we will look at is the *index buffer*. The index buffer provides the very useful ability to define primitives by referencing the vertex data stored in vertex buffers. More or less, the index buffer provides a list of indices that point into the list of vertices. Depending on the desired type of primitive (such as points, lines, and triangles) appropriately sized groups of indices are formed to define which vertices that primitive is made up of. Figure 2.10 depicts this operation visually.

The use of index buffers can potentially provide a significant reduction in the total number of vertices that need to be defined. Since each adjacent primitive definition can reference the same vertex data as its neighboring primitives, the shared vertices do not need to be repeated in the vertex buffer. In addition, this sharing of vertices allows multiple primitives to use the same output vertex from a vertex shader. We will discuss the vertex shader in more detail in Chapter 3, "The Rendering Pipeline," but for now, we just need to understand that after a vertex has been run through the vertex shader, the result can be

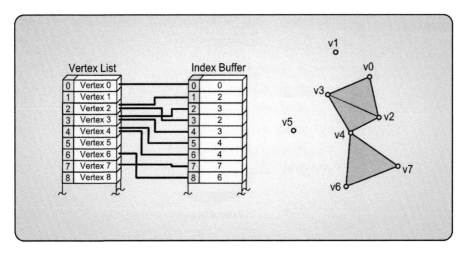

Figure 2.10. An index buffer creating triangles by referencing vertices.

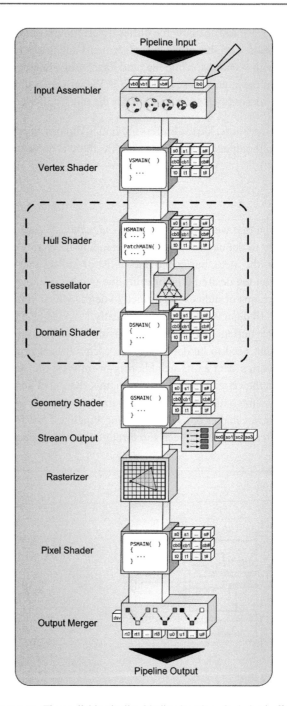

Figure 2.11. The available pipeline binding locations for index buffers.

cached and reused by multiple primitives. This can potentially reduce the amount of vertex shader processing required for a given model.

Using index buffers. Since the index buffer specifies which vertices are to be used in the primitive setup operations, you must know ahead of time what primitive topology you will be using—otherwise you wouldn't know which order to put the indices into. This is typically determined well in advance, with the selection of the rendering algorithm and the geometry loading routines. During the pipeline configuration when preparing to perform a *draw* operation, the desired index buffer is bound to the *input assembler* stage, where it will be used to generate the input primitives for the pipeline. Since these buffers serve a very specific purpose, they are normally not bound to the pipeline in other locations. This binding location is shown in Figure 2.11.

Creating index buffers. When creating an index buffer, we follow the standard buffer creation process and fill in a **D3D11_BUFFER_DESC** structure. The description structure of an index buffer does not change very frequently from buffer to buffer, since this type of data would normally be defined once in a content creation program and then exported as is. The index buffer indices normally would not change after the application startup phase. However, if some new algorithm does require dynamic updating of an index buffer, it can also be created with the dynamic update properties discussed in the "Vertext Buffers" section of this chapter. This could be used in the draw call reduction scheme discussed in the "Vertext Buffers" section, where multiple sets of geometry are dynamically grouped together into a single set of vertex and index buffers. A typical creation sequence is provided in Listing 2.7.

```
ID3D11Buffer* CreateIndexBuffer( UINT size,
                                 bool dynamic,
                                 D3D11_SUBRESOURCE_DATA* pData )
{
    D3D11_BUFFER_DESC desc;
    desc.ByteWidth = size;
    desc.MiscFlags = 0;
    desc.StructureByteStride = 0;
    desc.BindFlags = D3D11_BIND_INDEX_BUFFER;

    // Select the appropriate usage and CPU access flags based on the passed
    // in flags
    if ( dynamic )
    {
        desc.Usage = D3D11_USAGE_DYNAMIC;
        desc.CPUAccessFlags = D3D11_CPU_ACCESS_WRITE;
    }
    else
```

```
{
    desc.Usage = D3D11_USAGE_IMMUTABLE;
    desc.CPUAccessFlags = 0;
}

// Create the buffer with the specified configuration
ID3D11Buffer* pBuffer = 0;
HRESULT hr = g_pDevice->CreateBuffer( &desc, pData, &pBuffer );

if ( FAILED( hr ) )
{
    // Handle the error here...
    return( 0 );
}

return( pBuffer );
}
```

Listing 2.7. A method for creating index buffers depending on their usage.

The first item to specify is the size of the buffer in bytes. As can be seen in the code listing, the index buffer is always created with the D3D11_BIND_INDEX_BUFFER bind flag. When a static index buffer is desired, the usage flag is specified as D3D11_USAGE_IMMUTABLE, without any CPU access flags set. In this case, the intended contents of the buffer must be provided to the creation call with the D3D11_SUBRESOURCE_DATA structure. If a dynamic buffer is required, we would choose the D3D11_USAGE_DYNAMIC, along with the D3D11_CPU_ACCESS_WRITE CPU access flag.

Resource view requirements. Index buffers are bound directly to the input assembler stage, without the aid of a resource view. Because of this, there is no need to create a resource view to use index buffers.

Constant Buffers

The *constant buffer* is the first resource type that we have encountered that is accessible from the programmable shader stages and is subsequently used in HLSL code. A constant buffer is used to provide constant information to the programmable shader programs being executed in the pipeline. The term *constant* refers to the fact that the data inside this buffer remains constant throughout the execution of a *draw* or *dispatch* call. Any information that only changes between pipeline invocations, such as a world transformation matrix or object color, is supplied to the shader program in a constant buffer. This mechanism is the primary means of data transfer from the host application to each of the programmable shader stages. The type and amount of information contained within a constant buffer may vary from buffer to buffer. This depends completely on the data needed for each particular shader program, and is defined by a structure declaration in the shader program. The buffer

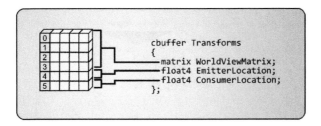

Figure 2.12. A sample constant buffer showing the contents of its structure.

can be tailored to more or less any combination of the basic HLSL types, as well as to structures composed of these basic types. We will cover the HLSL data types in more detail in Chapter 6, "High Level Shading Language," but we'll mention now that these types include scalars, vectors, matrices, arrays of these types, class instances, and combinations of each of these types within structures. A single combination of these types is depicted in Figure 2.12.

The constant buffer is somewhat different than the other buffers that we have discussed thus far. Both the vertex buffer and the index buffer define a basic data element and then repeat that element many times in an array-like fashion. The constant buffer defines a basic element, but it doesn't provide more than one instance of the element—the buffer is created large enough to fit the desired information, but not larger. This implies that the constant buffer implements a structure instead of an array.

Using constant buffers. Each of the programmable pipeline stages can accept one or more constant buffers. It then makes the information in the buffers available for use in the shader program. The data is accessed as if the structure contents were declared globally in the shader program. This means that the elements of each structure must have unique names at this pseudo-global scope. This ability provides a significant amount of flexibility for adding variation to any of the shader programs that are required for a particular rendering algorithm. It is important to note that a buffer created for binding as a constant buffer may not be bound to any other type of connection points on the pipeline. In practice, this is not a problem, since the contents of a constant buffer are already available to all of the programmable pipeline stages anyway. The locations on the pipeline that are available for binding constant buffers are highlighted in Figure 2.13.

The ability to use relatively large constant buffers does not mean that you should automatically create one large buffer for all of the variables that are needed in a shader program. Since constant buffers are updated between uses by the CPU, their contents must be uploaded to the GPU after any changes. Let's assume a hypothetical situation as an example. Suppose there are ten parameters needed by a shader program, but only two of them change between each time that an object is rendered. In this case, we would be writing all

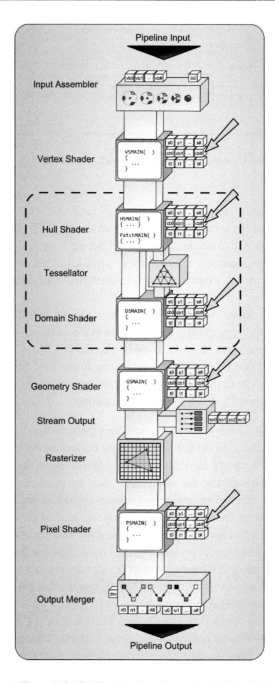

Figure 2.13. The available binding locations for constant buffers in the pipeline.

ten parameters to the buffer between every draw call, just to update the two varying parameters, since the entire buffer is always completely updated. Depending on the number of objects to render, these additional parameter updates can add up to a large amount of unnecessarily wasted bandwidth. If we instead create two constant buffers, where one of them holds the eight static parameters and the second one holds the two dynamic parameters, we can update only the changing parameters and significantly reduce the volume of the required data update for each frame. However, we would need to take care to ensure that the buffers are only updated when they really need to be!

Another potential reduction in updating constant buffers comes from the fact that these buffers only support read accesses from the shader programs. Since it is read-only, a single constant buffer can be simultaneously bound to multiple locations in the pipeline, without the possibility of causing memory access conflicts.

Creating constant buffers. Constant buffers also provide a number of different resource configurations for the best possible performance. Depending on how frequently a constant buffer will be updated by the CPU, one of two typical configurations is chosen. If constant buffers are updated many times throughout an application's lifetime, a dynamic buffer resource makes the most sense. Of course, if a constant buffer will contain some data that will not change throughout an application (such as a fixed back buffer size) it would be more efficient to create it as a completely static buffer, with immutable usage flags. Listing 2.8 demonstrates how a constant buffer is typically created.

```
ID3D11Buffer* CreateConstantBuffer( UINT size,
                                    bool dynamic,
                                    bool CPUupdates,
                                    D3D11_SUBRESOURCE_DATA* pData )
{
    D3D11_BUFFER_DESC desc;
    desc.ByteWidth = size;
    desc.MiscFlags = 0;
    desc.StructureByteStride = 0;
    desc.BindFlags = D3D11_BIND_CONSTANT_BUFFER;

    // Select the appropriate usage and CPU access flags based on the passed
    // in flags
    if ( dynamic && CPUupdates )
    {
        desc.Usage = D3D11_USAGE_DYNAMIC;
        desc.CPUAccessFlags = D3D11_CPU_ACCESS_WRITE;
    }
    else if ( dynamic && !CPUupdates )
    {
        desc.Usage = D3D11_USAGE_DEFAULT;
        desc.CPUAccessFlags = 0;
    }
    else
```

```
    {
        desc.Usage = D3D11_USAGE_IMMUTABLE;
        desc.CPUAccessFlags = 0;
    }

    // Create the buffer with the specified configuration
    ID3D11Buffer* pBuffer = 0;
    HRESULT hr = g_pDevice->CreateBuffer( &desc, pData, &pBuffer );

    if ( FAILED( hr ) )
    {
        // Handle the error here...
        return( 0 );
    }

    return( pBuffer );
}
```

Listing 2.8. A method for creating a constant buffer depending on its intended usage.

As with vertex and index buffers, the creation of dynamic constant buffers uses the D3D11_USAGE_DYNAMIC usage flag, in combination with the D3D11_CPU_ACCESS_WRITE flag, to allow the CPU to update the resource at runtime. For static contents, the D3D11_ USAGE_IMMUTABLE usage is used without any CPU access flags. One additional possibility is to have a buffer that is updated at runtime only by the GPU, such as when copying the number of elements in an Append/Consume buffer to a constant buffer with the ID3D11 DeviceContext::CopyStructureCount() method. In this case, we would require a default usage flag, but we would not set any of the CPU access flags. This allows the runtime to optimize the created resource for use on the GPU only.

There are several items of interest that are not shown in Listing 2.8. The first thing to notice is that the application typically defines a C++ structure that mirrors the desired contents of the constant buffer in HLSL. This allows the application updates to be applied to a system memory instance of this structure, which can then be directly copied into the buffer. We will explore the methods of updating the contents of resources later in this chapter. The second point of interest is that a constant buffer must be created with a ByteWidth that is a multiple of 16 bytes. This requirement allows for efficient processing of the buffer with the 4-tuple register types of the GPU, and it only appears as a requirement for constant buffers. This must be accounted for in the structure declaration within C/C++. The third interesting point is that the buffer description is not allowed to contain any other bind flags than the D3D11_BIND_CONSTANT_BUFFER. Once again, in practice this is not really a big restriction, since there typically are no situations where it would be desirable to bind a constant buffer to another location in the pipeline.

Resource view requirements. Although constant buffers are bound to the programmable shader stages and are accessible through HLSL, their contents are not interpreted with resource views in any way. They are specified exactly as they should be made available within HLSL, so there is no need to use a resource view.

HLSL constant buffer resource object. This is the first buffer type we have encountered that is directly accessible from within HLSL code in the programmable shader stages. In HLSL, constant buffers are resource objects that are declared with the **cbuffer** keyword. A sample declaration is provided in Listing 2.9.

```
cbuffer Transforms
{
        matrix WorldMatrix;
        matrix ViewProjMatrix;
        matrix SkinMatrices[26];
};

cbuffer LightParameters
{
        float3 LightPositionWS;
        float4 LightColor;
};

cbuffer ParticleInsertParameters
{
        float4 EmitterLocation;
        float4 RandomVector;
};
```

Listing 2.9. Various declarations of a constant buffer from within HLSL.

Each of the variables declared within this cbuffer structure can be directly used within the HLSL shader program as if it were declared at a global scope. The name of the constant buffer, as specified in the listing above, is used by the host application to identify the buffer by name, and to load the appropriate contents into it. However, the buffer name is not used within HLSL. As with the constant buffer name, the names and types of the individual elements of the constant buffer are also accessible through the shader reflection API.[1] This series of methods can be used to determine the name and type of each subparameter, to allow the application to know which information to insert into the buffer at runtime.

Buffer/Structured Buffer Resources

Our next type of buffer is referred to by two different names, depending on what type of data it holds. A *standard buffer resource* means a buffer whose elements are one of the

---

[1] The shader reflection API is discussed in more detail in Chapter 6: High Level Shading Language.

built-in data types. In this way, a standard buffer resource is similar to an array of values. Each value is stored at a unique array location and can be referenced by the index of its location. This allows the resource to be easily accessed by many different instances of a shader program simultaneously on the GPU. Since each element is uniquely identified, the developer can easily structure the program to avoid any memory collisions. Figure 2.14 graphically represents a buffer resource.

Very similar to the buffer resource is a structured buffer resource. The only difference between the two is that a structured buffer allows the user to define a structure as the basic element, instead of one of the built-in data types. This makes mapping the data of a particular processing problem to a resource relatively simple. If the developer can define a suitable structure in C++, a corresponding structure in HLSL can be defined, and the buffer will contain an array of these structures, which can be used by the programmable shader stages as a resource object.

The structured buffer is intended to provide a flexible memory resource for simplifying custom algorithm development. Since the data format within the structure can use any of the available types within HLSL, it can be customized to suit a particular scenario. The available structure formation is quite similar to that of the constant buffer, except that the structured buffer provides an array of the desired structures, instead of a single instance like the constant buffer. This concept is depicted graphically in Figure 2.15. In this figure, a hypothetical structure is shown that could represent the data for a particle in a particle system.

Here we see that the particle structure uses multiple variables, and that the complete particle system is contained within the structured buffer. A very interesting point to note is that these buffer types are the first resources that we have discussed that can be used for writing, as well as reading, by the programmable pipeline stages. We will explore the details of how to use this functionality throughout the remainder of this section.

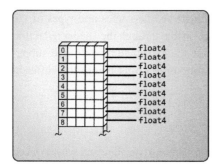

Figure 2.14. A depiction of a standard buffer resource.

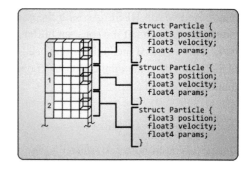

Figure 2.15. A structured buffer used to contain data from a particle system.

Using buffers/structured buffers. Due to this ability to provide a large amount of structured data, the buffer and structured buffer are great options for larger data structures that are to be accessed by the programmable shader stages. These buffers are the first resources that are attached to the pipeline with resource views instead of being directly bound. These buffers are available with read access to all of the programmable pipeline stages, and read/write access is also possible in the pixel and compute shader stages, depending on which resource view is used. This means that these resources can potentially provide a means of communication between the various pipeline stages, as well as between the individual processing elements within the same pipeline stage.

A pipeline stage's resource access capabilities are determined by which type of resource view is used to bind the resource to the pipeline. Unlike the constant buffer, the buffer/structured buffer must be bound to the pipeline through a resource view, which means that it must be created with an appropriate bind flag for binding with either a shader resource view, an unordered access view, or both. The shader resource view allows binding to all of the programmable pipeline stages, while the unordered access view is only allowed to be bound to the pixel and compute shader stages. Since an unordered access view is required to perform write accesses to a resource, the available binding locations for UAVs effectively determine where these write accesses can be performed. The available connection points for both shader resource views and unordered access views are shown in Figure 2.16. Like a constant buffer, a buffer/structured buffer can be bound to multiple locations when used for read-only uses. This is done by binding it to the pipeline with a shader resource view. However, when an unordered access view is used, it may only be bound to a single location. This is enforced by the runtime, which will print error messages if the resource is bound to multiple locations for writing. The use of an unordered access view also excludes simultaneous use of shader resource views, due to the read-after-write issues that it could introduce.

Creating buffers/structured buffers. As we have seen in the other buffer resource creation discussions, it must be determined which usage scenario the buffers/structured buffers will experience. There are in general three different types of usage patterns for these resources. The first case is that the data within the buffer will be static throughout the lifetime of the application. An example of this would be to use the buffer/structured buffer resource to hold precalculated data, such as precomputed radiance transfer data or some other form of a lookup table. The second case is when data needs to be loaded periodically into the buffer by the CPU. This is a likely scenario in many general purpose computation systems, where the GPU is being used as a co-processor to the CPU and is continually feeding the GPU with new data to process. In both of these cases, the GPU is only allowed to read the buffer data and is not able to write to the resource.

The final case arises when the buffer contents are updated in some way by the GPU itself. A common example of this can be found in situations where the GPU is performing

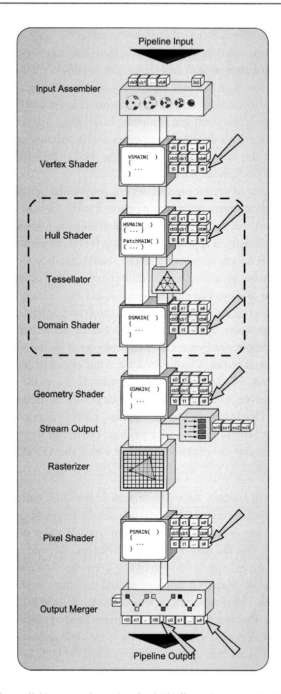

Figure 2.16. The available connection points for the buffer and structured buffer resource types.

some type of physical simulation that is later used for rendering a representation of the results of the simulation. In this case, the GPU will require read and write access to the buffer so that it can read the contents, perform some updates, and then write the results back into the buffer. Each of these three scenarios will require different settings for the usage, CPU Access, and bind flags used when creating the buffer.

In addition to these flags, we must also provide some additional size information when using structured buffers. All types of buffers have required the specification of the ByteWidth parameter, which indicates the desired size of the buffer in bytes. Structured buffers require us to also specify the size of the structure element being used. This information is used as the step size when the resource is accessed by one of the programmable pipeline stages. The other requirement for using structured buffers is to indicate that the buffer resource will indeed be used as a structured buffer. This is done by setting the D3D11_RESOURCE_MISC_BUFFER_STRUCTURED miscellaneous flag. A typical sample creation is shown in Listing 2.10.

```
ID3D11Buffer* CreateStructuredBuffer( UINT count,
                                      UINT structsize,
                                      bool CPUWritable,
                                      bool GPUWritable,
                                      D3D11_SUBRESOURCE_DATA* pData )
{
    D3D11_BUFFER_DESC desc;
    desc.ByteWidth = count * structsize;
    desc.MiscFlags = D3D11_RESOURCE_MISC_BUFFER_STRUCTURED;
    desc.StructureByteStride = structsize;

    // Select the appropriate usage and CPU access flags based on the passed in flags
    if ( !CPUWritable && !GPUWritable )
    {
        desc.BindFlags = D3D11_BIND_SHADER_RESOURCE;
        desc.Usage = D3D11_USAGE_IMMUTABLE;
        desc.CPUAccessFlags = 0;
    }
    else if ( CPUWritable && !GPUWritable )
    {
        desc.BindFlags = D3D11_BIND_SHADER_RESOURCE;
        desc.Usage = D3D11_USAGE_DYNAMIC;
        desc.CPUAccessFlags = D3D11_CPU_ACCESS_WRITE;
    }
    else if ( !CPUWritable && GPUWritable )
    {
        desc.BindFlags = D3D11_BIND_SHADER_RESOURCE |
                         D3D11_BIND_UNORDERED_ACCESS;
        desc.Usage = D3D11_USAGE_DEFAULT;
        desc.CPUAccessFlags = 0;
    }
    else if ( CPUWritable && GPUWritable )
```

```
{
    // Handle the error here...
    // Resources can't be writable by both CPU and GPU simultaneously!
}

// Create the buffer with the specified configuration
ID3D11Buffer* pBuffer = 0;
HRESULT hr = g_pDevice->CreateBuffer( &desc, pData, &pBuffer );

if ( FAILED( hr ) )
{
    // Handle the error here...
    return( 0 );
}

return( pBuffer );
}
```

Listing 2.10. A method for creating a structured buffer.

We split the size information into a *count* of the number of structures to include in the buffer, and a `structsize` parameter to indicate the size of each structure in bytes. These are then used to calculate the overall byte width of the buffer. In addition, each of the three usage scenarios discussed above is represented in the code listing; they differentiate themselves with the CPU access, usage, and bind flags.

Resource view requirements. We have already seen that the only resource views that can be used with a buffer or structured buffer are the shader resource view and the unordered access view. In fact, these are the only ways to bind these buffers to the pipeline. For standard buffer types, the format must be provided directly in the SRV description structure. For structured buffer resources, the format should be set to **DXGI_FORMAT_UNKNOWN**, since there won't be a DXGI format available to match all possible structure types. Instead, the format is derived from the HLSL structure declaration in the shader program, and the size of the elements is provided by the application when the buffer resource is created.

The shader resource view description structure for buffer resources requires specification of the beginning element, and of the number of elements to include in the view. This may select the entire range of elements in the buffer, or it can be used to select a subset of the total resource. This effectively maps the selected data to the [0, width-1] range when viewed from HLSL, with all other accesses considered out of bounds. These values are specified in the D3D11_BUFFER_SRV structure, as shown in Listing 2.11.

```
ID3D11ShaderResourceView* CreateBufferSRV( ID3D11Resource* pResource )
{
    D3D11_SHADER_RESOURCE_VIEW_DESC desc;
```

```
// For structured buffers, DXGI_FORMAT_UNKNOWN must be used!
// For standard buffers, utilize the appropriate format.
desc.Format = DXGI_FORMAT_R32G32B32_FLOAT;

desc.ViewDimension = D3D11_SRV_DIMENSION_BUFFER;
desc.Buffer.ElementOffset = 0;
desc.Buffer.ElementWidth = 100;

ID3D11ShaderResourceView* pView = 0;
HRESULT hr = g_pDevice->CreateShaderResourceView( pResource, &desc,
                                                  &pView );

return( pView );
}
```

Listing 2.11. A method for creating a shader resource view for a buffer resource.

By adjusting the ElementOffset and ElementWidth parameters, a single large buffer resource can be used to contain a collection of smaller datasets, each with its own SRV, to provide access to them. By using the shader resource view to restrict access to a particular subset of the resource and creating an appropriate number of shader resource views, the application can effectively control which data is visible to each invocation of a shader program by using a particular shader resource view. It is also possible to use more than one view to simultaneously select different ranges of a buffer.

The unordered access view uses the same element range selection and format specification, and also provides an additional set of flags that can be specified for special-use scenarios. These flags enable the unordered access view to be used for *append/consume buffers* and *byte address buffers*, both of which are specialized ways to use buffer resources through unordered access views. The specific details of how these function are discussed in more detail in the following two sections. Listing 2.12 demonstrates how to create an unordered access view.

```
ID3D11UnorderedAccessView* CreateBufferUAV( ID3D11Resource* pResource )
{
    D3D11_UNORDERED_ACCESS_VIEW_DESC desc;

    // For structured buffers, DXGI_FORMAT_UNKNOWN must be used!
    // For standard buffers, utilize the appropriate format.
    desc.Format = DXGI_FORMAT_R32G32B32_FLOAT;

    desc.ViewDimension = D3D11_UAV_DIMENSION_BUFFER;
    desc.Buffer.FirstElement = 0;
    desc.Buffer.NumElements = 100;

    desc.Buffer.Flags = D3D11_BUFFER_UAV_FLAG_COUNTER;
    //desc.Buffer.Flags = D3D11_BUFFER_UAV_FLAG_APPEND;
    //desc.Buffer.Flags = D3D11_BUFFER_UAV_FLAG_RAW;
```

```
    ID3D11UnorderedAccessView* pView = 0;
    HRESULT hr = g_pDevice->CreateUnorderedAccessView( pResource, &desc,
                                                        &pView );

    return( pView );
}
```

Listing 2.12. A method for creating an unordered access view for a buffer or structured buffer resource.

The listing shows the setting of the `D3D11_BUFFER_UAV_FLAG_COUNTER` flag, with the append and raw flags commented out. This flag is used to provide a counter for the buffer resource that is accessible through HLSL. This counter is operated by using the methods of the HLSL buffer resource object, as described in the following section.

HLSL structured buffer resource object. Since this type of buffer resource is available in the programmable shader stages, it is necessary to understand how it can be accessed through HLSL. Fortunately, the resource object declaration and usage are fairly straightforward. A structured buffer requires that the corresponding structure be defined in the shader program. Once a suitable structure is available, the structured buffer is declared with a template-like syntax. A sample declaration is shown in Listing 2.13, taken from the fluid simulation demo described in Chapter 12.

```
// Declare the structure that represents one fluid column's state
struct GridPoint
{
    float  Height;
    float4 Flow;
};

// Declare the input and output resources
RWStructuredBuffer<GridPoint> NewWaterState    : register( u0 );
StructuredBuffer<GridPoint>   CurrentWaterState : register( t0 );
```

Listing 2.13. A declaration in HLSL of a standard buffer, a structured buffer, and read/write standard and structured buffers.

In this code listing, we see that there are actually two different ways to declare a structured buffer—with `StructuredBuffer<>` and with `RWStructuredBuffer<>`. These two forms represent the differences between a structured buffer bound to the pipeline with a shader resource view (for read-only access) and with an unordered access view (for read and write access), respectively. This is also indicated by the register statements, with the read/write resource using a u# register and the read-only resource using a t# register. The choice of the declaration type will determine which type of resource view the application will have to use to bind the resource to the programmable shader stage for use in this shader program.

Once the object is declared, we can see that there is a simple way to interact with the structured buffer from within HLSL. The individual elements are accessed with an array like syntax, using brackets to indicate which index to use. This operates very similarly to C++, and the individual members of a structure can be accessed with a dot operator. When the resource is bound with a shader resource view, the elements can only be read. However, with the unordered access view, they can also be written to, using the same syntax. When a subset of a buffer resource is selected with the resource view, its elements appear to be remapped, such that the first element in the subrange is accessed with index 0, and each subsequent element is accessed with an incremented index.

These HLSL resource objects also provide a `GetDimensions()` method that returns the size of a buffer, or the size of the buffer and the structure element size for structured buffers. These can be used by the shader programs to implement range checking. It is also possible to use a counter variable that is built into a `RWStructuredBuffer` if the UAV is created with the `D3D11_BUFFER_UAV_FLAG_COUNTER` flag (as described above in the "Creating Buffers/Structured Buffers" section of this chapter). If a UAV is created with this flag set, the HLSL program can use the `IncrementCounter()` and `DecrementCounter()` methods. These can be used to implement customized data structures by providing a total count of the elements that have been stored in the structured buffer. Using this functionality requires setting the UAV format to be `DXGI_FORMAT_R32_UNKNOWN`.

Append/Consume Buffers

The *append* and *consume* buffer type provide a special variation of a structured buffer resource. In essence, these are both buffers that require unordered access views, but they provide new access methods in HLSL to implement a stack-like behavior. They simply use special unordered access view creation flags that provide a new method for accessing their contents. More specifically, these access methods are used to push elements into the buffers (using the `append()` HLSL method) and pull elements out of them (using the `consume()` HLSL method). In this way, a resource can be used to accumulate data, as well as to distribute data. The order in which elements are added to the buffer is not important, but the number of elements added is. The unordered access view maintains an internal count of the number of items that have been added or removed from the buffer. This allows the runtime and driver to perform more efficiently in some cases, since the ordering of the elements within the buffer does not need to be synchronized between individual GPU threads of execution.

Using append/consume buffers. This functionality can best be explained with an example. In this scenario, a shader program can be used to implement a GPU-based particle system. In fact, this is one of the example algorithms provided later in this book. The particle system would use two buffer resources to hold its particle information. One buffer would hold the current state of the particles, and the other would be used to receive the updated particles. When running, a compute shader program would read one particle per thread from the current state buffer using the consume functionality, then perform the update procedure on the particle, and finally, append the resulting particle state to the output

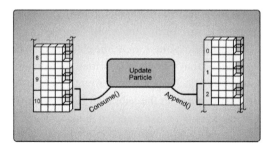

Figure 2.17. A particle system implementation using "append" and "consume" buffers.

buffer. Since each particle is updated in isolation from the others, the particle order within the buffer is completely irrelevant to the updating procedure. In addition, since the buffers maintain a count of the number of elements contained in the buffers, the update procedure can add or remove particles as needed. The buffers' internal counter tracks the total number of elements that are currently residing in the buffer. Because of this, there is no need to synchronize between GPU threads for accessing the correct number of particle elements. This particle system example is depicted in Figure 2.17.

Creating append/consume buffers. To create a buffer that can be used as an append or consume buffer, we require a slightly more restrictive creation process. Since the buffers are intended for read/write access by the programmable shader stages, we require the **D3D11_USAGE_DEFAULT** usage flag to be used. In addition, the bind flag must include the **D3D11_BIND_UNORDERED_ACCESS** flag to allow use of an unordered access view and will typically also include the **D3D11_BIND_SHADER_RESOURCE** flag as well. This also indicates that the CPU is not capable of directly accessing the buffer contents, due to the default usage flag. The remainder of the configurations can be selected as we have already seen in the "Buffer/Structured Buffer Resources" section. An example creation method is shown in Listing 2.14.

```
ID3D11Buffer* CreateAppendConsumeBuffer( UINT size,
                                         UINT structsize,
                                         D3D11_SUBRESOURCE_DATA* pData )
{
    D3D11_BUFFER_DESC desc;
    desc.ByteWidth = size * structsize;
    desc.MiscFlags = D3D11_RESOURCE_MISC_BUFFER_STRUCTURED;
    desc.StructureByteStride = structsize;

    // Select the appropriate usage and CPU access flags based on the passed
    // in flags
    desc.BindFlags = D3D11_BIND_SHADER_RESOURCE | D3D11_BIND_UNORDERED_ACCESS;
    desc.Usage = D3D11_USAGE_DEFAULT;
    desc.CPUAccessFlags = 0;
```

```
    // Create the buffer with the specified configuration
    ID3D11Buffer* pBuffer = 0;
    HRESULT hr = g_pDevice->CreateBuffer( &desc, pData, &pBuffer );

    if ( FAILED( hr ) )
    {
        // Handle the error here...
        return( 0 );
    }

    return( pBuffer );
}
```

Listing 2.14. A method for creating a buffer resource to be used as an append/consume buffer.

One can see that this is more or less one of the configurations that was already available for creating a buffer/structured buffer resource, with the exception that it is mandatory to use the default usage flag. This means that the append/consume buffer usage does not require a specially created buffer resource, but instead, it is the unordered access view that must be specially configured to provide the additional functionality. The UAV used to bind the buffer resource to the pipeline as an append/consume buffer must be created with the D3D11_BUFFER_UAV_FLAG_APPEND flag specified in its description structure.

Resource view requirements. When creating the resource view to be used with the append/consume buffer, there are a few differences from the previously discussed resource views. The format parameter of the resource view must be set to **DXGI_FORMAT_R32_UNKNOWN** to use the append/consume functionality. In addition, the unordered access view must be created with the **D3D11_BUFFER_UAV_FLAG_APPEND** for both append and consume buffers. These requirements are the only additional settings for creating an unordered access view for using a buffer resource as an append/consume buffer.

As we have mentioned in previous buffer discussions, it is possible to simultaneously use multiple resource views on the same resource. Unfortunately, this is not true for the unordered access views. An entire buffer resource is considered to be composed of a single subresource. The unordered access views allow read and write access to their resources, and are not allowed to have more than one writable resource view attached to a single subresource at the same time. This means that, for example, it is not possible to use a single buffer with two unordered access views to simultaneously append and consume from different regions of the buffer. In this case, the append and consume UAVs would need to be attached to separate buffer resources

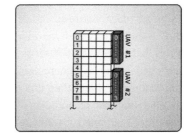

Figure 2.18. A single buffer resource being used by multiple unordered access views.

to be used at the same time. Figure 2.18 shows graphically how a buffer can be referenced simultaneously by more than one resource view.

HLSL append/consume buffer resource objects. Once the buffers have been created with the appropriate resource views, they can be bound to the pipeline and used from within a shader program. The syntax for declaring append and consume buffers is provided in Listing 2.15.

```
struct Particle
{
    float3 position;
    float3 velocity;
    float  time;
};

AppendStructuredBuffer<Particle>   NewSimulationState : register( u0 );
ConsumeStructuredBuffer<Particle>  CurrentSimulationState : register( u1 );
```

Listing 2.15. The declaration of append and consume buffers in HLSL.

Probably the next most frequently used methods for append and consume buffers are the actual `append()` and `consume()` resource object methods. These operate as would be expected, with the desired value being returned in the case of `consume()`, and being pushed into the buffer in the case of `append()`. We can also see in the listing that the `GetDimensions()` method is also available on the append and consume buffers. This can be used from within HLSL to read the number of elements that are available in the buffer, and to subsequently ensure that data is not appended to the buffer too many times, which would result in loss of the additional data.

Byte Address Buffers

Still another variation of the buffer resource type is available. The *byte address buffer* provides a much more raw memory block for the HLSL program to manipulate. Instead of using a fixed structure size to determine where to index within the resource, a byte address buffer simply takes a byte offset from the beginning of the resource and returns the four bytes that begin at that offset as a 32-bit unsigned integer. Since the returned data is always retrieved in four-byte increments, the requested offset addresses must also be a multiple of four. However, other than this minimum size requirement, the HLSL program is allowed to manipulate the resource memory as it sees fit. The returned unsigned integer values can also be reinterpreted as other data types by some of the type-converting intrinsic functions.

It may not be immediately obvious what the utility of such a buffer would be. Since the HLSL program can interpret and manipulate memory contents as it sees fit, there is no requirement for each data record to be the same length, as we have seen in the various types of structured buffers. With variable record lengths, an HLSL program can be written to implement almost any data structure that will fit within the bounds of the resource. It would be equally simple to create a variable-element-sized linked list as it would be to create an array-based binary tree. As long as the program implements the accessing semantics for the data structure, there is complete freedom to use memory as desired. This is a very powerful feature indeed, and it opens the doorway for a very large class of algorithms that either were impossible to implement on the GPU before, or were unwieldy to implement.

Using byte address buffers. As mentioned above, byte address buffers are intended to allow the developer to implement custom data structures in buffer resources. The use of the memory block is then, by definition, open to interpretation by the algorithm that it is being used in conjunction with. For example, if a linked-list data structure will be used to store a list of 32-bit color values, each link node will consist of a color value followed by a link to the next element. When the first element is added, the color value is written into the memory location at offset 0, and the link to the next element is initialized to -1, since the next element has not been added yet. When another element is to be added to the list, the HLSL program would start indexing the memory at location 0 and read two 32-bit elements worth of data—one for the color value and one for the link to the next element. If the link is set to -1, the program has found the tail of the list and can add another element at the end. This simple scenario is shown in Figure 2.19.

With these basics of a linked list available, the program can insert or remove nodes in the same way as in traditional CPU-based programs. This provides a completely generic way to use the resources from within the GPU. However, the parallel nature of the GPU can also introduce some complexity into the use of these resources. Since the memory access patterns are defined by the HLSL program, it is necessary to ensure that memory is accessed in a coherent and safe manner between multiple threads of execution. We will discuss the details of this type of synchronization in Chapter 5, "The Computation Pipeline," but there is a range of techniques available to the developer to ensure thread-safe access to the resources.

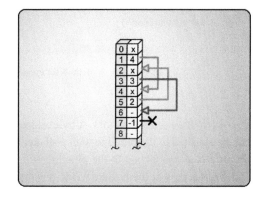

Figure 2.19. A linked-list implementation using byte address buffers.

Creating byte address buffers. Creating a byte address buffer is very similar to the process we have seen for the previous buffer types. It must also indicate the usage scenario that it will be used for, and at what locations it will be bound to the pipeline with its bind flags. A byte address buffer must also be created with the **D3D11_RESOURCE_MISC_ BUFFER_ALLOW_RAW_VIEWS** miscellaneous flag. An example method for creating such a resource is provided in Listing 2.16.

```
ID3D11Buffer* CreateRawBuffer( UINT size,
                               bool GPUWritable,
                               D3D11_SUBRESOURCE_DATA* pData )
{
    D3D11_BUFFER_DESC desc;
    desc.ByteWidth = size;
    desc.MiscFlags = D3D11_RESOURCE_MISC_BUFFER_ALLOW_RAW_VIEWS;
    desc.StructureByteStride = 0;

    // Select the appropriate usage and CPU access flags based on the passed
    // in flags
    desc.BindFlags = D3D11_BIND_SHADER_RESOURCE | D3D11_BIND_UNORDERED_ACCESS;
    desc.Usage = D3D11_USAGE_DEFAULT;
    desc.CPUAccessFlags = 0;

    // Create the buffer with the specified configuration
    ID3D11Buffer* pBuffer = 0;
    HRESULT hr = g_pDevice->CreateBuffer( &desc, pData, &pBuffer );

    if ( FAILED( hr ) )
    {
        // Handle the error here...
        return( 0 );
    }

    return( pBuffer );
}
```

Listing 2.16. A method for creating byte address buffer resources.

Resource view requirements. A byte address buffer can be used in two different ways. It can be attached to the pipeline with a shader resource view for read-only access to the buffer, or it can be attached to the pipeline with an unordered access view that provides read and write access to the buffer. While the shader resource view provides read-only access, it is available to all of the programmable shader stages. The unordered access view is only available in the compute and pixel shader stages. The formats for the resource views must be the **DXGI_FORMAT_R32_TYPELESS** in both cases. When an unordered access view is used, it must also be created with the **D3D11_BUFFER_UAV_FLAG_RAW** to indicate that it would supply access to data as a byte address buffer.

HLSL byte address buffer objects. Once the buffers have been created and appropriate resource views are available for binding the resources to the pipeline, the HLSL program must declare the appropriate resource object to interact with the buffers. There are two different objects that can be declared, each of which corresponds to the type of access that the HLSL code will have to the resource. If the buffer is bound to the pipeline with a shader resource view, the corresponding HLSL object will be declared as a *ByteAddressBuffer*. If it is bound to the pipeline with an unordered access view, the corresponding HLSL object will be declared as a *RWByteAddressBuffer*. Samples of these byte address buffer type declarations are provided in Listing 2.17.

```
ByteAddressBuffer   rawBuffer1
RWByteAddressBuffer rawBuffer2;
```

Listing 2.17. Declaring byte address buffers in HLSL.

In general, the methods of these objects require an address to specify the location within the buffer at which to perform the given operation. These addresses must be 4-byte aligned, and always work with the `uint` data type. If other data types are required, the returned data can be reinterpreted using one of the conversion intrinsic methods.[2] Each of these two objects support the various *Load* methods that can be used to retrieve data from the buffer resource, and up to four 32-bit memory locations can be read at a time. However, the read/write byte address buffer also provides a large array of methods for manipulating the contents of the buffer. There are the simple *Store* analogs to the *Load* methods, in addition to a host of atomic instructions for updating the resource simultaneously from many threads.

Indirect Argument Buffers

The final buffer configuration that we will inspect is the *indirect argument buffer*. As we will see in many cases throughout this book, Direct3D 11 has taken many steps toward making the GPU more useful and able to operate on its own. The indirect argument buffer is one of these steps. It is used to provide parameters to the rendering and computation pipeline invocation methods from within a resource, instead of having those parameters directly passed by the host program. The concept behind allowing this type of pipeline control is to allow the GPU to generate the needed geometry or input data resources in one execution pass, and to then process or render that data in a following pass, without the CPU having any knowledge of how many primitives are being processed. Even though the CPU must still initiate the pipeline execution and pass the indirect argument buffer reference, this scenario reduces the CPU's role to a simple broker between GPU passes.

Using indirect argument buffers. Indirect argument buffers can be used in several different pipeline execution methods. We haven't covered the pipeline execution in detail yet,

---

[2] The conversion intrinsic methods are discussed in Chapter 6, "High Level Shading Language."

but we can still discuss how these buffers operate without intimate knowledge of what each parameter means. The available indirect argument pipeline execution methods are listed below.

▪ DrawInstancedIndirect(ID3D11Buffer *pBufferForArgs,
   UINT AlignedByteOffsetForArgs)

▪ DrawIndexedInstancedIndirect(ID3D11Buffer *pBufferForArgs,
   UINT AlignedByteOffsetForArgs)

▪ DispatchIndirect(ID3D11Buffer *pBufferForArgs, UINT AlignedByteOffsetForArgs)

Each of these methods appends the *Indirect* term at the end of one of the standard methods, and simply replaces the arguments of the method with a reference to a buffer resource and an offset into the buffer, which identifies where the parameters are located in it. For example, the standard `DrawInstanced()` method takes four `uint` parameters: `VertexCountPerInstance`, `InstanceCount`, `StartVertexLocation`, and `StartInstanceLocation`. These arguments are passed directly in the API call to instruct the runtime and driver what to draw and what options are desired. The `DrawInstancedIndirect()` method passes these values to the runtime and driver in a buffer resource, instead of directly passing them. For this to work, the application must ensure that there are four consecutive `uint` values stored within the buffer at the location indicated by the offset argument passed with the buffer reference. Including an offset argument in these methods allows a single buffer to be used for many different pipeline invocations, and makes it possible to store the parameters simultaneously in different locations. Figure 2.20 shows how this scheme operates. It is also important to note that the offset into the buffer must be 4-byte aligned to allow the individual values to stride the standard 32-bit variable sizes.

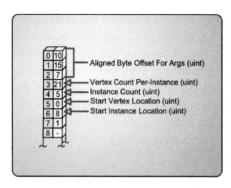

Figure 2.20. The indirect arguments buffer resource being used to execute the rendering pipeline.

There are many ways to update the data in the indirect argument buffer. Any pipeline output method or CPU manipulation technique can be used. As long as the data is available in the buffer, it can be used to execute the pipeline. For example, one possibility is to use the `AppendStructuredBuffer` resource type in a compute shader to fill it with some GPU-generated content. The number of items within the buffer could be copied to the indirect argument buffer with the `ID3D11DeviceContext::CopyStructureCount()` method, and the geometry could finally be drawn with the `ID3D11DeviceContext::DrawInstancedIndirect()` method, with a single instance provided. This would allow the GPU-generated content within the `AppendStructuredBuffer` to be rendered directly, without the CPU ever knowing how many items exist within the buffer itself.

Creating indirect argument buffers. Creating a indirect argument buffer is quite similar to the other creation methods we have seen before. It also follows the same usage control semantics that we have seen. In general, if we want to be able to generate the data on the GPU and then subsequently use the buffer as input to the pipeline, we must select the default usage flag to provide both read and write access to the GPU. In addition, to indicate to the runtime that this buffer will be used for indirect arguments, the **D3D11_RESOURCE_MISC_DRAWINDIRECT_ARGS** miscellaneous flag must be specified. The **ByteWidth** parameter should be a 4-byte multiple, since the input arguments to the *draw* and *dispatch* methods are all 32-bit variables. A sample method for creating these buffers is provided in Listing 2.18.

```
ID3D11Buffer* CreateIndirectArgsBuffer( UINT size, D3D11_SUBRESOURCE_DATA* pData )
{
    D3D11_BUFFER_DESC desc;
    desc.ByteWidth = size;
    desc.MiscFlags = D3D11_RESOURCE_MISC_DRAWINDIRECT_ARGS;
    desc.Usage = D3D11_USAGE_DEFAULT;
    desc.StructureByteStride = 0;
    desc.BindFlags = 0;
    desc.CPUAccessFlags = 0;

    // Create the buffer with the specified configuration
    ID3D11Buffer* pBuffer = 0;
    HRESULT hr = g_pDevice->CreateBuffer( &desc, pData, &pBuffer );
    if ( FAILED( hr ) )
    {
        // Handle the error here...
        return( 0 );
    }

    return( pBuffer );
}
```

Listing 2.18. A sample method for creating a buffer resource to be used as an indirect arguments buffer.

Resource view requirements. An indirect argument buffer is used in two different phases. First, it is loaded by the GPU or CPU with the data that will eventually represent the parameters to a pipeline invocation call. Second, it is passed as input to a *draw/dispatch* call by the CPU. The first case may require a resource view, while the second case does not.

When the buffer is to be filled by the GPU, it can be updated with the stream output functionality, written to as a render target, or modified with an unordered access view. If the buffer will be updated by the CPU, one of the standard methods for modifying its contents can be used, such as the `ID3D11DeviceContext::UpdateSubresource()` method, or updated with the `ID3D11DeviceContext::CopyStructureCount()` method to read the hidden counter value from a buffer. In general, these techniques for writing to the buffer are discussed in their respective areas of Chapter 2 and Chapter 3 and won't be repeated here. However, if it is possible to get the data into the buffer, it can subsequently be used to execute the pipeline.

HLSL indirect argument buffer objects. The indirect argument buffers are used as input to the pipeline execution methods, and are hence not directly used in HLSL. As mentioned above, there are a variety of methods for getting the parameter data into the buffer, some of which will require resource views and must be written directly from HLSL, and some which do not. The particular technique used will dictate if the data can be written to the buffer through an HLSL resource object.

2.2.2 Texture Resources

As we have seen throughout the last few sections, there are a myriad of different buffer resources, and many different ways to configure them for various uses. However, the buffers are only half of the resource story. Textures provide a different concept for resources based on the evolutionary roots of the GPU as a rendering co-processor. The term *texture* refers to a memory resource that is similar in nature to an image, or image-like. This is a very loose definition, because there is a range of different texture types, with varying dimensions, topologies, and multisampling characteristics. These make the term *image-like* seem somewhat out of place, since some textures don't resemble a traditional two-dimensional image at all. The common link between these texture resources and an image is the concept of a *pixel element*, which is the same across all texture types. The pixel (or *texel*, as it is referred to in textures) is the smallest data element that all textures are comprised of. Each pixel is represented by up to four components. The format of the components varies, depending on the function of the texture resource.

While texture resources are still blocks of memory that are made available to the GPU (as are buffer resources), additional dedicated hardware is available for certain texture operations, which can provide greater efficiency than a corresponding operation on a buffer. One example of this would be texture-filtering hardware in a GPU. It is simply faster to interpolate between two neighboring data elements with a texture resource than it is with a buffer resource, because of the additional hardware. In addition, some functions are naturally better suited to textures, such as render targets or depth stencil targets. Since buffers and textures have different qualities, the appropriate resource type must be chosen for a particular situation. In addition, textures support the ability to use mip-maps, which are essential for achieving good performance and visual quality when rendering texture-mapped geometry.

We will explore the texture resource types in more detail throughout this chapter. We begin with a general discussion about textures, and some specific details about their uses and available operations. This is followed by a more detailed look at using, creating, and manipulating each type of texture resource, both in C++ and in HLSL.

Common Texture Properties

As mentioned above, texture resources are available in 1D, 2D, and 3D versions. The specific memory layouts that they use are distinct from one another, due to their differences in dimension. These are similar to the differences you would expect to see in C/C++ in layout between arrays of different dimensions. As an indication of how these layouts are organized from a usage perspective, consider the graphical representations shown in Figure 2.21. Even with distinct layouts, the texture resources share many other properties. We will explore some of these common properties before moving on to examine each texture type.

As shown in Figure 2.21, textures are arranged in according to the familiar X, Y, and Z axes. This figure also includes additional axis descriptions, named U, V, and W. These indicate the texture addressing scheme used by Direct3D 11, in which the total size of a texture (regardless of how large or small it is) in each of the X, Y, and Z directions is mapped into a range of [0,1]. This is done to provide a convenient method for sampling a texture. If you sample a one-dimensional texture at address $u=0.5$, then regardless of whether the texture has 16 elements, or 1600 elements, you will receive the element that is at the halfway point of the texture. This is a very important consideration when performing sampling, and will be considered further when we discuss samplers in the "Sampler State Object" section.

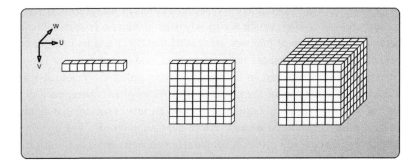

Figure 2.21. A visualization of the various texture configurations.

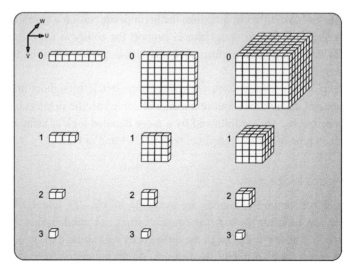

Figure 2.22. A visual representation of texture mip-maps.

Texture resource mip-maps. All of the various types of textures share the concept of *mip-maps*. A mip-map quite simply represents a lower-resolution version of a texture. It is common for a texture to contain several mip-map levels, with each successive level providing a half scale copy of the texture contents. Using mip-maps allows for lower-resolution texture lookups in situations where the highest levels of resolution will not add useful information, or could even contribute to rendering artifacts. A common example of this is when a textured model is rendered far from the camera, and a texture applied to the model appears to sparkle when the model moves. This is caused when the sampling pattern of the rendered image is at a much lower resolution than the texture content's sampling pattern. By lowering the effective texture resolution by looking up an appropriate mip-map level instead, the two sampling patterns can be brought much closer together, virtually eliminating this shimmering effect.

Figure 2.22 visualizes these mip-maps for each of the three texture types. Because mip-mapping also reduces the effective size of the resources being used, it significantly reduces *texture cache thrashing*, which occurs when the texture cache is continually forced to miss because large chunks of memory are loaded that don't end up being used, which ultimately leads to very high memory request latencies, and degrades performance.

Each mip-map is considered to be a *subresource* within the resource itself. Each mip-map level reduces the size of the resource by a factor of two along each dimension. The number of mip-map levels is limited by the lowest level containing a single texture element, which is logical, since you can't reduce the resolution of a 1×1 texture. We will see later in this chapter how subresources can be selected when creating resource views, as well as for manipulating resource contents.

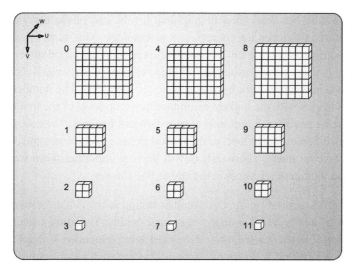

Figure 2.23. A texture resource with texture slices.

Texture resource arrays. Several types of texture resources can also be created as an array of textures. For example, a single 2D texture can be created as an array of 2D texture subresources. This allows a single resource to contain what are referred to as many individual *texture slices*. This is shown in Figure 2.23.

When a resource is created as an array, each of the individual texture slices can be selected with a resource view and used as if it were a standalone resource. This allows the texture resources to be used as a higher-level data structure that encapsulates multiple different contents in each texture slice. For example, having access to multiple texture slices can simplify some forms of *texture atlases*, which are essentially a texture resource that contains multiple individual textures. Texture atlases can then be used for multiple object renderings, which allows them to be rendered consecutively without modifying the pipeline state in between draw calls. Having access to an array-based resource significantly simplifies this type of technique.

Texture arrays can be created for 1D and 2D texture types, as well as the *texture cube* type (which is a form of 2D texture array), but which is not allowed for 3D textures. In practice, this limitation is not really a problem, since having an array of 3D textures would essentially be representing a 4D data resource. The memory consumption of such a resource would grow extremely quickly with the resource size in each dimension. However, if such a data structure is needed, the application can simply use multiple individual 3D texture resources or can simply create a 3D texture where the third dimension is specified as a multiple of basic record size and can manually be indexed by a shader program, or implemented with a series of resource views that select appropriate subregions of the resource.

Subresources. With our discussion of mip-map levels, and of the array elements of a resource, we have introduced the concept of a *subresource*. This name is used to identify complete subportions of a resource. For example, every mip-map level of every element of a texture array is a unique subresource. To aid in selecting a particular subresource, each one is given a subresource index by which it can be identified. The numbering of subresource indices begins with the highest-resolution mip-map level of the first element of an array, and increments for each mip-map level within that element. The count continues on the highest-resolution mip-map level of the second element of the array, and continues until all resources have an index. These indices will be very important when we consider the methods for manipulating resources at the end of this chapter.

Texture 2D multisampled resources. Another option is available for two-dimensional texture resources, and is aimed at improving image quality. Two-dimensional textures can be created as *multisample textures*, and can be used to implement *multisample anti-aliasing* (*MSAA*). The idea behind this resource type is that for each pixel, multiple subsamples are actually stored in a pattern within the pixel boundaries. When rendering is carried out, these subsamples are used to determine at a subpixel level the amount of coverage that a pixel should receive from a given primitive being rasterized. This lets each pixel be generated from a number of subsamples, which improves the quality of geometry edges by effectively increasing the sampling rate of the render target. An example pixel with a subsample pattern is shown in Figure 2.24.

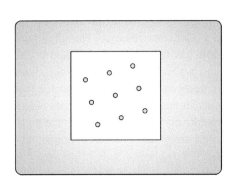

Up to 32 subsamples can be selected by the application, depending on hardware support. In the "Texture 2D" section we will see how to create a texture with a valid number of samples for a given hardware configuration. Care must be taken when using MSAA, since the amount of memory consumed by the resources in question is effectively multiplied by the number of subsamples used. Further details about why MSAA is useful, as well as information on how to use these resources, is discussed in Chapter 3 in the "Rasterizer," "Pixel Shader," and "Output Merger" sections.

Figure 2.24. An example subsample pattern within a pixel.

Texture 1D

The first texture resource type that we will examine is the 1D texture. These textures are arranged along a single axis, with each element being comprised of one of the basic DXGI formats provided in the DXGI_FORMAT enumeration. The various subresource configurations of a 1D texture are shown in Figure 2.25. These texture resources can be created

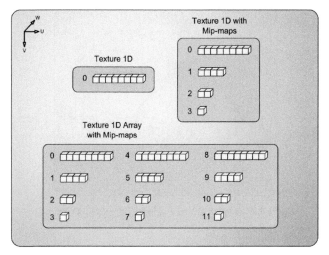

Figure 2.25. A visualization of a 1D texture resource.

with or without mip-maps, or as an array of 1D textures with or without mip-maps. The subresource index is also visible in Figure 2.25.

Using 1D textures. One-dimensional textures are commonly used to implement lookup tables. For example, if a very complex function can be replaced by a 1D lookup texture, it is possible to reduce the number of arithmetic instructions needed by replacing the function with a texture lookup instruction instead. Depending on the workload present in a given shader program, this can be beneficial to the performance of the shader. In addition, since the GPU has dedicated hardware for filtering texture data with samplers, the filtering process is performed more or less for free, from the shader program's point of view. It doesn't need to perform any additional arithmetic to receive a filtered texture value. Figure 2.26 demonstrates a complex trigonometric function that can be replaced by a 1D texture with the values indicated in the image. In this type of usage, the texture can use a single component format, such the **DXGI_FORMAT_R32_FLOAT** format.

Another common use of 1D textures is to implement a color visualization scale. In this scenario, a 1D texture resource is

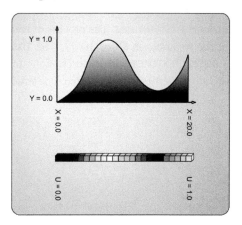

Figure 2.26. A visualization of replacing a complex function with a simple lookup-table, which is implemented as a 1D texture.

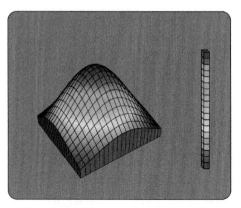

Figure 2.27. Implementing a color scale with a 1D texture.

populated with color values, where each color represents a scalar data value range. For example, when the value is in a normal range, it can be colored blue and progress to "hotter" colors as the input value approaches a more critical range. To implement such a color scale within a shader program, the particular scalar property value that is being visualized is mapped to the [0,1] range and is then used to sample the appropriate color in the 1D texture. This process is depicted in Figure 2.27.

A 1D texture resource can also be used in a computational context within the compute shader. With the general purpose computation capabilities of the compute shader, any 1D data array can be stored in a texture. If the data can be stored as a texture—that is, if it doesn't require a structure to represent its basic element—it can take advantage of the sampling abilities of the GPU, as described above, with little or no additional computational cost.

Creating 1D textures. The process of creating a 1D texture follows the same process as all of the resources in Direct3D 11. The **ID3D11Device::CreateTexture1D()** method takes a **D3D11_TEXTURE1D_DESC** description structure pointer, which specifies all of the desired texture properties of the resource. The structure and its members are provided in Listing 2.19.

```
struct D3D11_TEXTURE1D_DESC {
    UINT        Width;
    UINT        MipLevels;
    UINT        ArraySize;
    DXGI_FORMAT Format;
    D3D11_USAGE Usage;
    UINT        BindFlags;
    UINT        CPUAccessFlags;
    UINT        MiscFlags;
}
```

Listing 2.19. The D3D11_TEXTURE1D_DESC structure and its members.

The Width parameter alone specifies the size of the texture, because this is a one-dimensional texture. It represents the number of texture elements that will be included in the texture. The desired number of mip-map levels is specified in the MipLevels member. The most important consideration for the number of mip-maps is that a value of 1 provides just the top level mip-map, while a value of 0 will create a complete mip-map chain all the way down to a single element. Any value greater than 1 will specify the number of mip-map levels, until the single element mip-map level is reached. The ArraySize member of this method indicates how many texture elements to create in the array. Each texture slice in the array will contain the number of mip-maps specified in the MipLevels member. The format of the texture is logically specified in the Format member. This must specify one of the DXGI_FORMAT types, which provide a large variety of different precisions, component counts, and data types. The next three members of this structure define the usage characteristics, as described at the beginning of this chapter. However, the MiscFlags member allows for some customized resource behavior for special uses. The available flags that are usable with a Texture1D resource are listed below.

- D3D11_RESOURCE_MISC_GENERATE_MIPS

- D3D11_RESOURCE_MISC_RESOURCE_CLAMP

The generate mips flag allows the Direct3D runtime to populate the available mip-map levels with data based on the current values within the top level. To use this functionality, the resource must be used as a render target (and thus created with the D3D11_BIND_RENDER_TARGET bind flag) to fill the top mip-map level, and then the lower mip-map levels can be filled with a call to the ID3D11DeviceContext::GenerateMips() method. This method takes a shader resource view of the texture resource and will fill in the mip-map levels indicated in the view. Of course, to be able to use the resource in a shader stage, it must also be created with the shader resource bind flag (D3D11_BIND_SHADER_RESOURCE).

The resource clamp flag is used to indicate that a texture resource will use the minimum LOD functionality. This is a way for the application to use the device context to set the maximum detail level of a texture that will be used. The specification of the maximum detail level is performed with the ID3D11DeviceContext::SetResourceMinLOD() method, which takes a pointer to the resource and a float parameter that can have a value between 0 and 1. This functionality allows the GPU to page out unused portions of a resource so that more memory can be devoted to resources that are more likely to be used. This setting could be used to implement a resource level of detail that varies as a function of the distance from the camera, where the top mip-map levels will not be used when an object is very far from the camera.

Resource view requirements. A 1D texture resource can use all four types of resource views and cannot be bound to the pipeline without a resource view. The usage semantics associated with each of these resource views that we have seen for buffers also hold true for textures. However, depending on the resource dimensions and the type of resource view being used, a subset of the resource can be specified to be visible to the resource view. This provides the ability to manipulate only portions of a resource at a time and also allows multiple views to reference different subresource regions of a resource. The subresource selections are described below for each of these cases.

Texture1D shader resource view. For standard, non-array-based 1D texture resources, a shader resource view can be used to select a range of mip-map levels and make them accessible to the programmable shader stages. This gives the developer the option of either using a complete resource, or only using a subset of it. In general, the subset range of mip-map levels will be treated the same as a full resource by shader programs, since they are not aware of what the resource behind the resource view actually looks like. The parameters used to determine the subset are specified in the **D3D11_TEX1D_SRV** structure, which is used as the union parameter from the **D3D11_SHADER_RESOURCE_VIEW_DESC** structure (which we have seen in the "Resource Views" section earlier in this chapter). These parameters are shown in Listing 2.20.

```
struct D3D11_TEX1D_SRV {
    UINT MostDetailedMip;
    UINT MipLevels;
}
```

Listing 2.20. The contents of the D3D11_TEX1D_SRV structure.

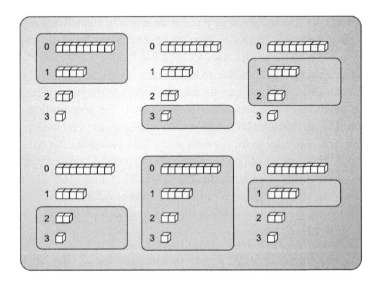

Figure 2.28. Several examples of mip-map ranges that can be selected.

The starting mip-map level is specified by the `MostDetailedMip` parameter, and the number of levels that follow it is specified in the `MipLevels` parameter. A few possibilities for mip-map level selection are graphically demonstrated in Figure 2.28.

For array-based 1D texture resources, a shader resource view references a subset of mip-map levels in each texture slice in the same way as mentioned above for non-array resources, but it can also select a subrange of texture slices to read those mip-map levels from. The range selections are made with the **D3D11_TEX1D_ARRAY_SRV** structure in the **D3D11_SHADER_RESOURCE_VIEW_DESC** structure. These parameters are shown in Listing 2.21.

```
struct D3D11_TEX1D_ARRAY_SRV {
    UINT MostDetailedMip;
    UINT MipLevels;
    UINT FirstArraySlice;
    UINT ArraySize;
}
```

Listing 2.21. The members of the D3D11_TEX1D_ARRAY_SRV structure.

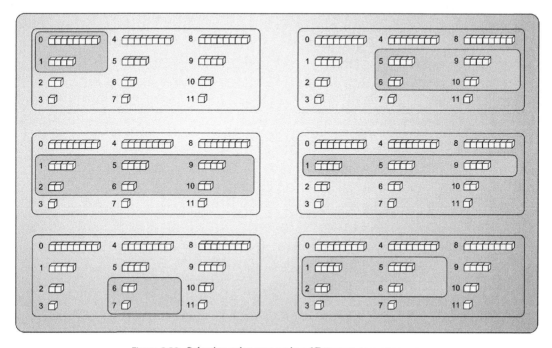

Figure 2.29. Selecting subresources in a 1D texture array resource.

As seen in Listing 2.21, the `MostDetailedMip` and `MipLevels` parameters are the same as they were for non-array resources. However, the texture slices are selected with the additional `FirstArraySlice` and the `ArraySize` parameters, which specify the starting slice and how many subsequent slices will be included in the view. This is depicted graphically in Figure 2.29.

Texture1D unordered access view. The unordered access view for non-array 1D texture resources specifies a single mip-map level. In practice, the restriction to a single mip-map level is not really a hindrance, since it is possible to create additional unordered access views to access other mip-map levels. This solution only partially overcomes the issue, since there are a limited number of UAV slots, but an increased number of mip-maps could be processed in multiple passes to overcome this limit. The desired mip-map level is specified in the **D3D11_TEX1D_UAV** structure within the **D3D11_UNORDERED_ACCESS_VIEW_DESC** structure.

```
struct D3D11_TEX1D_UAV {
    UINT MipSlice;
}
```

Listing 2.22. The members of the D3D11_TEX1D_UAV structure.

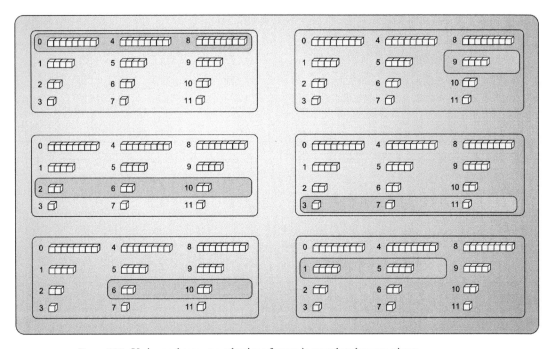

Figure 2.30. Various subresource selections for use in unordered access views.

For array-based 1D texture resources, the unordered access view still specifies a single mip-map level, but it also specifies an array subset in the same way as the render target view did. These parameters are specified in the `D3D11_TEX1D_ARRAY_UAV` structure within the `D3D11_UNORDERED_ACCESS_VIEW_DESC` structure. This range is depicted graphically in Figure 2.30 for a variety of different unordered access views.

```
struct D3D11_TEX1D_ARRAY_UAV {
    UINT MipSlice;
    UINT FirstArraySlice;
    UINT ArraySize;
}
```

Listing 2.23. The members of the D3D11_TEX1D_ARRAY_UAV structure.

Texture1D render target view. When a 1D texture resource is used as a render target, the render target view only specifies a single mip-map level. This is due to the way that a render target is written to, which does not allow for multiple mip-map subresources to be written to simultaneously. This mip-map level selection is performed with the **D3D11_TEX1D_RTV** structure within the **D3D11_RENDER_TARGET_VIEW_DESC** structure, as shown in Listing 2.24.

```
struct D3D11_TEX1D_RTV {
    UINT MipSlice;
}
```

Listing 2.24. The members of the D3D11_TEX1D_RTV structure.

When a render target view is used with a 1D texture array resource, it exposes a single mip-map level of a subrange of the array elements of the resource. This selection pattern is the same as that of the unordered access view with array-based resources. The subresource selection is performed with the **D3D11_TEX1D_ARRAY_RTV** structure within the **D3D11_RENDER_TARGET_VIEW_DESC** structure, as shown in Listing 2.25.

```
struct D3D11_TEX1D_ARRAY_RTV {
    UINT MipSlice;
    UINT FirstArraySlice;
    UINT ArraySize;
}
```

Listing 2.25. The members of the D3D11_TEX1D_ARRAY_RTV structure.

Texture 1D depth stencil view. In both the array and non-array **Texture1D** resources, the depth stencil view uses the same subresource regions as indicated for the render target views. This makes sense, since the render target size and dimension are always required to match that of the depth stencil target. The two structures used to select these subresource regions are shown in Listing 2.26.

```
struct D3D11_TEX1D_DSV {
    UINT MipSlice;
}
struct D3D11_TEX1D_ARRAY_DSV {
    UINT MipSlice;
    UINT FirstArraySlice;
    UINT ArraySize;
}
```

Listing 2.26. The members of the D3D11_TEX1D_DSV and D3D11_TEX1D_ARRAY_DSV structures.

HLSL objects. One-dimensional texture resources may be accessed from within the HLSL programs that are executed in the programmable shader stages. However, they interact with the resources through resource objects that must be declared in the HLSL source file. These objects essentially form the connection between the shader program and the data that is exposed through the resource view used to bind a resource to the pipeline. There are two possibilities for binding resources to a pipeline stage so that they can be accessed from HLSL, either using a shader resource view or an unordered access view. The shader resource view provides read-only access to resources. If a resource is bound with a shader resource view, its resource object in HLSL will be declared as either a **Texture1D** or a **Texture1DArray**, depending on whether the resource is array based or not. The unordered access view allows read and write access to the resource data that it exposes. To indicate this difference in usage, its resource objects in HLSL are declared with the same names, but prepended with an *RW* to indicate their read/write nature. Listing 2.27 demonstrates several example declarations in HLSL using these various resource objects.

```
Texture1D<float> tex01;
Texture1DArray<uint3> tex02;
RWTexture1D<float4> tex03;
RWTexture1DArray<int2> tex04;
```

Listing 2.27. Several resource object declarations for use with 1D texture resources.

When the texture is declared in HLSL, it specifies a template-style parameter to indicate what format the resource must be compatible with. This is used to ensure that when a resource view is bound to the pipeline for a particular resource object declaration, it contains the appropriate format. The usage of HLSL resource objects is described in more detail in Chapter 6.

Texture 2D

The second texture type that we will explore is the 2D texture. Since this texture type is closely related to a standard 2D image layout, it is typically one of the most widely used types of texture. This resource is organized as a two dimensional grid of elements, where each element is a member of the DXGI_FORMAT enumeration. The various subresource forms of a 2D texture are shown in Figure 2.31.

As seen in Figure 2.31, these textures support the use of mip-map levels and texture arrays in a similar manner as the 1D texture did. However, they also support multisampled resources to implement MSAA. This is the only texture resource type that allows multisampling, and it is hence a very important resource type.

Using 2D textures. Two-dimensional textures are the primary texture resource used in Direct3D 11. As such, they have many different uses within a real-time rendering application. The first and most visible use is that of a render target. The Direct3D 11 rendering pipeline is the backbone of any real-time rendering software, and a render target is needed to receive the results of all of the calculations performed in it. In addition, during this rendering process, 2D textures are typically used for the depth/stencil target as well, since the render and depth/stencil targets are required to match in size and dimension. Once a complete scene rendering has been performed, the contents of the render target are displayed in the desired output window for viewing onscreen. We will see this usage pattern many times throughout the remainder of the book.

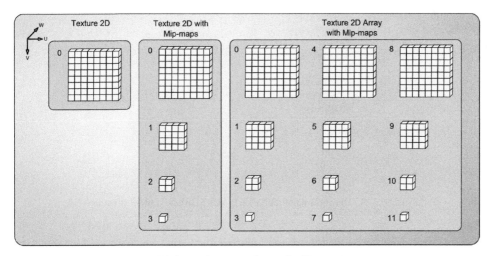

Figure 2.31. Various subresource forms of a 2D texture resource.

Outside of the standard render and depth targets, 2D textures are commonly used for applying surface properties to geometry being rendered. For example, if a brick wall is being rendered, the color of the bricks is typically supplied in a 2D texture that is read during the rendering process. The surface properties supplied don't necessarily need to be limited to the color of the surface, either. It is very common in modern rendering systems for surface normal vectors to be modified with what is called a *normal map*. This is a texture that supplies a 3D vector to represent a surface normal, where the *X*, *Y*, and *Z* coordinates are stored in the R, G, and B channels of the texture. Still other properties are encoded into textures, such as surface glossiness, displacement, and emissivity. It is fairly common to hear these types of textures referred to as *maps*. For example, the textures mentioned above would be referred to as *gloss maps*, *displacement maps*, or *emissivity maps*.

In addition to the per-object uses described above, it is also becoming increasingly important to perform post-processing of a scene rendering to increase the quality of the output presented to the user. In these situations, the output rendering is used as an input to various algorithms that process the image further or combine the image with other information to enhance it in some way. The algorithms in Chapter 10 demonstrate how this can be performed.

Creating 2D textures. The creation of a 2D texture is quite similar to the 1D texture case. Along with the addition of the extra dimension specification, an extra structure parameter is added to determine the multisample level to be used in the texture. The texture is created with the **ID3DDevice::CreateTexture2D()** method, which takes a pointer to the **D3D11_TEXTURE2D_DESC** structure, as shown in Listing 2.28.

```
struct D3D11_TEXTURE2D_DESC {
    UINT              Width;
    UINT              Height;
    UINT              MipLevels;
    UINT              ArraySize;
    DXGI_FORMAT       Format;
    DXGI_SAMPLE_DESC  SampleDesc;
    D3D11_USAGE       Usage;
    UINT              BindFlags;
    UINT              CPUAccessFlags;
    UINT              MiscFlags;
}
struct DXGI_SAMPLE_DESC {
    UINT Count;
    UINT Quality;
}
```

Listing 2.28. The contents of the D3D11_TEXTURE2D_DESC structure.

As mentioned above, this structure is quite similar to that which is used for 1D texture creation. The size of the texture is specified with the Width and Height parameters, and the number of mip-map levels and array slices are specified in the `MipLevels` and `ArraySize` elements, respectively. The texture format also operates in the same manner as for the 1D texture and must be set to a member of the `DXGI_FORMAT` enumeration. The `SampleDesc` parameter is a structure that describes the multi-sampling features that will be included with the 2D texture. It contains two members: `Count` and `Quality`. The `Count` parameter specifies the number of subsamples to include in the texture, while the `Quality` parameter indicates the pattern and algorithm used to resolve the subsamples into a single pixel value.

The quality level is a vendor-specific metric, and it may or may not support more than one quality level. Since support for various quality levels is left up to the GPU manufacturer, the application must query what the available quality level is for a particular texture format and sample count. This is performed with the `ID3D11Device::CheckMultisample QualityLevels()` method. An example usage of this method is shown in Listing 2.29.

```
UINT NumQuality;
HRESULT hr = m_pDevice->CheckMultisampleQualityLevels(
                        DXGI_FORMAT_R8G8B8A8_UNORM, 4, &NumQuality );
```

Listing 2.29. A demonstration of how to check the available quality level of a texture format and multi-sample count.

In this example, we can see that the desired format is `DXGI_FORMAT_R8G8B8A8_ UNORM` and the desired number of samples is 4. The available quality level is returned in the `NumQuality` variable. If the result is 0, that particular format and sample count are not supported. If the value is 1 or greater, any quality level can be used, up to the returned value in the 2D texture description structure described above. It is important to note that if n is returned, that the value of n-1 should be used in the sample description structure!

The `Usage`, `BindFlags`, and `CPUAccess` parameters are selected according to the options specified in the beginning of this chapter and define where and how the texture resource can be used.

Finally, we have `MiscFlags`. The flags that correspond to 2D textures are listed below:

- D3D11_RESOURCE_MISC_GENERATE_MIPS

- D3D11_RESOURCE_MISC_RESOURCE_CLAMP

- D3D11_RESOURCE_MISC_TEXTURECUBE

The first two flags have the same behavior as seen in the 1D texture case. The mips generation flag allows for automatically filling in the mip-map levels of a texture if its top level has been written to as a render target, while the resource clamp flag provides a mechanism to manually control which mip-map levels must reside in video memory.

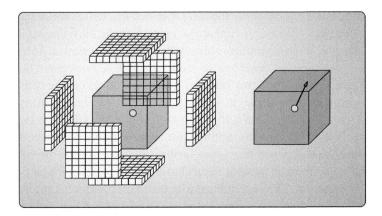

Figure 2.32. A texture 2D resource used as a cube map, and a demonstration of how it is sampled.

The texture cube flag is unique to the 2D texture type. This flag indicates that a 2D texture that is created with six texture slices can be used to generate all six faces of a cube map simultaneously. A *cube map* is a collection of six textures that can be sampled with a three-component vector positioned at the cube center, returning the texture element that is intersected by that vector. This is depicted in Figure 2.32. Intrinsic sampling functions are available to use this particular type of texture resource object, which we will see in more detail later in this section.

The other parameter for creating the texture is the initial data to load into the resource. This is provided with the D3D11_SUBRESOURCE_DATA point, which represents an array of these structures, with one structure for each subresource. The data layout for initializing the resource is provided Figure 2.33. If the texture is multisampled, the initial data parameter must be NULL, since these types of resources are not allowed to be initialized.

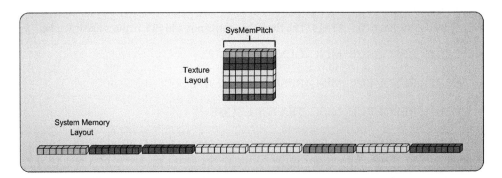

Figure 2.33. The layout of the subresource data to be applied to the initial contents of a 2D texture.

Resource view requirements. Like the 1D texture resource, the 2D texture resource can also be used with all four resource view types and can specify a subset of a resource to be exposed by the resource view. The 2D texture provides a large variety of resource creation options and also has a correspondingly large number of resource view configurations. We examine each of these possibilities below.

Texture2D shader resource view. For non-array, non-multisampled 2D texture resources, the shader resource view can select a range of mip-map levels from the resource. This is essentially the same selection mechanism that we have seen for 1D texture resources, except that each mip-map level is a 2D subresource, instead of 1D. This is specified in the **D3D11_TEX2D_SRV** structure within **D3D11_SHADER_RESOURCE_VIEW_DESC**, as shown in Listing 2.30.

```
struct D3D11_TEX2D_SRV {
    UINT MostDetailedMip;
    UINT MipLevels;
}
```

Listing 2.30. The members of the D3D11_TEX2D_SRV structure.

The starting mip-map level is specified in the `MostDetailedMip` parameter, and the maximum number of mip-map levels to use is specified in the `MipLevels` parameter.

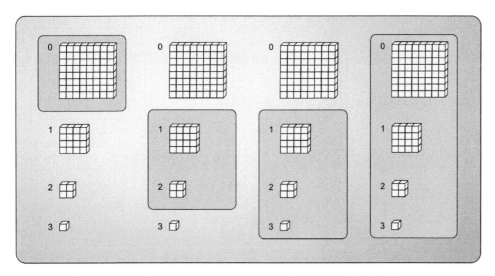

Figure 2.34. The various shader resource view subresource selections available for 2D texture resources.

Figure 2.34 demonstrates some example resource regions selected within a standard 2D texture resource.

The 2D texture array resources also select their ranges in the same way that their 1D resource counterparts do. That is, they select a range of mip-map levels, as well as an array element range. This is specified in the D3D11_TEX2D_ARRAY_SRV structure within the D3D11_SHADER_RESOURCE_VIEW_DESC structure shown in Listing 2.31.

```
struct D3D11_TEX2D_ARRAY_SRV {
    UINT MostDetailedMip;
    UINT MipLevels;
    UINT FirstArraySlice;
    UINT ArraySize;
}
```

Listing 2.31. The members of the D3D11_TEX2D_ARRAY_SRV structure.

The mip-map level range is selected in the MostDetailedMip and MipLevel parameters, as seen above. In addition, the array element range is selected with the FirstArraySlice and ArraySize parameters. A few example ranges are shown in Figure 2.35.

Multisample 2D texture resources can also be used with shader resource views. Since multisampled textures do not contain mip-maps, there is no sub-range of the standard resource to select. This is reflected in the D3D11_TEX2DMS_SRV structure by its lack of parameters. However, the multisampled 2D texture array resources do allow the selection

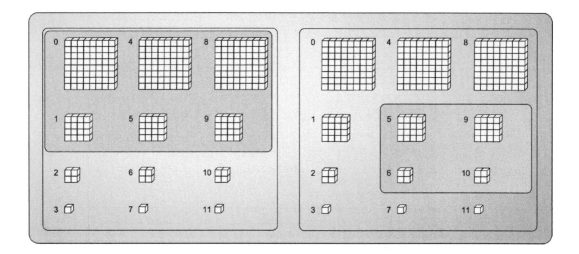

Figure 2.35. The various shader resource view subresource selections available for 2D texture array resources.

of a subrange of the elements of the texture array. This is done with the D3D11_TEX2DMS_
ARRAY_SRV structure within the D3D11_SHADER_RESOURCE_VIEW_DESC structure, as
shown in Listing 2.32.

```
struct D3D11_TEX2DMS_SRV {
    UINT UnusedField_NothingToDefine;
}
struct D3D11_TEX2DMS_ARRAY_SRV {
    UINT FirstArraySlice;
    UINT ArraySize;
}
```

Listing 2.32. The members of the D3D11_TEX2DMS_SRV and D3D11_TEX2DMS_ARRAY_SRV
structures.

As we have seen before, the array range is selected with the FirstArraySlice and the
ArraySize parameters. A few examples of this range selection are shown in Figure 2.36.

There are two additional special uses for 2D texture array resources when they are
created with the D3D11_RESOURCE_MISC_TEXTURECUBE miscellaneous flag. These shader
resource views can be used to provide cube map access to six array elements, where each
element represents one face of the cube map. For single cube map shader resource views,
the D3D11_TEXCUBE_SRV structure is used. In this case, the texture array resource must
have six elements, and a subrange of each face's mip-map levels can be specified. This is
demonstrated in Listing 2.33.

```
struct D3D11_TEXCUBE_SRV {
    UINT MostDetailedMip;
    UINT MipLevels;
}
```

Listing 2.33. The members of the D3D11_TEXCUBE_SRV structure.

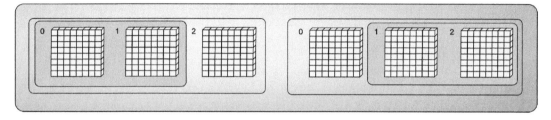

Figure 2.36. The various shader resource view subresource selections available for 2D multisampled texture array
resources.

For arrays of cube maps to be specified in the shader resource view, the D3D11_ TEXCUBE_ARRAY_SRV structure is used. The members of this structure allow for selecting the mip-map level range to be used, similar to the single cube map version. However, they also allow for specification of the first texture array element to be used, and then the number of cube maps that will be visible through the shader resource view. The number of cube maps is multiplied by six (for the number of faces on the cube) to determine how many array elements are needed. This is shown in Listing 2.34. Several example resource subregions are shown in Figure 2.37.

```
struct D3D11_TEXCUBE_ARRAY_SRV {
    UINT MostDetailedMip;
    UINT MipLevels;
    UINT First2DArrayFace;
    UINT NumCubes;
}
```

Listing 2.34. The members of the D3D11_TEXCUBE_ARRAY_SRV structure.

Texture2D unordered access view. The unordered access view can also be used to bind 2D texture resources to the pipeline. Two different subresource configurations can be used, which follow the same range selection paradigm as seen for the 1D texture resources. Non-array-based 2D texture resources can only specify a single mip-map level to be exposed to the unordered access view. This is performed with the **D3D11_TEX2D_UAV** structure within the **D3D11_UNORDERED_ACCESS_VIEW_DESC** structure, which is shown in Listing 2.35.

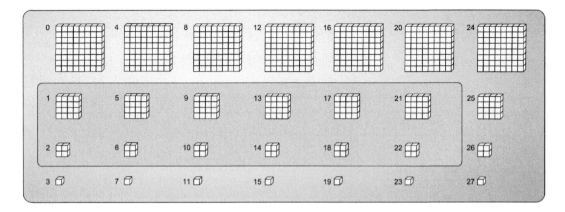

Figure 2.37. Various subresource regions of 2D texture arrays for use as texture cubes with shader resource views.

```
struct D3D11_TEX2D_UAV {
    UINT MipSlice;
}
```

Listing 2.35. The members of the D3D11_TEX2D_UAV structure.

The mip-map level is selected with the `MipSlice` member. The difference from the shader resource view subresource is that only a single mip-map level is allowed. This is shown in Figure 2.38.

The array-based 2D texture resources are also able to be exposed to the programmable shader programs with an unordered access view. This allows a single mip-map level to be specified over a subrange of the array elements. These selections are made with the `D3D11_TEX2D_ARRAY_UAV` structure in the `D3D11_UNORDERED_ACCESS_VIEW_DESC` structure, as shown in Listing 2.36.

```
struct D3D11_TEX2D_ARRAY_UAV {
    UINT MipSlice;
    UINT FirstArraySlice;
    UINT ArraySize;
}
```

Listing 2.36. The members of the D3D11_TEX2D_ARRAY_UAV structure.

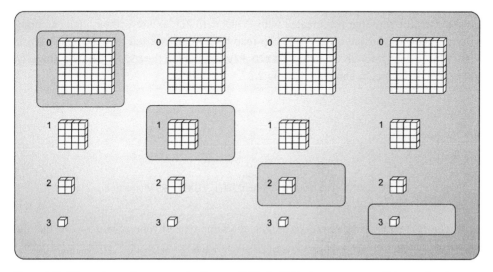

Figure 2.38. The various subresource selections available for 2D texture resources within unordered access views.

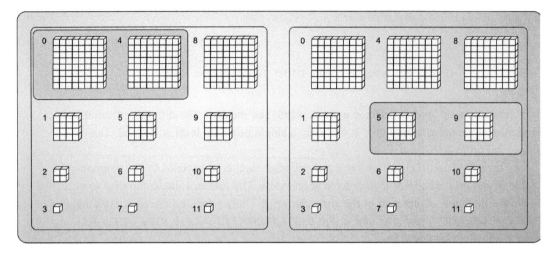

Figure 2.39. The unordered access view subresource selection for 2D texture array resources.

The array range selection is performed in the familiar manner, with the `First ArraySlice` marking the beginning of the range, and the `ArraySize` specifying how many slices to use. For clarification, this subresource selection is demonstrated in Figure 2.39.

Texture2D render target view. As we have seen in the "Using 2D Textures" section, one of the primary uses for a 2D texture is as a render target. Four different configurations are allowed for these resources to be used with a render target view, depending on which options the texture resource was created with. A standard 2D texture without multisampling or array elements can specify a single mip-map level to be exposed with the render target view. This is done through the **D3D11_TEX2D_RTV** structure in the **D3D11_RENDER_TARGET_VIEW_DESC** structure, as shown in Listing 2.37.

```
struct D3D11_TEX2D_RTV {
    UINT MipSlice;
}
```

Listing 2.37. The members of the D3D11_TEX2D_RTV structure.

The 2D texture array resources provide the same type of subresource options that the unordered access view does, allowing a single mip-map level to be specified over a selectable range within the texture array elements. This is done through the `D3D11_TEX2D_ARRAY_RTV` structure within the `D3D11_RENDER_TARGET_VIEW_DESC` structure, as shown in Listing 2.38.

```
struct D3D11_TEX2D_ARRAY_RTV {
    UINT MipSlice;
    UINT FirstArraySlice;
    UINT ArraySize;
}
```

Listing 2.38. The members of the D3D11_TEX2D_ARRAY_RTV structure.

The multisampled 2D texture resource's primary responsibility is to receive the results of rendering operations as render targets. The render target views are the primary
means of attaching these resources to the pipeline. The subresource selection structures are
shown in Listing 2.39 for both the non-array and array versions of the resource.

```
struct D3D11_TEX2DMS_RTV {
    UINT UnusedField_NothingToDefine;
}
struct D3D11_TEX2DMS_ARRAY_RTV {
    UINT FirstArraySlice;
    UINT ArraySize;
}
```

Listing 2.39. The members of the D3D11_TEX2DMS_RTV and D3D11_TEX2DMS_ARRAY_RTV
structures.

As you can see from Listing 2.39, there are no subresource selections possible for
standard 2D multisampled texture resources. However, a 2D multisampled texture array
allows the option of selecting a subrange of array elements. These selections are shown
graphically in Figure 2.40.

Texture2D depth stencil view. The depth stencil view provides the same options that the
render target views provide, making matching the render target and depth stencil target
views relatively simple. The individual structures used in creating the resource views are
provided in Listing 2.40 for reference.

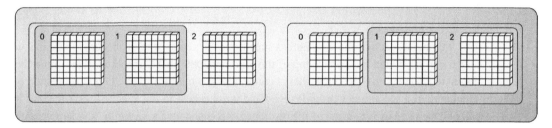

Figure 2.40. Various render target view subresource selections for a 2D multisampled texture array resource.

```
struct D3D11_TEX2D_DSV {
    UINT MipSlice;
}
struct D3D11_TEX2D_ARRAY_DSV {
    UINT MipSlice;
    UINT FirstArraySlice;
    UINT ArraySize;
}
struct D3D11_TEX2DMS_DSV {
    UINT UnusedField_NothingToDefine;
}
struct D3D11_TEX2DMS_ARRAY_DSV {
    UINT FirstArraySlice;
    UINT ArraySize;
}
```

Listing 2.40. The various depth stencil view subresource selection structures.

HLSL objects. As we have seen in the 1D texture resource case, it is possible to interact with texture resources in HLSL by declaring an appropriate resource object. This resource object must then be matched with either a shader resource view or an unordered access view with an appropriate resource attached to it. When a shader resource view is used, the resource object allows only read access to the resource contents. This is reflected in the types of resource objects that can be declared for use with shader resource views. The read-only resource types are listed below.

- Texture2D

- Texture2DArray

- Texture2DMS

- Texture2DMSArray

- TextureCube

- TextureCubeArray

Of course, each of these resources specifies a particular type of shader resource view that it can be bound to. For example, the Texture2DMS HLSL resource object must be bound to a shader resource view that is connected to a multisampled 2D texture resource. The unordered access view provides read and write access to the attached resource, but it can only be used with a subset of the resource objects shown above for the shader resource views. The available resource objects are shown below.

- RWTexture2D

- RWTexture2DArray

As you can see, only the standard 2D texture resource and its array based counterpart can be used with unordered access views. This leaves out both multisampled resources and cube map textures, although the resources of the latter can still be manipulated, as with an RWTexture2DArray. An example HLSL declaration for each of these resource objects is shown in Listing 2.41.

```
Texture2D<float>         tex01;
Texture2DArray<int3>     tex02;
Texture2DMS<float4, 4>   tex03;
Texture2DMS<float2, 16>  tex04;
TextureCube<float3>      tex05;
TextureCubeArray<float>  tex06;
RWTexture2D<uint3>       tex07;
RWTexture2DArray<float3> tex08;
```

Listing 2.41. Example resource object declarations in HLSL for the various types of 2D texture resources.

As can be seen in these declarations, the multisampled texture resource objects specify their format and can also optionally specify the number of subsamples that are used in the resource. This sample count used to be required in Direct3D 10, but since Direct3D 10.1, the count specification has become optional. Now it is possible to query the number of samples through the HLSL GetDimensions() method, making the direct specification unnecessary. This ensures that any methods that access the subsamples only try to use the appropriate number of samples. The semantics surrounding the usage of each of these resource objects will be covered in more detail in Chapter 6.

Texture 3D

The third texture type continues the trend and implements the expected increase in dimension to provide a 3D texture type. This texture type is organized similarly to the previous two textures, except that a third axis is added to the texture. It is essentially a 3D grid of texture elements, with each element consisting of one of the DXGI_FORMAT enumeration types. The various types of 3D texture resources are depicted in Figure 2.41.

As you can see from Figure 2.41, the 3D textures allow for single textures and mipmapped textures. Texture arrays are not supported in 3D textures. In addition, multi-sampling cannot be used with a 3D texture.

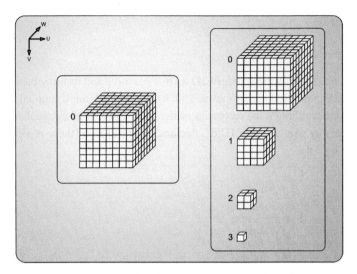

Figure 2.41. The various 3D texture resource configurations.

Using 3D textures. The use cases for 3D textures are typically somewhat more specialized than those for the previous two texture types. Due to the nature of a 3D representation, this texture type requires a significant amount of memory to hold all of its data, and hence it is typically either limited in resolution or is used in cases where it is absolutely mandatory to have a full 3D representation. Of course, as the amount of memory in GPUs continues to increase, this situation will continually improve. At first 3D textures seem to be almost exactly like a 2D texture array, and in some ways they are very similar. The single biggest difference is that a texture sample performed an a texture array will only sample + filter from the specified texture slice, while a sample performed on a 3D texture will filter between two adjacent slices. This means that on a 3D texture, a single trilinear filtered sample will perform 16 memory fetches!

Even with the current performance and size limitations, many interesting and creative uses have been found for 3D texture resources. The storage of voxel data is perhaps the prototypical example. A *voxel* is an extension of a 2D pixel (as a picture element) to 3D (as a volume element). This data structure stores a scalar value at each point within a 3D grid. The data in the grid can then be used to extract isosurface information to produce a model. For example, we could extract a surface from the model where the scalar value is equal to 0.5. This is commonly done with the well-known marching cubes algorithm (Lorensen, 1987) or some variant of it. An easy-to-visualize example of such a voxel-based data set is the results of a magnetic resonance imaging (MRI) scan, where the biological material density of a patient is represented as a scalar value at each point within the 3D grid.

Other increasingly popular uses are global illumination lighting solutions. In these scenarios, light propagation is calculated over a complete scene volume, which is represented by a 3D texture. Since the 3D grid is limited in resolution, this provides a lower calculation cost than trying to perform ray-tracing, which more or less entails tracing individual paths of light through a scene. Many recent research papers have been dedicated to these types of algorithms, such as seen in (Kaplanyan, 2010).

Creating 3D textures. As we have also seen in the 1D and 2D texture cases, the creation of the 3D texture is performed by passing a texture description structure to one of the device methods. The **ID3D11Device::CreateTexture3D()** method is used for 3D textures. It takes as input a pointer to a **D3D11_TEXTURE3D_DESC** structure, which is shown Listing 2.42.

```
struct D3D11_TEXTURE3D_DESC {
    UINT        Width;
    UINT        Height;
    UINT        Depth;
    UINT        MipLevels;
    DXGI_FORMAT Format;
    D3D11_USAGE Usage;
    UINT        BindFlags;
    UINT        CPUAccessFlags;
    UINT        MiscFlags;
}
```

Listing 2.42. The members of the D3D11_TEXTURE3D_DESC structure.

This structure is quite similar to the 1D and 2D versions. In a 3D texture, its dimensions are specified in the Width, Height, and Depth parameters. The number of mipmap levels and the element format are declared in the same manner as before, with the MipLevels and Format parameters. The Usage, BindFlags, and CPUAccessFlags parameters are used to determine the resource usage patterns, as described at the beginning of this chapter. Finally, the MiscFlags parameter provides the following optional flags for 3D textures:

- D3D11_RESOURCE_MISC_GENERATE_MIPS

- D3D11_RESOURCE_MISC_RESOURCE_CLAMP

We have seen both of these flags before. The first flag indicates the ability to auto-generate mip-maps from a render target texture, and the second flag allows for capping the highest resolution mip-map level, which allows the driver to manage the GPU memory more flexibly.

Resource view requirements. The 3D texture is the only texture resource that is not available for use with all resource view types. Specifically, it can be used for shader resource views, unordered access views, and render target views, but cannot be used in conjunction with a depth stencil view. At first consideration, this would seem to violate the rule that a depth stencil target always matches the size, dimension, and subsample count of the render target being used. However, for render target views, the 3D texture is actually treated like a 2D texture array, where each depth level of the texture acts like a texture slice. In this case, a depth stencil view would use a 2D texture array as its required resource type, allowing the render and depth target types to match. We will take a closer look at each of the available resource view configurations below.

Texture3D shader resource view. A single type of sub-resource specification is available for using texture 3D resources within a shader resource view. This is provided in the **D3D11_TEX3D_SRV** structure within the **D3D11_SHADER_RESOURCE_VIEW_DESC** structure. This structure is shown in Listing 2.43.

```
struct D3D11_TEX3D_SRV {
    UINT MostDetailedMip;
    UINT MipLevels;
}
```

Listing 2.43. The members of the D3D11_TEX3D_SRV structure.

As seen in Listing 2.43, a subset of the total mip-map levels can be chosen for use with a shader resource view, effectively hiding the other mip-map levels from the resource

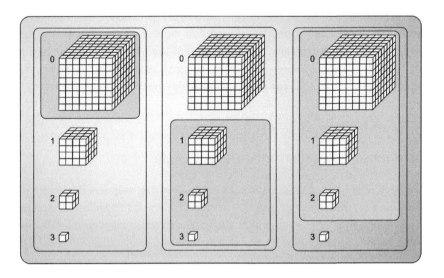

Figure 2.42. Various 3D texture subresource selections for use with shader resource views.

view. We have seen these mip-map level range selections before, with the most detailed mip-map level specified in the `MostDetailedMip` parameter, and the number of levels to be used specified in the `MipLevels` parameter. A visualization of the mip level subset range is shown in Figure 2.42.

Texture3D unordered access view. The unordered access view also only supports one form of 3D texture access, which supplies a single mip-map level and allows for the selection of a depth range. This is performed with the **D3D11_TEX3D_UAV** structure within the **D3D11_UNORDERED_ACCESS_VIEW_DESC** structure, as shown in Listing 2.44.

```
struct D3D11_TEX3D_UAV {
    UINT MipSlice;
    UINT FirstWSlice;
    UINT WSize;
}
```

Listing 2.44. The members of the D3D11_TEX3D_UAV structure.

This structure allows for the mip-map level to be specified in the `MipSlice` parameter, while the range of depth slices is selected with `FirstWSlice` and `WSize`. This makes it fairly simple to use a single 3D texture resource for multiple different uses. This special range selection is visualized in Figure 2.43.

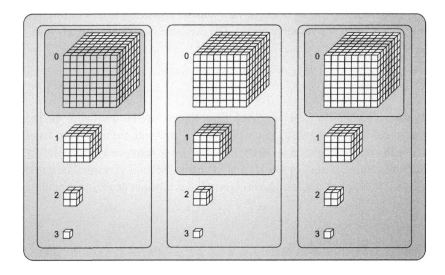

Figure 2.43. A 3D texture subresource selection for use with an unordered access view.

Texture3D render target view. The 3D texture resources can be used as a render target, although they are actually interpreted as 2D render target arrays. In this case, each depth layer is interpreted as a separate array element. The subset of the resource to be bound with the render target view is specified in the **D3D11_TEX3D_RTV** structure within the **D3D11_RENDER_TARGET_VIEW_DESC** structure, as shown in Listing 2.45. This selection range is exactly the same as the one shown for the unordered access views above, so we will not repeat the description here.

```
struct D3D11_TEX3D_RTV {
    UINT MipSlice;
    UINT FirstWSlice;
    UINT WSize;
}
```

Listing 2.45. The members of the D3D11_TEX3D_RTV structure.

Texture3D depth stencil view. As mentioned above, the 3D texture cannot be used with the depth stencil view. Instead, a 2D texture array resource should be used in conjunction with a 3D texture render target.

HLSL objects. The 3D texture objects are used in shader programs in the same way that we have seen for 1D and 2D texture types. There are significantly fewer resource objects available for declaration and use, which corresponds to the reduced selection of resource view configurations. The shader resource view and the unordered access views can both be bound to a single matching resource object. The available objects are listed below.

- Texture3D

- RWTexture3D

By now it should be clear which of these objects can be used with the read-only shader resource view (Texture3D), and which can be used with the unordered access view (RWTexture3D). The declaration syntax for each of these resource types is similar to the other examples we have seen in the 1D and 2D cases. Several examples are provided in Listing 2.46. The detailed usage of these objects is described in depth in Chapter 6.

```
Texture3D<float4> tex01;
RWTexture3D<uint3> tex02;
```

Listing 2.46. Sample HLSL resource object declarations for 3D texture resources.

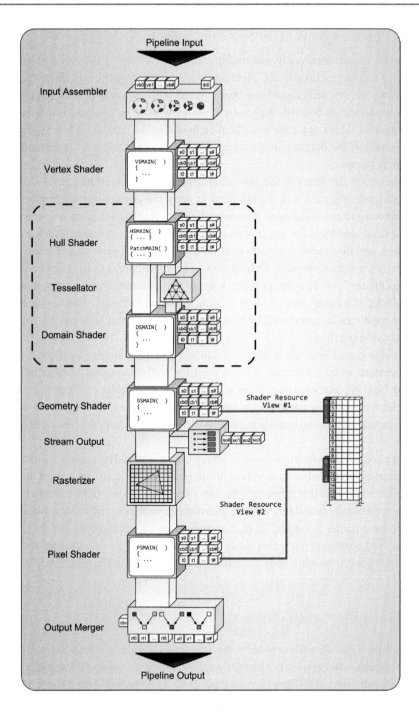

Figure 2.44. Using the same resource with different resource views in separate passes.

Multiple Resource Views of the Same Resource

It is possible to simultaneously use multiple resource views of the same resource. For example, to bind a resource to the pipeline with a render target view, three things must happen. First, the resource must have been created with appropriate bind flags for being used as a render target. Second, the render target view must be created for referencing the desired resource. Third, the view must then be bound to the pipeline. More specifically, it must be bound to the output merger stage, since that is the only place that render target views can be bound to.

However, if you want to use that same resource as an input to a pixel shader on a subsequent rendering pass, you do not need to copy the contents of the render target to another memory resource. Instead, you would ensure that the resource was created with the appropriate bind flags for use as a shader resource, create a shader resource view for the resource, and then use that view to bind the resource to the pixel shader stage, where it is then used by a shader program. This concept of multiple connections is shown in a block diagram in Figure 2.44. This brings us back to our earlier comment about resources—they are just blocks of memory that are used as input or output to the pipeline, or for both. The use of the resources is determined by how they are created, and how and where they are connected to the pipeline.

It is also possible to use multiple resource views of a single resource at the same time. In general, when a resource is being used for read access, any number of resource views can be used. For example, if we wanted to write a shader program that used one shader resource view for each mip-map level of a 2D texture resource, we could just create separate resource views and bind them individually to the pipeline, as if they were separate resources.

There are different restrictions when trying to use multiple resource views simultaneously when they have read or write access privileges. If a resource view is used for writing, any other resource view that references the same subresource cannot be bound to the pipeline. This is due to the possibility that the subresource contents may change while another portion of the pipeline is reading them, resulting in unpredictable behavior. In general, if you try to bind a resource view that reads from a resource that is also bound with a writable resource view, a warning and/or an error will be reported in the debug output window.

2.2.3 Sampler State Objects

The final objects that we will consider in this section are `sampler state objects`. A sampler state object is used to specify all of the parameters that are used when a shader program samples a texture resource. The process of sampling a texture involves looking up several pixels of texture resource and combining them with a mechanism selected by the developer, to reduce several types of common artifacts, as well as to increase the performance of texture lookup operations. All major GPU designs include additional hardware functionality to implement these sampling operations, so using them is generally much faster than trying to emulate the same sampling process from within a programmable shader.

Options that can be specified with a sampler state object include the texture coordinate addressing mode used to determine the location to be sampled, the type of filtering to be performed during the sampling process, several LOD parameters, a border color to line the texture with, and a comparison function that can be used to modify the data returned by the sampling functions. This provides quite a range of different types of sampling that can be performed, and these options can have quite a potent effect on the performance of a texture sample operation. It is thus important to properly configure the object for the intended use case of the texture being sampled.

Sampler state objects are actually not resources in the same sense as buffers and textures, but they are managed like resources in the programmable pipeline, so we are including them here. We will explore them in more detail in the following sections by first considering how they are created and then briefly reviewing how they are used in the programmable shader stages.

Creating Sampler State Objects

Before looking at the details of how a sampler object is used in a shader program, we first need to understand what exactly it is capable of doing. This is best described by looking at the process used to create the sampler object, since all of its various options must be specified during the creation process. Like the other object types we have seen throughout this chapter, a sampler state object is created by filling out a description structure and then passing that structure to a device method. In this case, we use the ID3D11Device::Create SamplerState() method to generate a ID3D11SamplerState object. The description structure is shown in Listing 2.47.

```
struct D3D11_SAMPLER_DESC {
    D3D11_FILTER               Filter;
    D3D11_TEXTURE_ADDRESS_MODE AddressU;
    D3D11_TEXTURE_ADDRESS_MODE AddressV;
    D3D11_TEXTURE_ADDRESS_MODE AddressW;
    FLOAT                      MipLODBias;
    UINT                       MaxAnisotropy;
    D3D11_COMPARISON_FUNC      ComparisonFunc;
    FLOAT                      BorderColor[4];
    FLOAT                      MinLOD;
    FLOAT                      MaxLOD;
}
```

Listing 2.47. The D3D11_SAMPLER_DESC structure and its members.

The first parameter is perhaps the most visible of all. The `Filter` structure member specifies the type of filtering that will be performed during the sampling process, in several different scenarios. Texture minification, magnification, and mip-mapping are all different types of sampling operations to which different filter qualities can be applied.

Texture minification occurs when a texture appears far enough from the viewer that its individual pixels (often referred to as *texels*, which is short for *texture pixels*) are smaller than the output render target pixels. Because the texels from the texture are smaller than the screen pixels, the texture can appear to sparkle or pop as multiple texels pass through a pixel from frame to frame. If an appropriate filtering mechanism is chosen, this effect can largely be reduced, or even eliminated. *Texture magnification* occurs in the opposite situation. When a texture is viewed up close, a single texel will cover multiple screen pixels, resulting in a "blocky" appearance. In some cases, this is the desired behavior, but in other cases a better quality image can be produced if the texture is filtered to reduce the visibility of the large texels. The final texture sampling situation to consider is *mip-mapping*. We have already discussed mip-mapping in the discussion of texture resources earlier in this chapter. Essentially, a mip-map provides a multi-resolution representation of the texture resource, and it is possible to specify the type of filtering that is performed during the mip-map sampling process.

The description structure has one filter parameter, which indicates that all three of these filtering scenarios are specified with a single enumeration. The available filtering modes are shown in Listing 2.48. As you can see, each enumerated filter type specifies how it will sample the texture under the three conditions mentioned above—MIN corresponds to texture minification, MAG corresponds to texture magnification, and MIP corresponds to texture mip-mapping.

```
enum D3D11_FILTER {
    D3D11_FILTER_MIN_MAG_MIP_POINT,
    D3D11_FILTER_MIN_MAG_POINT_MIP_LINEAR,
    D3D11_FILTER_MIN_POINT_MAG_LINEAR_MIP_POINT,
    D3D11_FILTER_MIN_POINT_MAG_MIP_LINEAR,
    D3D11_FILTER_MIN_LINEAR_MAG_MIP_POINT,
    D3D11_FILTER_MIN_LINEAR_MAG_POINT_MIP_LINEAR,
    D3D11_FILTER_MIN_MAG_LINEAR_MIP_POINT,
    D3D11_FILTER_MIN_MAG_MIP_LINEAR,
    D3D11_FILTER_ANISOTROPIC,
    D3D11_FILTER_COMPARISON_MIN_MAG_MIP_POINT,
    D3D11_FILTER_COMPARISON_MIN_MAG_POINT_MIP_LINEAR,
    D3D11_FILTER_COMPARISON_MIN_POINT_MAG_LINEAR_MIP_POINT,
    D3D11_FILTER_COMPARISON_MIN_POINT_MAG_MIP_LINEAR,
    D3D11_FILTER_COMPARISON_MIN_LINEAR_MAG_MIP_POINT,
    D3D11_FILTER_COMPARISON_MIN_LINEAR_MAG_POINT_MIP_LINEAR,
    D3D11_FILTER_COMPARISON_MIN_MAG_LINEAR_MIP_POINT,
    D3D11_FILTER_COMPARISON_MIN_MAG_MIP_LINEAR,
    D3D11_FILTER_COMPARISON_ANISOTROPIC,
    D3D11_FILTER_TEXT_1BIT
}
```

Listing 2.48. The available filter types in the D3D11_FILTER enumeration.

The available types of filtering for each of these scenarios are point sampling, linear sampling, and anisotropic sampling. *Point sampling* simply returns the texel that the input texture coordinates happen to fall on. This is the least expensive form of sampling, but it also produces the lowest image quality in most situations. *Linear sampling* provides a higher quality of sampling, in which the texels that surround an input texture coordinate are interpolated to find an approximated value somewhere between the selected texels. This produces smoother transitions between texels, but it requires additional texels to be read, in addition to performing some arithmetic on the sampled data before returning the result. The final option is to utilize *anisotropic sampling*, a much higher-quality sampling technique. When a texture is viewed at any angle that is not perpendicular to the camera's line of sight, anisotropic filtering performs a number of samples to determine the average color that is visible to the viewer and returns this combined color. The maximum number of samples is specified in the MaxAnisotropy parameter of the sampler state description structure. While this filtering mode produces the best results, it can be extremely expensive to use, due to the much higher bandwidth and computation needed to produce the final sample value.

In the first half of the D3D11_FILTER enumeration, we see all of the various combinations of filtering types that can be used. In the second half of the enumeration, we see a repeat of the list from the first half, except for the addition of COMPARISON in the value names. This indicates that each of the individual samples used in the sampling process will be used in the comparison function (which will be discussed shortly) prior to combining the result. This allows the result of the comparisons to be filtered, instead of filtering the texture first and then performing a comparison.

The next three options in the sampler description structure specify the texture addressing modes in the *U*, *V*, and *W* directions that correspond to the *X*, *Y*, and *Z* coordinates of the texture.[3] These modes let you specify what action to take when a texture coordinate that lies outside of the [0,1] range is used during sampling. A different addressing mode can be specified for each direction. You can have the sampler wrap the texture coordinates back around to 0 when they are greater than 1, effectively "wrapping" the texture over again. If the texture coordinate is greater than 2, the texture would be wrapped again, and so on. This allows many copies of a texture to appear on a textured surface. Another possible mode is the *mirror* addressing mode, which essentially flips the texture at every integer coordinate. This also allows many copies of the texture to be visible, but it flips the orientation for each consecutive copy of the texture. You can also specify a *clamp* mode, in which the texture coordinates are "clamped" to the [0,1] range. This effectively makes the boundary values of the texture appear at any location outside of the texture. An alternative to this mode is a *border addressing* mode, in which the sampler state can specify a replacement color to return when the coordinates are outside of the [0,1] range. The border color is specified in the BorderColor parameter of the sampler description object. Finally, you can use a *mir-*

---

[3] These coordinates are subject to the same dimensionality that was described for the texture coordinates. For example, a 1D texture does not have a notion of a *Y* or *Z* coordinate, while a 3D texture has all three.

ror once mode, in which the absolute value of the texture coordinates are used and then clamped in the same way as in the clamp texture addressing mode.

The `MipLODBias`, `MinLOD`, and `MaxLOD` parameters are all used to manipulate the mip-map level that is used during the sampling process. The `min` and `max` parameters specify the minimum and maximum levels that can be accessed during sampling, and the `bias` parameter provides a constant offset to add to the mip-map level selected by the sampling hardware. In all of these parameters, 0 is the most detailed mip-map level, so adding a positive bias will select a less detailed mip-map level.

The final member in the sampler state description structure is the `ComparisonFunc` parameter. This indicates what type of comparison should be performed on the texture samples prior to performing the filtering operation. The other value that is used in the comparison is supplied in the texture object sampling method call. The result is either a 1 if the comparison passes, or a 0 if it fails. The results of each individual comparison are then combined according to the chosen filtering method, and returned.

Using Samplers

To use a sampler object, it must first be created as detailed above and then bound to one of the programmable pipeline stages. Each of the programmable stages can simultaneously bind up to 16 different sampler objects, and the same sampler state object can be used in more than one programmable stage at once. After the state objects have been bound to a pipeline stage, the HLSL program that will run in that stage must declare an appropriate HLSL `SamplerState` object, which corresponds to a particular sampler slot. Then the shader program can pass the `SamplerState` object to one of the many sampling methods provided by the various texture resource types. By having to pass the `SamplerState` object as an argument in the sample methods, D3D11 allows multiple texture resources within a shader program to simultaneously use the same sampling object. Due to the large number of available resources that can be used in a shader program (up to 128), sharing these states becomes a significant benefit if the same set of options is needed many times.[4]

2.3 Resource Manipulations

We have seen throughout this chapter what resource types are available, their typical usage, how they are constructed, and how to declare them for use in HLSL. We also saw in the

---

[4] In previous versions of Direct3D, the sampler could only be used with a single texture at a time. This limited the overall number of textures that could be used in a shader program. Without this limitation, Direct3D 11 provides a significantly larger number of textures (or more specifically resources) that can be accessed in a shader.

"Resource Creation" section that resources are created with very specific usage and access patterns. Depending on what a resource will be used for, it can be advantageous to only allow it to be accessible by the GPU. In other cases, it may be necessary to be able to directly manipulate the resources in the host C/C++ application.

Direct3D 11 provides a number of different techniques to modify or manipulate the contents of a resource. We will explore each of these methods in detail in this section, and will also consider how each of these methods provides useful operations in different situations.

2.3.1 Manipulating Resources

The first group of methods that we will look at is used to modify the contents of a resource. As we will see shortly, it is also possible to read the contents of a resource with some of the methods.

Mapping Resources

The primary way to manipulate the contents of a buffer for both reading and writing from the CPU is the *Map / Unmap* method of the device context. If a resource has CPU read or write access flags set, an application can map the contents of the resource to system memory. The *map* and *unmap* method prototypes are shown in Listing 2.49.

```
HRESULT Map(
    ID3D11Resource *pResource,
    UINT Subresource,
    D3D11_MAP MapType,
    UINT MapFlags,
    D3D11_MAPPED_SUBRESOURCE *pMappedResource
);
void Unmap(
    ID3D11Resource *pResource,
    UINT Subresource
);
```

Listing 2.49. The method prototypes of the map and unmap methods.

The resource and subresource to be mapped are specified in the pResource and Subresource parameters, respectively. Next, the mapping type is specified in the MapType parameter, which indicates if the resource will be read or written, and how any writing will be performed. The members of the D3D11_MAP enumeration are shown in Listing 2.50.

```
enum D3D11_MAP {
    D3D11_MAP_READ,
    D3D11_MAP_WRITE,
    D3D11_MAP_READ_WRITE,
    D3D11_MAP_WRITE_DISCARD,
    D3D11_MAP_WRITE_NO_OVERWRITE
}
```

Listing 2.50. The members of the D3D11_MAP enumeration.

The first three values in the enumeration are fairly self-explanatory. If the resource is to be read, it must have been created with the D3D11_CPU_ACCESS_READ flag, and if the resource is to be written to, it must have been created with the D3D11_CPU_ACCESS_WRITE flag. When reading a resource, its contents will be available in the D3D11_MAPPED_SUBRESOURCE structure pointed to by the pMappedResource parameter. When writing to a resource, the same D3D11_MAPPED_SUBRESOURCE structure is used by the application to know where to write the desired resource data to. When both reading and writing, the contents of the resource are made available and can then be overwritten by the application, thus "mapping" the contents of the resource to system memory. The call to unmap the resource will make any written changes take effect in the resource.

The D3D11_MAP_WRITE_DISCARD and D3D11_MAP_WRITE_NO_OVERWRITE flags are used together to allow for dynamic resource usage. The D3D11_MAP_WRITE_DISCARD mapping type invalidates the contents of the resource when it is mapped, even if only a sub-resource is currently being mapped. The D3D11_MAP_WRITE_NO_OVERWRITE mapping type allows writing to a portion of a buffer that has not previously been updated since the last call to D3D11_MAP_WRITE_DISCARD. In this way, the contents of the resource that have been loaded into a dynamic resource may start to be used while other portions of the resource are still being filled up.

After the resource has been read or written from the mapped subresource structure, the resource must eventually be unmapped to allow the GPU to continue using it. This is done simply with the ID3D11DeviceContext::Unmap() method, which identifies the resource and subresource index to be unmapped.

An important consideration for the mapping of resources is that a complete subresource must be mapped at a time. This means that it is not possible to read/write a portion of a subresource, although the entire subresource can be mapped while only updating a smaller portion of it. Depending on the size of the subresource, this can involve a large amount of memory being transferred to and from system memory. In some cases this is unavoidable, but in other cases it would be beneficial to have a method for updating only a portion of the subresource. We will see such a method in the "Update Subresource" section.

A common example of using the map and unmap methods is to update the contents of a constant buffer before using it in a pipeline execution. In this case, the resource would be mapped with the D3D11_MAP_WRITE flag, since we are not interested in reading the contents of the constant buffer. Instead, we will just write the new data into the buffer. Once the data has been written to the system memory provided with the pMappedResource structure, the resource would be unmapped and could then be bound to the pipeline for use.

Update Subresource

The other mechanism that can be used to modify the contents of a resource is the device context's UpdateSubresource() method. This method differs from mapping of resources in that it only writes to the resource and does not allow reading its contents. The prototype of this method is shown in Listing 2.51.

```
void UpdateSubresource(
    ID3D11Resource *pDstResource,
    UINT DstSubresource,
    const D3D11_BOX *pDstBox,
    const void *pSrcData,
    UINT SrcRowPitch,
    UINT SrcDepthPitch
);
```

Listing 2.51. The update subresource method prototype.

In Listing 2.50 we can see that data to be written to the resource is given a pointer to its memory location. This is accompanied by the row pitch and depth pitch of the data block, measured in bytes. The destination is identified with a pointer to the resource and a subresource index. It is possible to write to only a portion of a subresource, by identifying it within a D3D11_BOX. While the source data is provided in terms of bytes and is thus format dependent, the destination is provided in indices and is not dependent on the format of the resource.

In contrast to methods for mapping a resource, the update subresource method can write to only a portion of a subresource. If only a small part of a large resource needs to be updated, the update subresource method can perform better under some conditions, since it would use less bandwidth.

An example of such a resource update could be changing the contents of a texture based on an action taken by the user. If the user selects an option that requires modifying a small portion of a texture, this can easily be performed using the update subresource method.

2.3.2 Copying Resources

There are many situations when the contents of one resource must be copied to another. This can be done for the needs of a particular algorithm, or to copy data to a second resource that has different access capabilities. In both cases, multiple methods are available to perform the copying. We will examine each of these methods and then present a few situations they can be employed in. These methods are distinct from those presented in the "Manipulating Resources" section, in that all of the data that is being moved is already located within a resource, while the previous methods were used to read or write data with the CPU.

Copy Resource

To copy the complete contents of one resource to another resource, the device context's copy resource method would be used. The prototype of this method is shown in Listing 2.52.

```
void CopyResource(
    ID3D11Resource *pDstResource,
    ID3D11Resource *pSrcResource
);
```

Listing 2.52. The copy resource method prototype.

This is a fairly simple method, which only takes pointers to the source and destination resource. The resources must be the same type and size and have compatible formats. Neither immutable resources nor depth stencil resources can serve as the destination. In addition, multisampled resources cannot be used as either the source or destination (we will see later in this section how to retrieve the contents of a multisampled resource).

Since this method copies the complete resource, it is most likely to be used to copy data to the second resource with the desired usage configuration. For example, if a compute shader is used to generate a data buffer and the application needs to save that buffer to the hard disk, it cannot directly map and read the buffer. This is because a buffer used for the unordered access view must have the default usage, which means it cannot be read by the CPU. This is true of all output from the pipeline—only resources with the default usage can receive output from the pipeline, and the CPU cannot directly read resources that have default usage. Instead, a second buffer resource would be created with the staging usage, and the contents of the default buffer would be copied to the staging buffer. Resources with the staging usage can be read by the CPU, which solves the problem of getting the data to the application.

An example of using the copy resource method would be to retrieve the contents of an append structured buffer that has been filled with data from the compute shader. Since the append structured buffer has the default usage flags, its contents would need to be copied to a secondary staging buffer using the CopyResource() method. The staging buffer could then be used by the CPU to read the contents using the map/unmap methods.

Copy Subresource Region

The next technique for copying resources is the device context's CopySubresourceRegion() method. This method allows a portion of a resource to be copied, as opposed to always copying the complete resource, as is required in the CopyResource() method described above. Listing 2.53 shows the prototype of the method.

```
void CopySubresourceRegion(
    ID3D11Resource *pDstResource,
    UINT DstSubresource,
    UINT DstX,
    UINT DstY,
    UINT DstZ,
    ID3D11Resource *pSrcResource,
    UINT SrcSubresource,
    const D3D11_BOX *pSrcBox
);
```

Listing 2.53. The copy subresource region method prototype.

Since this method allows copying a subset of one resource to another, the resources do not need to have the same size. However, the two resources must have the same type, and compatible formats. As with the source, the destination can be a subresource. This is specified by using a zero-based subresource index in the DstSubresource parameter. In addition, a destination offset can be specified in each of the three coordinates, to positioning the copied data. The source resource data is also determined with a subresource index, and the region to be copied is determined with a D3D11_BOX selection.

The ability to selectively copy portions of a resource to another resource can be used to create dynamic texture atlases, which combine several textures into a single larger texture. This can reduce the number of individual draw calls needed to render a scene (topics such as these are discussed in more detail in Chapter 3).

Copy Structure Count

One final method allows copying a portion of one resource into another resource, although this is a less direct copying technique. The device context's CopyStructureCount() method allows copying a buffer resource's hidden counter into another resource. The method prototype is shown in Listing 2.54.

```
void CopyStructureCount(
    ID3D11Buffer *pDstBuffer,
    UINT DstAlignedByteOffset,
    ID3D11UnorderedAccessView *pSrcView
);
```

Listing 2.54. The copy structure count method prototype.

As we have seen earlier in this chapter, there are two unordered access view description flags that create a hidden counter for managing the contents of a buffer. The first is the D3D11_BUFFER_UAV_FLAG_APPEND flag, which lets a buffer be used as an append/consume buffer. The second is the D3D11_BUFFER_UAV_FLAG_COUNTER flag, which directly adds the hidden counter for manipulation with the IncrementCounter() and DecrementCounter() methods in HLSL. The CopyStructureCount() method works by taking a pointer to the UAV that contains the counter as the source and then copies that value into a destination buffer at the offset specified in the DstAlignedByteOffset parameter.

The ability to copy the counter value into another buffer has uses in indirect rendering operations. When a buffer is filled with vertex data by a compute or pixel shader program, it can be used to render a model *indirectly* with a buffer resource that indicates how many primitives to render. The counter value can be placed at the proper location in the buffer by using the appropriate index when copying it. Then, the application simply passes the indirect buffer to the draw call. Indirect rendering has been discussed in the "Buffer Resources" section of this chapter, and is also covered in Chapter 3.

This functionality can also be used to stage the structure count data for eventual reading by the CPU. This is performed in the same way we described in the "Copy Resource" section.

2.3.3 Generating Resource Contents

There is also a pair of device context methods that can be used to generate the contents of a resource. We discuss both of these methods here.

Generate Mips

Earlier in this chapter, we saw the miscellaneous resource creation flag for indicating that a texture resource should be able to generate the lower-resolution mip-map levels when its top level is rendered. This is indicated with the D3D11_RESOURCE_MISC_GENERATE_MIPS flag. The generation of the mip-map level data is initiated with the ID3D11DeviceContext:: GenerateMips() method. The method prototype is shown in Listing 2.55. The method only takes a pointer to the shader resource view to be updated, whose attached resource must have been created with the flag mentioned above. In addition, the number of mip-map levels that will be generated is defined by the subresource range selected by the shader resource view.

```
void GenerateMips(
    ID3D11ShaderResourceView *pShaderResourceView
);
```

Listing 2.55. The generate mips method prototype.

Resolve Subresource

The other method that can be used to generate resource content is the `ID3D11DeviceContext` `::ResolveSubresource()` method. This is used to take a source multisampled texture resource and use the subsamples to calculate the final color value of the corresponding pixel in a non-multisampled destination resource. This is a required step before displaying the contents of a multisampled render target on the screen, and hence is a vital operation for high-quality rendered image output. In the case of a DXGI swap chain, it supports automatically performing the resolve function during the buffer swap triggered with the `Present()` method. This technically allows the application to skip the manual resolve process in some cases, but if a multisampled render target is used as an input to any further algorithms (such as any post-processing techniques) the render target will need to be resolved well in advance of any per-pixel algorithms being applied to it. The prototype of the method is shown in Listing 2.56.

```
void ResolveSubresource(
    ID3D11Resource *pDstResource,
    UINT DstSubresource,
    ID3D11Resource *pSrcResource,
    UINT SrcSubresource,
    DXGI_FORMAT Format
);
```

Listing 2.56. The resolve subresource method prototype.

This method allows for subresource indices to be provided for both the source and destination resource. This also means that a complete subresource must be resolved at the same time, and it does not allow portions of a subresource to be manipulated in isolation. In addition, a format identifier is passed to indicate the format to be used when the two resources have compatible `TYPELESS` formats.

3   The Rendering Pipeline

The *Direct3D 11 Rendering Pipeline* is the mechanism used to process memory resources into a rendered image with the GPU. The pipeline itself is made up of a number of smaller logical units, called *pipeline stages*. Data is processed by progressing through the pipeline one stage at a time and is manipulated in some way at each stage. By understanding how the individual stages of the pipeline operate, and the semantics for using them, we can use the pipeline as a whole to implement a wide variety of algorithms that execute in real time.

As GPUs have become increasingly powerful with each new architecture generation, the size and capabilities of the pipeline have expanded significantly. In addition, the complexity and configurability of each pipeline stage has steadily increased. The current rendering pipeline has fixed-function stages, as well as programmable shader stages. This chapter will first consider the differences between these types of pipeline stages. In particular, it will focus on the states that are required to make them function, as well as the processing they can perform. After clarifying this distinction, we will consider the higher-level details of how the pipeline is invoked, and how each pipeline stage communicates with its neighbors.

We will then explore each of the pipeline stages in detail. This includes the individual functions that each stage performs, a thorough discussion of how the stages are configured, and the general semantics that they bring along with them. With a firm understanding of what each individual component of the pipeline does, we can consider some of the higher-level functionalities that are implemented by groups of pipeline stages in various configurations. We will also discuss several high-level data processing concepts for the pipeline as a whole, including how to manage such a complex processing architecture. The rendering pipeline has evolved into a sophisticated set of APIs, which can be used to implement a wide variety of algorithms. After completing this chapter, we will have a deep and thorough understanding of how to use the rendering pipeline to develop efficient and interesting rendering techniques for use in real-time rendering applications.

3.1 Pipeline State

To understand how the pipeline operates, we need to look no further than its name. Data is submitted as input at one end of the pipeline, and then processed by the first pipeline stage. This data consists of vector-based variables with up to four components. After processing in the first stage is completed, the modified output data is passed on to the next stage. The next set of data is then brought into the first stage. This means that the first two stages are processing different pieces of data at the same time. This process is repeated until the complete pipeline is operating simultaneously on different portions of the input data. The pipeline architecture specifically allows multiple operations to be performed at the same time by different pipeline stages, which lets many specialized processes be carried out on an individual data item as it travels through the pipeline. Once a data item reaches the end of the pipeline, it is stored to an output resource, which can later be used as needed by the host application. This pipeline concept is a simple but powerful processing technique, similar in nature to an assembly line. Figure 3.1 shows how data is processed by a pipeline.

The task of the developer is to properly configure each stage of the pipeline to obtain the desired result when the data emerges from the end of the pipeline. The pipeline configuration process is performed by manipulating the state of each individual stage of the pipeline. By organizing the pipeline into stages, Direct3D 11 effectively groups related sets of states together and consolidates how they are manipulated. There are two different types of pipeline stages, the fixed function stages and the programmable shader stages. Both stage types share some common concepts regarding how data flows through them and how their states are manipulated by an application. The following sections will explore these two types of states and provide some general concepts of how to work with them.

3.1.1 Fixed Pipeline Stages

The *fixed-function pipeline* stages perform a fixed set of operations on the data passed to them. They perform specific operations, and hence provide a "fixed" scope of available functionality. They can be configured in various ways, but they always perform the same

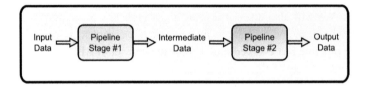

Figure 3.1. The Direct3D 11 rendering pipeline.

operations on data passed to them. A useful analogy to this concept is to consider how a regular function works in C++. You first pass the function a predefined list of parameters. It then processes the data and returns the result as its output. The body of the function represents the operations that the fixed function stage implements, while the input parameters are the available configurations and the actual input data. We can't change the function body, but we can control what options are used during processing of the input data.

Some examples of this type of fixed functions in previous generations of Direct3D include the ability to change the vertex culling order (which is now a portion of the rasterizer stage), selecting a depth test function (now a portion of the output merger stage), and setting the alpha blending mode (also a portion of the output merger stage). This focus on a single functional area is primarily due to performance considerations. If a particular pipeline stage is designed for a specific task, it can usually be optimized to perform that task more efficiently than a general purpose solution could.

In Direct3D 9, changing the state of these fixed function stages was performed by calling an API function for each individual setting that was to be changed. This required many API calls to be performed to configure a particular processing setup for each stage. Depending on the scene contents and the rendering configuration, the number of API calls could easily add up and begin to cause performance problems. Direct3D 10 introduced the concept of *state objects* to replace these individual state settings. A state object is used to configure a complete functional state with a single API call. This significantly reduces the number of API calls required to configure a state and also reduces the amount of error checking required by the runtime. Direct3D 11 follows this state object paradigm. The application must describe the desired state with a description structure, and then create a state object from it, which can be used to control the fixed-function pipeline stages.

The requirement to create a complete state all at once moves state validation from the runtime API calls to the state creation methods. If incompatible states are configured together, an error is returned at creation time. After creation, the state object is immutable and cannot be modified. Thus, there is no way to set an invalid state for the fixed function pipelines, which effectively removes the validation burden from the state-setting API calls. Overall, this system of using state objects to represent the pipeline state allows for a more streamlined pipeline configuration, with only minimal application interactions required.

3.1.2 Programmable Pipeline Stages

The *programmable pipeline stages* comprise the remaining stages of the pipeline. The word programmable here means that these stages can execute programs written in the High Level Shading Language (HLSL). Using the same C++ analogy from above, the programmable stages allow you to define the required input parameters *and* the body of the function. In fact, the programs that run in these stages are authored as functions in HLSL.

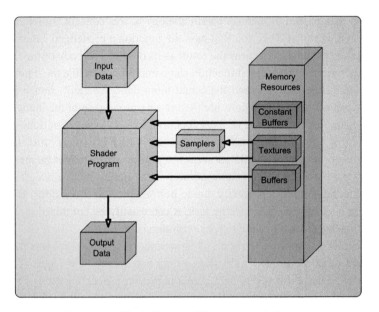

Figure 3.2. A block diagram of the common shader core.

The ability to execute a program lets these pipeline stages be used for a wide variety of processing tasks. This contrasts with the fixed function stages, which are intended for a very specific task and only offer a small amount of configurability. The programs that are executed in these programmable pipeline stages are commonly referred to as *shader programs*, a name inherited from the early steps of programmability, when pixel shader stage was initially used to modify how an object reacted to lighting. As more and more programmable stages were added to the pipeline, the name *shader* was adopted to refer to all of the programmable pipeline stages. We will use this term interchangeably with *programmable stages* throughout the book. In the following sections, we will first take a look at these programmable shader stages from a high level, and then dive into the details of their architecture, after establishing the basic concepts.

Common Shader Core

All of the programmable shader stages are built upon a common base of functionality, which is referred to as the *common shader core*. The common shader core defines the generic input and output design of the pipeline stage, providing a set of intrinsic functions that all of the programmable shader stages support, as well as the interface to the resources that a programmable shader can use. Figure 3.2 provides a visual representation of how the common shader core operates. As described above, the data flows into the top of the stage, is processed within the shader core, and then flows out as output at the bottom of the stage. The shader program that executes within these stages is a function written in HLSL for a specific purpose. While the data is being processed within the shader stage, the shader program has access to constant buffers, samplers, and shader resource views that are bound to that stage by the application. Once the shader program has finished processing its data, that data is passed out of the stage, and the next piece of data is brought in to start the process over again.

The configurability of the programmable pipeline stages is not limited to the processing performed within the shader programs. The input and output data structures for these stages are also specified by the shader program, which provides a flexible way to pass data from stage to stage. Some rules are imposed on these interfaces between stages, such as requiring certain types of data to be included in the interfaces of particular stages. A typical example of a required output is for one of the stages before the rasterizer to provide a position output that can be used to determine which pixels of a render target are covered by a primitive. Depending on which stages are passing the data, the parameters are either provided unmodified, or can be interpolated before being passed on to the next stage.

While all the programmable stages share a common set of functionality, each stage can also provide additional specialized features and functionalities unique to it. These are usually related to input and output semantics, and are hence only applicable to particular stages. The individual behaviors of each stage will be discussed in more detail later in this chapter as we describe each stage of the pipeline.

With so much flexibility, and a wide variety of different pipeline stages, there are many algorithms that do not require all stages to be active. Thus, it is possible to disable the programmable shader stage's by clearing its shader program. In addition, since there is a large amount of flexibility in the processing done in each stage, it is quite possible to perform some or all of the required work in different combinations of all of the programmable stages. This ability to choose where to perform a particular calculation can be used to our advantage. If a calculation can be done prior to any data amplification, it is possible to calculate the same data with far fewer operations. We will see a good example of this in the Chapter 8, "Mesh Rendering," where vertex skinning is performed prior to a model being tessellated, to reduce the number of vertices being skinned. (The details of what vertex skinning is can be found in Chapter 8.)

Shader Core Architecture

In the previous section, we have seen the general concept of how a programmable stage operates. It receives input from the previous stage, executes an HLSL program on it, and passes the results on to the next pipeline stage. However, this is just an overview description of what is really going on. The shader program that executes in a programmable stage is actually compiled from HLSL into a vector-register-based assembly language designed for use in specialized shader processor cores in the GPU. Even though all shader programs

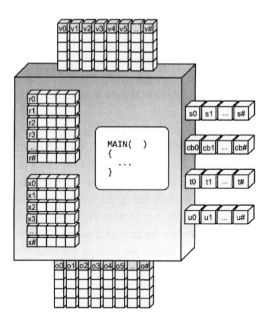

Figure 3.3. The assembly language view of the common shader core.

must be written in HLSL, they must still be compiled into this assembly byte code before being used in the rendering pipeline.[1,2]

We can learn a great deal of information from this assembly language. It defines a specific set of registers that can be used by the compiler to map an HLSL program to assembly language. The registers are generally four component vector registers, which can use individual components to provide scalar register functionality as well. There are registers for a shader core to receive its input data, temporary registers for performing computations, registers for interacting with resources, and registers for passing data out of the stage. The assembly language instructions use these registers to perform their respective operations. By understanding how the assembly programs can use these registers, we can gain a deep insight into how shader stages operate.

To get started, we will consider the common shader core overview shown in Figure 3.2, but this time we will view it from an assembly language viewpoint. Figure 3.3 shows the assembly version of the common shader core.

---

[1] The assembly program produced by compiling an HLSL program is not directly executed in the GPU. It is further processed by the video driver into machine-specific instructions, which can vary from GPU to GPU. Even so, the assembly language provides a common point of reference with which we can gain an insight into the operations of a shader processor.

[2] The details of the compilation process can be found in Chapter 6, "The High Level Shading Language."

As seen in Figure 3.3, the input to the shader core is provided in the v# registers.[3] Since they are providing the input to the stage, they are naturally read only. When a shader program is executed, its input data is available in the v# registers. After the data has been read, it can be manipulated and combined with other data, and any intermediate calculations can be stored in the r# and x#[n] registers. These are called *temporary registers*, and since they hold intermediate values, they are both readable and writable by a shader program. The texture registers (t#), constant buffer registers (cb#[n]), immediate constant buffer register (icb[index]), and unordered access registers (u#) are also available as data sources. These registers are used to provide access to the device memory resources, as described in Chapter 2, and they are all read only except for the unordered access registers. Finally, the calculated values that will be passed on to the next pipeline stage are written into the output registers (o#). When the shader program has terminated, the values stored in the output registers are passed on to the input registers of the next stage, where the process is repeated. A few other special purpose registers are only available in certain stages, so we will defer discussion of them until later in the chapter.

Typically, a developer does not need to inspect the assembly listing of a compiled shader program, unless there is a performance issue.[4] This makes understanding the details of how the assembly instructions operate less critical. Even so, it is still helpful to have a basic knowledge of the assembly-based world. For example, when developers define input and output data structures for a shader program, they must be aware of the limits on how many input and output vectors can be used for each stage. This is determined by the number of input and output registers available for that particular stage. Similarly, the available number of constant buffers, textures, and unordered access resources is limited by their respective registers, as well. This is very important information, and should be taken into consideration as we proceed through each of the pipeline stage discussions.

GPU Architectures

Even with a strictly defined assembly language specification, the actual GPU hardware is not required to directly implement the specification. There are many different architectural implementations, which can vary widely from one vendor to the next. In fact, even consecutive generations of GPU hardware from the same vendor can vary significantly from one another. This makes it incredibly difficult to predict how efficiently a given shader program will execute on a current or future GPU. Depending on the architecture of the GPU executing the program, one particular memory access pattern may be more efficient than another, but the opposite could be true with a different architecture.

---

[3] The # symbol indicates that multiple registers are available that are identified by an integer index. For example, v0 and v1 are the first two input registers available for use.

[4] Details about how to compile a shader and view its assembly listing are provided in Chapter 6.

Due to the huge amount of variation among implementations, it would be impractical to provide detailed information here. We invite the reader to explore this fascinating topic in more detail, with a good starting point available in (Fatahalian). However, we can still provide a general idea of how a GPU is organized, so that discussions later in the book will have a context to build upon. The GPU has evolved into a massively parallel processor, housing hundreds of individual ALU processing cores. These processors can run custom programs, and they can access a large bank of memory, which allows for high-bandwidth data transfers. Each GPU uses some form of a memory cache system to reduce the effective latency of memory requests, although the cache system is vendor specific and its low-level design is generally not disclosed. In the past, it has been necessary to understand the individual architectural details of a particular target GPU in order to attain the maximum available performance level. This trend will likely continue for some time into the future, due to the fact that GPUs are still evolving at a very rapid pace.

3.2 Pipeline Execution

The fixed function stages, together with the programmable stages, combine to provide an interesting and diverse pipeline with which we can render images. The process of executing the pipeline consists of configuring all of the pipeline stage states, binding input and output resources to the pipeline, and then calling one of the Draw methods to begin execution. All of these tasks are performed through the methods of the ID3D11DeviceContext interface. The number of pipeline executions performed to generate a rendered frame depends on the application and the current scene, but we can assume that at least one pipeline execution is performed for each frame.

Once the pipeline has been invoked with one of these draw methods, data is read into the first stage of the pipeline from the input memory resources. The amount of data read from the input resources depends on the type and parameters of the draw call used to invoke the pipeline. Each piece of data is processed and passed on to the next stage until it reaches the end of the pipeline, where it will be written to an output memory resource. Once all of the data from these draw calls have been processed, the resulting output resource (typically a 2D render target) can be presented to an output window or saved to a file.

Throughout this chapter, we will see how to configure each stage of the pipeline, as well as understand the differences between each of the draw methods. There are seven different draw methods that can be used to invoke the rendering pipeline. They are listed here for easy reference, and to demonstrate that a fairly wide variety of methods is available for invoking the rendering pipeline.

- `Draw(...)`

- `DrawAuto(...)`

- `DrawIndexed(...)`

- `DrawIndexedInstanced(...)`

- `DrawInstanced(...)`

- `DrawIndexedInstancedIndirect(...)`

- `DrawInstancedIndirect(...)`

Each of these methods instructs the pipeline to interpret its input data in a different way. Each method also offers different parameters to further configure how much, and which, input data to process. Once the pipeline execution has been initiated, the input data specified in the `draw` method is processed piece by piece as it is brought into the pipeline.

One of the most important considerations when planning the work that will be performed during one pipeline execution is that each of the stages (both fixed and programmable) has different input and output types. In addition, the semantics for using those types have relatively large performance implications. For example, in a simple rendering sequence we could submit four vertices to the pipeline as two triangles. This would lead to four vertex shader invocations, which would pass their output to the rasterizer stage, which produces a number of fragments. These fragments are then passed to the pixel shader, then finally to the output merger. If the triangles are close to the viewer, many fragments may be generated, causing correspondingly large number of pixel shader invocations. The number of vertices is in this case is constant, while the number of fragments depends on the conditions of the scene. The opposite situation can also occur, where the tessellation stages produce a variable number of vertices to be rasterized, but the object remains the same size onscreen, and thus the number of pixels remains constant. One must consider this when designing an algorithm, since it is critical to balancing the workload placed on the GPU.

During the actual execution of the pipeline on hardware, each individual pipeline stage must perform its calculations as quickly as possible on the available GPU hardware. In many cases the GPU can perform multiple types of computation at the same time in different groups of processors. In the example above, it would be possible for the GPU to process the first triangle and rasterize it to generate fragments. It could then use some of its processing cores to process pixel shader invocations, and some to perform the remaining vertex shader work. On the other hand, it could process all of the vertices first and then switch to processing all of the fragments afterward. There is no guarantee about which order the GPU will process the data in—except that it will respect the logical order presented in the pipeline.

3.2.1 Stage-to-Stage Communication

As data is processed by the individual pipeline stages, it must be transmitted from one stage to the next. To do this, each programmable stage defines its required inputs and outputs for its shader program. We will refer to these inputs and outputs as *attributes*, which consist of vector variables (from 1 to 4 elements). All of the elements in each vector consist of one of the scalar types (both integer and floating point types are available). We can also refer to these attributes as *semantics* or *binding semantics*, which is the name given to the text-based identifier for each input/output attribute. In general, it should be clear from the context in which they are used when we are referring to these types of stage-to-stage variables by either of these names.

The input attributes to a shader program define the information required for that program to execute. Quite literally, the input attributes are the input parameters to the shader program function. Likewise, the function's return value defines its output attributes. Thus, each previous stage must produce output parameters that match the required input parameters for the next stage. These input and output attributes are implemented with the input and output registers we discussed in the "Shader Core Architecture" section. The brief example in Listing 3.1 provides a sample declaration for two shader functions, with each function defining its inputs and outputs. Notice that the output from the first function matches the input from the second function.

```
struct VS_INPUT
{
    float3 position : POSITION;
    float2 coords : TEXCOORDS;
    uint vertexID : SV_VertexID;
};

struct VS_OUTPUT
{
    float4 position : SV_POSITION;
    float4 color : COLOR;
};

VS_OUTPUT VSMAIN( in VS_INPUT v )
{
    // Perform vertex processing here...
}

float4 PSMAIN( in VS_OUTPUT input ) : SV_Target
{
    // Perform pixel processing here...
}
```

Listing 3.1. Two shader functions defining their input and output attributes which must match.

In the declaration of each function's input and output, a structure definition is used to group all of the inputs and outputs together. Each attribute of the input and output structures has an associated semantic, a textual name to link two data items together between stages. The semantic is located after the name of the attribute, following a semicolon. You can also see the type declarations for each of the components to the left of their names, just as you would expect from a similar declaration in C/C++.

You can see from this example that the output from the first shader function nearly matches the input of the second function, with the exception of the vertexID parameter. The semantic listed next to this particular parameter is SV_VertexID. In addition to the user-defined semantic attributes shown in Listing 3.1, there is another group of parameters that can be used by a programmable shader in its input/output signature. These are referred to as *system value semantics*. They represent attributes that the runtime either generates or consumes, depending on where they are declared, and which system value semantics are being used. These system value semantics provide helpful information for use in HLSL programs, and are also used to indicate required values to the pipeline. System values are always prefixed with SV_ to indicate that they have a special significance and are not user defined. We will see each of the system values, and how they are used in each of the pipeline stages, as we progress through the pipeline.

Understanding how data is supplied to the pipeline, and seeing how each pipeline stage's states influence the type of processing that occurs, are the key to successfully using the rendering pipeline. In the following sections we will inspect each of the pipeline stages in great detail, in the order in which they appear within the pipeline.

3.3 Input Assembler

The *input assembler* stage is the first stop in the rendering pipeline. It is a fixed function stage that is responsible for putting together all of the vertices that will be processed further down the pipeline—hence the name. As the entry point of the pipeline, the input assembler must create vertices that have the attributes required in the next pipeline stage, the vertex shader. The process of assembling vertices from one or more vertex buffer resources can actually cover a large number of different configurations, as we will see in detail later in this section. The application provides the input assembler a road map to constructing the vertices with a vertex layout object, which also allows for flexible strategies on the application side when creating vertex buffer resources. The input assembler's location in the pipeline is highlighted in Figure 3.4.

In addition to constructing the input vertices, the input assembler also determines how those vertices are connected to one another by specifying the topology of the geometry being rendered. This identifies the types of primitives or control points that define

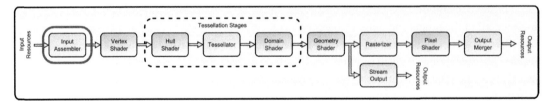

Figure 3.4. The input assembler stage.

how the individual vertices should be grouped together later in the pipeline. By specifying the topology of the geometry being rendered, the input assembler can have a significant impact on how the same set of input vertices is interpreted by the rest of the pipeline. If a geometric object is composed of triangles, but it is rendered as a set of points, the resulting rendering will be quite different, even though the input vertex buffers are exactly the same.

The input assembler is also the enabler that allows several different types of rendering operations to be performed by the rendering pipeline. The rendering pipeline supports standard draw calls, indexed drawing, instanced drawing, indirect drawing, and several combinations of these various operations—all of which require different input resource configurations. The input assembler works in conjunction with its input resources and the information provided in the draw call to convert these various input configurations into a format that is usable by the rest of the pipeline. In many cases, the input assembler is the only stage that executes differently for different draw calls. It effectively hides the data submission details from the rest of the pipeline, and instead, provides its output data in a consistent format. Since the input assembler forms the primary link between an application's model data and the rendering pipeline, it serves a very important function. We will investigate this pipeline stage in more detail to understand precisely how to properly use it, and what options are available for invoking a pipeline execution.

3.3.1 Input Assembler Pipeline Input

Prior to passing data into the pipeline, the input assembler must be connected to the appropriate resources to acquire the input geometric data. This data consists of vertex data and index data, both of which are supplied to the pipeline in the form of buffer resources. The following two sections describe how to bind these buffer resources to the input assembler.

Vertex Buffer Usage

Having primary responsibility to correctly assemble vertices for use further down the pipeline, the input assembler requires access to vertex data from the application. This data is stored in one or more vertex buffer resources and can be organized in a variety of different

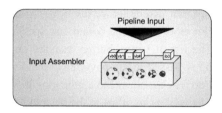

Figure 3.5. An overview of the available input slots of the input assembler.

layouts. Conceptually, we can view the input assembler as having several input slots, which can be filled with vertex buffer resources. Figure 3.5 shows this visualization.

Each slot can be filled with a vertex buffer resource that contains one or more attributes of a complete vertex. For example, one vertex buffer can contain the position data for the vertices, and a second vertex buffer can contain all of the normal vector data. It is also possible to provide all of the vertex data within a single vertex buffer in an array of structures. The developer can choose how to best organize the input data. There are currently 16 input slots available for the application to configure, which allows for fairly flexible storage options for the needed vertex data. With the individual attributes stored separately, it is possible to dynamically decide which vertex components will be needed for a particular rendering. This can potentially reduce the bandwidth required for reading the vertex data, by eliminating unused attributes from the vertices. We will discuss the details of how to store various data in multiple buffers later in this chapter.

To bind the vertex buffers to the input assembler stage, we use the device context interface. As we saw in Chapter 1, the device context is the primary interface to the complete pipeline. The process to bind several vertex buffers to the input assembler with the `ID3D11` `DeviceContext::IASetVertexBuffers` method is shown in Listing 3.2. Each of the variables declared in the listing will be filled with its appropriate buffer references, strides, and offsets, before the set vertex buffer method is called.

```
UINT StartSlot;
UINT NumBuffers;
ID3D11Buffer* aBuffers[D3D11_IA_VERTEX_INPUT_RESOURCE_SLOT_COUNT];
UINT aStrides[D3D11_IA_VERTEX_INPUT_RESOURCE_SLOT_COUNT];
UINT aOffsets[D3D11_IA_VERTEX_INPUT_RESOURCE_SLOT_COUNT];

// Fill in all data here...

pContext->IASetVertexBuffers( StartSlot, NumBuffers, aBuffers, aStrides,
                              aOffsets );
```

Listing 3.2. How to bind several vertex buffers to the input assembler stage.

The index of the first input assembler vertex buffer slot to begin binding to, and how many subsequent buffer slots should be bound, are specified in the StartSlot and NumBuffers parameters. The list of vertex buffer resource pointers must be assembled into a contiguous array, and is passed in the ppVertexBuffers parameter. The final two parameters, pStrides and pOffsets, allow the application to specify the per-vertex step size for each vertex buffer, as well as an offset to the desired location to begin using each vertex buffer. Each of these array elements will start being used at the StartSlot index and will be used to fill NumBuffers number of slots. This means that the arrays must be appropriately sized and filled in, or the function will access an invalid memory location. In addition, to bind the vertex buffers to the input assembler, the buffers must not be bound for writing at another location in the pipeline. This is a reasonable requirement, since it would lead to seemingly unpredictable results, with buffers being read from and written to at the same time.

Index Buffer Usage

When taken on its own, vertex data is interpreted by the input assembler in the order that it appears within the vertex buffers. This means that if the vertex data is being used to render triangles, the first three vertices define the first triangle, followed by the next three vertices for the next triangle, and so on.[5] This can lead to duplicated vertex data in cases where two adjacent triangles actually reference the same vertex. In fact, this is usually the case—most triangle meshes have a significant number of adjacent triangles. When the individual vertex data is specified more than once in the vertex buffer, significant amounts of memory can be wasted. This inefficiency can be eliminated with the use of index buffers. An index buffer is used to define the vertex order for each primitive by providing index offsets into the vertex buffers. In this case, each vertex can be specified once, and then an index that points to it can be used for any triangle that it is a part of. With indexed rendering, the three vertices to be used in a triangle are specified by three consecutive indices in the index buffer.

To use indexed rendering, an application must bind an index buffer to the input assembler. There is only one slot available for binding an index buffer, and it is only used during one of the indexed draw calls to specify the order of the vertices used to generate primitives. To bind the index buffer to the input assembler, the application must use the ID3D11DeviceContext::IASetIndexBuffer method. Listing 3.3 demonstrates how to perform this binding operation. This method provides three simple parameters: a pointer to the index buffer, the format that the indices are stored in (either 16- or 32-bit unsigned integers), and finally an offset into the index buffer to begin building primitives from.

---

[5] The ordering of vertices and indices will vary for different primitive types. These are described in more detail later in this section.

```
ID3D11Buffer* pBuffer = 0;
UINT offset = 0;

// Set pBuffer to the desired index buffer reference here, and
// set the offset to an appropriate location in the buffer.

m_pContext->IASetIndexBuffer( pBuffer, DXGI_FORMAT_R32_UINT, offset );
```

Listing 3.3. How to bind an index buffer to the input assembler stage.

Once a vertex or index buffer is bound to the pipeline, it can be unbound by setting a NULL pointer value to the corresponding input slot. This is important to remember, especially when reconfiguring the pipeline in between draw calls. If a multiple vertex buffer configuration is used in a previous draw call, and the new draw call uses fewer vertex buffers, the unused slots should be filled with NULL to unbind the unused buffers. This is why the vertex buffer binding method always uses an array sized for the maximum number of vertex buffers and initializes them to NULL values before filling the desired configuration.

3.3.2 Input Assembler State Configuration

After the appropriate resources have been bound to the vertex and index buffer slots, there are two other configurations that must be set in the input assembler before it can be used. The first is the Input Layout object, which is used by the input assembler to know which input slots to read the per-vertex data from to build complete vertices. The second parameter is the primitive topology that should be used by the input assembler to determine how vertices are grouped together into primitives. These two states, in combination with the draw method that is used to execute the pipeline, determines how vertex and primitive data are interpreted by the pipeline. We will explore each of these states in detail and then consider how they interact with the available draw methods to produce different input configurations.

Input Layout

The Input Layout object can be thought of as a recipe that tells the input assembler how to create vertices. Every vertex is composed of a collection of vector attributes, each with up to four components. With up to 16 vertex buffers available for binding, the input assembler needs to know where to read each of these components from, as well as understanding what order to put them in the final assembled vertices. The Input Layout object provides this information to the input assembler.

To create an Input Layout object, the application must create an array of D3D11_INPUT_ELEMENT_DESC structures, with one structure for each component of a desired vertex configuration. The members of D3D11_INPUT_ELEMENT_DESC are shown in Listing 3.4. We will examine what each of these structure members represents, and how they define the needed information for the input assembler to do its job.

```
struct D3D11_INPUT_ELEMENT_DESC {
    LPCSTR                      SemanticName;
    UINT                        SemanticIndex;
    DXGI_FORMAT                 Format;
    UINT                        InputSlot;
    UINT                        AlignedByteOffset;
    D3D11_INPUT_CLASSIFICATION  InputSlotClass;
    UINT                        InstanceDataStepRate;
}
```

Listing 3.4. The D3D11_INPUT_ELEMENT_DESC structure members.

Per-vertex elements. The first parameter, **SemanticName**, identifies the textual name of a vertex attribute, which must match the corresponding name provided in the vertex shader program. Each input to a vertex shader must have a semantic defined in its HLSL source code. That semantic name is used to match the vertex shader input to the vertex data provided by the input assembler. The **SemanticIndex** is an integer number which allows a **SemanticName** to be used more than once. For example, if more than one set of texture coordinates is used within a single vertex layout, each set of coordinates could use the same **SemanticName**, but would use increasing **SemanticIndex** values to differentiate between them.

The next structure member is the Format of the component. This specifies what data type, and how many elements are contained in the attribute. The available formats range from 1 to 4 elements, with both floating point and integer types available. The next parameter is the InputSlot, which indicates which of the 16 vertex buffer slots this component's data should be read from. The AlignedByteOffset indicates the offset to the first element in a vertex buffer for the items described by this D3D11_INPUT_ELEMENT_DESC. This tells the input assembler where in a vertex buffer to begin reading input data from.

Per-instance elements. The final two structure members declare vertex functionality regarding instancing. The instancing **draw** methods essentially use a single **draw** call to submit a model to the pipeline. This model is then "instanced" multiple times, but with a subset of the vertex data specified at a per-instance level, instead of a per-vertex level. This is frequently used when many copies of the same object must be rendered at different places throughout a scene. The vertex format will include a transformation matrix in its definition, which is only incremented to the next value for each instance of the model. Then the transformation matrices for all of the instances of the model are provided in a vertex buffer and bound to the input assembler. When the **draw** call is performed, the vertex data is submitted to the vertex shader one copy at a time, with the transformation matrix being updated between instances of the model.

If a vertex component is providing per-instance data, it must specify that with the InputSlotClass parameter. In this case, there is also an option to increment to the next element in the vertex buffer only after a certain number of instances have been submitted to the pipeline. This is specified in the InstanceDataStepRate and provides a rough control for the frequency with which a per-instance parameter is changed. Since this is specified at the individual attribute level, it is possible to have different attributes to advance to the next data member at different rates. After all of the desired vertex attributes have been identified, and an array of description structures has been filled in, the application can create the ID3D11InputLayout object with the ID3D11Device::CreateInputLayout() method. Listing 3.5 demonstrates how this process is performed.

```
// Create array of elements here for the API call.
D3D11_INPUT_ELEMENT_DESC* pElements = new D3D11_INPUT_ELEMENT_DESC[elements.
    count()];

// Fill in the array of elements (this will vary depending on how they are
// submitted to this method).
for ( int i = 0; i < elements.count(); i++ )
   pElements[i] = elements[i];

// Attempt to create the input layout from the input information (the
// compiled shader byte code storage will also vary by the data
// structures used).
ID3DBlob* pCompiledShader = m_vShaders[ShaderID]->pCompiledShader;
ID3D11InputLayout* pLayout = 0;

// Create the input layout object, and check the HRESULT afterwards.
HRESULT hr = m_pDevice->CreateInputLayout( pElements, elements.count(),
    pCompiledShader->GetBufferPointer(), pCompiledShader->GetBufferSize(),
                                &pLayout );
```

Listing 3.5. The creation of an input layout object.

Here we can see the array of D3D11_INPUT_ELEMENT_DESC structures being allocated and then being filled in from an input container object. The first two parameters to ID3D11 Device::CreateInputLayout() provide the array of descriptions and the number of elements that exist in the array. The third and fourth parameters are a pointer to the shader byte code that results from compiling the vertex shader source code, and the size of that byte code, respectively. This is used to compare the input array of element descriptions against the HLSL shader that it will be used to supply data to. Comparing these two objects while creating the input layout lowers the amount of validation required at runtime when a shader is used with a particular input layout.

Primitive Topology

The final state that must be set in the input assembler is the *primitive topology*. We can view the work of the input assembler as two tasks—first, it must assemble the vertices from the provided input resources, using the input layout object; and second, it must organize that vertex stream into a stream of primitives. Each primitive uses a small sequential group of vertices (with the sequence determined either by their order of appearance in the vertex buffers, or by the order of the indices in the index buffer for indexed rendering) to define the components that make it up. Common examples of primitive topology are triangle lists, triangle strips, line lists, and line strips. In the example of a triangle list, a primitive is created for every three vertices in the vertex stream.

The primitive topology setting tells the input assembler how to build the various primitives from the assembled vertex stream. The number of vertices used in each primitive, as well as how they are selected from the assembled vertex stream, is a function of the primitive topology setting. The available primitive types are provided in Listing 3.6, along with a demonstration of how to set the primitive topology with the `ID3D11DeviceContext::IASetPrimitiveTopology()` method.

```
enum D3D11_PRIMITIVE_TOPOLOGY {
    D3D11_PRIMITIVE_TOPOLOGY_UNDEFINED,
    D3D11_PRIMITIVE_TOPOLOGY_POINTLIST,
    D3D11_PRIMITIVE_TOPOLOGY_LINELIST,
    D3D11_PRIMITIVE_TOPOLOGY_LINESTRIP,
    D3D11_PRIMITIVE_TOPOLOGY_TRIANGLELIST,
    D3D11_PRIMITIVE_TOPOLOGY_TRIANGLESTRIP,
    D3D11_PRIMITIVE_TOPOLOGY_LINELIST_ADJ,
    D3D11_PRIMITIVE_TOPOLOGY_LINESTRIP_ADJ,
    D3D11_PRIMITIVE_TOPOLOGY_TRIANGLELIST_ADJ,
    D3D11_PRIMITIVE_TOPOLOGY_TRIANGLESTRIP_ADJ,
    D3D11_PRIMITIVE_TOPOLOGY_1_CONTROL_POINT_PATCHLIST,
    D3D11_PRIMITIVE_TOPOLOGY_2_CONTROL_POINT_PATCHLIST,
    D3D11_PRIMITIVE_TOPOLOGY_3_CONTROL_POINT_PATCHLIST,
    D3D11_PRIMITIVE_TOPOLOGY_4_CONTROL_POINT_PATCHLIST,
    D3D11_PRIMITIVE_TOPOLOGY_5_CONTROL_POINT_PATCHLIST,
    D3D11_PRIMITIVE_TOPOLOGY_6_CONTROL_POINT_PATCHLIST,
    D3D11_PRIMITIVE_TOPOLOGY_7_CONTROL_POINT_PATCHLIST,
    D3D11_PRIMITIVE_TOPOLOGY_8_CONTROL_POINT_PATCHLIST,
    D3D11_PRIMITIVE_TOPOLOGY_9_CONTROL_POINT_PATCHLIST,
    D3D11_PRIMITIVE_TOPOLOGY_10_CONTROL_POINT_PATCHLIST,
    D3D11_PRIMITIVE_TOPOLOGY_11_CONTROL_POINT_PATCHLIST,
    D3D11_PRIMITIVE_TOPOLOGY_12_CONTROL_POINT_PATCHLIST,
    D3D11_PRIMITIVE_TOPOLOGY_13_CONTROL_POINT_PATCHLIST,
    D3D11_PRIMITIVE_TOPOLOGY_14_CONTROL_POINT_PATCHLIST,
```

```
        D3D11_PRIMITIVE_TOPOLOGY_15_CONTROL_POINT_PATCHLIST,
        D3D11_PRIMITIVE_TOPOLOGY_16_CONTROL_POINT_PATCHLIST,
        D3D11_PRIMITIVE_TOPOLOGY_17_CONTROL_POINT_PATCHLIST,
        D3D11_PRIMITIVE_TOPOLOGY_18_CONTROL_POINT_PATCHLIST,
        D3D11_PRIMITIVE_TOPOLOGY_19_CONTROL_POINT_PATCHLIST,
        D3D11_PRIMITIVE_TOPOLOGY_20_CONTROL_POINT_PATCHLIST,
        D3D11_PRIMITIVE_TOPOLOGY_21_CONTROL_POINT_PATCHLIST,
        D3D11_PRIMITIVE_TOPOLOGY_22_CONTROL_POINT_PATCHLIST,
        D3D11_PRIMITIVE_TOPOLOGY_23_CONTROL_POINT_PATCHLIST,
        D3D11_PRIMITIVE_TOPOLOGY_24_CONTROL_POINT_PATCHLIST,
        D3D11_PRIMITIVE_TOPOLOGY_25_CONTROL_POINT_PATCHLIST,
        D3D11_PRIMITIVE_TOPOLOGY_26_CONTROL_POINT_PATCHLIST,
        D3D11_PRIMITIVE_TOPOLOGY_27_CONTROL_POINT_PATCHLIST,
        D3D11_PRIMITIVE_TOPOLOGY_28_CONTROL_POINT_PATCHLIST,
        D3D11_PRIMITIVE_TOPOLOGY_29_CONTROL_POINT_PATCHLIST,
        D3D11_PRIMITIVE_TOPOLOGY_30_CONTROL_POINT_PATCHLIST,
        D3D11_PRIMITIVE_TOPOLOGY_31_CONTROL_POINT_PATCHLIST,
        D3D11_PRIMITIVE_TOPOLOGY_32_CONTROL_POINT_PATCHLIST
}

// Specify the type of geometry that we will be dealing with.
m_pContext->IASetPrimitiveTopology( primType );
```

Listing 3.6. The available primitive topology types, and an example of how to set the topology type with the device context interface.

As you can see from Listing 3.6, there are quite a variety of available primitive types. The first nine entries in the list should look familiar if you have used Direct3D 10 before, since they have been in use prior to Direct3D 11. However, the remainder of the list provides a series of different control point patch lists, each with a different number of points to be included in the control patch, up to a maximum of 32. These primitive types have been introduced to support the new tessellation stages that were introduced in Direct3D 11. To better understand how all of these primitive topologies organize a stream of vertices, we will create an example with an ordered series of vertices and then examine how each topology type would create primitives with this vertex stream. Figure 3.6 shows the sample set of vertices, each numbered according to its location in the vertex stream.

Point primitives. The *point list primitive topology* is the simplest of all of the primitive types. Vertices are grouped into single-vertex primitives, meaning that the stream of vertices produces an equal-sized stream of primitives. Therefore, the output group of primitives will look identical to that shown in Figure 3.6.

Line primitives. The *line list primitive topology* is only slightly more complex, with every pair of vertices producing a line primitive. This essentially indicates that the two vertices comprising a line must be included in the vertex buffer for every line to be rendered. The line strip primitive topology provides a more compact representation of the line primitives,

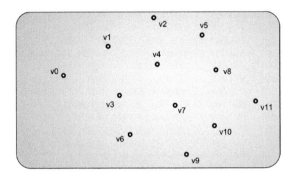

Figure 3.6. A number of vertices to be used to create a variety of primitive types.

but it can only represent a connected list of lines. The input assembler produces the first line from the first two vertices in the vertex stream. Every vertex after the first two defines a new line, where the other vertex of the line is the previous vertex in the stream. This provides a more dense representation of vertices in situations where the lines to be drawn are connected end to end. Both of these primitive topologies are shown in Figure 3.7.

Triangle primitives. The *standard triangle primitive* **topologies** follow the same paradigm as their line primitive counterparts. The triangle list topology creates a triangle primitive from every three vertices in the vertex stream. The triangle strip creates a triangle primitive out of the first three vertices in the stream, then a new triangle primitive is created for every subsequent vertex, using the prior two vertices in the stream to define the remaining portion of the triangle. As with the line strip primitives, the triangle strip primitive topology can only represent connected "strips" of triangles. These primitive topologies and their ordering are shown in Figure 3.8.

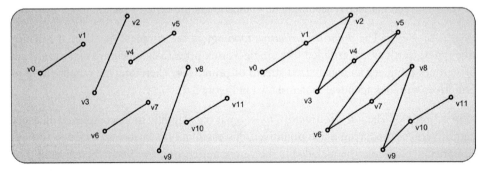

Figure 3.7. Line list and line strip primitives created from our input vertices.

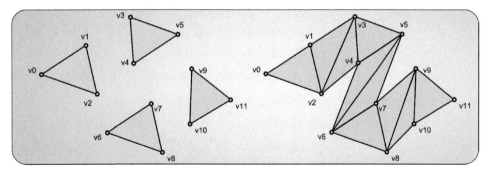

Figure 3.8. Triangle list and triangle strip primitives created from our input vertices.

Primitives with adjacency. In some situations, it is desirable to have adjacent primitive information available for use in the pipeline. This means that every primitive will be provided to the pipeline with access to the primitives immediately neighboring it. There are four different primitive topologies that provide adjacency information: *line lists with adjacency*, *line strips with adjacency*, *triangle lists with adjacency*, and *triangle strips with adjacency*. These representations provide significantly more information to the pipeline for processing at the primitive level, at the expense of a higher memory and bandwidth cost per primitive. Each of these primitive topologies is shown in Figures 3.9 and 3.10.

Control point primitives. To facilitate the use of higher-order primitives with the tessellation pipeline, additional primitive topology selections are provided. There is one primitive type for each number of control points, ranging from 1 to 32. This provides the ability to perform a very large number of different control patch schemes. One sample control patch primitive is shown in Figure 3.11.

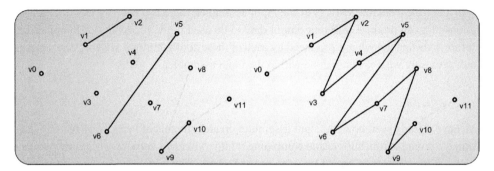

Figure 3.9. Line lists and line strip primitives with adjacency created from our input vertices.

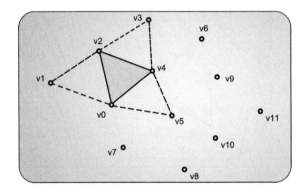

Figure 3.10. Triangle list primitive with adjacency created from our input vertices.

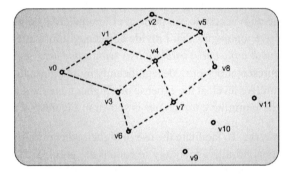

Figure 3.11. Control patch primitives created from our input vertices.

3.3.3 Input Assembler Stage Processing

With all of the available settings of the input assembler, there is a wide variety of configurations that can produce different output data to be used in the remainder of the pipeline. Before investigating what is produced by each of these configurations, we will take a closer look at exactly what is being done within the input assembler.

Vertex Streams

We have already seen how the input assembler creates streams of vertices by reading data from the bound vertex buffers and combining it into individual vertices. This vertex stream produces the vertices in one of two ordered sequences. The vertices are either left in their existing order from their vertex buffers, or they are rearranged into the order specified by the indices in the index buffer. The determination of which order is used comes from the type of draw call used to invoke pipeline execution. Regardless of where the ordering comes from, the input assembler produces a stream of vertices to represent the input geometry.

Primitive Streams

The vertex stream is further refined into a stream of primitives. Various portions of the pipeline are intended to operate on only vertices (vertex shader), on both vertices and primitives (hull shader and domain shader), or only on complete primitives (geometry shader). In each of these cases, the stream of primitives is shaped by the selected primitive topology and the type of draw call used to initiate pipeline execution.

Draw Call Effects

So how exactly are the vertex and primitive streams that result from a draw call influenced by the type of call used? This can best be demonstrated by examining the output from the input assembler for each of the types of draw calls. The following sections provide information about each of the classes of draw calls, and what effects they have on the resulting data streams.

Standard draw methods. The simplest **draw** methods to consider are the **Draw()** and **DrawAuto()** methods. Both of these methods trigger the input assembler to assemble vertices based on its input layout settings. This creates a stream of vertices, which is then passed into the pipeline. The primitive information generated by these **draw** calls is determined by the primitive topology currently specified in the input assembler, and the construction of primitives is dependent on the order of the vertices found in the vertex buffers. This process can be considered the basic vertex and primitive construction process.

Indexed draw methods. The first variation on the basic **draw** calls is the addition of **indexed rendering**. As we have already discussed, indexed rendering uses the indices of an index buffer to determine which vertices are used to construct primitives instead of simply using the vertex order from the vertex buffers. The vertex stream remains unchanged, with the exception that the vertices within it are chosen by the indices of the index buffer. In addition, the primitive stream creates its primitives with the index buffer ordering, instead of the vertex buffer ordering. Several **draw** calls support indexed rendering, including **DrawIndexed()**, **DrawIndexedInstanced()**, and **DrawIndexedInstancedIndirect()**.

Instanced draw methods. In addition to indexed rendering, Direct3D 11 also allows **instanced rendering**. We discussed instanced rendering briefly while describing the available members of the **D3D11_INPUT_ELEMENT_DESC** structure. If a vertex component is declared as a per-instance component, and the pipeline is invoked with one of the instanced

draw methods, the vertex and primitive streams are created in largely the same ways that were described for basic rendering and indexed rendering. However, the vertex and primitive streams are repeated for each instance of the object, except that the per-instance vertex components are updated for each complete instance. The number of instances is determined by the parameters passed to the **draw** call, and the per-instance vertex components are taken from one or more vertex buffers bound to the input assembler.

Overall, the result of using instanced rendering methods is that the vertex and primitive streams are multiplied by the number of instances that are being rendered. There are four different instanced draw calls: `DrawInstanced()`, `DrawIndexedInstanced()`, `DrawInstancedIndirect()`, and `DrawIndexedInstancedIndirect()`.

Indirect draw methods. In addition to the basic, indexed, and instanced rendering methods, there is another type of **draw** call: **indirect rendering**. This technique doesn't actually modify the vertex and primitive streams, but it should be discussed here for a complete view of the available **draw** methods. Instead, indirect rendering methods allow a buffer resource to be passed as the **draw** call parameters. The buffer contains the required input information that would normally be passed by the application. The indirect rendering methods are **DrawInstancedIndirect()** and **DrawIndexedInstancedIndirect()**. As an example, the number of vertices, the starting vertex, the number of instances, and the starting instance of the **DrawInstancedIndirect()** method would be contained within a buffer resource, instead of being directly passed to the function.

The purpose of these calls is to allow the GPU to fill in a buffer, which then can control how a draw sequence is performed. This shifts control of the draw call to the GPU instead of the application and provides the first steps for making the GPU more autonomous in its operations. However, indirect rendering operations don't modify the construction of the vertex and primitive streams—they only modify how the draw method parameters are passed to the runtime.

Mixed rendering draw methods. As we have seen, each class of rendering operation is not mutually exclusive of the others. The names of the **draw** methods typically include several of these rendering techniques, and provide a variety of different ways to execute the pipeline. In each of these mixed cases, the resulting vertex and primitive streams are a mixture of the independent rendering types.

3.3.4 Input Assembler Pipeline Output

User Defined Attributes

Along with an understanding of what the input assembler can produce, and how to manipulate its output, we need to consider how the output streams interact with the remainder of the pipeline. For the input assembler to produce output that is compatible with the desired vertex shader, the array of D3D11_INPUT_ELEMENT_DESC structures used to create the input layout object must match the declared input signature of the vertex shader. Therefore, the individual vertex attributes must match what is expected by the vertex shader, including the data type and semantic name. Listing 3.7 shows a sample vertex shader program in HLSL.

```
struct VS_INPUT
{
    float3 position : POSITION;
    float2 tex : TEXCOORDS;
    float3 normal : NORMAL;
};

VS_OUTPUT VSMAIN( in VS_INPUT v )
{
    ...
}
```

Listing 3.7. A sample vertex shader input declaration.

For this example, the application would need to provide vertex data that consists of a POSITION semantic attribute with three floating point elements, a TEXCOORDS semantic component with two floating point elements, and a NORMAL semantic component with three floating point elements. Creating an input layout object helps the developer ensure that the vertex data and the vertex shader signature match one another. Since the vertex data is supplied by the application, it can be created or loaded in the proper format to be used with a particular vertex shader. These inputs are then used by the vertex shader to perform its own calculations, and it will then also define a number of output attributes, which it will write out to be used by the next active pipeline stage. This process is repeated for each subsequent pipeline stage, with each stage declaring its output signature and then supplying its output data in them. Some typical examples of input data produced by the input assembler include, but are certainly not limited to, the following:

- float3 position—the position of the input vertex

- float3 normal—the normal vector of the input vertex

- float3 tangent—the tangent vector of the input vertex

- float3 bitangent — the bitangent vector of the input vertex

- uint4 boneIDs—four-bone IDs of the vertex used for skinning

- float4 boneWeights—four-bone weights of the vertex used for skinning

The input assembler can also produce three different system value semantics: SV_ VertexID, SV_PrimitiveID, and SV_InstanceID. These system value semantics are not supplied by the user, but are instead generated by the input assembler as it creates the output data streams. The SV_VertexID system value is an unsigned integer, which uniquely identifies each vertex in the assembled vertex stream. It is first available in the vertex shader, and provides a simple way to differentiate between vertices later in the pipeline. The SV_PrimitiveID provides a similar unique identifier for each primitive in the primitive stream. It is first available in the hull shader stage, since the vertex shader doesn't use primitive-level information. Finally, the SV_InstanceID uniquely identifies each instance of the geometry in an instanced draw call. It is first available for use in the vertex shader stage. Together, these three system values provide a fairly extensive method for identifying each of the data elements that the input assembler produces.

3.4 Vertex Shader

The first programmable shader stage in the rendering pipeline is the vertex shader. As described above, the programmable shader stages execute a custom function written in HLSL. In the case of the vertex shader stage, the vertex shader program is a function that is invoked once for each vertex in the vertex stream produced by the input assembler. Each of the input vertices is received as the argument to the vertex shader program, and the processed vertices are returned as the result of the function. Each vertex shader invocation is executed in complete isolation from the others, with no communication between them possible. Since it is executed once for each input vertex, there will be a one-to-one mapping of input vertices to output vertices from the vertex shader stage. The location of the vertex shader within the pipeline is highlighted in Figure 3.12.

In addition, each invocation of the vertex shader is not aware of the primitive stream produced by the input assembler. Its sole purpose is to process vertex data and leave any higher-level processing to the stages further down the pipeline. With no knowledge of primitives, the vertex shader operates in the same way for all different topology types. Regardless of whether it is processing geometry that was submitted as points, lines,

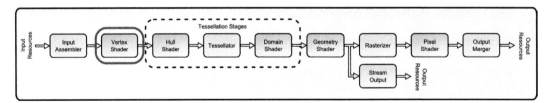

Figure 3.12. The vertex shader stage.

triangles, or patches, the individual vertices are all treated the same by the vertex shader. The processed output from the vertex shader is combined with the primitive information generated by the input assembler stage later in the pipeline. This simplifies the job of the vertex shader stage, allowing it to perform the exact same operations regardless of the primitive type.

The vertex shader is located between the input assembler stage and the hull shader stage. This means that it processes geometry data immediately after it is inserted into the pipeline, and immediately before any tessellation is performed. Some examples of operations that are typically performed in the vertex shader are the application of transformation matrices to input geometry, performing bone-based animation transformations (also known as *vertex skinning* [6]), and performing per-vertex lighting calculations. We will investigate more precisely what the vertex shader stage is typically used for, and what tools it has at its disposal throughout the remainder of this section. We begin by considering what data the vertex shader receives as input followed by the types of processing that it performs, and then finally consider what data it produces as output.

3.4.1 Vertex Shader Pipeline Input

Since the vertex shader is located directly after the input assembler in the pipeline, it naturally receives its input from there. All of the work done to configure the input assembler's input layout is intended to make the created vertices match the format expected by the vertex shader stage's current program. This is why compiled shader byte code is required as an input when creating an ID3D11InputLayout object—to ensure that the assembled vertices will match what is needed to execute the vertex shader program.

As indicated in the description of the input assembler's output, the vertex shader input can also use system value semantics in addition to the attributes of the assembled vertices. The two system value semantics available as inputs to the vertex shader are SV_VertexID and SV_InstanceID. Both of these provide useful information to the vertex shader program, and the developer only needs to include them in the vertex shader input signature to acquire them. For example, the SV_VertexID can be used as an index into a lookup table for providing pseudo-random values to the shader. Similarly, the SV_InstanceID can also be used as an indexing mechanism. However, it is only updated for each instance generated by the input assembler, so any lookup table values would be applied uniformly to all of the vertices of an instance. This could be used to introduce variation between instances of a mesh to make each instance appear more unique.

We have already described how these system value semantics are available to the vertex shader as inputs. They can also be provided to later stages in the pipeline, but they

---

[6] Vertex skinning is described in detail in Chapter 8, "Mesh Rendering."

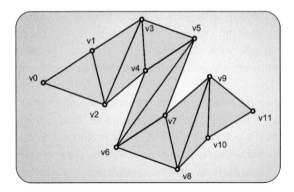

Figure 3.13. A depiction of vertices of a triangle strip that belong to multiple primitives.

must be passed through the vertex shader output if they are to be used later on. This is a reasonable restriction—if the system values are generated in the input assembler, they are directly available to the vertex shader. If that shader decides not to propagate the system values further down the pipeline, they would not simply reappear in a later pipeline stage that needs them. The developer must ensure that any system value semantics that are needed later on are passed from stage to stage until they are used.

One system value semantic that is not available to the vertex shader is the SV_PrimitiveID attribute. SV_PrimitiveID is produced by the input assembler and provides an identifier for each primitive in the primitive stream. Vertices are processed by the vertex shader stage one at a time, regardless of the primitive type being rendered. In addition, a single processed output vertex produced by the vertex shader can be use in more than one primitive, such as in triangle strips, or as seen in our discussion of indexed rendering. This makes having a unique primitive ID for an input vertex attribute illogical, since it can be used in multiple primitives. Figure 3.13 shows a very simple set of geometry that is submitted as a triangle strip, which highlights this situation.

When geometry such as this is processed by the pipeline, each vertex would be processed individually by the vertex shader stage. If the SV_PrimitiveID were allowed in the vertex shader, which primitive index would be assigned to vertex *v3*? It belongs to multiple primitives, and hence it would not be clear which primitive ID a shared vertex should use.

3.4.2 Vertex Shader State Configuration

As a programmable shader stage, the vertex shader implements the common shader core functionality. This means that it provides a standard set of resource interface methods that allow an application to provide the shader program access to the required resources. As with all pipeline manipulations, all of the methods that can change the vertex shader stage's state belong to the `ID3D11DeviceContext` interface. We will review each of these available resources and see how they are used in the context of the vertex shader stage.

The Shader Program

The first, and probably the most important state of the vertex shader stage is the shader program itself. The process for compiling a shader and creating a shader object from it is described in detail in Chapter 6, so we will assume here that a compiled shader has been used to create a valid vertex shader object. An application can, and normally does, create more than one vertex shader object for use in rendering various objects. To configure the vertex shader stage with the desired shader program, the application must use the `ID3D11Device Context::VSSetShader` method. Listing 3.8 shows an example of how this is performed.

```
// pNextShader must be initialized to the desired shader object.

ID3D11VertexShader* pNextShader = pShader1;
m_pContext->VSSetShader( pNextShader, 0, 0 );
```

Listing 3.8. Setting a vertex shader program.

The method to set the desired shader object requires three arguments. The first is a pointer to the shader object to bind to the stage. The second and third parameters are used to pass an array of shader class instances to be used in the shader program. These class instances are objects that can be used to provide a portion of a shader program with an external object, and are intended to reduce the number of individual shader objects needed to support a wide variety of similar rendering scenarios.[7] There is also a corresponding `ID3D11DeviceContext ::VSGetShader` method, which can be used to retrieve the shader and class instance objects that are currently bound to the vertex shader stage. This can be useful to read an existing shader object setting, so that it can be restored after an interim operation is performed.

Constant Buffers

The next configuration that the application can provide to the vertex shader program is an array of constant buffers. Constant buffers are described in detail in Chapter 2. These are essentially buffer resources that the application loads with parameters that are made available to the shader program, which can use the data directly within the HLSL code. This is used to communicate parameters to the shader program that will remain the same throughout a pipeline execution—hence the name *constant* buffers. The process for filling a constant buffer with data is also described in detail in Chapter 2, so we will focus here on

---

[7] These class instances are described briefly in Chapter 6, but are not discussed in great detail. The reader is referred to the DXSDK documentation for further guidance on how to use these objects.

how to bind these resources to the vertex shader stage. Listing 3.9 shows how to bind one or more constant buffers to the vertex shader stage.

```
ID3D11Buffer* cbuffers[D3D11_COMMONSHADER_CONSTANT_BUFFER_API_SLOT_COUNT];

// Fill each element of the array with buffers here...

pContext->VSSetConstantBuffers( 0, count, cbuffers );
```

Listing 3.9. Binding an array of constant buffers to the vertex shader stage.

Listing 3.9 shows how to set multiple constant buffers with a single device context method. In fact, this example shows how to set all of the available constant buffer slots. This can be an advantageous technique if all of the elements of the ID3D11Buffer* array are initialized to NULL. Then any constant buffers that were previously attached to this stage are automatically detached when the NULL value is bound in its place. Figure 3.14 depicts the constant buffers slots in the vertex shader before and after such a "complete" state setting. The array of constant buffer pointers that are currently set in the vertex shader stage can also be retrieved with a corresponding ID3D11DeviceContext::VSGetConstant

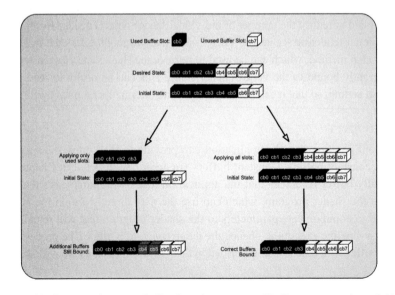

Figure 3.14. Unbinding unused constant buffers by using an array of buffer resource pointers initialized to NULL.

Buffers method. Once again, this can be useful for reinstating an existing setting after an interim operation is performed.

Shader Resource Views

For the vertex shader program to be able to access the various types of read-only resources that are available to it, the resources must be bound to the vertex shader stage with a shader resource view. Both buffer resources and texture resources are bound to the pipeline in exactly the same manner, using the `ID3D11DeviceContext::VSSetShaderResources` method. As with constant buffers, one or more shader resource views can be set simultaneously with the same method call. Listing 3.10 demonstrates how this method is used.

```
ID3D11ShaderResourceView*
ShaderResourceViews[D3D11_COMMONSHADER_INPUT_RESOURCE_SLOT_COUNT];

// Fill the array with the desired resource views...
pContext->VSSetShaderResources( start, count, ShaderResourceViews );
```

Listing 3.10. Binding an array of shader resource views to the vertex shader stage.

The array of shader resource views operates in much the same fashion as described for the constant buffers. Once a shader resource view has been set in one of the slots, it will remain bound in this location until it is replaced by either another shader resource view or a NULL pointer. The technique of clearing out all unused shader resource views with NULL in between pipeline executions takes on more importance when applied to resources. If a resource that is dynamically generated by the GPU is used as a shader resource in another rendering pass, it is first written to by the pipeline in the first pass, and then read in the shader program in the second pass. Writing to the resource is done with either a render target view or an unordered access view, while reading is performed with a shader resource view. If a resource is bound for reading with a shader resource view and is mistakenly left bound to the vertex shader stage, any attempt to bind the resource for writing will generate an error, since a resource cannot be simultaneously read and written by multiple views. This situation is depicted in Figure 3.15. The simplest way to ensure that this does not happen is to clear all unneeded shader resource views from the vertex shader stage each time the pipeline is configured for a new rendering effect.

Samplers

The final configuration that the application can supply to the vertex shader stage is *sampler objects*. Samplers provide the ability to perform various types of filtering on textures

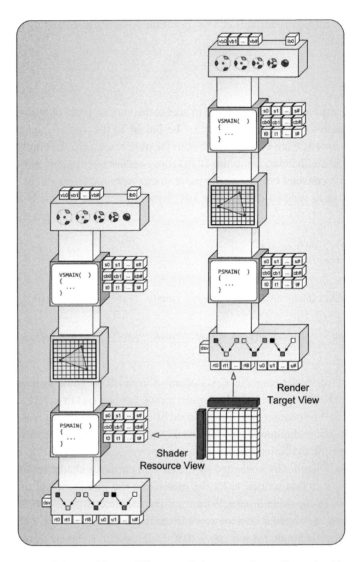

Figure 3.15. A resource being used in two different rendering passes for reading and writing.

when they are read from memory.[8] This often leads to significant performance and/or quality increases for the sampled values, since the GPU typically has additional specialized hardware specifically intended to accelerate this process. Listing 3.11 demonstrates how ID3D11SamplerState instances are bound to the pipeline. Once again, we see the familiar

---

[8] Samplers were described in detail in Chapter 2.

method of simultaneously setting an array of sampler states to reduce the number of interactions with the API required to configure the pipeline for rendering.

```
ID3D11SamplerState* SamplerStates[D3D11_COMMONSHADER_SAMPLER_SLOT_COUNT];

// Fill in the sampler states here...
pContext->VSSetSamplers( start, count, SamplerStates );
```

Listing 3.11. Binding an array of sampler state objects to the vertex shader stage.

Overall, we can group these four configurations together and consider them to be
the available vertex shader stage states. Configuring the vertex shader requires the proper
shader program, along with all the constant buffers, shader resources, and samplers that are
called for in the shader program. Since each vertex shader program can have significantly
different uses and requirements, it is possible to have many different combinations of these
states.

3.4.3 Vertex Shader Stage Processing

We now know what data the vertex shader can receive as input data from the input assembler, and which resources can be bound by the host application. We also know that the vertex shader program provides custom processing of individual vertices, regardless of which
topology the pipeline has been invoked with. So, what types of operations are performed
in the vertex shader program? Are certain types of operations better suited to this stage's
semantics? We will consider what operations the vertex shader stage is typically used for
and will comment on how best to use this stage of the pipeline.

Geometric Manipulation

The vertex shader is traditionally used to perform geometric manipulations on input data
before it is rasterized later on in the pipeline. This is because the vertices of a model hold
the information that represents the geometric surface of the object being rendered. Since
all vertices are passed through the vertex shader early in the pipeline, this is a logical place
to perform geometric manipulations. A few examples can clarify this topic fairly easily.

The prototypical operation normally performed in the vertex shader is to apply transformation matrices to vertex positions. This operation modifies the input position attribute
of the vertices to reflect the placement of the geometry within the scene by performing a
matrix multiplication of the vertex position with a transformation matrix supplied by the

application in a constant buffer.[9] In addition, any vertex attributes that are sensitive to the model's coordinate system must also be transformed into the new coordinate system with a similar matrix multiplication. Both the position of the vertices and their normal vector define the physical representation of the model, which is modified by the transformation matrices. By performing this calculation in the vertex shader, we can be certain that all of the model's geometry will be converted to the desired coordinate space before further processing is performed further down the pipeline.

Another operation frequently performed in the vertex shader stage is *vertex skinning*.[10] This performs a different type of coordinate transformation on a model, in which bones are hierarchically linked to one another within a system of bones, and each vertex is assigned to one or more bones. These bones move different parts of the model in different ways, for example, to simulate how your bones move your skin when you move your arm. Like the standard transformations example above, this operation modifies the physical structure of the model to provide the desired pose. It is also appropriate to perform this operation in the vertex shader, before further processing.

Vertex Lighting

Geometric manipulation and vertex skinning can be efficiently performed in the vertex shader, since they can be performed directly on the data that the vertices represent—the physical shape, structure, orientation, and location of the geometry being rendered. Of course, these are not the only types of operations that can be performed in a vertex shader. Because this stage is programmable, many special types of processing can be performed in the vertex shader and then passed to later stages. A common example of this type of operation is *per-vertex lighting*. Per-vertex lighting calculates the amount of light reflected from a vertex, normally using some form of simplified lighting model (Hoxley). The lighting data can be stored as a three- or four- component floating point attribute in the output vertex format, where a value of 1.0 means that a full amount of the color is available and a 0.0 represents the complete absence of that color.

These per-vertex lighting values are passed down the pipeline and eventually arrive at the rasterizer stage, which interpolates them as attributes between each vertex in the primitive. These interpolated values are then applied to each pixel generated by the rasterizer, and are eventually used to determine the final color written to the render target. By performing these calculations in the vertex shader, lighting quantities are calculated only at a sparse set of points (at the vertices), instead of at every pixel that a model generates. The generated values are interpolated between the vertices, making the reduced resolution of the calculation less noticeable, as long as the vertices are not too far apart on the screen.

---

[9] A basic introduction to transformation matrices, as well as coordinate spaces, is provided in Chapter 8. "Mesh Rendering."

[10] Vertex skinning is also discussed in detail in Chapter 8, "Mesh Rendering."

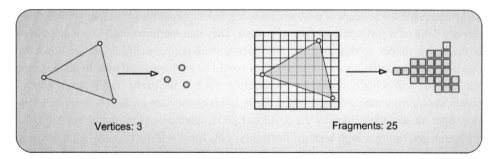

Figure 3.16. The difference between per-vertex and per-pixel calculations in a pair of triangles.

Generic Per-Vertex Calculations

Many other types of calculations can follow a model similar to the one described above for lighting. If a calculation can be performed in the vertex shader stage and an interpolated version of the data can be used at the per-pixel level, then it makes sense to perform the calculations in the vertex shader where they will only be performed once per vertex. If the calculations were instead performed after rasterization, they would be performed many more times. Figure 3.16 shows the difference in calculation frequency between a per-vertex calculation and a per-pixel calculation.

Mathematically speaking, if the calculation is a linear combination of its inputs, then the results of the per-vertex calculation will be identical to those of the per-pixel calculations, because of the interpolation between vertex attributes. Even if a calculation is not linear, per-vertex calculations can still provide a fairly good approximation of the full per-pixel calculations. The effectiveness of this approximation depends on the type of calculation, as well as on how large the final rasterized primitive will appear in the final render target. If the number of pixels generated between vertices is very small, any inaccuracy of the interpolated values is less likely to be noticed. This also means that if an input set of geometry is created to have a larger number of vertices, the interpolation effects will be less noticeable, at the expense of an increased number of vertices to process. There is no simple rule to determine the required level of detail for a particular model. This is a function of the available processing power and the desired image quality. As a small indication of what is coming later in this chapter, the new tessellation stages are aimed at striking a balance in this regard, by dynamically generating an appropriate number of vertices, only where they are needed (where they are visible). This reduces the overall vertex processing costs, while still producing small screen-space-sized primitives, which effectively raises the available image quality for the same amount of required processing.

Control Point Processing

One other way to use the vertex shader stage is to process data that is actually not vertices, but rather control points of a higher-order primitive or control patch. These control points still encapsulate the concept of position, and perhaps of orientation, even though they don't directly define the geometric surface of a mesh. They can therefore easily be manipulated in the vertex shader. In this case, the processed control points would be passed down the pipeline to the tessellation stages, where they would be evaluated and used to generate the vertices that will actually define the geometric surface of the mesh. The types of control points, the information contained within them, and how they are evaluated to produce vertices later on, are all decided by the developer and implemented in programmable shader programs, providing a high level of flexibility. This topic will be revisited again several times throughout the book (especially in Chapter 4, "The Tessellation Pipeline," as well as in several of the sample algorithm chapters).

Vertex Caching

Earlier in this chapter, we discussed how the vertex shader program is invoked once for each vertex. While the vertex shader stage is effectively executed once per vertex, individual processed vertex results may be shared among more than one primitive. For example, if a vertex is shared by two triangle primitives in a triangle strip, it can be processed once, and its result can be used in both triangle primitives later in the pipeline, after the vertex shader stage. Using any of the "strip" primitive topology types will allow this type of vertex reuse. In addition, when using indexed rendering even the "list" primitive topology types can define multiple primitives which share the same vertices. This can significantly lower the number of vertices to be processed when compared to rendering techniques that cannot share vertices. As an example, consider the geometry shown in Figure 3.17. In each of these three cases, different primitive topologies are used to define the required model, with varying numbers of vertices to be processed.

One additional consideration regarding vertex caching is that this operation is hardware dependent. The size of the cache (and even its very existence) can vary significantly among different GPUs. Therefore, it is a best practice to ensure that any shared vertices are always referenced as closely together as possible in the vertex or input. In the case of the

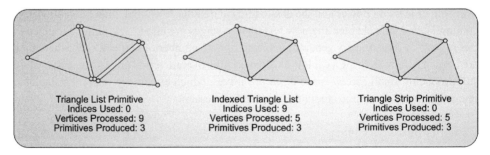

Figure 3.17. Varying sets of geometry with different topologies, and the resulting number of required vertices to be processed.

"strip" primitive topology types, this is automatically done, since each primitive uses the vertices immediately before it to create new primitives. In the case of indexed rendering, the indices that reference the same vertex should be located as closely together as possible, to ensure that the cached vertex can be reused.

3.4.4 Vertex Shader Pipeline Output

When deciding what information to include in the output vertex structure, there are some considerations to take into account regarding how the remainder of the pipeline will be used. Figure 3.18 shows a block diagram of the rendering pipeline.

The vertex shader stage is followed by the group of tessellation stages (the hull shader, tessellator, and domain shader stages), then the geometry shader, and then the rasterizer stage. Depending on which of these stages are active between the vertex shader and rasterizer stages, different requirements must be met. If the vertex shader is connected directly to the rasterizer stage, it must produce the final clip space position of each of the vertices in the SV_Position system value semantic. If the tessellation stages (all three are either active or inactive together) or the geometry shader stage is active, then it is optional for the vertex shader to produce the clip space position. However, the last active stage immediately before the rasterizer stage must provide the SV_Position system value semantic.

Similarly, if the tessellation stages are active, the vertex shader stage must provide control points to the hull shader stage. The distinction between an output vertex and a control point really depends only on the tessellation scheme being implemented, but in both cases the data is still produced by the vertex shader. If the tessellation stages are disabled and the geometry shader is enabled, the vertex shader output is sent directly to the geometry shader. In this scenario, the geometry shader must supply the SV_Position semantic to the rasterizer, although it could be calculated in the vertex shader and then passed to the geometry shader. The wide variety of options for producing just one position calculation underscores the flexibility that the pipeline provides, and the corresponding freedom for the developer to implement an algorithm in the most advantageous method available.

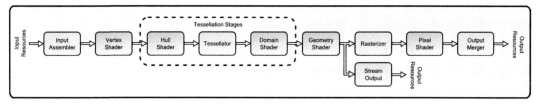

Figure 3.18. A block diagram of the rendering pipeline.

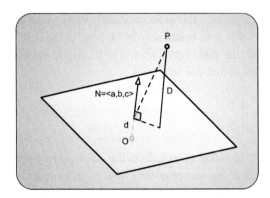

Figure 3.19. Calculating the distance from a point to a plane.

System Values

Besides these stage-based considerations, two new system values are available for the vertex shader stage to write to: `SV_ClipDistance[n]` and `SV_CullDistance[n]`. Both are available for writing in the stages before the rasterizer, where they are used to perform two different types of operations: *clipping* and *culling*. Clipping and culling for the pipeline are implemented in the rasterizer, and are discussed in more detail in the "Rasterizer" section of this chapter. However, we will describe these operations briefly here to explain the functionality provided in these two system values.

We begin with a brief mathematical introduction to point-to-plane distance calculations. In general, the shortest distance from a point to a plane can be found by taking the dot product of the point's position and the normalized normal vector of the plane, minus the shortest distance from the plane to the origin of its coordinate space. This result is easily obtained when the familiar plane equation is available for a plane. Equation (3.1) shows the equation that provides this property, where a, b, and c are the components of the normal-length normal vector, and d is the shortest distance from the origin to the plane. This equation produces a scalar result, which can take three different value ranges. It can be a positive value if the point is on the side of the plane that the normal points into, a zero value if the point is exactly on the plane, or a negative value if the point is on the side of the plane pointing away from the normal vector. Figure 3.19 shows this calculation for a plane and a point:

$$D = ax_1 + by_1 + cz_1 - d. \tag{3.1}$$

To conservatively cull a complete primitive from further processing, all of the vertex positions must be completely outside of at least one of the six planes defining clip space.[11] Each of the distances between the vertex positions and each of the planes can be calculated as described above. If all of the vertices of a primitive result in negative distance values for any of the six planes, the primitive can be discarded since it would not be visible. The SV_CullDistance[n] system value semantics behave in much the same way as the form of conservative culling mentioned above. Each attribute declared with this system value represents the distance to a culling plane. When these system values are interpreted by the rasterizer stage, it will eliminate any primitive whose vertices all have a negative value in the same register. An example of this type of system value usage is shown in Listing 3.12. If the same component of the "clips" attribute were negative for all of the vertices of a primitive, it would be culled without being rasterized at all. This arrangement allows for the results of up to four different culling equation results to be stored for this attribute.

```
struct VS_OUT
{
    float4 position : SV_Position;
    float4 clips    : SV_CullDistance;
};
```

Listing 3.12. A sample output vertex structure that uses a four-component SV_CullDistance system value semantic.

The SV_ClipDistance system value semantic operates in a similar fashion, and defines the distance to a clipping plane. If one or more vertices of a primitive have negative values in this attribute, its value will be used to determine which portion of the current primitive should be clipped. The actual clipping can be performed before or after rasterization—this is an implementation-specific detail that the developer can't know in advance. Regardless of how the primitive is clipped, the result is that none of the fragments generated after rasterization will contain a negative value in the interpolated SV_ClipDistance system value semantic. This differs from the SV_CullDistance system variable, which performs the culling test on the entire primitive instead of trying to "clip" the primitive to only have a positive SV_ClipDistance attribute.

The maximum number of these clip and cull system values is a total of two attributes, with up to four components, each of which can be declared in any combination of SV_ClipDistance and SV_CullDistance. This means that a total of eight planes can be either clipped against (if two float4s are used as SV_ClipDistance) or culled against (if two float4s are used as SV_CullDistance) or some combination of the two.

---

[11] We discuss clip space and its properties in more detail in the rasterizer and pixel shader sections of this chapter.

3.5 Hull Shader

Our next stop in the pipeline brings us to the new tessellation stages. These three stages work together to implement a flexible and configurable system. The stages, in order of appearance in the pipeline, are the hull shader stage, the tessellator stage, and the domain shader stage. Of these stages, the hull shader and domain shader stages are programmable, while the tessellator stage performs fixed-function operations. The hull shader stage can be thought of as performing the setup operations that prepare the input primitives for further processing. It instructs the Tessellator stage how finely to split up the input geometry, and provides the processed control point primitives[12] for the domain shader stage, which will fill in the freshly tessellated vertices with data. Thus, the hull shader plays a very important role in configuring the overall tessellation scheme. The location of the hull shader within the pipeline is highlighted in Figure 3.20.

The responsibilities of the hull shader stage are divided into two separate HLSL functions. The first is the hull shader program. The hull shader program is invoked once for each control point needed to create an output control patch. It receives input control *patch* primitives constructed from the control points produced by the vertex shader, and must create one processed output control *point* for each invocation of the hull shader program. Unlike the vertex shader, the hull shader program is aware of both the input and output primitive topology of the geometry being passed through it. In fact, *every* invocation of the hull shader has access to all of the input control points that belong to its input control patch. With access to the complete input data, the hull shader stage can also statically expand or reduce the number of control points in the control patch before the control patch is passed to the next stage in the pipeline. This change in control point count is statically declared with a special attribute in HLSL, and all control patches passed in to the hull shader will be reduced or expanded in the same way. This is the first time we have seen a pipeline stage that can perform an amplification of the primitive data stream passing through it.

The second HLSL function that performs operations within the hull shader stage is referred to as a patch constant function. In contrast to the hull shader program, the patch

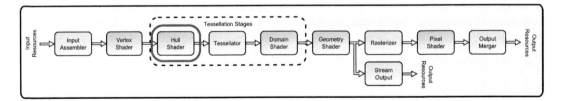

Figure 3.20. The hull shader stage.

---

[12] The control patch primitive types are described in the "Primitive Topology" section earlier in this chapter.

constant function is only executed once for each complete control patch. This function's primary purpose is to define several tessellation factors, which tell the tessellator stage how finely to tessellate the input primitives. It can also calculate additional attributes that are constant for a complete control point patch, which are then supplied to the domain shader later in the pipeline. The hull shader stage is the only programmable pipeline stage that is required to provide two functions in HLSL.

Throughout the remainder of this chapter, we will further explore how this combination of functions allows the hull shader stage to enable a wide variety of tessellation algorithms. We begin by examining the typical inputs to the stage, and what types of states are available for configuration. This is followed by a discussion of typical processing operations that are suitable for the hull shader. Finally, we will look at the data produced by the hull shader stage, and where in the pipeline it is passed on to.

3.5.1 Hull Shader Pipeline Input

The hull shader stage is situated immediately after the vertex shader stage in the pipeline. Its input data is a set of processed vertices produced by the vertex shader, which is combined with the primitive stream information produced by the input assembler stage to form control patch primitives. Strictly speaking, the hull shader program operates on complete control patches, which are made up of control points, and not vertices. However, there is little distinction between control points and vertices. Both vertices and control points define a position, along with some additional attributes, so there is not much difference in their contents. The only difference between control points and vertices is actually determined by the primitive topology specified in the input assembler stage. If the pipeline configuration for a particular draw call will use the tessellation stages,[13] it must declare one of the control point patch list primitive types. If an equivalent topology is passed instead (a triangle list topology in the place of a three-point control patch list, for example) the runtime will issue an error. Because of this, we can assume that if the tessellation stages are active, we can say that the vertex shader processes control points. If the tessellation stages are not active, we say that the vertex shader as processes vertices.

As mentioned above, the hull shader program is executed once for each of the desired output patch's control points. Every invocation of the hull shader program has full knowledge of all of the control points in the current input control patch. These points are received as a pipeline input attribute, which is declared in a template-like fashion. Two template parameters are declared in the attribute: the structure of the data produced by the vertex shader, and the number of points in the input control patch. The number of points in the control patch must match the number specified in the input assembler's topology type.

---

[13] The geometry shader stage is also capable of receiving control patches, as well as the tessellation stages.

Listing 3.13 shows a sample attribute declaration for a hull shader program for a three-point control patch list topology.

```
HS_CONTROL_POINT_OUTPUT HSMAIN( InputPatch<HS_CONTROL_POINT_INPUT, 3> ip,
                                uint i : SV_OutputControlPointID,
                                uint PatchID : SV_PrimitiveID )
{
    HS_CONTROL_POINT_OUTPUT output;

    // Insert code to compute Output here.
    output.WorldPosition = ip[i].WorldPosition;

    return output;
}
```

Listing 3.13. A sample hull shader program, demonstrating its attribute declaration.

Listing 3.13 also shows a pair of system value attributes. The particular control point that a hull shader invocation is processing is indicated in the SV_OutputControlPointID system value, which must be declared as an input attribute. This system value semantic provides a single-component unsigned integer that identifies the current hull shader invocation control point within the specified output control patch size. By using this attribute, the hull shader program can determine which portion of the input control points to use when generating the output control point. The second system value is the SV_PrimitiveID, which provides a unique single-component unsigned integer identifier for each control patch. This is the first pipeline stage where this system value semantic is available, since the vertex shader has no concept of primitives.

The patch constant function runs only once for a complete control patch primitive that is passed into the hull shader stage. Like the hull shader program, it has access to the complete set of input control points, and all their attributes. It can also use the SV_PrimitiveID system value, but not the SV_OutputControlPointID system value semantic. This makes sense, because this function operates once for the entire patch—it would have no use for an identifier of individual control points.

3.5.2 Hull Shader State Configuration

Since the hull shader stage is one of the programmable shader stages and has access to the same set of resources that have been discussed for the vertex shader. They are covered here for completeness, but in fact, the names of the methods used to set and get the resources are simply updated to indicate which pipeline stage should be affected by the method, so we won't repeat the code listings here. The method names are provided here for reference.

- `ID3D11DeviceContext::HSSetShader()`

- `ID3D11DeviceContext::HSSetConstantBuffers()`

- `ID3D11DeviceContext::HSSetShaderResources()`

- `ID3D11DeviceContext::HSSetSamplers()`

In addition to the resources that can be bound to the hull shader stage by the application, a number of other function attributes can be specified in the HLSL code. These are also described below.

Shader Program

The hull shader program is compiled in the same way as any of the other HLSL programs.[14] The distinction is that the hull shader program must be preceded by the patch constant function, and immediately preceded by a list of function-level attributes. These items must all be located within the same HLSL file and compiled into shader byte code, along with the main hull shader program. The resulting byte code is then used to create an instance of the `ID3D11HullShader` object, which is then bound to the pipeline through the device context.

Constant Buffers

The primary means of communicating data to the hull shader program is the constant buffer mechanism. This operates in precisely the same way shown for the vertex shader stage. The same the recommendations regarding the persistence of setting of constant buffers between pipeline executions applies here.

Shader Resource Views

Another possible source of read-only data for the hull shader program is the shader resource views. These provide read-only access to a wide array of resources, as discussed in Chapter 2. Resources provide a larger pool of available memory to the hull shader than constant buffers do, but they can also potentially be slower to access.

Samplers

Once again, samplers provide a mechanism for performing filtering operations on texture resources. The hull shader stage has access to this functionality, just as discussed for the vertex shader.

---

[14] See Chapter 6, "The High Level Shading Language," for details about how to compile and create shader objects.

Function Attributes

To control the tessellation scheme used in the three tessellation stages, several *function attributes* are required in the HLSL code of the hull shader. These attributes are different than the input and output pipeline data flow attributes passed between pipeline stages. Instead, these are individual statements that perform a specific configuration of the tessellation system. The attributes must be located above the hull shader program in the HLSL source code, since they are used to validate the shader program itself. The list of attributes is provided in Listing 3.14. Each attribute will be described when its subject area is discussed throughout this section. We list them here, followed by a brief description.

```
// Specifies the 'domain' of the primitive being tessellated.
[domain("tri")]

// Specifies the method to tessellate with.
[partitioning("fractional_even")]

// Specifies the type of primitives to be created from the tessellation.
[outputtopology("triangle_cw")]

// Specifies the number of points created by the Hull Shader program (which
// is also the number of times the hull shader program will execute).
[outputcontrolpoints(3)]

// Specifies the name of the patch constant function.
[patchconstantfunc("PassThroughConstantHS")]

// Specifies the largest tessellation factor that the patch constant
// function will be able to produce.  This is optional and provided as a
// hint to the driver.
[maxtessfactor(5)]
```

Listing 3.14. Function attributes of the hull shader program.

3.5.3 Hull Shader Stage Processing

The processing tasks intended for the hull shader stage are split between the two HLSL functions that an application must provide. Since the hull shader program creates the output control points that are passed down the pipeline, its primary responsibility is to read the input control patches that were produced earlier in the pipeline, and translate them to the desired output control patch format. The patch constant function is responsible for determining the necessary tessellation factors according to a developer-defined metric. In the following sections, we will examine some higher-level concepts that these functions may be used to perform.

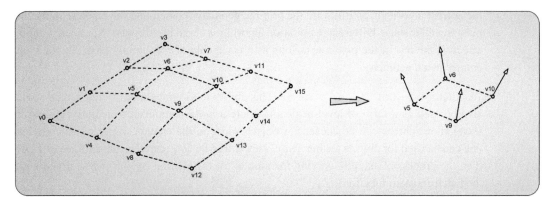

Figure 3.21. Changing the number of control points in the control patches passed on to the domain shader stage.

Hull Shader Program

As mentioned above, the hull shader operates once for each output control point that needs to be produced for the selected control patch output size. The number of output control points is configured by the application by specifying it in the `outputcontrolpoints` function attribute. The system value semantic input attribute `SV_OutputControlPoint` identifies which output control point is being created in each invocation of the hull shader program. Its values range from 0 to n-1, where n is the number of output control points defined in the `outputcontrolpoints` function attribute. This function attribute is the mechanism by which the hull shader can expand or reduce the number of points in the output control patch. To produce these output control points, the hull shader program receives the complete set of control points that are included in the input control patch primitive as an input attribute. This concept is visualized in Figure 3.21 for a hull shader that receives a 16-point control patch and produces a four-point control patch.

Tessellation algorithm setup. This output control patch is the data that will later be used in the domain shader stage. Because of this configuration, the hull shader can be thought of as somewhat of a data setup stage for the tessellation system. As we just saw, the format of the input control patches can be modified, expanded, or reduced in the hull shader. This functionality can be used to produce a control patch type that is compatible with the desired tessellation algorithm implemented in the domain shader stage. With this in mind, the hull shader could be used to consume a standard control patch primitive type (perhaps mandated by the content creation tool chosen for a given project) and then produce the required output patch format for the desired type of tessellation algorithm. This acts as an isolation mechanism for the tessellation system, since a standard input can be provided, and the output from the hull shader stage will always appear in the appropriate format for the current algorithm, as if it were the original geometry passed into the pipeline in the input assembler stage. Different tessellation algorithms could be swapped in and out without affecting the rest of the pipeline, as long as it is capable of consuming the available input control patch geometry.

Tessellation algorithm level of detail. This ability to swap tessellation algorithms could also be used as a form of level-of-detail, where a more complicated or more aggressive tessellation scheme can be chosen for objects close to the camera, while a less complex one can be used for objects farther away. This would be considered a coarser form of LOD than the process of actually varying the amount of tessellation (which is performed with the patch constant function).

Generic control point calculations. The hull shader program can also be used for a variety of other processing tasks not related to tessellation. Since it is the last stage before tessellation actually occurs, it is a good candidate for performing calculations that will be shared by more than one of the tessellated points. This can effectively pull the calculations back to an area of the pipeline before the data amplification of the tessellation procedure, saving significant processing costs.

Tessellation Factor Calculations

The idea behind the patch constant function is that it will calculate two tessellation factors, which are written to special system value semantics. These two system values semantics tell the fixed function tessellator stage how finely to tessellate the primitive being generated from the input control patch. The number of tessellation factors required is influenced by the `domain` function attribute mentioned above. The type of primitive that is conceptually being split up by the tessellator stage is specified in this attribute. Depending on which type of domain is specified (isoline, triangle, or quad) the number of required edge tessellation factors will vary.

The patch constant function itself is declared with another attribute—`patchconstantfunc`. This simply identifies which HLSL function should be executed for the patch constant function. The remaining function attributes listed above are used to configure execution of the tessellation stage. The `partitioning` attribute configures the type of tessellation to perform, the `outputtopology` attribute defines the type of output primitives that will be created by the tessellation, and the optional `maxtessfactor` attribute tells the driver the maximum amount of tessellation it should expect, so that it can allocate an appropriate amount of memory to hold the results.

These attributes and their available values will be discussed in detail in Chapter 4, "The Tessellation Pipeline." However, the important concept to take away from these configurations is that the patch constant function and the function attributes specified with it are used to determine precisely what the tessellator stage produces. In many ways this is similar to the setup functionality discussed for the hull shader program. The function attributes configure what will be tessellated (with the `domain` attribute), how it will be tessellated (with the `partitioning` attribute), and what format its output will be in (with the `outputtopology` attribute). The patch constant function then produces the tessellation factors for each patch, which instruct the tessellator stage how finely to chop up the chosen domain primitive. If the hull shader program sets up the data of the tessellation algorithm, the patch constant function sets up the actual tessellation machinery.

Fine level of detail. The calculation of the tessellation factors can be based on any number of different criteria. For example, it is possible to modify the amount of tessellation based on the distance of the input control patch to the viewer. It is of course better to have higher tessellation closer to the viewer than when a patch is farther away, to reduce the number of required computations. However, in other cases this approach is either not appropriate or not ideal. For example, when an object is relatively smooth, with only a few sharp details, it may be better to base the tessellation factors on some measure of the amount of change in a surface. This could either be precalculated and stored in the control points, or perhaps stored in a texture for lookup later on. It is also possible (and in fact, more probable) to combine more than one heuristic approach for this problem. Examples of this type of analysis are presented in Chapter 9, "Dynamic Tessellation."

Image quality. Many other types of properties should be considered when determining the desired tessellation density. For example, if a control patch is positioned on the silhouette edge of a mesh, it will improve the image quality to tessellate the patch more than if it were perpendicular to the view direction. Another very good example of improving image quality with the use of varying tessellation levels is in the generation of dual-paraboloid shadow maps, or environment maps.[15] The paraboloid projection is known to produce artifacts when it is generated with triangles that cover large areas of the paraboloid map. This happens because each vertex is transformed with the paraboloid projection, but the triangle primitives that the vertices are used to produce are linearly rasterized. This situation is depicted in Figure 3.22.

The situation becomes more problematic as a primitive covers more and more texels of the paraboloid map. A simple method to alleviate this issue is to calculate the screen-space area of a triangle in the patch constant function and then increase the tessellation factors according to this screen-space size. The tessellated points can then be properly projected in the domain shader stage to minimize the effects of this issue.

Generic control patch calculations. In addition to tessellation factors, the patch constant function can also produce user-defined attributes that will be applied to all tessellated points from the current control patch. This means that any data that can be shared by all

---

[15] Further details about paraboloid maps can be found in Chapter 13, "Multithreaded Samples."

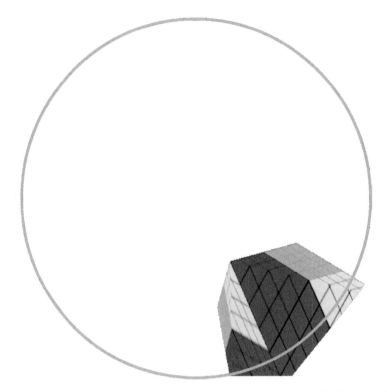

Figure 3.22. Artifacts from linear rasterization when used in projections that don't preserve lines.

control points of a control patch (such as a complete patch normal vector or a screen-space derivative of the patch position) can be calculated once and used many times later. Once again, this reduces the number of computations required to perform a calculation.

3.5.4 Hull Shader Pipeline Output

We have already seen that the results of executing the hull shader program are the output control points of the control patch to be used in tessellation. These control points are its primary way to pass information down the pipeline. Ultimately, these control points will be consumed by the domain shader stage, where they will be used to generate positions for each of the newly tessellated points from the tessellator stage. Requirements for the attributes of the output of the hull shader are very flexible. There are no mandatory system values that need to be written. Instead, it is up to the developer to supply the information needed for processing in later stages. But although the output requirements are very flexible, some common attributes are usually included. The positions of the control points are typically provided by the hull shader program, which are typically either copied from the input control points or extracted from them. In addition, any material properties not read out of a resource would normally be included in the control points. These may include per control point colors, shininess, or material identifiers, just to name a few.

In contrast, the patch constant function's output is somewhat more predefined. It must produce the appropriately sized set of tessellation factors for use by the tessellator stage. These are written to output attributes with the system value semantic of SV_TessFactor and SV_InsideTessFactor. In addition, any patch constant information is also passed as output attributes. These attributes are not sent to the tessellator stage, since it doesn't perform custom processing, but are instead passed directly to each invocation of the domain shader program as input attributes.

3.6 Tessellator

The tessellator stage is the fixed function pipeline stage of the tessellation system. It is located somewhere between the hull shader stage and the domain shader stage, and this location defines the source and destination of its input and output. When the tessellation stages are used in the rendering pipeline, they are all used together—they cannot operate individually. As such, the tessellator stage plays an important role in implementing any tessellation algorithm. The location of the tessellator stage is highlighted in Figure 3.23.

The job of the tessellator stage is to convert the requested amount of tessellation, which is determined in the tessellation factors of the hull shader stage patch constant function, into a group of coordinate points within the current "domain." The domain in which these points are generated can be an isoline, triangle, or a quad—whichever is specified in the domain function attribute of the hull shader program. This should be somewhat surprising, since the operation described above makes no use of control points or control patches. Instead, the tessellator stage simply generates a set of points that indicate where vertices should be created within the specified domain. These points are later passed to the domain shader stage, where they are rejoined with the control points produced by the hull shader

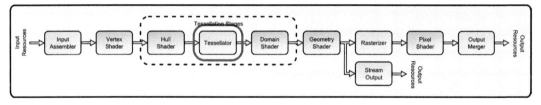

Figure 3.23. The tessellator stage.

program to give a physical meaning to the points. Understanding how the tessellator stage operates, and why it is implemented in this counterintuitive way, will provide a solid foundation upon which to build tessellation algorithms.

3.6.1 Tessellator Stage Pipeline Inputs

To understand the tessellator stage, we will begin by considering the type of data that flows into it. The vertex shader stage receives assembled vertices/control points from the input assembler stage, and the hull shader stage receives control patches created from the output control points from the vertex shader stage (along with primitive information from the input assembler). This basic concept is depicted in Figure 3.24.

In both of these cases earlier in the pipeline, the data that flows into the stage is processed in some fashion and then passed to the next stage. This is not the case in the tessellator stage. Instead, it receives the tessellation factors produced by the hull shader stage patch constant function. These factors are simple floating point numbers that indicate how heavily to tessellate the different parts of the domain object. Regardless of how the tessellation factors are calculated in the hull shader stage, if two control patches produce the same set of tessellation factors, they appear identical to the tessellator stage. Since the tessellator stage cannot distinguish between the two sets of input, it will produce precisely the same output in both situations.

Strictly speaking, all of the other pipeline stages would produce the same output if given two sets of identical input. The reason the tessellator is unique is that it is more likely to encounter the same input values from multiple different control patches, due to the "flattening" of the input data to simple factors. This is a profound difference from the other pipeline stages. This will be important to consider as we further explore the type of processing performed in the tessellator stage.

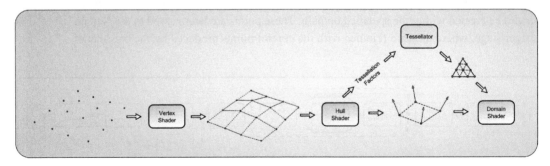

Figure 3.24. A representation of the data flowing through the pipeline.

3.6.2 Tessellator Stage State Configuration

The tessellator stage is not directly configured by the application, but by the attributes specified in the hull shader program (which *is* configured by the application). The available set of configurations is the collection of function attributes mentioned in the "Hull Shader" section. Since most of these function attributes are used to configure the tessellator stage (even though they are specified in the hull shader stage), we will take a closer look at what they are intended to do.

Domain Attribute

We begin with the domain attribute. We have already mentioned that the tessellator stage does not actually tessellate control patches, but tessellates a *domain* object. The type of domain specified in this attribute can have one of three values: an `isoline`, `tri`, or `quad`. At a high level, we can say that the tessellator stage determines points within this domain that need to be realized by the domain shader stage. This attribute is mandatory, since the type of domain is also used to validate the number of tessellation factors produced by the patch constant function.

Partitioning Attribute

The next attribute is the partitioning attribute. This can take on one of four values: `integer`, `fractional_even`, `fractional_odd`, or `pow2`. Each value identifies a different scheme for how to split up the specified domain. As we will see in more detail in Chapter 4, "The Tessellation Pipeline," there is a significant difference in the chosen tessellation points for each of these partitioning schemes. This attribute is also mandatory.

Output Topology Attribute

Once the specified domain is split up according to the specified partitioning scheme, the tessellator stage also needs to know how to assemble primitives from these individual points. The available options for this attribute are `triangle_cw`, `triangle_ccw`, and `line`. This attribute is also mandatory; otherwise, the primitives produced by the tessellation stage could have significantly different results than intended. For example, if the incorrect triangle winding were produced, the rendered geometry would appear inside out!

Max Tessellation Factor Attribute

The final attribute that configures the tessellation stage is `maxtessfactor`. This attribute is an optional hint to the driver about the maximum possible amount of data amplification. Using this upper limit to the tessellation factors, the driver can efficiently preallocate enough memory to receive the results of the tessellation operations.

3.6.3 Tessellator Stage Processing

So what exactly is the operation that the tessellator stage performs with the tessellation factors and configurations mentioned above? The simple answer is that it selects a series of points within the given domain and then generates primitive data that creates an output primitive from these points later in the pipeline. However, the details about how this is performed and about how to control the produced output are not quite as clear. In this section we will take a closer look at the tessellation process and what is being performed by the tessellator stage.

Sample Locations

The tessellator stage is more of a data generation stage than a data processing stage. Once a domain type has been specified, the inner tessellation factor and edge tessellation factors specify how much tessellation to perform. The tessellation itself occurs in two steps. First, tessellation points are chosen that will produce an appropriate number of triangles. If the edge tessellation factors are higher than the interior factors, more points will be chosen closer to the edges of the domain than in the middle. The inverse is also true—if a higher tessellation factor is received for the interior, more points will be chosen in the inner portion of the domain. This is depicted in Figure 3.25 for the quad domain.

The selected points are identified only by a set of coordinates within the current domain. At first, this is somewhat surprising when one considers what the tessellator stage is doing. These coordinates are the only data that the tessellator produces, and they only provide locations within a generic domain shape that should become vertices later, based on the control patch produced by the hull shader. This is what was meant by calling the tessellator stage a *data generation stage* instead of a *data processing stage*. It takes the domain configuration and the set of tessellation factors, and then generates all of the required sample locations. The amount of output data will typically be larger than the amount

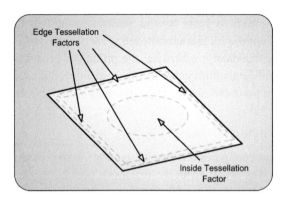

Figure 3.25. The regions of a quad that are affected by tessellation factors.

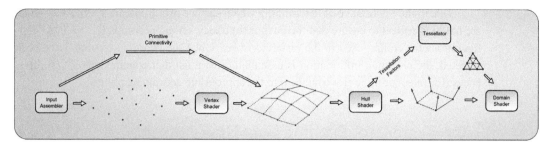

Figure 3.26. Primitive data flowing through the pipeline, up to through the tessellation stages.

of input data, simply because this stage only consumes a maximum of five floating point values for tessellation factors.

Primitive Generation

The second step in the tessellator stage is to generate the required primitive information needed to use the sampled locations as renderable geometry later in the pipeline. You may wonder why the primitive data must be generated in this stage, since a primitive topology was specified in the input assembler stage. In fact, it is correct that the input assembler creates primitive connectivity information that skips the vertex shader stage and is passed to the hull shader stage when the tessellation system is active. However, when the tessellation stages are active, the input primitive from the input assembler stage must be one of the control patch types. This specifies the connectivity of the vertices as control points and defines a control patch, instead of a primitive type that can be rasterized directly. Figure 3.26 visualizes the distinction in how primitive data flows through the pipeline up to the end of the tessellation system.

We have already seen that the type of output primitive is specified by the `outputtopology` attribute. When lines are being generated, they have no front or back side, so it isn't important in what order the vertices are arranged in within the primitive. However, when triangles are produced, it certainly matters in which order the vertices are specified. As we will see in the rasterizer stage, triangles facing away from the current viewpoint are discarded and do not influence the final rendered image. Which direction a triangle faces is determined by taking a cross product of the two vectors created by the edges touching the first vertex. This is demonstrated in Figure 3.27.

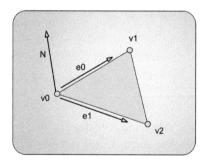

Figure 3.27. Determination of the direction a triangle is facing.

Due to the importance of determining which surface primitives are visible and which are not, it is critical to ensure that the `outputtopology` setting is compatible with the configurations for culling used in the rasterizer stage. Once the coordinate points have been selected, the primitive information is essentially just a list that contains either two references (for lines) or three (for triangles) to the appropriate coordinate points.

3.6.4 Tessellator Pipeline Data Flow Outputs

Once the tessellation process has been completed for a particular invocation of the tessellator stage, the results are passed through a system value semantic to the domain shader stage. The `SV_DomainLocation` system value provides one domain coordinate point to the domain shader stage at a time, and the domain shader program is invoked once for each of these output points. Before the points are processed in the domain shader, they are simple coordinates in the current domain.

Since the primitive topology information created for the freshly tessellated geometry is not used by the domain shader stage, it is conceptually passed beyond the domain shader stage to the geometry shader stage. The geometry shader stage is aware of complete primitives, and hence can make use of the information. One important note to keep in mind is that there is *no* primitive topology with adjacency information available for creation in the tessellator stage. This means that when the tessellation stages are active, the geometry shader stage can't receive a primitive type with adjacency. This limits the available algorithms that can be implemented in the geometry shader program, but typically, any calculations that would require adjacency information can be incorporated into either the hull or domain shader programs, since they *do* have various levels of adjacency information available to them. If the geometry shader stage is not active, the primitive information and vertices created in the domain shader stage are passed directly to the rasterizer stage, where they further processed and used to generate pixels.

3.7 Domain Shader

The domain shader stage is the final stop in the tessellation system. It is a programmable pipeline stage, which receives input from both the hull shader stage and the tessellator stage, and produces output vertices, which can be further processed by the rasterization stages later in the pipeline. With this responsibility, the domain shader stage can be considered the heart of the tessellation algorithm, which is set up by the hull shader and by the granularity of the produced geometry set up by the tessellator stage. The location of the domain shader stage is highlighted in Figure 3.28.

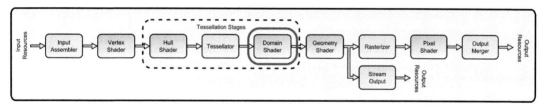

Figure 3.28. The domain shader stage.

3.7.1 Domain Shader Pipeline Input

The domain shader stage is required to produce vertices to complete the tessellation process and produce output geometry. It is invoked once for each coordinate point produced by the tessellator stage. To create these vertices, it receives the complete control patch produced by the hull shader stage. This can vary from 1 to 32 control points, but it must match the number of output points declared in the hull shader program's `outputcontrolpoints` attribute. The control patch is declared here in a similar manner to how it was used in the hull shader program, where the individual control points are provided in a template style attribute. The structure type for each control point is the first argument, and the number of points in the control patch is the second argument. Listing 3.15 provides a sample domain shader program input signature, which demonstrates this template style attribute.

```
struct HS_CONTROL_POINT_OUTPUT
{
    float3 WorldPosition        : POSITION;
};

struct HS_CONSTANT_DATA_OUTPUT
{
    float Edges[3] : SV_TessFactor;
    float Inside   : SV_InsideTessFactor;
};

[domain("tri")]
DS_OUTPUT DSMAIN(
        const OutputPatch<HS_CONTROL_POINT_OUTPUT, 3> TrianglePatch,
        float3 BarycentricCoordinates : SV_DomainLocation,
        HS_CONSTANT_DATA_OUTPUT input )
{
    // ...
}
```

Listing 3.15. A sample domain shader declaration, demonstrating the input attribute declarations.

As seen in Listing 3.15, the input control patch is declared as a `const` input parameter, since it can't be modified. Due to this read-only access, any modifications or additional data that must be included in the control patch for the domain shader program to create the output vertices must be provided by the hull shader program before this stage.

In addition to the control patch, the domain shader program also receives the domain location that it is supposed to implement from the tessellator stage. This data is provided as the `SV_DomainLocation` system value semantic. The format of this attribute varies, depending on which domain type is specified, with a `float2` used for `isoline` and `quad` domains, and a `float3` for `tri` domains. These values represent where within the domain this domain shader invocation should generate vertex data. In this respect, the coordinate location defines a sampling point over a virtual surface determined by the function that the domain shader program implements. We will investigate this concept further in the processing portion of this section.

The final set of input data that the domain shader program receives is the constants that were produced by the patch constant function of the hull shader stage. The data contained within this structure varies, according to what is calculated in the patch constant function, but it will remain the same for each domain shader program invocation for a given control patch.

To provide a high-level view of what the domain shader program is required to do, we can examine its input from the pipeline. All of the input control patch data is provided to every invocation of the domain shader program. In addition, the data received from the patch constant function is also constant across a complete control patch. This means that the only input that varies between invocations of a domain shader for a complete control patch is the domain location. This makes sense, since the number of points produced by the tessellator stage will vary, but the control patch data input doesn't change with a varying tessellation level. The separation of these two functionalities in the hull shader stage provides a clear distinction in this regard.

3.7.2 Domain Shader State Configuration

As a programmable pipeline stage, the domain shader stage can use the standard set of configurations available to all programmable stages. Again, these configurations are set by the application through the `ID3D11DeviceContext` interface with the familiar method names, with the exception that the beginning of the method names use `DS` to indicate that they will affect the domain shader stage. Since these methods and their usage characteristics have already been discussed, we won't repeat the code listings here. Further details on using these methods can be found in the "Vertex Shader" section. The methods are listed here for reference.

- `ID3D11DeviceContext::DSSetShader()`

- `ID3D11DeviceContext::DSSetConstantBuffers()`

- `ID3D11DeviceContext::DSSetShaderResources()`

- `ID3D11DeviceContext::DSSetSamplers()`

3.7.3 Domain Shader Stage Processing

The primary job of the domain shader stage is to create vertices using the generated set of coordinate points from the tessellator stage, by using the control patch and patch constants produced by the hull shader stage. The most important task for creating these vertices is determining their position, since this will directly affect how the final tessellated geometry will be rasterized. With this in mind, we will take a closer look at how the position is calculated in the domain shader program and will provide some high-level concepts for thinking about this process.

The data held in the points within the control patch can represent a wide variety of different types of higher-order surfaces. The control points can be used to implement Bézier curved surfaces (Akenine-Moeller, 2002) just as easily as they can represent normal vertices, as used in curved point triangles (Vlachos, 2001).[16] The only requirements are that the hull shader program must produce the expected format of data, and that the domain shader must implement a compatible algorithm for generating an output position for each coordinate point.

In a more generalized view, we can think of the control patch as a set of parameters which define a virtual surface. This surface must be valid over the entire domain of the control patch, since this is the region in which all of the input coordinate points will reside. Then we can think of each of the coordinate points produced by the tessellator stage as a selected location where we want to sample the virtual surface. This can be visualized with a simplified case, as shown in Figure 3.29, where we see an isoline domain, its control points, and the virtual surface that it represents. The selected sampling points along the isoline domain indicate the locations that are sampled along the virtual surface.

The important message to take from this concept is that the virtual surface remains the same regardless of where the tessellator-generated points are located or how many there are. This can help us break the overall

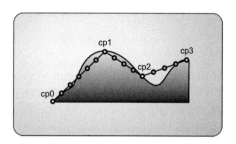

Figure 3.29. A depiction of how a control patch defines the parameters of a virtual surface, and the tessellator-generated points that can be considered as sampling locations for this surface.

---

[16] Bézier curved surfaces are discussed in more detail in Chapter 4, while curved point triangles are discussed further in Chapter 9.

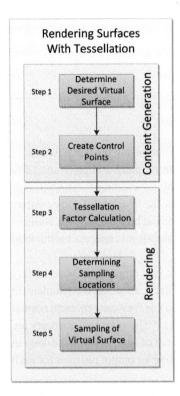

Rendering Surfaces With Tessellation

Content Generation

Step 1 — Determine Desired Virtual Surface

Step 2 — Create Control Points

Rendering

Step 3 — Tessellation Factor Calculation

Step 4 — Determining Sampling Locations

Step 5 — Sampling of Virtual Surface

Figure 3.30. The process of constructing geometry from a virtual surface at the specified sampling points.

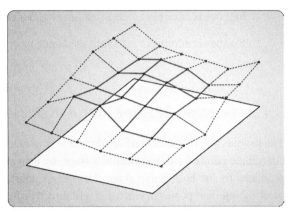

Figure 3.31. The set of control points that will affect a single quad of the surface.

tessellation algorithm design task into a few different components. First, our control patches must define the surface that we actually want. This is typically done at the content creation phase if the object is pre-created by an artist. Second, our tessellation factors must be calculated from the control patch to ensure that our desired surface is adequately sampled and therefore it appears correct in the current viewing conditions, while minimizing the number of vertices required to achieve that visual quality. Finally, when provided with a set of coordinates, our algorithm must be able to properly sample the virtual surface from the control patch data. These steps are shown in a block diagram in Figure 3.30.

Bézier Curves

With these steps in mind, we can consider a traditional tessellation algorithm and see how it would fit into this whole tessellation paradigm. We will consider one of the most well-known surface representations—the Bézier surface. The general concept of a Bézier surface is that the desired surface is shaped by moving a grid of control points. The number of points can vary, but for our example we will assume a 4×4 grid of control points to define the surface shape for a quad portion of the surface. This setup is shown in Figure 3.31.

Once the surface has been edited into the desired shape, the control points are stored as vertices in a vertex buffer, and an index buffer is used to define groups of control points as control patches. Next, when the object is going to be rendered, the input assembler builds the control points as vertices and passes them to the vertex shader. At the same time,

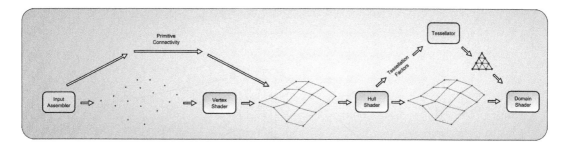

Figure 3.32. Our 4×4 Bézier surface as it is passed through the input assembler, vertex shader, and hull shader stages.

it determines which control points belong to a particular control patch, based on the indices provided in an index buffer and its primitive type setting (a 16-point control patch in this case). The control patch points are read by the vertex shader and passed to the hull shader. The 4×4 group of control points is then read by the hull shader and passed to the domain shader in the same 4×4 control patch configuration. The process up to this point represents step one in our block diagram from Figure 3.30. Figure 3.32 depicts this process.

Next, the patch constant function must calculate the needed tessellation amount to sufficiently represent our Bézier surface in the given viewing conditions. This can take into account the size of the control patch, the distance from the current view point, or any number of other metrics. The required tessellation factors are passed to the tessellation stage, which converts that data into a series of coordinates within the domain. Those points are then passed to the domain shader stage. This represents step two in our tessellation process, and is depicted in Figure 3.33.

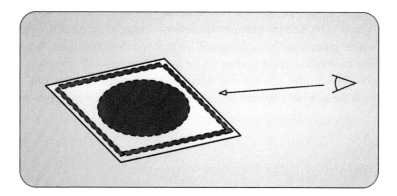

Figure 3.33. Determining the required edge factors for a particular quad, based on the current viewing conditions.

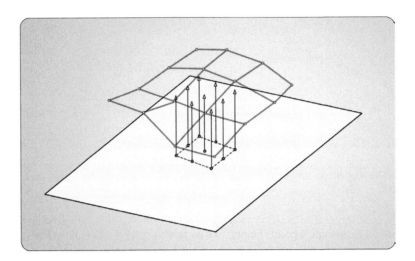

Figure 3.34. The generation of vertex locations in the domain shader for our 4×4 Bézier surface example.

Finally, the domain shader program must be able to receive the control patch, along with one coordinate point for each invocation of the domain shader, and produce a vertex that represents the Bézier surface. This is basically performed by evaluating the equations for Bézier surfaces, where each of the control points contributes to the calculated location of the vertex. The amount of the contribution is determined by the coordinates, which more or less specify the proximity to each of the control points where we are evaluating. Once this location has been calculated, the position is produced by the domain shader and is passed down the pipeline for further processing. This is shown in Figure 3.34.

Of course, the domain shader will normally need to calculate more than just the position of each of the tessellated vertices. One attribute that will be required most of the time is a normal vector to be used for lighting calculations further in the pipeline. However, this follows the same process we have discussed for position determination. The only difference is that we must calculate the normal vector in the domain shader program, in addition to the position. This is also true for any other attribute needed for rendering. It must be calculated from the available information, even if it is simply read from per-control point attributes.

3.7.4 Domain Shader Pipeline Output

After the required attribute data is calculated, the newly created vertices are returned by the domain shader program and passed to the next stage. The position output from this stage is typically written in the `SV_Position` system value semantic, which must be present in the input to the rasterization stage. In addition, any other per-vertex attributes required to determine the final pixel color are also added to the domain shader output.

It is possible that the geometry shader stage can be either active or inactive, depending on the desired pipeline configuration. If it is active, output from the domain shader is passed to the geometry shader stage, where it is consumed as complete primitives (remember that the primitive information in this case was generated at the tessellator stage). Otherwise, the output is passed directly to the rasterizer stage, where it is also consumed as primitives.

3.8 Geometry Shader

The geometry shader is the final pipeline stage that can manipulate the geometry being passed through the pipeline before it is rasterized. It is a programmable stage and has several unique capabilities not found in any other stage, including the ability to programmatically insert/remove geometry in the pipeline, the ability to pass geometry information to vertex buffers through the stream output stage, and the ability to produce a different primitive type than is passed into it. This provides some very interesting use cases for this stage, which are outside of the traditional pipeline model, ranging from saving processed geometry data to a file, pipeline operation debugging, and of course, rendering operations. The geometry shader stage also operates on complete primitives, including adjacent primitive information, which provides the additional ability to analyze and test various aspects of each primitive and perform customized calculations, depending on the geometry and its immediate neighbors. The location of the geometry shader stage in the pipeline is highlighted in Figure 3.35.

The geometry shader stage receives a list of vertices that represent the input primitives. It is then free to pass these vertices to an output stream, where they are then re-interpreted

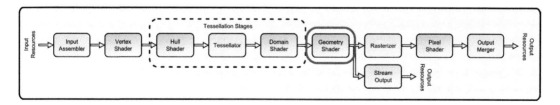

Figure 3.35. The geometry shader stage.

as primitives as they are passed to the next stage in the pipeline. Depending on the type of stream object declared for receiving this output, the streamed vertices will be used to construct different primitives. In addition, up to four output streams can be declared to use in conjunction with the stream output stage, making it quite simple to produce geometry intended for rasterizing as well as geometry to save into a buffer resource. This streaming model is the heart of the operation of the geometry shader, and we will see throughout this section how it can be used in various situations.

3.8.1 Geometry Shader Pipeline Input

The geometry shader stage operates on complete primitives composed of an array of vertices. Depending on which type of primitive is being passed, the array will vary in size, from a single vertex all the way up to six vertices for a triangle with adjacency information. The input to the geometry shader program declares the type of primitive that it will receive in the input attribute definition. A sample geometry shader function signature is shown in Listing 3.16.

```
[instance(4)]
[maxvertexcount(3)]
void GSScene( triangleadj GSSceneIn input[6],
              inout TriangleStream<PSSceneIn> OutputStream )
{
    PSSceneIn output = (PSSceneIn)0;

    for ( uint i = 0; i < 6; i += 2 )
    {
        output.Pos = input[i].Pos;
        output.Norm = input[i].Norm;
        output.Tex = input[i].Tex;

        OutputStream.Append( output );
    }

    OutputStream.RestartStrip();
}
```

Listing 3.16. A sample geometry shader program, demonstrating how to declare an input stream attribute.

In this listing, the input attribute represents a triangle with adjacency, which produces four triangles out of six vertices. The order of the vertices within this input array depends on which type of primitive is being passed to the function. The various primitive types, their vertex counts, and the order that their vertices are provided in, are listed in Table 3.1. In addition, each of these primitive types is shown in Figure 3.36 for easy reference.

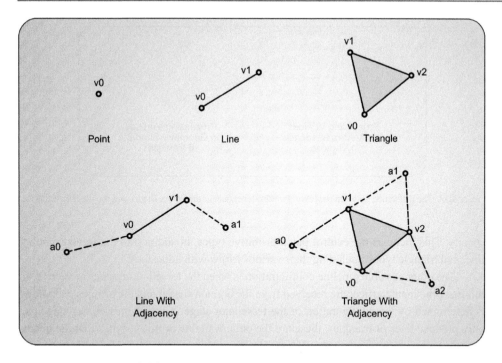

Figure 3.36. The various primitive arrangements that are available to be received by the geometry shader.

Keyword	Primitive Type	Number of Vertices	Vertex Ordering
point	point list	1	[V0]
line	line list, line strip	2	[V0,V1]
triangle	triangle list, triangle strip	3	[V0,V1,V2]
lineadj	line list w/adj., line strip w/adj.	4	[A0,V0,V1,A1]
triangleadj	triangle list w/adj., triangle strip w/adj.	6	[V0,A0,V1,A1,V2,A2]

Table 3.1. A table for determining how to evaluate the geometry shader input vertex stream.

With the primitive information available to it, the geometry shader program can inspect the geometry, perform visibility tests, or calculate high-level information for the primitive as a whole. There are two different possible pipeline configurations that can send primitive data to the geometry shader stage. If tessellation is inactive (the hull and domain shader programs are set to NULL), the input vertices are received directly from the vertex shader, with the primitive connectivity specified by the input assembler. In this case, the primitive topology can be any of the primitive topologies that the input assembler can

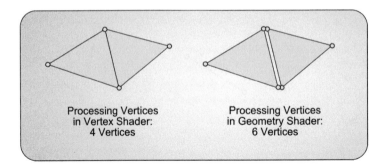

Figure 3.37. The difference between processing vertex-level data in the vertex shader and the geometry shader.

specify. This includes the control patch primitive types, in addition to the standard point, line, and triangle types, including their various forms with adjacency.

The other possible pipeline configuration is when the tessellation stages are active. In this case, the input vertices are received from the domain shader and the primitive topology is determined by the configuration of the tessellator stage. Since the tessellator stage can only produce lines or triangles, these are the only available primitive types that the geometry shader can receive when tessellation is active. This is why primitives with adjacency are not supported when the tessellation stages are active.

Regardless of the pipeline configuration, the geometry is received as a list of vertices representing the primitives present at this point in the pipeline. This has some performance implications, since any of the rendering modes that reduce the amount of vertex processing will not produce the same savings in the geometry shader. For example, if a triangle strip with four vertices is processed in the vertex shader, four invocations of the vertex shader will process the complete set of geometry. However, if each vertex is processed in the geometry shader instead, all three vertices will be submitted to each geometry shader invocation—which is the number of primitives being processed. In this case, there would be two geometry shader invocations, and each would receive the three vertices of the triangle being processed. Therefore, processing the vertices in the vertex shader requires only four sets of calculations, while the same operations performed in the geometry shader will require six. This difference is shown in Figure 3.37.

We can also see from Listing 3.16 how a stream output object is declared for a geometry shader. This object is actually the mechanism that the geometry shader uses to pass its output out of the stage, as opposed to the function returning its output values directly. Even so, the stream is passed as an input argument and hence will be discussed in this input discussion. The syntax for declaring a stream output object begins with the `inout` keyword, which specifies that the object is both an input and an output. This is followed by one of three possible stream types. The stream output object can be used to output a point list, a line strip, or a triangle strip; each of these is represented by the available stream declarations. The argument can be declared as a `PointStream<T>`, `LineStream<T>`, or `TriangleStream<T>`, where each `T` represents the format of the vertex structure that will be passed into the stream. We will further discuss the mechanics of how these output streams must be used in the "Geometry Shader" section.

One final input consideration is the use of two system value semantics, one that we have seen before and one new one. The `SV_PrimitiveID` system value attribute is available to uniquely identify each of the primitives that are passed to the geometry shader. We have previously seen this system value in the hull shader stage, where it can be used to uniquely identify control patches. Since the geometry shader is invoked once per primitive, this system value semantic also uniquely identifies the geometry shader invocations, as well.

The other system value semantic that the geometry shader can use as an input is `SV_GSInstanceID`. This system value semantic works similarly to `SV_InstanceID`, which we saw in the vertex shader stage, except that the instances are actually instanced in the geometry shader stage itself, instead of in the input assembler stage. It is possible to declare a static number of instances to create for each primitive that the geometry shader receives. The `SV_GSInstanceID` is used to differentiate among the individual primitive instances and take the appropriate action in the shader program. This system value must be declared as a standalone input to the geometry shader, since it can't be declared in one of the input streams. Since the identifier indicates which geometry shader invocation is being processed, it makes sense to not allow this to be a per-vertex streaming attribute. The technique to enable this instancing mechanism is discussed in the "Geometry Shader State Configuration" section, while the use of the functionality is discussed in more detail in the "Geometry Shader Stage Processing" section.

3.8.2 Geometry Shader State Configuration

Like all of the other programmable pipeline stages, the geometry shader stage provides the usual group of common shader core resource configurations that can be manipulated by the application with the `ID3D11DeviceContext` interface. There are some key differences in how geometry shader objects are created, due to the geometry shader stage's close relationship to the stream output stage, which we will cover briefly in this section, with a more detailed examination in the "Stream Output" stage section.

Geometry Shader Program

The geometry shader object can be created with one of two methods. Which one must be used depends on whether the stream output stage will be used or not.[17] If the stream output stage will not be used, the geometry shader object is created in the same way that we have

---

[17] The particular details of how the geometry shader writes data to the stream output stage is covered in the "Geometry Shader Stage Processing" section.

already seen for the other programmable stages using the `ID3D11Device::CreateGeometry Shader(...)` method. However, if the stream output stage will be used, the shader object must be created with the `ID3D11Device::CreateGeometryShaderWithStreamOutput(...)` method. This method takes several additional parameters, which configure the stream output stage and the way that data will be streamed to the buffers attached to it. The details of how to configure these parameters for the stream output version will be discussed in the "Stream Output" stage section of this chapter. Simply keep in mind for now that there are two different ways to create the shader object, depending on whether the stream output will be used.

Once the geometry shader object has been created, it can be set in the geometry shader stage with the usual `ID3D11DeviceContext` method. Similarly, the constant buffers, shader resource views, and samplers required for the geometry shader program are manipulated in the standard common shader core methods. They are listed here for reference.

- `ID3D11DeviceContext::GSSetShader()`

- `ID3D11DeviceContext::GSSetConstantBuffers()`

- `ID3D11DeviceContext::GSSetShaderResources()`

- `ID3D11DeviceContext::GSSetSamplers()`

Function Attributes

The geometry shader stage also supports two function attributes that must be declared prior to the geometry shader function in the HLSL source file. The first attribute is the `maxvertexcount` parameter, which allows the developer to specify just that—the maximum number of vertices that an invocation of the geometry shader will emit into its output stream. This is required to ensure that the geometry shader doesn't output an erroneous number of vertices if it has a logic error. It also lets the GPU properly allocate memory for the number of vertices it expects to be produced by each invocation of the geometry shader. This attribute is mandatory and must be provided for the geometry shader function to compile properly. An example of this function attribute can be seen in Listing 3.16.

There is a limit to how much data that can be produced by a single geometry shader invocation. The number of `scalar` values may not exceed 1024. This means that the number of scalar values used in a vertex structure must be summed, and then multiplied by the maximum number of vertices provided in the `maxvertexcount` attribute. This value must be less than or equal to 1024. We will see later that up to four different streams can be used simultaneously when using the stream output functionality, but for the purposes of this maximum output calculation, we can simply take the largest vertex size of all of the output streams being used and multiply that by the `maxvertexcount` attribute.

The second function attribute allowed by the geometry shader is the `instance` attribute. This attribute activates the geometry shader instancing mechanism, in which the primitives passed into the geometry shader are duplicated as many times as specified in the instance attribute declaration. The geometry shader can then declare the `SV_GSInstanceID` (as described above) to provide a unique identifier for the instance. The maximum number of instances that can be created in this way is 32, which allows for significant data amplification. This instancing technique is aimed at simplifying geometry shader programs that need to process geometry for several different outputs, such as when more than one viewpoint is being rendered simultaneously. For example, when generating a cube map[18] with a single rendering pass, geometry shader instancing can be used to generate the six copies of the primitive data, which can then choose the appropriate transformation matrix with the `SV_GSInstanceID` system value semantic.

3.8.3 Geometry Shader Stage Processing

Geometry Shader Process Flow

Before we investigate the possible operations that can be performed in the geometry shader, we must first clarify how it uses the stream objects to produce its output. An example geometry shader was shown in Listing 3.16, and we will reference it throughout this discussion. As we have seen in the previous sections, the geometry shader program receives input primitives as an array of vertices. The ordering of the vertices will vary, depending on the type of input primitive that the pipeline produces prior to the geometry shader stage, as shown in Table 3.1 and Figure 3.36, respectively.

Once the vertex data is available within the geometry shader, some calculations are performed to either modify the input vertices or create entirely new ones. The sample code in Listing 3.16 receives a triangle with adjacency as its input, and then simply passes the main triangle through to its stream output object, ignoring the adjacency information. The method used to generate the output primitives is to call the stream output object's `Append()` method, which takes an instance of the output vertex structure as its argument. With three different stream types available (one for triangle strips, one for line strips, and one for point lists) a series of vertices that are appended to a stream will create different primitives. For example, if a triangle stream is used, and 5 vertices are appended to it, then a total of 3 triangles will be created in the "strip" vertex.[19] Each additional vertex will generate a new triangle using the previous two vertices to complete the primitive. The same methodology exists for the line strip stream object, which would create an additional line primitive for each additional vertex appended after the first one.

---

[18] Cube maps are briefly described in the "Using 2D Textures" section of Chapter 2, "Direct3D 11 Resources."

[19] This is the same "strip" ordering principle that was explained in the "Input Assembler" section of this chapter.

If the triangle/line strip needs to be restarted to create detached groups of geometry, the geometry shader can call the `RestartStrip()` method of the output stream. This will reset the primitive generation to allow another strip to be started, which is not connected to the previously streamed geometry. While it is more efficient to produce geometry in a single strip, if the required output geometry must reside in multiple disconnected sections, then the strip must be restarted. This process of producing primitives by appending vertices to the output stream continues until the geometry shader has completed, which marks the end of this particular invocation of the shader program.

Multiple Stream Objects

One of the new features in Direct3D 11 is the ability of the geometry shader program to use up to four different stream output objects. When properly configured, these objects can be used not only to pass pass primitive data to the rasterizer stage, but to send their data to the stream output stage, where it can be stored in buffer resources. Since the buffer resource is accessible to the application,[20] this means that the stream of pipeline data is directly available for reading or storing in a file. This is a powerful facility, which can be used for some interesting data analysis scenarios. To declare multiple streams in the geometry shader program, you simply include additional stream objects as input parameters to the function. Listing 3.17 demonstrates this technique. These streams are used in the same way that the individual output stream is, with the `Append()` and `RestartStrip()` methods. Each output stream can be written to independently from one another—they don't have to have the same output frequency.

```
void MyGS( InVertex verts[2],
           inout PointStream<OutVertex1> myStream1,
           inout PointStream<OutVertex2> myStream2 )
{
    OutVertex1 myVert1 = TransformVertex1( verts[0] );
    OutVertex2 myVert2 = TransformVertex2( verts[1] );
    myStream1.Append( myVert1 );
    myStream2.Append( myVert2 );
}
```

Listing 3.17. A geometry shader that declares more than one output stream, with different vertex structure types.

Normally, when the geometry shader uses only a single output stream, the results of that stream are passed on to the rasterizer stage, where they are processed and split into

---

[20] Techniques for manipulating resources, including how to read their contents from C/C++, is provided in Chapter 2: Direct3D 11 Resources.

fragments. When there are multiple streams used in the geometry shader program, only one of the declared streams can be selected to be passed on to the rasterizer stage through the stream configuration set by the application when compiling the geometry shader program. The remaining streams must be configured to be streamed to a buffer (one buffer per stream) attached to the stream output stage. The stream that is passed to the rasterizer can optionally be streamed to a buffer as well. Details about how to configure these streams from the application will be covered in the "Stream Output" stage section. Figure 3.38 shows multiple stream outputs from the geometry shader.

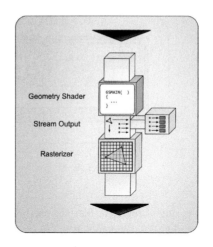

Figure 3.38. Multiple stream outputs from the geometry shader.

A very important consideration when using multiple output streams is that all stream types must be point streams. This means that all vertices passed to a buffer resource are collected individually, which also means that it is not possible to produce triangles or lines that share vertices. In this scenario, a given set of geometry may consume more memory than would normally be used with line strips or triangle strips. However, if the algorithm in the geometry shader produces a known pattern of vertices, it may be possible to use a preloaded index buffer to reference the vertices as the desired primitives. In addition to this restriction, if a stream is sent to the rasterizer stage, it will only appear as points. Because of this, most geometry shader programs with multiple output streams don't rasterize any of the streams unless point rendering is desired.

Another possible configuration is to use a single stream output object (which can use any of the three stream types) and pass it to both the rasterizer and the stream output stage. Data sent to the stream output buffer is still stored as non-indexed geometry, so it causes the same memory consumption increases as mentioned above. However, if the object is being rasterized, it is possible to use triangles and lines instead of just points. These are all options for the developer, and each situation may require using a different combination of these streams to be used.

The final consideration when using multiple output streams is the scalar value output limits imposed on each geometry shader invocation. We have already seen that there is a limit of 1024 scalar values, and that this limit applies to the total number of scalars passed into an output stream. In the case of multiple streams, we take the maximum vertex size and use this to determine the maximum number of vertices that can be passed into the output streams. There is only one set of output registers in the stage, and the registers are shared among all of the output streams. The maximum vertex size is chosen, since

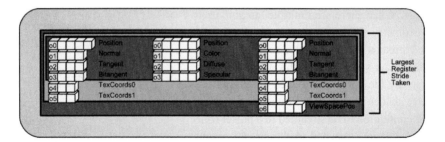

Figure 3.39. Multiple vertex structures using the same output registers.

the output registers that are used to receive the output vertex data will overlap with one another. This is depicted in Figure 3.39.

Since the output frequencies of the streams are not required to be equal, there is no easy way to ensure that dynamic branching won't produce an execution path that streams out the largest vertex for all vertex outputs. Therefore, the easiest way to ensure that only 1024 scalars are output is to determine at compile time the "worst case" vertex output and only allow a valid number of those vertices to be streamed. Any other vertices that are streamed out will only reduce the total number of scalars produced, ensuring that the limit of 1024 scalars is observed.

Primitive Manipulations

Another interesting result of the geometry shader's process flow is that while primitives must be passed into the stage, they may or may not be passed to the next stage. This lets the geometry shader selectively discard unneeded geometry, reducing the amount of geometry for the rasterizer to process, and allowing for special rendering operations that require only a subset of a model to be rendered. A good example of this is the rendering of a model's silhouette to indicate that it is highlighted.

Shadow Volumes

Shadow volumes are a shadowing technique that extracts a model's silhouette as viewed from a light source. The extracted silhouette is expanded outward away from the light, and this expanded geometry is rasterized into the stencil buffer[21] as an indication of which pixels are inside of a shadow volume. This is the heart of the algorithm—the extracted silhouette geometry forms a volume. If the viewer only sees one side of the shadow volume at a particular pixel, it means that that particular point in the scene is inside of the shadow and would not be illuminated by the light. If the viewer can see both sides

---

[21] The stencil buffer and its operation are described later in this chapter, in the output merger stage section.

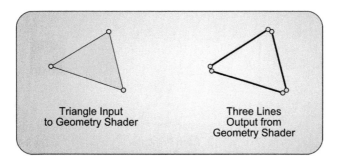

Figure 3.40. Splitting a triangle into three lines to provide a customized wireframe rendering.

of the shadow volume at that pixel, then that point in the scene lies on the far side of the shadow volume and hence should be illuminated.

Before the geometry shader existed, shadow volume extrusion had to be performed on the CPU and then passed to the GPU to be rasterized into the stencil buffer. However, the geometry shader can easily detect when a triangle is located at the silhouette by comparing its face normal vector to the adjacent triangle face normal vectors. If the current triangle is facing the camera, and at least one adjacent triangle is facing away from the camera, then the edge that the two triangles share should be extruded as a part of the shadow volume. If a triangle is not a portion of the silhouette, the geometry shader can discard it by simply not outputting its vertices. In this way, the geometry shader reduces the workload downstream in the pipeline by eliminating unneeded geometry from further processing.

Point Sprites

Since the geometry shader receives primitives in its input array of vertices and it can declare the type of output stream it wants, there are no restrictions regarding the input and output primitive types. In combination with the geometry shader stage's ability to produce a variable number of vertices, this allows for complete control over primitive type conversion. For example, a triangle passed into the geometry shader can be converted to three lines representing the edges of the triangle. This is a simple way to implement a wireframe rendering scheme. The creation of the individual triangle edges is shown in Figure 3.40.

Another popular primitive conversion technique is to convert point primitives into a pair of triangles that form a quad. These quads can then have a texture applied to them, which effectively converts point geometry into a sprite. This process is typically referred to as *creating point sprites*, and is frequently used to give particles in a particle system a more compelling appearance. The process of creating point sprites is demonstrated in Figure 3.41. An example of this technique is used in Chapter 12 to add textures to the particles in the particle system example.

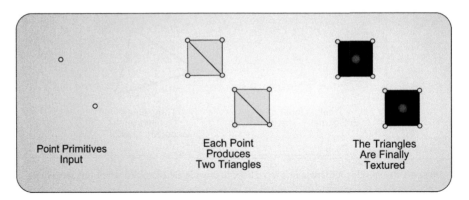

Figure 3.41. The generation of point sprites from point primitives.

Instancing Geometry

The final feature of the geometry shader that we will examine is its ability to perform instancing. As mentioned above, the number of instances of each primitive to create is statically declared with the `instance` attribute. The individual instances can then be identified by the `SV_GSInstanceID` system value semantic. This is a fairly generic ability, which can be applied to many different situations. For example, one point sprite can be instanced and offset from another to amplify the number of particles that appear in the final rendering, while keeping the number of particles in the system to a smaller amount.

Another good example of how to use this functionality is to simultaneously create geometry for multiple render targets. This is especially effective when the geometry for two render targets must have different transformation matrices applied to them, as is the case in cube-mapped environment mapping or dual-paraboloid environment mapping. Separate instances of a primitive are created, which can be transformed and sent to the appropriate render target. This would eliminate repeating all the processing that would have been required prior to the geometry shader stage if two full rendering passes were needed.[22]

3.8.4 Geometry Shader Pipeline Output

We have discussed many ways to use the output vertex streams of the geometry shader, and noted that these output streams are the mechanism by which the geometry shader passes its results down the pipeline. However, we have not discussed in detail what the vertices that are appended to the output streams can contain. Since the geometry shader stage occurs immediately before the rasterizer stage, it must produce vertices with the `SV_Position`

---

[22] This geometry shader instancing technique is implemented in the sample program for Chapter 13, which uses the dual-paraboloid environment map method.

system value semantic. In addition, the contents of this attribute must contain the post-projection clip space position of the vertex in order to be rasterized properly. Stages prior to the geometry shader stage could use this attribute with other forms of the position (such as object space, world space, or view space positions), but the geometry shader must pass the final clip space position to the rasterizer stage.

In addition to the position of each vertex, the geometry shader can take advantage of a pair of system values that have not been available prior to this stage in the pipeline. The first of these system values is SV_RenderTargetIndex. This system value provides an unsigned integer value that identifies the slice of a render target that this primitive should be rendered into. As described in Chapter 2, it is possible to create 2D texture resources that contain an array of textures. Each of these textures is referred to as a slice of the texture. If a render target is bound for output to the pipeline and contains multiple slices, then the geometry shader can dynamically decide which render target to apply a particular primitive to. This method of selecting the render target to be applied to goes hand in hand with the single-pass environment mapping examples described earlier in this section. The SV_GSInstanceID value would be used to determine which copy of a primitive is passed into each render target by copying its value into SV_RenderTargetIndex. Of course, the bound render target must have an appropriate number of slices to accommodate the range of instances specified in the geometry shader. We will discuss binding render targets to the pipeline in more detail in the "Output Merger" stage section. The use of this system value semantic is depicted in Figure 3.42.

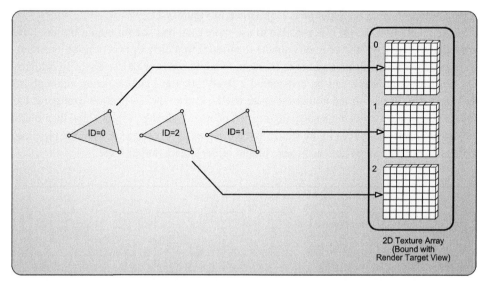

Figure 3.42. Using the geometry shader to write to multiple texture slices simultaneously with the "SV_RenderTargetIndex" system value semantic.

The second system value that the geometry shader can use is the `SV_Viewport` `ArrayIndex` attribute. This attribute operates in a similar way to `SV_RenderTargetIndex`, except that it determines which viewport it will be applied to, instead of the texture slice. Viewports reside in the rasterizer stage, and hence have not been discussed yet. However, we can generalize a viewport to represent a subregion within a render target. Multiple viewports can be bound simultaneously, with each representing different portions of a render target. The usefulness of these viewports is that different representations of a scene can be rendered into each viewport. This is commonly done for split-screen rendering, where each side of the screen shows one player's view of the scene. The geometry shader stage's instancing abilities can also be used to provide an index for use as the viewport array index, effectively allowing multiple sets of geometry to be passed to multiple viewports.

3.9 Stream Output

In the geometry shader stage, we saw how its stream output object is used to produce a stream of primitives that is passed on to the rasterizer stage, where they are ultimately used to modify the contents of a render target later in the pipeline. However, this is not the only use of the output streams. We have also seen how the stream can be sent to the stream output stage, where the output primitives can be streamed into a buffer resource for later use. This is the sole purpose of the stream output stage—to provide a mechanism for connecting the geometry shader and output buffer resources to hold the vertex data. The location of the stream output stage in the pipeline is highlighted in Figure 3.43.

As mentioned above, it is possible to use more than one stream output buffer at the same time. This allows the geometry shader to produce four different versions of its output, and it provides a significant amount of freedom to implement a wide variety of algorithms. The stream output stage can be considered a fixed-function pipeline stage, although no real processing is performed within the stage itself. Instead, the application configures the stream output stage by using a specially created geometry shader program, and then binds the desired buffer resources to the stage. Since it serves as an exit point from the pipeline, there is no pipeline stage attached to the output of the stream output stage.

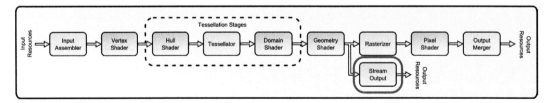

Figure 3.43. The stream output stage.

3.9.1 Stream Output Pipeline Input

The stream output stage can only receive information from the geometry shader stage in the form of the output streams declared in the geometry shader program. Up to four streams are available to the geometry shader to pass data to the stream output stage, and each stream can receive a different vertex structure as its input. This allows multiple variations of the same geometry to be streamed simultaneously. For example, the positions, normal vector, tangent and bitangent vectors, or other optional per-vertex data, can be split into multiple streams and then selectively bound to the input assembler stage later on, providing a very flexible method of binding only the vertex data that is required for a particular effect.

The restrictions on these vertex structures are that a maximum of 128 scalars is allowed for each structure, and that a geometry shader invocation may not produce more than 1024 scalars. Thus, if four output streams are used, and each uses a structure of 128 scalar vertices, a total of 512 scalars for all four streams can be written twice. In practice, this should be a sufficient amount of data to stream, since it is performed for each primitive that the geometry shader receives.

It is also possible to produce completely disjoint streams of data as well. For instance, if an algorithm requires that all front faces must be rendered separately from the back faces, then the front and back faces for a set of geometry can be passed to two different streams. This provides a simple mechanism for generating these mixed sets of geometry from a standard geometry format.

3.9.2 Stream Output State Configuration

The stream output stage requires two different types of configurations. The first is that the application must bind the appropriate number of buffer resources to this stage, through the `ID3D11DeviceContext::SOSetTargets(...)` method. This method operates in a similar way to other resource binding functions, and it allows up to four buffers to be bound simultaneously from within the same method invocation. Listing 3.18 demonstrates how this method is used. There is also a corresponding `Get` method that can be used to retrieve references to the buffer resources that are currently bound to the stage.

```
ID3D11Buffer* pBuffers[1] = { pBuffer1 };
UINT aOffsets[1] = { 0 };

pContext->SOSetTargets( 1, pBuffers, offset );
```

Listing 3.18. Binding buffer resources to the stream output stage.

Each buffer resource that will be bound for output from the stream output stage must be created with the D3D11_BIND_STREAM_OUTPUT bind flag, in addition to any other bind flags that indicate how the buffer will be used for vertex buffer usage. The buffer resource references that are stored in this stage persist between pipeline executions, meaning that they must be managed by the developer. This is the same principle that other resource-binding functionalities use, and the same management techniques can be employed. As we have seen before, it can be beneficial to either always bind all four buffer entries, with the unused slots receiving a NULL pointer, or to have the current state of the slots managed by the application, modifying only the appropriate slots as needed.

The second configuration required for the stream output stage to operate is the specification of what data will be streamed to which output slot. This information is provided to the ID3D11Device::CreateGeometryShaderWithStreamOutput(...) method in the form of an array of D3D11_SO_DECLARATION_ENTRY structures. The elements of this structure are specified in Listing 3.19.

```
struct D3D11_SO_DECLARATION_ENTRY {
    UINT    Stream;
    LPCSTR  SemanticName;
    UINT    SemanticIndex;
    BYTE    StartComponent;
    BYTE    ComponentCount;
    BYTE    OutputSlot;
};
```

Listing 3.19. The members of the D3D11_SO_DECLARATION_ENTRY structure.

In this structure, we can see the information required for each piece of every vertex output by the geometry shader that will end up in one of the stream output buffers. One of these structures must be provided for each output attribute that will be streamed out. The first argument, Stream, identifies which stream output object the data will be coming from in the geometry shader. The SemanticName attribute provides the semantic that is defined in the geometry shader for the attribute to be streamed, and the SemanticIndex attribute provides an index for multiple attributes that share the same semantic name to be uniquely identified. This is the same type of semantic index that we have seen for vertex buffer element declarations.

The StartComponent and ComponentCount arguments determine which portion of a four-component attribute will be streamed. StartComponent can have a value of {0,1,2,3}, which corresponds to the register components {x,y,z,w} respectively. The ComponentCount simply indicates how many of these components to stream. For example, if StartComponent is 1 and ComponentCount is 2, the $\{y,z\}$ components would be streamed. The final argument for this structure is OutputSlot, which determines which stream output stage buffer resource will receive the streamed data. The valid values range from 0 to 3, to correspond with as many as four output buffers.

Once an application has properly filled in an array of these structures to define the information to be streamed out, the geometry shader object can be created. Listing 3.20 shows the method used to create this special version of the shader object.

```
HRESULT CreateGeometryShaderWithStreamOutput(
    const void *pShaderBytecode,
    SIZE_T BytecodeLength,
    const D3D11_SO_DECLARATION_ENTRY *pSODeclaration,
    UINT NumEntries,
    const UINT *pBufferStrides,
    UINT NumStrides,
    UINT RasterizedStream,
    ID3D11ClassLinkage *pClassLinkage,
    ID3D11GeometryShader **ppGeometryShader
);
```

Listing 3.20. The CreateGeometryShaderWithStreamOutput() method of the ID3D11Device interface.

We are the most interested in seeing which of these parameters is different than the normal geometry shader object creation method. The first new parameter, pSODeclaration, is the pointer to an array of the declarations described above. This is followed by NumEntries, which indicates how many entries are in the array. The next two parameters, pBuffer-Strides and NumStrides, provide the vertex strides of each of the vertices that will be streamed to each buffer. This allows each buffer to use a different vertex format to store the desired information. The final new parameter is RasterizedStream, which provides the index of the output stream that should be sent to the rasterizer stage. If none of the streams will be rasterized, this parameter should be set to the constant D3D11_SO_NO_RASTERIZED_STREAM. Whichever stream is specified for rasterization, it can still provide data to one of the stream output buffers as well.

When a geometry shader object is created with this method and then subsequently bound in the geometry shader stage, it activates the stream output stage. Then when a pipeline execution begins, any data passed into an output stream in the geometry shader is matched up with the output configuration based on its semantic name and semantic index. Once a match has been found, that data is streamed to the appropriate buffer to create the completed output vertices.

3.9.3 Stream Output Stage Processing

Automated Drawing

There are two general ways to make use of the stream output functionality. The first is to actually use the resulting buffer resource as an input to the input assembler stage for supplying the vertex data for a pipeline execution. In this case, a special draw call is used to execute the pipeline. The buffer resource must be bound to the input assembler stage slot 0, and the corresponding input layout object must also be configured by the application. Then the ID3D11DeviceContext::DrawAuto() method will inspect the contents of a streamed-out buffer that was bound to the input assembler stage. It then provides the appropriate number of vertices and primitives to the input assembler to render the entire contents of the buffer, with no interaction from the application. In this case, the geometry shader object produces the streamed output vertices[23] and the primitive connectivity information is determined by the input assembler's primitive type configuration. The primitive type should be chosen to be compatible with the type of primitive stream used to stream the data out. If it is then rendered with the DrawAuto method, the complete rendering is performed without the application having knowledge of the contents of the buffer. This is another way in which GPUs are becoming more autonomous and can operate somewhat independently of the CPU.

The utility of this operation may not be immediately obvious. After all, if the geometry reaches the geometry shader, we could just as easily pass its output stream to the rasterizer stage and use it directly—what benefit is there to streaming a mesh out and then re-rendering it? The important point to consider is that the complete set of operations that have been performed between the input assembler and the geometry shader are included in the output stream generated by the geometry shader. Any vertex transformations performed in the vertex shader can be saved, allowing any further rendering passes requiring the same geometry to skip the vertex transformations and continue directly to the next stage. When an expensive vertex-level operation is performed, such as vertex skinning, this can result in a significant reduction in the number of calculations performed during a rendering pass.

This is also especially useful for algorithms that require multiple rendering passes when tessellation has been used. If a model's geometry is first transformed in the vertex shader, and then tessellated in the tessellation stages, and finally streamed out in the geometry shader stage, the results are available for future rendering passes without having to perform all of those calculations again. The tessellation stages can be quite expensive to compute if a very high tessellation level is used, which means that caching the tessellated mesh in this way can save a significant number of calculations. The processing savings are depicted in Figure 3.44.

---

[23] If either a triangle stream or a line stream is used to produce the output vertex data, the strips that are passed into these streams are converted into lists of primitives. This provides a consistent data format for use with the DrawAuto() method.

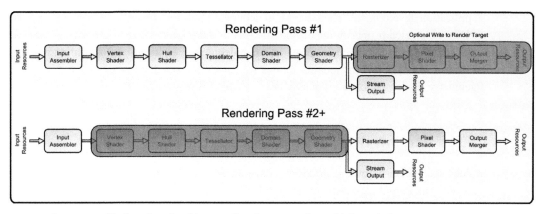

Figure 3.44. The benefits of caching tessellated geometry for multiple rendering passes.

Stream Output as a Debugging Tool

Another potential use for the stream output stage is to tap into the pipeline data at the geometry shader stage and retrieve a portion of the data for debugging purposes. This can be helpful for inspecting intermediate values that would normally not be available in the final rendered output. In this case, two different buffer resources are needed. The first is the stream output buffer, which must be created with the default resource usage flag. The application cannot access this type of buffer directly, since it is not possible to create buffers that can be written to by the pipeline and also read from the CPU. Instead, we create a second buffer with the staging usage flag, and then copy the streamed data from the default usage buffer to the staging usage buffer. The application can then map the staging buffer and read its contents.

Once the contents are accessible to the application, they can be accessed by using a structure that identifies the layout of the streamed vertices. This is typically known well in advance, since the stream output configurations are needed before executing the pipeline, anyway. The number of primitives written to the buffers can be retrieved with a pipeline query object of the type D3D11_QUERY_DATA_SO_STATISTICS (more details about how to perform a pipeline query can be found in Appendix B). This allows the application to know at what point in the buffer the pipeline-generated data ends, and where the default or invalid contents of the buffer begin.

With the buffer data accessible in an evaluated form, the application can use the data in whatever way is needed. This can include writing the data to a file for offline inspection, analyzing data, such as finding the axis-aligned bounding box of the new data, or even printing statistics to the screen in real time for a live visualization of the model data being passed through the geometry shader. These types of techniques would typically not be used for a user-facing application, but they can provide a number of debugging options during application development. The other main debugging tool for Direct3D 11 applications, the PIX tool, is used primarily for offline analysis of data. This makes streaming of pipeline data an appropriate method for providing live debugging information. In addition, the use of the stream output functionality allows for saving intermediate values from within the geometry shader, which are typically not available for reading in PIX.

3.9.4 Stream Output Pipeline Output

Since the stream output stage doesn't have another stage attached to its output, we can consider its output as consisting of the data streamed to the buffers. This is just one of several ways to extract data from the pipeline, but depending on the configuration of the pipeline, it can be used as the primary output. For example, the pipeline could be used as a transformation and tessellation accelerator for a software-based rendering system. In such a scenario, the geometry that will be rendered would be transformed and tessellated by the GPU and then would be streamed out and read back to the CPU, where it could be used in the software renderer. The geometry would never be rasterized by Direct3D 11, but could instead be rendered with a ray-tracing engine or some other high-quality techniques not available on the GPU.

3.10 Rasterizer

The geometry shader stage marks the final point in the pipeline that deals strictly with geometric data. Regardless of whether it receives data directly from the vertex shader stage, the domain shader stage, or the geometry shader stage, the geometric pipeline data is fed into the rasterizer stage. The location of the rasterizer stage is highlighted in Figure 3.45.

The primary purpose of this stage is to convert the geometric data into a regularly sampled representation that can later be applied to a render target. This sampling process

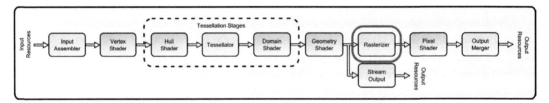

Figure 3.45. The rasterizer stage.

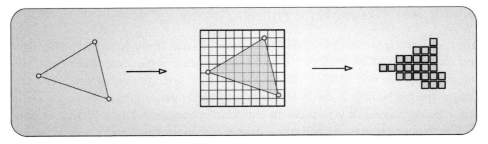

Figure 3.46. A triangle that has been rasterized.

is referred to as *rasterization*, and it maps individual primitives to a format appropriate for storage in a texture resource. The result of rasterization is to produce a number of *fragments* that approximate the original geometry. This is the first step toward generating individual pixel data that will contribute to an output image. Figure 3.46 shows what a rasterized triangle looks like.

In addition to performing the rasterization operation, the rasterizer stage also performs a number of additional operations before fragments are generated. The first operation is *primitive culling*, which eliminates primitives that will not contribute to the output rendering due to their location within clip space. By eliminating primitives prior to rasterization, the overall number of primitives to process is reduced, which increases the efficiency of the stage.

The second operation is the clipping of primitives. This process "clips" primitives to the portion of clip space that can affect the rendered output. This essentially cuts up primitives that are partially in the viewable area into primitives that are completely within it. The portion outside of this region is not needed since it won't produce fragments that can affect the output render target. The clipped primitives are then mapped to the active viewport. The viewport is a structure that describes a subset of the current render target to receive the processed primitives. Finally, the *scissor test* is applied during the rasterization process. This essentially allows the application to specify a rectangular region where rasterization is allowed to occur. Any generated fragments that fall outside of the scissor rectangle are discarded instead of being passed further down the pipeline. This also improves overall pipeline efficiency when only a subregion of the rasterized output will be used.

With all of these operations to perform, the rasterizer stage contributes a large amount of functionality to the processing of pipeline data. It can have a profound impact on the performance of the pipeline as a whole, and hence it is important to understand how all of the processes mentioned above work, and how to influence their operation. The output from the rasterizer stage sets up the rest of the pixel-based pipeline stages, which makes this a data amplification point.

3.10.1 Rasterizer Stage Pipeline Input

The rasterizer stage receives individual primitives as its input from the pipeline. The primitive data consists of the vertices that define its geometric shape. Each vertex must, at a minimum, contain one attribute with the SV_Position system value semantic. This is a four-component homogeneous coordinate that represents the position of the vertex in clip space. Before moving on to consider the individual operations that the rasterizer performs on these input primitives, we will take a short excursion to clarify what clip space is, and how geometry is transformed into this coordinate space. Perhaps more importantly, we will consider the properties of this coordinate space, and how it is used in the context of rasterizing a scene.

Clip Space and Normalized Device Coordinates

The term *clip space* refers to a post-projection coordinate space that can be used to identify what portion of a scene will be visible under the current viewing conditions. This space is essentially a result of the current projection transformation, which is typically either a perspective or orthographic projection. We are primarily concerned with perspective projections, and will focus on them throughout this section. However, many of the same concepts apply to orthographic projections as well, so the following discussion is valid in both cases.[24]

We mentioned above that the position data supplied in the SV_Position system value semantic are homogeneous coordinates consisting of four components. In general, the first three components of this position specify the X, Y, and Z coordinates of the vertex in the 3D space in which it currently resides. The fourth component, which is referred to as the W coordinate, always has a value of 1 before a projection transformation is applied to it. Thus, the position coordinates are always in a form such as $[X,Y,Z,1]$. After a projection transform is applied to these points, the resulting form will depend on the type of projection that was performed. If a perspective projection is used, the resulting point produces a W-value that is equal to the Z-component prior to the projection[25] while the Z component is a scaled version of itself, based on the projection transformations' near and far clipping planes. Equations (3.2) and (3.3) show the results of the projection transformation on an input Z (and W) component based on the projection matrix shown in Equation (8.7). Here we assume that the input coordinates are denoted by the subscript 1, and that the post-projection

---

[24] A general discussion of the various transformations that a set of geometry pass through is provided in Chapter 8, "Mesh Rendering." This includes a discussion of the perspective projection.

[25] This assumes that the projection matrix used is the same as described in the DXSDK documentation, and also as described in Chapter 8.

coordinates are denoted with subscript p. The following equations are used:

$$z_p = \frac{z_f(z_1 - z_n)}{(z_f - z_n)} \; ; \qquad (3.2)$$

$$w_p = z_1 . \qquad (3.3)$$

The result of Equation (3.2) may not be obvious at first inspection, but we can gain some insight into this equation by plotting z_p after the divide by W as a function of the input Z-coordinate. This is demonstrated in Figure 3.47. Here we can see that the post projection Z-value can reside in one of three general areas. If the input Z-coordinate is less than the near clipping plane, Z_n, the resulting Z_p has a negative value. If the input Z-coordinate is between Z_n and Z_p, the resulting Z_p has a value between 0 and Z_f. Any input Z-coordinate that is greater than Z_f produces a Z_p that is greater than Z_f.[26]

At some point after projection, these post-projection coordinates must be rehomogenized by dividing by the W-coordinate. By dividing all of the coordinates by the W-coordinate, we ensure that the post-divide W-coordinate will be 1. We can also consider what happens to our post divide Z-coordinate by simply dividing the results discussed above by Z_1. When Z_1 equals Z_n, the result is 0. When Z_1 equals Z_p, the result is 1. Thus, after the divide by W we have a valid depth range of values between 0 and 1, which correspond to the input values between Z_n and Z_f. We won't repeat the calculations here, but the X- and Y-coordinates produce a similar behavior, with each of them producing valid values in the -1 to 1

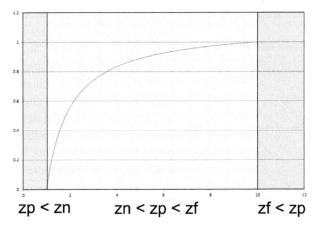

zp < zn   zn < zp < zf   zf < zp

Figure 3.47. The post-projection Z-coordinate of a position.

[26] All of these equations assume a left-handed coordinate system, where Z is positive in the direction of view. This is the standard coordinate system used in Direct3D.

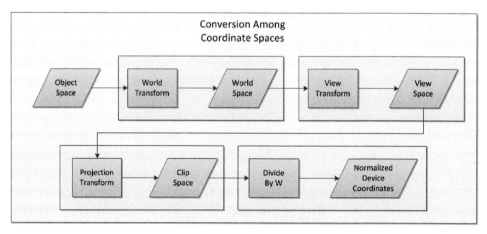

Figure 3.48. The change from one coordinate space to another.

range in the post-W-divide range. Thus, we have two different ways to express the post-projection coordinates—before and after the W-divide. We will refer to the pre-divide coordinates as residing in clip space, and we will refer to the post-divide coordinates simply as normalized device coordinates. This distinction is very important, since both coordinates describe the same point, but their values will vary significantly between the two. The various coordinate spaces and the operations that produce them are shown in Figure 3.48.

As we noted above, the position received by the rasterizer stage in the SV_Position system value semantic should be the result of the projection transformation, and hence should be in clip space. The rasterizer stage performs the W-divide later in its sequence of processing, so the shader program that performs the projection transformation doesn't need to manually execute the division.[27] In many cases, it can often be easier to visualize the post-projection coordinates in terms of the normalized device coordinates when considering whether a particular vertex will be visible or not. In these normalized device coordinates, a vertex must reside within the cube between the points [1,1,1] and [-1,-1,0]. This range of coordinates is depicted in Figure 3.49.

As we can see in Figure 3.49, the positive X-axis points to the right, the positive Y-axis points upward, and the positive Z-axis points

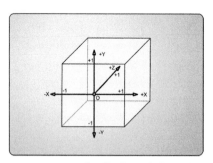

Figure 3.49. A visualization of the region where vertices are within the current viewing area.

---

[27] This doesn't mean that the shader can't perform the W-divide if it needs to for further calculations. If so, the rasterizer stage will simply divide the other coordinates by 1, which will of course produce the same result.

into the scene. This region is often referred to as the *unit cube*, although in reality it is not a complete unit cube, due to its range in the *Z* direction, but it is still commonly referred to this way. We will explore clip space and normalized device coordinates further in the "Rasterizer Stage Processing" section.

Clip and Cull Distances

Now that we have an understanding of how the position data received by the rasterizer should look, we can consider some of the other inputs it may receive. The next two input attributes we will look at were initially discussed in the vertex shader section. The system value attributes, SV_ClipDistance and SV_CullDistance are both consumed in the rasterizer stage as additional user-defined inputs to the culling and clipping operations. These allow the culling and clipping processes to be manipulated with customized data provided by the programmable shader programs. We will discuss these in more detail later in the "Rasterizer Stage Processing" section.

Viewport Array Index

In addition to the clip and cull distances, the rasterizer stage can also receive the SV_ViewportArrayIndex system value semantic. This is an unsigned integer that specifies which viewport definition structure the rasterizer should use when rasterizing a primitive. The mechanics of how these individual viewports affect the output will be described in detail in the "Rasterizer Stage Processing" section.

Render Target Array Index

We also have seen that the geometry shader can specify the SV_RenderTargetArrayIndex system value semantic to indicate which texture slice of a render target array to write to. This is simply an unsigned integer value that selects the texture slice to target. We don't implement any sample programs that use this system value semantic in the book, but there is a very good Direct3D 10 sample in the DXSDK that demonstrates this technique being used, called CubeMapGS. In addition, a further description of how this is used can be found in (Zink).

Additional Input Attributes

The final group of inputs that can be sent to the rasterizer stage is actually a wide variety of different types of information. Any additional attributes in the vertex structure that are received by the rasterizer represent data that the developer has defined at each vertex of the input primitive. In the case of a triangle, there are three instances of an attribute available for each triangle at each of its vertices. Since the rasterization process generates fragments that represent a regular sampling pattern between these vertices, the three vertex attribute values must be interpolated to find an appropriate intermediate value to pass on to each fragment. This is performed in the actual rasterization process, and will be discussed later in this section.

3.10.2 Rasterizer Stage State Configuration

With such a large number of different functionalities available in the rasterizer stage, there are a correspondingly large number of states to configure in order to control its operation. These configurations are controlled by the application through three different sets of methods, which can be used either to query the existing state or to overwrite the existing state with a new one.

Rasterizer State

The primary configuration of the rasterizer stage is performed with the rasterizer state object. This object, represented by the `ID3D11RasterizerState` interface, is an immutable state object that is validated when it is created. As is the case with all objects in Direct3D 11, the object is created through a device method, the `ID3D11Device::CreateRasterizer State(...)` method in this case. Like other state objects, this method takes a pointer to a description structure that provides the desired configuration values. Listing 3.21 provides a description declaration, followed by the creation of a rasterizer state object. This is followed by reading the existing state in the current device context, and then setting the new state that we just created.

```
D3D11_RASTERIZER_STATE rs;

rs.FillMode = D3D11_FILL_SOLID;
rs.CullMode = D3D11_CULL_BACK;
rs.FrontCounterClockwise = false;
rs.DepthBias = 0;
rs.SlopeScaledDepthBias = 0.0f;
rs.DepthBiasClamp = 0.0f;
rs.DepthClipEnable = true;
rs.ScissorEnable = false;
rs.MultisampleEnable = false;
rs.AntialiasedLineEnable = false;

ID3D11RasterizerState* pState = 0;

HRESULT hr = m_pDevice->CreateRasterizerState( &rs, &pState );
```

Listing 3.21. The specification and creation of a rasterizer state object.

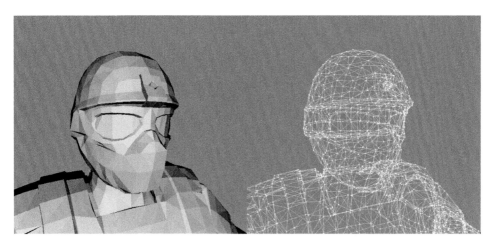

Figure 3.50. The difference between wireframe and solid fill modes. Model courtesy of Radioactive Software.

We will examine each of the elements of the rasterizer state individually. The first setting is the `FillMode` parameter, which determines how primitives are rasterized. The available options are `solid fill` or `wireframe`. In solid fill, fragments are generated between the vertices of the primitive to completely fill its interior, while in wireframe mode only the edges of a primitive are rasterized. In practice, this only affects triangle primitives since lines only consist of a single edge, and points only consist of a single point, so the fill modes are equivalent for them. Figure 3.50 shows a simple example of the difference between these two modes for a triangle.

The second setting in this structure is the `CullMode`. The cull mode controls the cull operation of the rasterizer stage. This can be used to enable culling of primitives that are facing the viewer (*front facing*) or facing away from the viewer (*back facing*). An option is also provided to disable the culling operation entirely. Determining if a triangle is front facing or back facing is done by examining the order that its vertices arrive in. If the vertices are ordered such that traversing them in order produces a clockwise trip around the triangle on the render target, it has *clockwise* vertex ordering. Otherwise, the triangle is said to have *counter-clockwise* winding. The `FrontCounterClockwise` setting determines if the front-facing triangles should be considered clockwise or counter-clockwise. In effect, this setting is used with the `CullMode` to select the target of the culling operation.

The next three settings control an optional feature to provide a depth bias to the fragments generated by the rasterizer. Some algorithms require an object to be rendered in two different ways to two separate render targets to achieve a particular effect. *Shadow mapping* is probably the most common example of this, where an object is rendered from a light source's perspective to generate a map of what objects in the scene are visible to the light. This rendering produces what is called a *depth map* for the light, as it determines

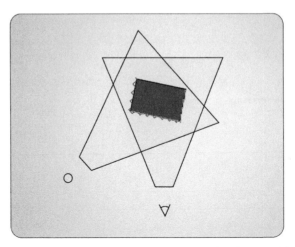

Figure 3.51. A mismatch between a scene rasterized from the light point of view and from the camera point of view.

the distance from the light in each of its pixels. In the subsequent rendering pass, an object is rendered from the viewer's perspective and the depth map is sampled from the first to determine if it can be "seen" by the light and hence would receive light from it. In theory, this is a very sensible way to determine which objects are in shadow and which are not. But in reality, this method can introduce many artifacts, due to differences in the effective sampling patterns of the two rendering passes.[28] This is demonstrated in Figure 3.51, which displays an overhead view of an object and how it is rasterized by both the light's depth map and by the scene rendering.

To combat this type of effect, the rasterizer stage provides a method to introduce a depth bias into the generated fragments. The `DepthBias` and `SlopeScaledDepthBias` settings provide a constant and a slope-based offset depth value, respectively. These two parameters apply a depth bias in one of two ways, depending on the type of depth buffer being used. Equation (3.4) and Equation (3.5) provide the two depth calculations.

$$\texttt{Bias = (float)DepthBias * r + SlopeScaledDepthBias * MaxDepthSlope;} \tag{3.4}$$

$$\texttt{Bias = (float)DepthBias * 2 (exponent(max z in primitive) - r) +}$$
$$\texttt{SlopeScaledDepthBias * MaxDepthSlope;} \tag{3.5}$$

---

[28] Many, many, many algorithms have been devised to improve the produced image quality when using shadow maps. A search in academic literature will easily return more than 100 different papers on the topic.

The first technique for calculating the depth bias is used when a depth buffer with a unorm format is used, or if no depth buffer is bound to the pipeline. The second method is used when a floating point depth buffer is used. In the first equation, the r parameter represents the minimum value that can be represented that is greater than 0, and the MaxDepthSlope is the greatest slope in depth found over the primitive. In the second equation, the r parameter represents the number of mantissa bits in the floating point type of the depth buffer. In either case, this bias value is calculated and then clamped to the DepthBiasClamp parameter of the rasterizer state description. Finally, the bias value is added to the z value of a vertex after clipping and before interpolation setup is performed. The effect of adding a bias can effectively shift the rasterized surface to ensure that small differences in calculation can't produce artifacts.

The remaining four settings of the rasterizer state description are all Boolean parameters that switch various functionalities on or off. We will explore each of these parameters in more detail in the "Rasterizer Stage Processing" section, but we make a brief mention of them here. The first of these settings is the DepthClipEnable parameter. This enables or disables clipping of primitives, based on their depth values. This essentially clips the geometry to the near and far clipping planes. Next is the ScissorEnable parameter, which enables or disables the use of the scissor test. The scissor test is used to cull any fragments that are generated outside of the scissor rectangle. The third Boolean parameter is multisampleEnable, which essentially toggles whether the rasterizer performs multiple coverage tests for MSAA render targets. And finally, AntialiasedLineEnable enables or disables the use of anti-aliasing when a line primitive is rasterized.

Viewport State

In addition to the main rasterizer state, the application must also provide at least one viewport for use in rendering operations. The viewport structure consists of six floating point parameters that identify the subregion of the current render target that will be rasterized into. Listing 3.22 shows the contents of this structure.

```
struct D3D11_VIEWPORT {
    FLOAT TopLeftX;
    FLOAT TopLeftY;
    FLOAT Width;
    FLOAT Height;
    FLOAT MinDepth;
    FLOAT MaxDepth;
}
```

Listing 3.22. The members of the D3D11_VIEWPORT structure.

As seen in Listing 3.22, the viewport structure defines a region that the normalized device coordinates will be mapped to in order to convert from the unit-sized coordinates of the unit cube to the pixel based coordinate system of a render target. As an example, if the entire render target should be covered by the viewport, the top and left values will be 0 and the width and height parameters will be the width and height of the render target being used. The `MinDepth` and `MaxDepth` parameters are used to scale the range of depth values being used to a subset of the complete regular range.

The application can either provide a single viewport, or it can specify multiple viewports simultaneously. If multiple viewports have been set, the viewport that is used for a particular primitive rasterization is determined by the value of the `SV_ViewportArrayIndex` system value semantic. Binding of viewports is performed with the `ID3D11DeviceContext::RSSetViewports` method. Unlike the state arrays that we have discussed previously, any viewports that are not set in the most recent set call will be cleared. In addition, there is the usual `Get` counterpart method for retrieving the currently set array of viewports.

Scissor Rectangle State

The final configuration available in the rasterizer stage is the ability to specify an array of scissor rectangles. These rectangles are used during the scissor test to specify a particular region of the render target for which fragments can be generated. This effectively eliminates any fragments that would have been generated for a region outside of the current scissor rectangle, which can reduce unnecessary computation. If multiple scissor rectangles are bound, the rectangle used for a given primitive is determined by the same `SV_ViewportArrayIndex`. This ensures that the viewport and scissor rectangle that are used together are a matching pair. The array of scissor rectangles is also managed in the same way as the array of viewports. It can be bound using the `ID3D11DeviceContext::RSSetScissorRects()` method, making only the scissor rectangles that were specified in the most recent set call remain active.

3.10.3 Rasterizer Stage Processing

All of the processes performed in the rasterizer stage are vital components for efficiently converting geometric data into image-based data suitable for storage in a render target. This section explores in greater detail what each of these processes encompasses, and attempts to provide some insight into how they can be used in common real-time rendering contexts. Figure 3.52 shows a block diagram of the functional operations performed within the rasterizer stage. These processes may be implemented in a different order in the actual GPU hardware, but we will conceptually consider them in the shown sequence.

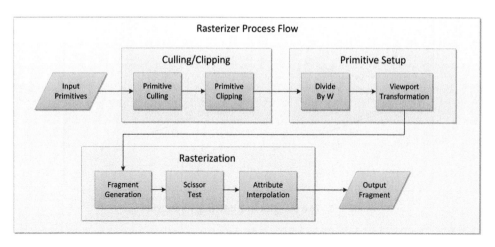

Figure 3.52. The functional operations performed by the rasterizer stage.

As can be seen in Figure 3.52, the rasterizer receives complete, individual primitives as its input. At this point in the pipeline, any primitive topologies with adjacency are converted to standard primitive topologies. If the primitives were passed into the pipeline as strips, those strips are broken apart and issued to the rasterizer stage individually. These primitives are then processed in a sequential pipeline fashion, with the process ending with the production of fragments, which are then passed to the pixel shader stage. The following sections explore in more detail what is performed in each of these functional blocks.

Culling

The first operation performed on the incoming primitives is culling. This operation is intended to remove primitives that will not contribute to the final rendered image from further processing, to increase the efficiency of the pipeline. There are two different types of culling: *back face culling* and what we will refer to as *primitive culling*.

Back face culling. The first form of culling we will examine is back face culling. As its name indicates, this operation is only applied to triangle primitives, since they are the only basic primitive with a concept of a face. However, triangles are typically the most frequently used primitive, and hence this process is quite important to most rendering sequences. When the vertices of the input primitive are received by the rasterizer stage, the winding of the vertices is determined. Figure 3.53 shows two different triangles, where the vertex winding is either clockwise or counter-clockwise. One technique for performing this check is described in the "Vertex Shader Pipeline Output" section, but the hardware implementation may vary.

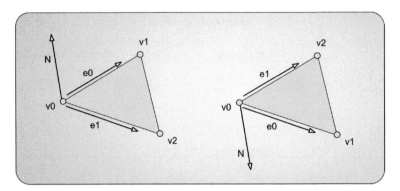

Figure 3.53. Two triangles with differing vertex windings.

We have seen that in the rasterizer state configuration, there are two different settings for the culling operation: CullMode and FrontCounterClockwise. Specifying which vertex winding represents a front-facing primitive depends on the input data introduced to the pipeline in the input assembler stage. The digital content creation tool that was used to generate the input geometry may use either winding, but it is common to have the option to reverse the winding, to make the model compatible with the convention of the end user's application. Reversing the vertex winding order is performed by swapping two of the vertices from their current locations in the primitive's list of vertices.

Once the convention of what represents the front face of a primitive has been specified by the FrontCounterClockwise parameter, the CullMode parameter determines which faces, if any, should be culled. Culling is used to reduce the amount of work required by the rasterizer stage by eliminating triangles that are facing away from the current view point. If a triangle faces away from the current view, it is considered to be back-facing. In traditional opaque rendering algorithms, back faces do not contribute to the final image and should be eliminated. In fact, roughly half of the primitives that enter this stage will be removed, since only half of a model can be visible in a single view within a frame. In this case, the CullMode parameter should be set to D3D11_CULL_BACK.

At the same time, there are many algorithms that require rendering geometry multiple times with different cull modes to achieve different effects. For example, a simple technique for finding the thickness of an object at every pixel is to first render it with front faces being culled and writing to the red channel of the render target, while using a blending state that selects the maximum value when writing to the render target.[29] A second rendering pass is performed with back faces being culled and writing to the green channel of the render target, while using a blending state that selects the minimum value when writing to the render target. After these two passes, the render target contains the farthest point in the red

---

[29] Blending will be covered in more detail in the output merger section of this chapter.

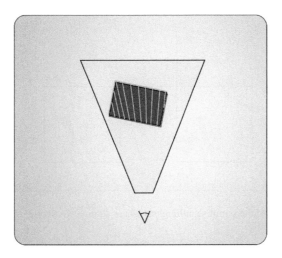

Figure 3.54. A depiction of the minimum and maximum depths of an object, enabled by using varying culling modes.

channel, and the nearest point in the green channel. Together, the difference between these two values provides an estimate of the thickness of an object.[30] This concept is depicted in Figure 3.54. In algorithms such as this, it is necessary and useful to cull triangles with one orientation in one pass, and then cull the opposite orientation in the second pass. A number of special effects can be achieved by manipulating the culling order.

Primitive culling. The second form of culling that is performed in the rasterizer stage tests if there are any primitives that reside completely outside of the unit cube in normalized device coordinates. This can be conservatively determined by testing if all of the vertices of a primitive are outside of the same clipping plane, where the clipping planes are defined by the faces of the unit cube. This operation can be performed very efficiently when performed in clip space,[31] since the planes are axis-aligned and are always located at the same distance from the origin. For example, to test if the three vertices of a triangle are all outside of the top plane of the unit cube, the test simply consists of checking if the Y-component of their clip space positions is greater than its W-component. The test can be performed by taking the difference of the W-component minus the Y-component. If all of the results are negative, the primitive can be safely discarded from further processing. This

---

[30] There are other, newer techniques that can perform this depth calculation in a single pass, but the example is still a valid situation where multiple rendering passes are required with switched culling order.

[31] Which space this is performed in is up to the hardware implementation, and may or may not be performed in clip space.

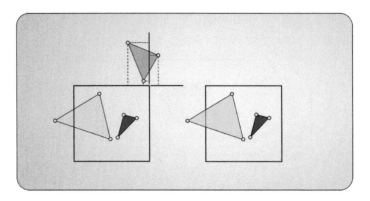

Figure 3.55. A depiction of the primitive culling test where all three vertices of a triangle must be outside of the same plane to be culled.

check is performed for each of the sides of the unit cube. This culling process is depicted in Figure 3.55.

In addition to this standard form of culling, the developer can also use custom culling algorithms. We have seen the system value attribute SV_CullDistance in the vertex shader section. This system value semantic will cull a primitive if all of its vertices arrive at the rasterizer stage with negative values in the same attribute component marked with the SV_CullDistance semantic. The value can be generated with a user-defined plane equation, performing a distance test and then storing the scalar result in the SV_CullDistance attribute. In practice, there is no limitation imposed on the type of calculation used to generate the culling value. If there is a more appropriate test to be applied, it can be implemented as needed and the result stored in the same way. For example, when generating a paraboloid environment map, geometry could be culled that exists in the opposite hemisphere from the one being generated at the moment (paraboloid maps are discussed in more detail in Chapter 13).

Primitive Clipping

The next operation performed by the rasterizer stage is primitive clipping. This takes each primitive that has survived the culling operations and tests if it is located fully inside of the unit cube, or if it is only partially inside. If the primitive resides completely outside of the unit cube, it can be discarded outright without any further processing. If it is partially inside, then the primitive is split into new primitives that reside entirely within the unit cube, and the exterior primitives are discarded. This procedure is depicted in Figure 3.56. Since the unit cube corresponds to the view frustum of the current projection matrix, this operation is also referred to as frustum clipping.

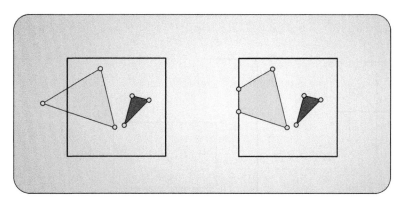

Figure 3.56. A depiction of the clipping process.

The implementation of this process is hardware dependent, but it will generate new vertices at the boundary of the unit cube, to allow the new primitives to be created. An important point to consider during this process is that these generated vertices receive interpolated attribute values from the original vertices, which will then be used later to perform the actual rasterization process.

Like the culling mechanism, the primitive clipping hardware can also be used by the developer to perform customized per-fragment clipping. By using the SV_ClipDistance system value attribute, each vertex can specify its location relative to a user-defined clipping function. This can be used to implement custom clipping planes, or some other function that will provide an appropriate clipping result. These attributes are interpolated across the primitive, and a per-fragment clipping test is performed. If the interpolated attribute value is negative, then the fragment is discarded. Multiple clipping functions can be implemented by using more than one component of a vertex attribute. For example, if a vertex attribute is declared as a float4 type, then the results of the four different clipping functions can be specified with one result in each component. This allows for a significant number of tests to be performed simultaneously.

Homogenous Divide

Once a primitive has passed through the culling and clipping operations, the *homogenous divide* is performed. This process simply divides the projected points by their *W*-components, as we discussed earlier, producing homogenous coordinates of the form [*X/W*, *Y/W*, *Z/W*, 1]. This format is referred to as *normalized device coordinates*, since the size of each coordinate has been scaled by the *W*-coordinate. Prior to this conversion, the coordinates are still in clip space, which is the form produced by the projection matrix. The input clip space points may or may not have a *W*-coordinate value of 1.

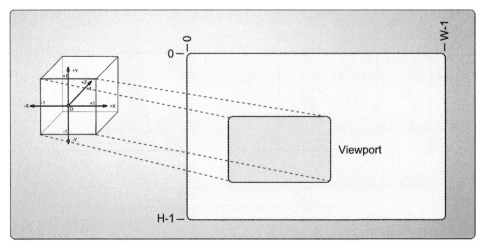

Figure 3.57. The mapping of the unit cube in normalized device coordinates to a render target's screen space coordinate system.

Viewport Transformation

The final step before the actual rasterization takes place is the *viewport transformation*. The vertices of the primitive currently reside in normalized device coordinates, where the X- and Y-coordinates range between [-1,1] and the Z coordinate ranges between [0,1]. The viewport provides the information needed to map the primitive from normalized device coordinates to pixel coordinates. The viewport data structure provides an offset (with the TopLeftX and TopLeftY parameters) and a scale (with the Width and Height parameters) for positioning the primitives within the desired region of a render target. This mapping process is depicted in Figure 3.57.

After the mapping has been performed, the new coordinates will have a range of [TopLeftX,TopLeftX+Width] for the X component, and [TopLeftY,TopLeftY+Height] for the Y component, where width and height are the dimensions of the render target area to be rendered to. It is also possible to scale the Z-component of the position as well, to manipulate the depth values used later in the pipeline. As we have mentioned previously, multiple viewports can be used to achieve effects such as split screen rendering, where the scene is rasterized into two different regions of a render target to indicate what is visible from each player's current view point. However, this could also be used to only render a scene to a region that isn't covered by a user interface element. For example, it is common in real-time strategy (RTS) games to have a large user interface element at the bottom of the screen. If this user interface is not transparent, then it can significantly reduce the size of the region being rasterized into, and subsequently reduce the number of fragments/pixels that need to be generated. In general, if there is a reason to restrict the region that is being rendered with a viewport, it is a good practice to do so!

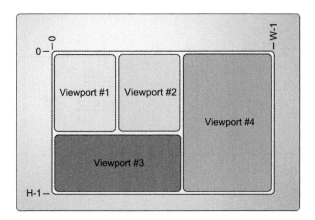

Figure 3.58. Multiple viewports within a single render target, which allows rasterizing of geometry to a particular region of the render target.

There are two different ways that the desired viewport can be selected by the application. If the input signature to the rasterizer stage (as determined by the declared output from the previous stage) contains the SV_ViewportArrayIndex system value semantic, then this value is used to select from the currently bound array of viewports. If this system value semantic is not present in the input signature, then the system defaults to the viewport located at index 0. Therefore, if only a single viewport is bound and no further manipulation of the viewport is needed, the developer can simply exclude SV_ViewportArrayIndex from the rasterizer input to select the primary viewport. Figure 3.58 demonstrates the use of multiple viewports and how they can be employed to select multiple regions for rasterization.

After the viewport transformation has been completed, the vertex position is split into its X- and Y-coordinates, which are used to identify the region of the render target that is covered by a primitive, and its Z-coordinate, which is a depth value that is used later in the depth buffering system to determine the visibility of individual fragments.

Rasterization

The final operation in the rasterizer stage is to perform the actual rasterization of the primitives that reach this point in the pipeline. The primary purpose of the rasterizer is to convert the geometric data that it receives into discretely sampled data that approximates the geometric data within the render target. This can be thought of as a sampling process, where the geometry can be considered as continuous data, while the output fragments are a digital representation of the geometry at regularly spaced intervals. A number of sample primitives are shown in Figure 3.59.

The rasterization process consists of two different tasks, fragment generation and attribute interpolation. We will discuss each of these items individually, and then

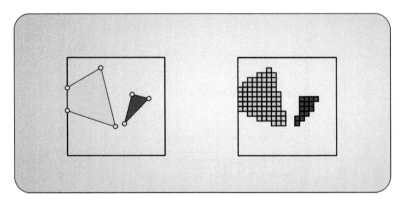

Figure 3.59. Various rasterized geometric primitives.

also look more closely at some of the fine details of using this process with multisample anti-aliasing (MSAA) enabled.

Fragment generation. The first step in rasterization is to determine which pixels of the render target are considered to be covered by the current primitive. This operation is performed by using a set of rasterization rules, which vary depending on the type of primitive that is being rasterized, as well as on the type of render target. If a render target supports MSAA, the rasterization process is different than if a standard render target is used. We will discuss the implications of MSAA later in this section, after the standard concepts have been introduced. With a grid of pixels representing the render target, the rasterizer stage must determine which of the pixels are "covered" by a primitive. Quotation marks were used here because there are many cases in which a pixel location is selected for rasterization but it is not completely covered by the primitive.

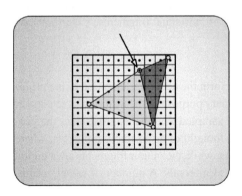

Figure 3.60. Pixels at the boundaries of multiple primitives.

To understand how this can happen, we first need to explore how fragment generation is performed. We will begin with the triangle rasterization process. Each pixel of the render target uses a point at the center of the pixel as the key for rasterization. Since it is at the center of the pixel, and each pixel is 1 unit large (in screen space), the center of the pixel is 0.5 units by 0.5 units from the edge of the pixel boundary. When a pixel center is completely covered by a primitive, it is clearly selected to produce a fragment. Since the pixel center can also land on one of the boundary edges of a primitive, some additional rules must be used in testing to ensure that all pixels in two adjacent triangles are selected, and that no pixels that are selected more than once by neighboring primitives. Figure 3.60 demonstrates some of these situations.

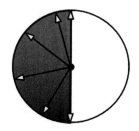

The triangle rasterization rules specify that any points that fall onto a `top edge` or `left edge` of a primitive are selected for rasterization from that primitive. A top edge is any perfectly horizontal edge, while a left edge is any edge that faces to the left at any angle. If you consider the normal vectors of the range of edges that will allow selection of a pixel, the angles that will produce a fragment when an edge resides on a pixel center would appear as shown in Figure 3.61.

Figure 3.61. The angular range of primitive edge normal vectors to select a pixel for rasterization.

Line rasterization is somewhat different from triangle rasterization. It can follow one of two paths—either `aliased` or `anti-aliased`. Aliased line rasterization uses a fairly simple algorithm to determine which pixels are covered by a line, while anti-aliased line rasterization performs a more sophisticated algorithm to determine how much of a pixel is covered by a line, and then multiplies the output color for that pixel by this coverage factor. Selection of aliased or non-aliased line rasterization is determined by the `AntialiasedLineEnable` parameter of the rasterizer state. We will examine the details of aliased rasterization, while leaving the anti-aliased lines. The anti-aliased line algorithm implementation is hardware specific, and hence it does not make sense to try to interpret in great detail here.

Aliased line rasterization is performed as follows. Instead of using a point-based algorithm, the process starts by inscribing a diamond shape on each pixel. This is shown in Figure 3.62. If a line's slope falls within the range $-1 <= slope <= 1$, it is considered to be an *x-major* line. In this case, if the line itself intersects the bottom left or bottom right edge or the bottom corner of the diamond, then the pixel is selected for fragment generation. If

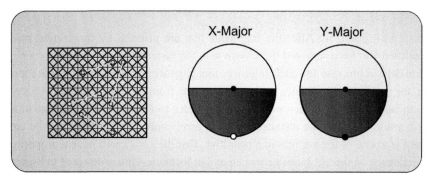

Figure 3.62. Some examples of line rasterization, with the two diamond selection possibilities depicted at the right.

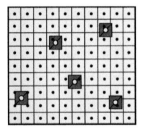

Figure 3.63. Several examples of point rasterization as a pair of triangles.

the line has a slope outside of the range defined above, then it is considered a *y-major* line. In this case, the same test is performed, with the exception that the right corner is also included in the test. These intersection images are shown in the right side of Figure 3.62.

The final primitive type that can be rasterized is the *point primitive*. Point primitive rasterization actually uses the same rules as triangle rasterization, by conceptually expanding itself into two triangles that form a 1×1 square around its point position. These triangles are then rasterized exactly as if they were triangle primitives. Several examples of this pattern are shown in Figure 3.63.

Scissor test. In all of these cases, the primitives must pass the *scissor test*, in addition to fulfilling the rasterization rules. As we have seen in the "Rasterizer State Configuration" section, the application can configure the scissor test by binding an array of scissor rectangles to the rasterizer stage. The rasterizer stage then selects the appropriate scissor rectangle, based on the same criteria used to select the viewport. If the **SV_ViewportArrayIndex** is declared as one of the input attributes for the rasterizer stage, then it is used to select from the array. Otherwise, the rasterizer defaults to the first entry. This selection mechanism ensures that the viewport index and scissor rectangle index are always the same, which allows the application to trivially reference pairs of these objects together.

The scissor test works by comparing the fragment's X- and Y-components against the scissor rectangle. If the fragment falls outside of the rectangle, it is culled and will not contribute to the render target. It is also important to note that the exact location of the scissor test in the sequence of operations of the rasterizer may vary, depending on the hardware implementation. This may have some performance implications, due to a variable location that fragments are culled from. However, the application can be sure that the pixels outside of the scissor rectangle will not be written to the render target.

Attribute interpolation. After the fragments that are affected by the current primitive are identified and have survived the scissor test, the rasterizer stage must determine what attribute data to produce for each fragment that it generates. To calculate these attribute values, the rasterizer interpolates each input attribute from its input primitive vertices. The input attribute value from each primitive contributes to the interpolated fragment output attribute value based on the distance from the pixel center. This means that the closer a fragment is to one of the vertices in a primitive, that the vertex will have a proportionally larger influence on the attributes generated at that location. This is depicted in Figure 3.64 with a simple example. In addition to the general attributes being interpolated, the depth at each fragment is also interpolated and will be used later in the pipeline in the depth test.

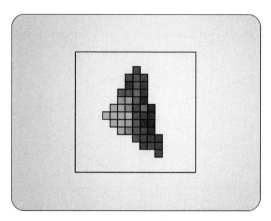

Figure 3.64. Several pixels of a rasterized triangle, which demonstrate how pixels are affected by their proximity to each of the vertices.

By default, the type of interpolation used to generate these per-fragment attributes is a perspective correct linear interpolation. However, the interpolation mode can be modified by adding an interpolation modifier before the type declaration of a pixel shader input attribute. The available interpolation modes are listed in Table 3.2, with a brief description of their behavior.

Interpolation Mode	Description
linear	Provides linear, perspective-correct interpolation. The interpolation is based on the center of the pixel.
centroid	Provides linear, perspective-correct interpolation. The interpolation is based on the centroid of the covered area of the pixel.
nointerpolation	Provides no interpolation. Attributes are passed as constants.
noperspective	Provides linear interpolation without accounting for perspective effects. The interpolation is based on the center of the pixel.
sample	Provides linear, perspective-correct interpolation. The interpolation is based on the MSAA sample location, instead of the center of the pixel.

Table 3.2. The available attribute interpolation modes.

Multisampling Considerations

Direct3D 11's rasterization pipeline is an efficient means of rendering 3D geometry with performance suitable for real-time applications. However one of the fundamental drawbacks of rasterization is that the image produced can suffer from noticeable *aliasing* artifacts. Aliasing is a term from the field of signal processing that refers to the effect that

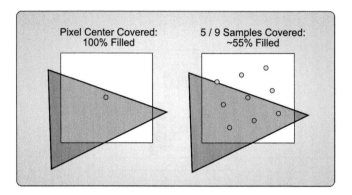

Figure 3.65. Rasterization sample points for non-MSAA and MSAA Rendering.

occurs when a continuous signal isn't sampled at a high enough sampling frequency to reproduce the original signal. If we were to describe rasterization in such terms, the "signal" would be the vector representation of our 3D geometry, while our "sampling frequency" is the X and Y resolution of the render target. In more general terms, we can say that aliasing occurs because our geometry must ultimately be rendered to a discrete grid of rectangles, using a binary coverage test. A grid of rectangles can never perfectly represent edges that are not completely horizontal or vertical, so a jagged "stair step" pattern occurs. This type of aliasing artifact is often referred to as *edge aliasing*, since it occurs at triangle edges. Another type of aliasing results from the fact that a pixel shader is only executed once for each pixel, thus discretely sampling the surface color resulting from the material and BRDF properties. This type of aliasing is referred to as *shader aliasing*. These artifacts can be very displeasing to the eye if the resolution of the image is small relative to the size of the display.

The classic signal processing approach to combat aliasing is to sample at a higher frequency, and then apply a low-pass filter. In rasterization, this roughly equates to rendering to a higher-resolution render target, averaging adjacent pixels, and using the result of the average as the pixel values for a normal-sized render target. In real time 3D graphics, this process is referred to as *super sampling anti-aliasing*, or *SSAA*. Increasing the resolution increases the sampling frequency used for both rasterization and pixel shading, thus effectively reducing the appearance of both types of aliasing. However, the additional performance cost of doubling or quadrupling the render target resolution often makes the technique prohibitively expensive. Because of this, Direct3D 11 includes a more simplified and optimized form of anti-aliasing known as *multisample anti-aliasing* (*MSAA*).

The basic premise of MSAA is that the resolution used for rasterization and the depth/stencil test is increased by an integral factor, but the pixel shader is still executed at the normal resolution of the render target. This allows it to effectively deal with edge aliasing, but not with shader aliasing. Since the depth and stencil tests are performed at a higher resolution, the depth-stencil buffer must store multiple subsamples per-pixel, causing the memory footprint to increase by the MSAA factor. The same applies to the render target, since it must store individual color values for each sample before they can be resolved to a single value. Because of this, the MSAA sample count must be explicitly specified when creating the render target and depth stencil resources, so that the runtime can allocate a sufficient amount of memory. Also, the MSAA sample count must be exactly the same for render targets and depth-stencil buffers that are simultaneously used for rendering. Additional details for creating the resources needed for MSAA can be found in Chapter 2.

Rasterizer interaction with MSAA. When MSAA is enabled for rendering, the first modification to the pipeline is in the rasterization stage. Normally, during rasterization a triangle is tested for coverage using a single point for each pixel as outlined above. In Direct3D 11, this point is located in the exact center of the pixel, or 0.5 units from the upper-left corner in both the X and Y directions. This point is also used for interpolating all vertex attributes output from the previous stage (either the vertex shader, domain shader, or geometry shader, depending on which stages are in use), including the depth used for performing the depth test. When MSAA is enabled, multiple sample points within a pixel are used to test for coverage. The location of these sample points is implementation-specific, although typically these samples are arranged in a rotated grid pattern. An example of the sampling points for MSAA and non-MSAA rasterization are shown in Figure 3.65.

When a depth stencil target and MSAA render target have been bound to the pipeline, the rasterizer behavior is determined by the `multisampleEnable` member of the rasterizer state. This parameter essentially toggles whether the rasterizer performs multiple coverage tests for MSAA render targets. If it is disabled and an MSAA target is bound, then only a single coverage test is performed at the pixel center. If the coverage test passes, the pixel shader result is written to all subsamples. If it does not pass, then none of them are written to. This selectable MSAA operation allows a portion of a scene to be rendered using the multi-sampling functionality, while other portions of the scene that don't require anti-aliasing (such as 2D sprites for a user interface) can be rendered in the normal fashion. This produces a consistent appearance for these scene elements that don't use anti-aliasing, and also improves performance by reducing the number of coverage tests that are needed.

Figure 3.66. The difference between a rendering with and without MSAA enabled.

The difference between primitives rasterized with and without MSAA active is shown in Figure 3.66. You can clearly see the difference in how the edges of the primitive are significantly smoother in the MSAA version (shown on the left) than the non-MSAA version (shown on the right).

3.10.4 Rasterizer Stage Pipeline Output

Fragment Generation

Throughout the rasterizer stage discussion, we have seen each of the processes that are used to convert a primitive into a set of fragments. Several different tests, such as culling, are intended to reduce the number of primitives that need to be processed. In addition, there are also several tests, such as clipping, viewport transform, and the scissor test, that potentially reduce the number of fragments generated by a primitive. It appears that there have been significant amounts of effort put into reducing the number of fragments generated by the rasterizer stage. Why would so many tests be aimed at reducing the output from this stage?

The reason is that the rasterizer stage has a 1-to-many relationship between its input and its output. For each primitive submitted to the rasterizer as input, many fragments are potentially generated. Due to this general data amplification, any processing performed after this point in the pipeline will be executed many more times than processing performed earlier. Therefore, it is clearly better to eliminate unused or unnecessary primitives before they are split into fragments, and it is also better to eliminate fragments in the rasterizer stage before they need to be processed later on. This idea can also be extended to general algorithm design, as well. When you have a choice between performing a calculation before or after rasterization, it is typically more efficient to perform the calculation before generating many fragments. In many cases, even if the calculations done before the rasterizer are only approximations of those that would be done afterward, this can produce a large efficiency gain at a minimal loss of image quality.

Fragment Data

With this in mind, we can consider what type of data stream is produced by the rasterizer stage. We have discussed in detail how each fragment is generated when its location within a render target is covered by a primitive. When it is determined that a location is covered by a primitive, each of the attributes that are passed into the rasterizer are interpolated, using the center of the pixel as the input to the interpolation function.[32] The depth value is also

---

[32] For MSAA rasterization, the sample locations may be used for interpolation as well.

interpolated and passed along with each fragment for use in the depth test of the output merger stage.

The position of the fragment may or may not be added into the generated fragments. This depends on whether the pixel shader input signature declares one of its input attributes to have the SV_Position system value semantic. Prior to the rasterizer stage, the SV_Position value was used to indicate a version of the vertex position. When passed into the rasterizer, it must contain the clip space vertex location. However, the *fragment* location is provided in the output of the rasterizer. This can be used in the pixel shader to look up values in other texture resources, or for performing processing based on the screen location, such as fading the render target color to black at the outer edges of the screen.

3.11 Pixel Shader

After a primitive is converted into fragments by the rasterizer, the fragments are passed to the pixel shader stage. The pixel shader stage is the final programmable shader stage in the rendering pipeline. It processes each fragment individually by invoking its pixel shader program. Each pixel shader invocation operates in isolation—no direct communication is possible between individual fragments being processed.[33],[34] After the pixel shader has completed, the result is a processed output fragment, which is passed to the output merger stage. The location of the pixel shader within the pipeline is highlighted in Figure 3.67.

The pixel shader stage is responsible for the primary appearance of a rendered primitive. Prior to rasterization, processing was primarily focused on manipulating the number of primitives, their size, shape, and attributes. These are all geometric-style operations. During rasterization, these primitives are used to select which pixels are affected by them, and fragments are generated accordingly. The pixel shader can only receive the fragments that are passed to it—it is not possible to change the location of a fragment from within the

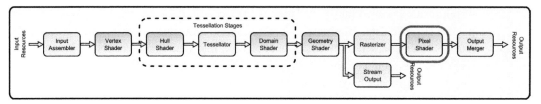

Figure 3.67. The pixel shader stage.

---

[33] The derivative instructions used data shared between multiple fragments.

[34] Unordered access views change this strict isolation, but there is still a fairly strong communication separation between pixel shader invocations.

pixel shader stage. Instead, the pixel shader determines how each of the selected fragments should appear, based on the fragment's attributes, and the information provided to the stage in various resources. This gives the pixel shader a relatively important role to play in the generation of images in real-time rendering.

The pixel shader stage also has some new abilities that were not available prior to Direct3D 11. It can use the new resource view type, the *unordered access view* (*UAV*), to perform read and write operations to the complete resource attached to the view.[35] This is a big departure from the traditional capabilities of the pixel shader stage, which could only write color and depth data into the pixel location that was passed to it. This provides the potential for a large number of algorithms to be implemented directly in the rendering pipeline, such as histogram generation, while the image is being rendered.

3.11.1 Pixel Shader Pipeline Input

The pixel shader stage receives its input fragment from the rasterizer stage. This means that the input attributes that the pixel shader program will use are produced by the rasterizer as well. We have already seen in the rasterizer section that it produces interpolated attribute data, based on the sampling location of a fragment within the primitive being rasterized, which is typically the center of the pixel (although sometimes from other sample locations). We also saw that various interpolation modifier keywords can be specified in the pixel shader program that will instruct the rasterizer to use a particular interpolation mode for each input attribute. These interpolation modes are useful in different scenarios, and must be chosen appropriately to ensure that the input attributes are calculated properly. We will explore these interpolation modes in more detail here.

Attribute Interpolation

The process of interpolating attributes requires three different pieces of information. The first is the data that will be interpolated—the attributes of the vertices of the primitive being rasterized. This includes both their positions and their attribute values. The second piece of information needed for interpolation is the location of the point that requires the interpolated attribute data. This more or less determines how much of each vertex attribute the interpolated value will receive. The final piece of information required is the interpolation technique that should be used. This final item performs the actual interpolation on the first two items described above. The interpolation mode is determined by interpolation modifier declared with the pixel shader input signature.

---

[35] A resource view can be used to expose only a subportion of a resource instead of the complete resource, as described in Chapter 2.

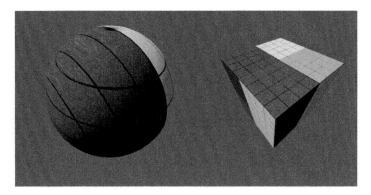

Figure 3.68. A sample of using the `linear` interpolation mode.

The linear mode. When the **linear** interpolation mode is used, the attributes that are generated for each fragment will be linearly interpolated, taking perspective effects into account. The need for perspective-correct interpolation arises from the fact that the perspective projection is not a linear operation, and this interpolation is performed after the projection. This means that standard linear interpolation can't be used between vertices in screen space, but instead, the depth of each vertex must also be taken into account.

The perspective effect can be accounted for in the interpolation process by dividing each vertex attribute by its depth from the viewer before interpolation. The depth value is found for each vertex as its W-value, after projection, but before the conversion to normalized device coordinates. The reciprocal of W is also calculated, and then interpolated, along with these modified attributes. After all of these values are interpolated for each rasterized fragment, the interpolated reciprocal of W is used to extract the original desired attribute, which will produce a perspective-corrected interpolation value. This is the most common interpolation mode, with texture coordinates on a three-dimensional model providing a perfect example of when they are needed. Perspective-correct interpolation is needed not only for texture coordinates, but is also needed if a position or direction vector (or any other linear attribute) from a linear space (such as world space or view space) comes into the post-projection stages. An example of the linear interpolation mode is shown in Figure 3.68.

The noperspective mode. When the **noperspective** interpolation mode is used, the interpolated attributes are strictly interpolated according to their two-dimensional position on the render target. In essence, you can consider the geometry as being projected onto the render target before the interpolation is performed. This interpolation mode implements standard linear interpolation between vertices based on their positions. The interpolation of attributes without perspective correction can be performed whenever the vertices being interpolated all are at the same depth, such as when constructing an onscreen user interface rendering. Figure 3.69 demonstrates the use of this interpolation mode.

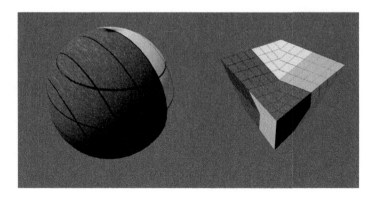

Figure 3.69. A sample of using the noperspective interpolation mode.

The nointerpolation mode. As its name implies, the **nointerpolation** mode does not perform any interpolation of the vertex attributes. This means that the attribute value of the first vertex of each primitive will be passed to all fragments for a given primitive, which produces a constant value over the entire surface of the primitive. This can be used to give a model a faceted appearance when this interpolation mode is used to modify lighting related attributes. In effect, not using interpolation will generally make the faces of the rendered geometry more noticeable, since the whole purpose of using interpolation is to hide the fact that the vertex-based geometry is approximating a smooth surface. This may be minimized when using the tessellation system, but could also be a good way to visualize how finely tessellated the final geometric results are. Figure 3.70 demonstrates the use of this interpolation mode.

The centroid mode. The **centroid** interpolation mode is intended to offer a more appropriate interpolation mode for special cases in MSAA rendering modes. When a pixel is

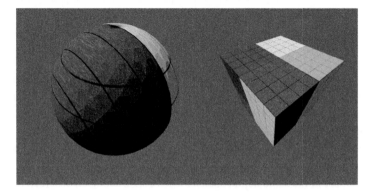

Figure 3.70. A sample of using the nointerpolation interpolation mode.

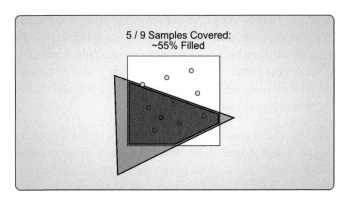

Figure 3.71. Interpolation point for centroid sampling.

partially covered by a primitive and MSAA is enabled, the rasterizer will produce a fragment for that pixel. However, if the primitive only covers a few samples within the pixel, but not the center of the pixel, then the normal **linear** interpolation mode would still use the center of the pixel as the input location to the interpolation function. Because of this, it is possible for the vertex attributes to be extrapolated beyond the true edge of the triangle, in cases in which only a portion of the sample positions are covered. For situations where extrapolation is undesirable, centroid sampling can be used on a per-attribute basis by adding the **centroid** modifier to the input variable declaration. When it is used, centroid sampling causes the affected attribute to be interpolated to a point that is guaranteed to be covered by the triangle. The location used is implementation-specific, but is generally located at the average of the covered sample points. An example of the centroid sampling mode is shown in Figure 3.71.

The sample mode. The **sample** interpolation mode provides interpolation at each of the sample points used in MSAA rendering mode. This is useful when the pixel shader is executed for each MSAA sample to provide attributes at each individual sample location. The topic of MSAA and how it relates to the pixel shader is discussed further in the following section.

Multisample Anti-Aliasing

Several additional system value semantics can be declared as inputs to the pixel shader to offer some interesting possibilities for controlling its behavior when MSAA is enabled. Direct3D 11 offers several ways in which the "standard" MSAA behavior can be altered. These changes are based on the pixel shader bound to the pipeline, and on how it declares its input and output attributes.

Per-sample execution. The first such modification is that pixel shaders can be run for each subsample, rather than once per pixel. This behavior is triggered when the pixel shader takes an input with the **SV_SampleIndex** system value semantic attached, which provides the index of the subsample currently being processed. It is also triggered if an input attribute is marked with the **sample** interpolation mode. We will discuss the usage of the **SV_SampleIndex** later in this section in more detail.

Subsample coverage. The second possible interaction with the MSAA system involves the use of the **SV_Coverage** system value semantic. When an input attribute is declared with this semantic, it provides an unsigned integer value, where each bit corresponds to one of the subsamples of the current pixel. This essentially indicates to the pixel shader which subsamples were detected as being covered by the rasterizer stage during rasterization. We will see later how this system value semantic can also be used as an output attribute as well to perform custom modifications to the selected subsamples that will be written to with the pixel shader's results. This lets custom coverage masks be implemented, which is commonly used for alpha-to-coverage algorithms.

Fragment Location

Two more system value semantics are accessible as inputs to the pixel shader to identify the location of a fragment, which is where it will ultimately be written to. These attributes can be used to aid the pixel shader program in processing a fragment.

Source position. When a pixel shader declares an input attribute with the **SV_Position** system value semantic, the pixel shader receives the four coordinate fragment positions produced by the rasterizer stage. In general, the most important parts of this position are the X- and Y- coordinates, which indicate the location of the fragment within the current render target. In most cases, these coordinates will indicate the pixel center location with the 0.5 pixel offset. This can be used to identify the location in other textures that need to be sampled when multiple textures are used to generate a pixel shader result. A common example of this is the use of multiple textures as a G-buffer for deferred rendering (see Chapter 11 for more details on deferred rendering).

In addition to the pixel center location, it is also possible to use this system value semantic to obtain the centroid position of the current fragment. As described above, the centroid location is guaranteed to be within the primitive boundary, and is normally used to ensure that the coordinates used for interpolation do not extend beyond the primitive's edge. The centroid value used in interpolation can be received if this semantic value is declared with the `centroid` interpolation modifier.

The depth value generated by the rasterizer stage is also available from within the `SV_Position` system value semantic. This is the normalized device coordinate depth, so its value will be in the range of [0,1]. This can be used to perform calculations within the pixel shader that use this depth information, and can also be used to modify the rasterizer-generated depth value with a customized depth value. We will see later in this section how the pixel shader can modify its depth value using the SV_Depth system value semantic.

Fragment destination. In addition to being able to receive the location that the fragment was generated from, the fragment's ultimate destination can also be declared as an input. If there is a standard, non-array render target bound for receiving output from the pipeline, the output position of the fragment is more or less described by the **SV_Position** semantic discussed above. However, if an array-based render target is used to generate multiple simultaneous renderings of a scene, the render target array index must also be used to determine where the fragment will finally be written to. This is commonly used to generate various forms of environment maps, as demonstrated in Chapter 13.

The pipeline determines which render target slice to apply a primitive to by interpreting the SV_RenderTargetArrayIndex system value semantic. This parameter can also be used as an input attribute to the pixel shader. This lets the pixel shader program perform selective processing of a fragment that depends on the render target slice that will ultimately receive the fragment.

Geometric Orientation

The final specialized input attribute that we can declare for input to the pixel shader is the SV_IsFrontFace system value semantic. This parameter provides an indication of the orientation of the primitive that generated the fragment. This is a Boolean attribute that is *true* when the fragment is generated by a front-facing primitive and *false* when it is generated by a back-facing primitive. You may recall that the rasterizer state can be used to determine which vertex winding defines the front-facing direction for a primitive, as well as which of these orientations are to be culled. If both orientations are allowed to be rasterized, this system value will specify which orientation was detected by the rasterizer stage.

Since point and line primitives have no notion of orientation, fragments generated from either of these primitive types will always produce a value of *true* for this input attribute. However, if a triangle primitive is rasterized in wireframe mode (which essentially generates lines from the edges of the triangles) the triangle orientation information is still used.

3.11.2 Pixel Shader State Configuration

With a clear understanding of what data the pixel shader program can receive, we will now consider what types of state configurations are available to the application. The pixel shader stage is a programmable stage, meaning that it has access to the standard common shader core functionality that we have seen throughout the various pipeline stages. Once again, we won't repeat any code listings here, since they are essentially the same as we have seen in the vertex shader section. These common methods are listed below easy reference.

- `ID3D11DeviceContext::PSSetShader()`

- `ID3D11DeviceContext::PSSetConstantBuffers()`

- `ID3D11DeviceContext::PSSetShaderResources()`

- `ID3D11DeviceContext::PSSetSamplers()`

In addition to the standard resources of the common shader core, the pixel shader also has access to the new unordered access views, which provides random access to read and write to resources directly from the pixel shader. Finally, a function attribute called `earlydepthstencil` can be used to force the depth stencil test to be performed prior to the pixel shader. We will investigate these unique configurations in more detail in the following sections.

Unordered Access Views

The pixel shader stage can use utilize unordered access views. As opposed to the shader resource views that provide *read-only* random access to resources, the unordered access views allow *read* and *write* random access. This lets the pixel shader program perform scatter writing operations, meaning that it can programmatically decide where to store the data it writes to a resource. Before having access to UAVs, the pixel shader could only write to the output render target and depth stencil render target in the location determined by the rasterizer stage when it generated the fragment.

The method for binding an unordered access view to the pipeline for use in the pixel shader is performed through the device context, just as it is for all other pipeline configurations. In this case though, the UAV is actually bound to the pipeline in the output merger stage instead of the pixel shader stage. This is done so that the render targets, depth stencil target, and any UAVs that will receive output from the pipeline are all bound for output from the pixel shader in the same method call. This also groups all possible pipeline output resources into a single pipeline stage, which somewhat simplifies the pipeline concept. In the output merger section we will see how resources are bound for use in the pixel shader with a UAV.

Early Depth Stencil Test

The other unique configuration for the pixel shader stage is performed from within the pixel shader program's source code. The `earlydepthstencil` function attribute can be declared before a pixel shader program to indicate that the depth and stencil tests should be performed before the pixel shader is executed (the depth and stencil tests will be covered in more detail in the output merger stage section). These tests are typically implemented to be executed as early in the pipeline as possible by the hardware implementation, but this attribute forces the tests to be performed early.

This is intended to improve the efficiency of a particular algorithm by eliminating fragments that would fail the depth test or the stencil test, and would thus be discarded. By doing so, any pixel shader invocations from fragments that would have failed the depth or stencil tests are prevented. This can be especially helpful when a scene has already been rasterized into the depth buffer and a second rendering pass is needed.

Another, more subtle situation that requires this attribute is when the rendering pipeline is only outputting data to a UAV through the pixel shader. In this case, the data is prevented from being written to the output merger, and hence the pipeline will never execute the depth or stencil tests. However, if these tests are still needed, it must be explicitly forced on with this function attribute.

3.11.3 Pixel Shader Stage Processing

We have now seen the types of information that the pixel shader stage receives from the rasterizer, and what additional resources it can be provided with to augment the available data. In this section, we will first consider some of the mechanics of how the pixel shader stage performs its duties. Next, we will further explore the pixel shader by considering how it can represent an object's material properties in a traditional rendering scenario. After this, we will look at some of the possible uses for the new UAV and consider some of the general operations that this new resource view type enables. Finally, we will complete this section with a discussion of how MSAA interacts with the pixel shader stage, and how to control the various MSAA functionalities to improve image quality while still retaining acceptable performance levels.

Pixel Shader Mechanics

We have seen how data is brought into the pixel shader stage. We can now consider some of the operations that the pixel shader will perform, and exactly what it must produce in a given pipeline configuration.

Pixel shader execution. Our conceptual operation model of the pixel shader stage is that when it is executing, each invocation processes a single fragment and then calculates an output color to pass to the output merger stage. Since the pixel shader operates after rasterization occurs, any calculations performed in the pixel shader are performed proportionally to the number of fragments that an object covers. Regardless of how complex the input geometry is, only the fragments within the viewport that the geometry covers are processed

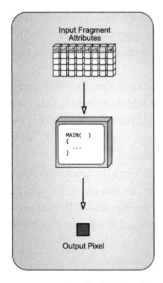

Figure 3.72. A pixel shader being executed to produce a single fragment output.

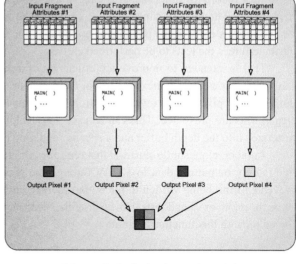

Figure 3.73. Many pixel shader invocations being executed simultaneously.

with the pixel shader. The concept of a single fragment being processed for each pixel shader invocation is depicted in Figure 3.72.

By following this model and restricting communication between threads, many different invocations of the pixel shader can be calculated simultaneously in parallel processing elements. Since each invocation receives its own data from its input fragment, and writes its own output to the location specified by the fragment, it does not rely on neighboring invocations. A schematic concept of this is shown in Figure 3.73.

At a lower hardware level, this is still mostly true, with the exception that pixel shader invocations are always performed in at least 2×2 groups of fragments. This is done to ensure that screen space derivative instructions can be calculated across the invocations in both the X- and Y-directions by performing a discrete difference. The calculation of these derivative instructions relies on the fact that multiple invocations are running at the same time, and if a variable is passed into one of the derivative instructions, then the GPU will use the difference of the neighboring invocations' variables to find how that variable's value is changing over screen space. In practice, this behavior is transparent to the application, and the boundary between pixel shader invocations remains.

Multiple render targets. We have already mentioned that the pixel shader calculates and outputs a color to pass to the output merger stage. However, this is not restricted to a single color. We will see in the output merger stage section that it is possible to bind up to eight

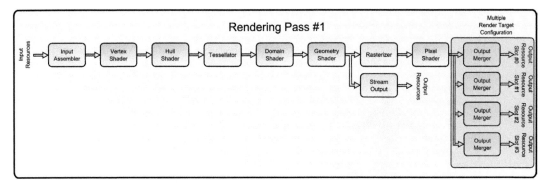

Figure 3.74. The difference between using MRTs and standard single render targets for filling several render targets with scene-based data.

different render targets for receiving output from the pipeline, as long as they meet certain size and format restrictions. A single pixel shader invocation calculates and outputs a color to write to each of these render targets. This ability to use more than one render target simultaneously is referred to as *multiple render targets* (*MRTs*) and provides the potential for improving the efficiency of a rendering algorithm.

If an algorithm requires multiple render targets to be filled with data produced from the scene's geometry, then it is possible to fill each render target one at a time using a traditional single render target configuration. This means that all of the calculations performed in the vertex shader, the tessellation stages, the geometry shader, and the rasterizer are repeated for each pass. The only calculations that are different between passes are those performed by the pixel shader. The use of MRTs allows all of the passes to be combined into a single one, after which the pixel shader writes to all outputs individually. This saves the cost of processing the geometric data multiple times, but still enables the same number of render targets to be filled with the desired information. Figure 3.74 visualizes the difference between using MRTs in this scenario and making multiple rendering passes instead. A very good example of the use of MRTs is provided in Chapter 11, "Deferred Rendering."

Modifying depth values. In addition to being responsible for writing the color values for a given fragment, the pixel shader also can output a new depth value to the **SV_Depth** system value semantic. If the pixel shader does not write the depth value, the depth generated in the rasterizer stage is passed to the output merger stage. However, in some cases it is more appropriate for the pixel shader to specify a depth value instead. A scenario where this would be useful is when using *billboards*. A billboard is commonly used to simulate more complex geometry with two triangles arranged to form a quad, which is then aligned to be perpendicular to the current view direction. A texture is applied to the quad, and the results

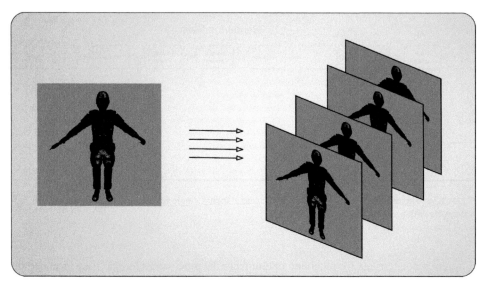

Figure 3.75. Using a billboard to simulate more complex geometry. Model courtesy of Radioactive Software.

of rendering just the two triangles still provide a significant amount of detail.[36] The concept of billboards is illustrated in Figure 3.75.

A billboard provides an increased amount of detail with very simple geometry, which is an efficient way to introduce complexity to a scene. However, since a billboard is essentially flat, when it is used in situations where it intersects other scene geometry, the illusion of the complex geometry is spoiled, because the billboard has a uniform depth across its surface. In this scenario, if the depth variations of the simulated geometry were included in the billboard texture (in the alpha channel for example), the pixel shader could write a modified depth value to the fragment, thereby reintroducing depth complexity to the billboard geometry. Then, when scene geometry intersects the billboards, they will actually appear more convincing, with partial occlusion rather than complete occlusion. An example of this type of depth modification is shown in Figure 3.76.

Conservative depth output. However, writing depth values from the pixel shader does have some drawbacks. Most modern GPUs implement an efficiency improvement technique referred to as *Hierarchical-Z Culling*, or *Hi-Z*. The concept behind this technique is for the GPU hardware to perform some simplified forms of occlusion tests to see if the current batch of geometry will be able to be seen in the final render target, or if it would appear behind another object in the rendering. If the geometry would be occluded, then it is simply

---

[36] Billboards are used in the particle system sample discussed in Chapter 12, "Simulations."

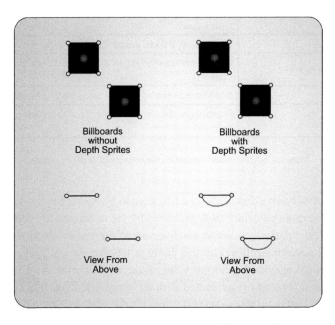

Figure 3.76. An overhead view of how depth can be added to a billboard to allow proper depth testing.

discarded before any further processing. This reduces the overall computational cost of the pipeline execution when there are many objects that overlap within a scene. The details about how this is implemented vary by GPU manufacturer, so they will not be covered in detail here. However, some of the assumptions made about the size of the geometry, which are used during this hierarchical testing, occur prior to the execution of the pixel shader. Therefore, when the depth value is modified in the pixel shader, Hi-Z may be unusable and its performance benefits can be lost.

To allow Hi-Z to remain active for a certain subset of algorithms that require depth output, Direct3D 11 introduces a new feature known as *conservative depth output*. This technique works by requiring pixel shaders to specify an inequality function along with the new depth value. The inequality effectively specifies an upper or lower bound on the depth output, which allows Hi-Z to continue to identify fragments that can be trivially rejected and thus don't require the pixel shader to be executed.

To have a pixel shader make use of conservative depth output, the shader program must assign one of four new system value semantics to the value used for outputting depth. These semantics each specify the inequality as part of the semantic name, and the depth value must satisfy that inequality relative to the interpolated depth value calculated by the rasterizer stage. If the pixel shader outputs a depth value that fails to satisfy the inequality, the runtime will automatically clamp the value to the appropriate minimum or maximum value. The four semantics are listed in Table 3.3.

Semantic	Description
SV_DepthGreater	Depth output must be greater than the interpolated depth value
SV_DepthGreaterEqual	Depth output must be greater than or equal to the interpolated depth value
SV_DepthLess	Depth output must be less than the interpolated depth value
SV_DepthLessEqual	Depth output must be less than or equal to the interpolated depth value

Table 3.3. System value semantics for conservative depth output.

Whether conservative depth allows a fragment to be rejected by Hi-Z depends on the inequality, the interpolated depth value produced by the rasterizer, the depth testing function specified in the depth-stencil state, and the current value in the depth buffer. For rejection to occur, all possible depth values that could satisfy the inequality must fail the depth test when compared against the current value in the depth buffer. As a simple example, let's suppose that a fragment is generated with an interpolated depth of 0.75, the current value in the depth buffer for that pixel is 0.5, and D3D11_COMPARISON_LESS is specified as the current depth comparison function. If a pixel shader without depth output is used, the Hi-Z hardware will determine that 0.75 is greater than the current depth value, and that

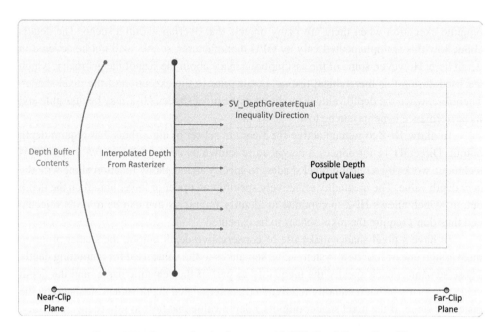

Figure 3.77. Conservative depth output with "SV_DepthGreaterEqual."

the fragment should be rejected. If a pixel shader with conservative depth output is used that specifies SV_DepthGreaterEqual for the depth output attribute, the Hi-Z hardware will still be able to reject the fragment. This is because the inequality guarantees that the depth output could only become larger, and consequently, it will always fail the depth test. Figure 3.77 illustrates this scenario.

Note that if SV_DepthLess or SV_DepthLessEqual were used instead of SV_DepthGreaterEqual, the Hi-Z unit would not be able to reject the fragment. This is because it would be legal for the pixel shader to output values less than the current depth buffer value of 0.5, causing the fragment to pass the depth test. In fact, if SV_DepthLess or SV_DepthLessEqual are used with D3D11_COMPARISON_LESS, there are no cases where the Hi-Z unit would be able to reject fragments. The same applies to using SV_DepthGreater or SV_DepthGreaterEqual with D3D11_COMPARISON_GREATER. Consequently, a conservative depth semantic with inequality direction *opposite* of the depth test direction should always be used for best performance.

A Small Example

Now we move on to consider how the pixel shader stage is used to implement a rendering model. The primary responsibility of this stage is to produce the output color that will be merged into the render target bound at the end of the pipeline. Thus, the ultimate decisions the developer must make are what type of calculation to perform, and what data inputs to use for that calculation. The types of algorithms can range from a single color output for all pixels that are rendered, all the way to a complete global illumination system that requires several simulation steps before the final rendering can be performed.

To explore how various algorithms can be implemented, we will first examine one simple rendering example and see how the pixel shader is used to formulate the rendered color output. Along the way, we will explain how this example scenario represents the basic methodology that the pixel shader operates with. By taking a high-level view of these concepts, we will be able to implement a wide variety of algorithms, including some that aren't even used to generate color data such as shadow maps. The details of the implementation aren't as important as the details of how processing is performed within the stage.

As mentioned above, the pixel shader is responsible for producing the final color values that will be merged into the render targets at the end of the pipeline. In this sample scenario, the color determination will be a product of the object's material properties, as well as of some environmental properties that the material interacts with, such as the presence of lights. An object with the same material properties can appear quite different when viewed in different lighting conditions, just as models with differing material properties can appear different even if they are being rendered in the same lighting conditions. Thus, both sets of properties must be made available to the pixel shader program to properly determine what color to produce.

Material properties. The material properties of an object determine how it looks at a basic level. Typical material properties include things like the base color of an object, some coefficients to describe how reflective the surface is, and perhaps a texture map that represents the fine surface color variations. We will assume that the pixel shader receives the vertex normal vector and texture coordinates as interpolated input attributes. The pixel shader program will calculate the color of the object's material with the function shown in Listing 3.23.

```
float4 PSMAIN( in VS_OUTPUT input ) : SV_Target
{
    // Determine the color properties of the surface from a texture
    float4 SurfaceColor = ColorTexture.Sample( LinearSampler, input.tex );

    // Return the surface color
    return( SurfaceColor * input.color );
}
```

Listing 3.23. A sample pixel shader to calculate the color of an object's material.

In a typical scene, many different objects need to be rendered, and each one will require at least partially different properties from the others. For this example, we will assume that the pixel shader is common among each of the objects being rendered. With this in mind, we can step through a few object rendering sequences and consider what is different between each one with respect to the pixel shader. We will assume there are three objects in the scene, called objects A, B, and C. For each of these objects to be rendered, the pipeline must be configured with the object properties prior to execution of the pipeline with one of the draw calls.

When the pipeline is configured to render object A, the object's texture is bound to the pipeline with a shader resource view, and its color is specified by binding a constant buffer to the pixel shader stage containing the appropriate color. Each of the generated fragments is passed to the pixel shader, where the texture is sampled and multiplied by the color supplied in the constant buffer, as shown in Listing 3.23. After this pipeline execution completes, the application would configure the pipeline for rendering object B. This would require binding a different texture and color constant buffer, followed by pipeline execution. Object C would follow the same pattern, but in this case, it would use the same texture as object B, and only a new color constant buffer would be needed.

This short sequence demonstrates that an object's material properties are typically supplied as resources external to the shader program itself. Both the color and texture of a model are controlled by the application manipulating the pixel shader stage's states, rather than by swapping the pixel shader program in and out. In addition, the geometry that is passed through the pipeline doesn't really matter to the material properties. In all three cases, the input geometry could have been exchanged between objects, and the resulting appearance of the material would have remained the same. This is because the pixel shader operates after the geometric data in the pipeline has been converted into a rasterized form. It processes data after rasterization, and so is not affected by changes in geometry. While keeping these properties in mind, we will next consider how to have our example objects interact with their environment by adding lighting to the example.

Lighting properties. By providing lighting information in a scene, we can significantly increase its quality. Lighting is a very fundamental part of how we see the physical world, and using it significantly improves generated scene renderings. Many different types of lighting representations are used in modern real-time rendering, but for the purposes of this sample, we will use a simple directional lighting model. The light will be described by a direction vector and a color, both of which will be provided in a second constant buffer. The amount of light that reaches a given surface is commonly approximated by taking the dot product of the surface normal vector and the vector representing the light's direction of travel. This produces a scalar value in the range of [0.0, 1.0] (as long as both the normal vector and the light vector are normalized) and can be used to scale the amount of light applied to a surface. A function that performs this operation is provided in Listing 3.24.

```
float4 PSMAIN( in VS_OUTPUT input ) : SV_Target
{
    // Normalize the world space normal and light vectors
    float3 n = normalize( input.normal );
    float3 l = normalize( input.light );

    // Calculate the amount of light reaching this fragment
    float4 Illumination = max(dot(n,l),0) + 0.2f;

    // Determine the color properties of the surface from a texture
    float4 SurfaceColor = ColorTexture.Sample( LinearSampler, input.tex );

    // Return the surface color modulated by the illumination
    return( SurfaceColor * input.color * Illumination );
}
```

Listing 3.24. A function that performs a simple lighting equation.

Now we can return to our sample renderings of objects A, B, and C. The rendering sequence will be repeated, except that the lighting information must be supplied to the pixel shader in a second constant buffer. Since each of these objects resides in the same scene, they all utilize the same light description. This means that the same constant buffer can be reused for all of the pipeline executions. To carry out the example, the pipeline is configured for each object in the same manner as before, with the addition that the lighting constant buffer is also bound. When the pixel shader executes, the object's material color is found in the same way as before, and the amount of light visible at each pixel location is calculated, based on the normal vector passed into the pixel shader as an input attribute and the light direction vector. The result of the lighting calculation is then used to modulate the color of the fragment. As each additional object is rendered, the process is repeated exactly with each of its material properties, and the resulting rendering now incorporates lighting in addition to material colors.

In this case, we see that the lighting information is the same throughout the scene and hence is applied to each of our three objects in the same way. Environmental data is the same for all objects that share the same environment. We also see that even though each of the three objects has a different material appearance, they all interact with the light in the same way, using the same calculation regardless of what color they are.

Generalizing the example. So what have we learned from this example, and how can we apply the results of these simple experiments to understand the general concept of using the pixel shader to implement rendering techniques? The pixel shader program is simply a function that takes a certain number of input arguments and produces a color. Some of the inputs are changed at every pipeline execution, such as material properties, and some of the inputs are changed once per rendered frame, such as the lighting properties. Still others won't change at all throughout an application's lifetime, such as the vertex normal vectors of a model.

The flexibility provided by the pixel shader becomes quite clear now. It does not really matter what the function is that calculates the color of a generated fragment. As long as it produces a color result that varies appropriately when the input is varied, the rendering model serves its purpose. Developing a different rendering model revolves around deciding what inputs should be used to calculate the output color, developing the function that carries out the mapping from input to output, and then producing the geometric content that fits into the rendering model. More complex rendering models may require more inputs, including the possibility of using dynamically generated inputs such as the result of additional rendering passes. However, the pixel shader itself always resolves to a way to convert the input data to an output color.

Using Unordered Access Views

The previous example demonstrated how the pixel shader can be used to implement a rendering model. In this section, we consider what kind of additional possibilities are made available by the inclusion of the unordered access views to the pixel shader. The UAVs allow the various resource types to be read or written at any location, by any invocation of the pixel shader. Before the addition UAVs, the pixel shader stage was restricted to only reading from resources, with the exception of writing its output color(s) and depth to the fragment location.

This new ability to write to any location of a resource can be used to extract additional information from the rendering process. For example, the generation of a histogram can be performed simultaneously with the rendering of the scene. A histogram provides a number of *bins*, where each bin represents the number of pixels that have a value that resides within a particular range. A histogram provides an indication of the distribution of the values in an image. By examining the output color of a pixel shader, the histogram bin that the color falls into can be determined. Then, a resource can be accessed using a UAV, and the appropriate bin value can be incremented. This would not have been possible prior to the introduction of UAVs.

Even more complex rendering algorithms can be implemented with the use of UAVs. Since read and write access to resource is available, generic data structures can be implemented using UAVs. This ability has been used to build linked list implementations out of GPU resources, which can then be used as building blocks for other algorithms.

Multisample Anti-Aliasing Considerations

Since the pixel shader operates very close to the final rendered image, several special pixel shader functionalities should be considered when using MSAA. With non-MSAA rendering, the coverage test performed for each pixel's sample point determines if the pixel shader stage should be executed for that pixel. With normal MSAA rendering, the pixel shader is still only executed once per pixel. Thus, the pixel shader will be executed as long as at least one of the sample points passes the coverage test. After the shader is executed, the value(s) output by the shader are written to the corresponding subsamples in the currently bound render target(s).

Sampling MSAA textures in shaders. In Direct3D 11, shader resource views can be created for MSAA textures, so that the textures can be bound as inputs to the various shader stages. This is done by setting **D3D11_SRV_DIMENSION_TEXTURE2DMS** as the **ViewDimension** member of the **D3D11_SHADER_RESOURCE_VIEW_DESC** structure passed to the **ID3D11Device ::CreateShaderResourceView** method. However, when the view is bound to the pipeline for a stage, that stage cannot sample the texture using the normal **Texture2D** object and its various methods. Instead, a **Texture2DMS** object must be declared. This object supports the **load** method, which allows you to specify both XY position and the index of the subsample that you want to retrieve. It also supports the **GetDimension** method for querying the XY dimensions and the number of subsamples in the texture, as well as the **GetSamplePosition** method for querying the pixel coverage sample point used for a specified subsample index. Using these methods in a pixel shader allows for custom resolve algorithms to be implemented, rather than relying on what the hardware provides.

Alpha-to-coverage. **Alpha-to-coverage** is a technique that modifies how a pixel shader's output is written to the render target. When enabled by setting the **AlphaToCoverageEnabled** member of the **D3D11_BLEND_DESC** structure to *true*, the output of the pixel shader is no longer written to the various subsamples solely based on the results of the rasterization stage coverage test. Instead, the mask generated by the coverage test is bitwise AND'ed with a mask generated from the alpha component of the value output from the pixel shader. This result is AND'ed with a screen-space dither mask, which causes the surface to be rendered with a fixed dithering pattern across the subsamples. The most common use of this feature is to implement a simplified version of transparency, where transparent surfaces are dithered based on their opacity, rather than blended. This provides a performance advantage, since the pixel output does not need to be blended with the render target contents, and also because it does not require transparent surfaces to be sorted based on their distance from the camera.

Using alpha-to-coverage doesn't explicitly require that MSAA be enabled. However, the quality is significantly improved when used in conjunction with MSAA, since it allows dithering to occur at the subsample level, rather than at the pixel level, providing a higher output image quality.

Pixel shader behavior modifications. Direct3D 11 offers several ways in which the "standard" MSAA behavior can be altered. These changes are based on the pixel shader bound to the pipeline, and how it declares its inputs and outputs.

Per-sample execution. The first such modification is that pixel shaders can be run for each subsample, rather than once per pixel. This behavior is triggered when the pixel shader takes an input with the **SV_SampleIndex** semantic attached, which provides the index of the subsample being currently shaded. This index corresponds to the index passed to **Texture2DMS.Load**, making it simple to retrieve the appropriate subsample from another MSAA render target. The per-sample behavior is also triggered if the "sample" modifier is attached to an input, which causes the attribute to be interpolated to the current sample point, rather than the pixel center.

Having the pixel shader run at a higher frequency means that shader aliasing can be reduced, in addition to edge aliasing, or that special behavior can be implemented for non-traditional rendering pipelines. However, it should be noted that having the pixel shader execute at sample frequency means that performance for those pixels is effectively reduced to that of supersampling, since the pipeline is essentially the same. Therefore, it is usually beneficial to only use per-sample shaders for a subset of the geometry rendered, or to use the stencil buffer to mask out pixels that don't require per-sample shading. For an example using per-sample pixel shaders, see Chapter 11, "Deferred Rendering."

Depth output. As described previously, the pixel shader allows a value to be manually specified for depth testing and depth writes, using the **SV_Depth** semantic. When MSAA is enabled and the shader is not executed per-sample, outputting to **SV_Depth** causes a single value to be used when testing depth for each sample point. This can cause MSAA to function incorrectly for pixels that are overlapped by multiple triangles.

Custom coverage masks. Under normal conditions, the value output from a pixel shader is based on the coverage mask generated by the coverage test performed in the rasterizer stage. However it is possible for pixel shaders to instead output their own coverage mask instead, using the **SV_Coverage** semantic. The value output is a *uint*, where each bit corresponds to a subsample in the render target. Setting a bit to 1 causes the output value to be written to a subsample, while setting it to 0 prevents it from being written. The obvious use of this semantic is for implementing a custom mask for alpha-to-coverage transparency, rather than relying on the fixed screen-space mask implemented in most hardware. Another possible use is for implementing MSAA is a *deferred renderer*, which is explored further in Chapter 11, "Deferred Rendering."

3.11.4 Pixel Shader Pipeline Output

Since the pixel shader stage sends its output to the end of the pipeline, the output merger stage, there are limitations on the output attributes that it can write to. The output merger can only process the color values and depth described above, and no additional information can be received. This is enforced by the shader compiler, which requires that any output attributes be limited to those system value semantics that represent color and depth, namely SV_Target[n] and SV_Depth.

The only exceptions to these attributes are the SV_DepthGreaterThan, SV_DepthLessThan, and SV_Coverage semantics. The first two provide the mechanism to continue to enable the hierarchical *z*-culling algorithm when the SV_Depth attribute is written to. Therefore, it is not useful to write either of these semantic attributes if the depth is not manually modified in the pixel shader. The SV_Coverage semantic makes it possible to use customized subsample coverage masks, as detailed above. Since it specifies a pattern to be used in an MSAA render target, it also should only be used in situations that use MSAA render targets.

3.12 Output Merger

The final stop in the pipeline is the output merger stage. This is a fixed function stage that receives color and depth results from the pixel shader stage, and then merges those results into the render targets bound to it for output. However, this stage provides significantly more functionality than simply writing color and depth values to resources. The output merger also performs visibility determination with the depth test, which implements the traditional Z-buffer algorithm. In addition, it can also perform a stencil test to precisely control which areas of the render target are written to. We will examine how these two

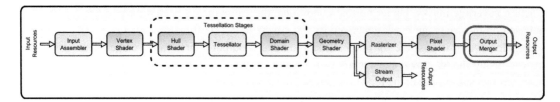

Figure 3.78. The output merger stage.

tests function in more detail later in this section. The location of the output merger stage is highlighted in Figure 3.78.

The output merger is also capable of modifying the color values that are passed on to it from the pixel shader through the blending function. This configurable functionality allows for a variety of different blending modes, which can be used to combine the current pixel shader results with the existing contents of the render target. This blending functionality, coupled with the ability to handle MSAA render targets, multiple render targets, and render target array resources makes the output merger play an important role in producing the final rendered image.

3.12.1 Output Merger Pipeline Input

Because it is the final stage in the pipeline, input to the output merger stage is the complete set of data produced by the pixel shader stage. The primary input for the output merger is the color values that were produced by the pixel shader program. The number of color values received depends on the output signature declared in the pixel shader, but in general, the number of output colors will match the number of render targets that have been bound to the output merger for an MRT configuration. There is one exception to this, which is when two colors are output from the pixel shader for use in dual-source blending. This concept is discussed further in the blending portion of the "Output Merger Stage Processing" section.

The second major input that the output merger stage receives is the fragment's depth value. This value is received either from the rasterizer stage, or the pixel shader stage if it is manually modified in the pixel shader program. These depth values are the input to the depth test, and will ultimately be used to update the depth buffer contents.

As described in the "Pixel Shader Pipeline Output" section, both of these values are output for each subsample when MSAA is used. In this case, an additional input system value semantic can be received from the pixel shader. The SV_Coverage semantic indicates which subsamples will be written to by the pixel shader output. This subsample selection process is performed automatically in the output merger stage and requires no additional effort by the application, but it is still important to understand how the output data will eventually end up in a render target.

3.12.2 Output Merger State Configuration

The output merger stage is controlled by a pair of state objects—the depth stencil state and the blend state. In addition to these, the output merger can also have resources bound to it, which ultimately represent the final output of the rendering pipeline. In this section, we will consider how each of these properties is configured, and what restrictions there are for handling each possible state and resource.

Depth Stencil State

Both the depth test and the stencil test are configured with the same state object, the `ID3D11DepthStencilState`. As with all resources in Direct3D 11, this object is created through the device interface. The depth stencil state is created with the `ID3D11Device::Create DepthStencilState()` method, which takes a pointer to a description structure that contains the desired state options. Listing 3.25 lists the members of the depth stencil state description.

```
struct D3D11_DEPTH_STENCIL_DESC {
    BOOL                        DepthEnable;
    D3D11_DEPTH_WRITE_MASK      DepthWriteMask;
    D3D11_COMPARISON_FUNC       DepthFunc;
    BOOL                        StencilEnable;
    UINT8                       StencilReadMask;
    UINT8                       StencilWriteMask;
    D3D11_DEPTH_STENCILOP_DESC  FrontFace;
    D3D11_DEPTH_STENCILOP_DESC  BackFace;
}
```

Listing 3.25. The D3D11_DEPTH_STENCIL_DESC structure and its members.

The first three parameters of the structure configure the depth test functionality, while the remaining parameters are used to configure the stencil test. Once the depth stencil state has been created, it is immutable and cannot be changed. If the application requires another state to be active, a separate state object must be created and bound to the pipeline in the place of the current state. This is accomplished through the `ID3D11DeviceContext::OMSet DepthStencilState()` method. As usual, there is a corresponding get method to retrieve the current state that is bound to the pipeline. The details of what these members do are discussed in more detail in the "Output Merger Stage Processing" section.

Blend State

The second state object that the output merger supports is the `ID3D11BlendState`. This blend state object controls how color values are blended together before being written to the output render targets. This stage object is created with the `ID3D11Device::Create` `BlendState()` method, which takes a pointer to a description structure that specifies what configuration the state object will represent. Listing 3.26 shows the contents of the blend state description.

```
struct D3D11_BLEND_DESC {
    BOOL                        AlphaToCoverageEnable;
    BOOL                        IndependentBlendEnable;
    D3D11_RENDER_TARGET_BLEND_DESC RenderTarget[8];
}
```

Listing 3.26. The D3D11_BLEND_DESC structure and its members.

The first parameter determines if the alpha value of the output color will be used to determine the subsample coverage instead of the rasterizer coverage values, or the `SV_Coverage` system value semantic if the pixel shader writes to it. This can be used to implement a form of transparent rendering without needing to first sort the objects in the scene. The second and third members of this structure are used in conjunction with one another. The `IndependentBlendEnable` parameter indicates if there will be a separate blending mode specified for each render target when multiple render targets are used. If the parameter is *true*, each of the eight elements in the `RenderTarget` array provides the blending state for its respective render target. If independent blend is *false*, the element at index 0 will be used for all render targets. Once the blend state object has been created, it is bound to the pipeline through the device context with the `ID3D11DeviceContext::OMSetBlendState()` method. And as always, there is a corresponding get method to retrieve the state currently bound to the pipeline. We will explore the details of the blending configurations in the "Output Merger Stage Processing" section.

Render Target State

The render target state of the output merger represents the actual resources that receive the results of all the rendering calculations performed throughout the pipeline. A relatively large number of different configurations can be used to receive this output, and of course there are also restrictions on what combinations of render targets can be used together. We will begin by discussing the various types of render targets that can be bound to the output merger and then see how they are manipulated by the application.

The output merger has eight render target slots and one depth stencil slot. Each render target is bound to the pipeline through a render target view (*RTV*), and the depth stencil target is bound through a depth stencil view (*DSV*). In a traditional rendering configuration, a single render target is bound with a single depth target. However, this is not a requirement.

The application may bind from 1 to 8 total render targets simultaneously for generating multiple versions of the same rasterized scene. These render targets must match in type (such as `Texture2D`, `Texture2DArray`, etc...) and size (including width, height, depth, array size, and sample counts), but may have different formats from one another. When more than one render target is bound, it is referred to as a multiple render target (*MRT*) configuration.

This arrangement also uses a single depth stencil target, even though there are multiple render targets in use. Since the depth stencil target is used to carry out the depth and stencil tests, it must be the same type and size as the resources bound as render targets, but its format will be one of the depth stencil formats. As indicated above, these size-matching requirements hold true for MSAA render targets as well. If an MSAA render target is bound for output, then the depth stencil target must also be an MSAA resource with the same number of subsamples. In addition, the array size of a resource must match between the render and depth stencil targets. You may recall from Chapter 2 that a resource can be created as an array resource, where a number of texture slices are created within the same resource. If a render target has six array slices to produce a cube-map, then the depth stencil target must also have six array slices.

One final configuration is also possible that doesn't use a render target at all. It is possible to bind only a depth stencil view to the pipeline. In this case, each render target slot is emptied by binding a NULL value to it, while the depth stencil view is bound as it is in a regular rendering configuration. This is commonly used to fill the depth stencil target with the depth information of the scene before proceeding with additional rendering passes.

MRT vs. render target arrays. A distinction should be noted between how MRTs and render target arrays differ. It is possible to create eight individual texture resources and then bind them to the output merger stage for use in an MRT configuration. It is also possible to create a single texture array resource that has eight texture slices and then bind that resource to the output merger for use in a single render target configuration. Even though they would provide the same number of effective render targets, there are several differences in the mechanics of using these configurations and in their capabilities. An MRT configuration uses more than one render target slot—one for each render target to be used. On the other hand, an array-based resource only occupies a single render target slot in the output merger. This produces another difference between the two configurations—MRT setups write to all of their render targets simultaneously, while array-based setups only write to one slice at a time. Since the pixel shader can write to all MRT render targets simultaneously, only a single pixel shader invocation is needed to write to all targets. Since the pixel shader can only write to a single render target in the array-based configurations, it would take one pixel shader invocation to write to each of its texture slices.

This situation seems to indicate that MRT configurations are a better choice, since this can reduce the number of pixel shader invocations, while still writing to the same

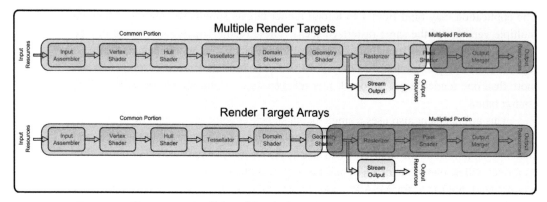

Figure 3.79. The conceptual splitting of the pipeline with MRTs and array-based resources.

number of render targets. However, render target array resources have several advantages, as well. In a sense, we can say that MRT configurations use the same rasterization for all of their render targets. This means that if a triangle is rasterized to the upper-right-hand corner of a render target, then it will appear in the same location for all of the render targets that are written to. Render target arrays each use an individual rasterization that can be customized, based on the `SV_RenderTargetArrayIndex` system value semantic. This allows different view transformations to be used for the geometry before rasterization, and thus also allows primitives to be rasterized to different locations within the render targets.

We can think of these two configurations as splitting the pipeline to ultimately write to multiple render targets, but they split it in different locations. The MRT configurations split the pipeline at the pixel shader stage, while array-based configurations split the pipeline prior to the rasterizer stage. Both configurations can be useful in different situations, depending on what type of pipeline output is needed. This difference is depicted in Figure 3.79.

There is still another facet to consider regarding the binding of render and depth stencil targets. We mentioned above that these resources are bound with resource views—RTVs and DSVs. Since a subresource of a resource can be specified by a resource view, it is possible to have a larger resource bound to the pipeline through a resource view that works with a smaller portion of that resource. When this is considered along with all of the other options and configurations mentioned above, the developer has a large variety of different options available to create specialized rendering algorithms.

Binding render targets. Both the render targets and the depth stencil target are bound to the output merger in a single device context method call. The **ID3D11DeviceContext ::OMSetRenderTargets()** method (shown in Listing 3.27) takes a pointer to an array of pointers to render target views and a pointer to a depth stencil view in addition to an integer specifying the number of targets in the array. The pipeline will hold a reference to the render targets after they are bound, and it will keep the reference until the render target is replaced with a NULL reference.

One other resource type can be bound to the output merger stage. We have seen in the section on the pixel shader stage that it can use using unordered access views (UAVs). However, the pixel shader stage cannot accept UAVs for binding. Instead, they are bound to the output merger stage in the same way that render targets are. When UAVs are used, they are bound with the `ID3D11DeviceContext::OMSetRenderTargetsAndUnordered AccessViews()` method. The first three arguments of this method are identical to the render-target-only version, while the remaining four are used to bind the UAVs. There are a total of eight UAV slots available, and the range of slots affected by this call is selected with the `UAVStartSlot` and the `NumUAVs` parameters. The `ppUnorderedAccessView` parameter is a pointer to an array of UAVs to be bound. Naturally, the number of UAVs in this array must match the number of views specified in the `NumUAVs` parameter.

The final parameter to this method is another pointer to an array of UINT values. These values provide the current buffer counter value for use in Append / Consume buffers. As described in Chapter 2, buffers used with a UAV contain an internal counter indicating the number of elements present in the buffer. This counter is hidden in the buffer and is maintained by the runtime. However, the value stored in these counters at the time of binding is controlled by what is passed into this array. This lets the application effectively reset the data in the buffer if desired, or if a value of -1 is passed, the current internal buffer counter value is maintained.

The total number of render targets and UAVs being bound must not exceed eight. However, any combination that adds up to eight or less is allowed—including using eight UAVs and no render targets. If there are no render targets bound to the pipeline, the UAVs represent the only outputs for the entire pipeline. Both methods for binding render targets to the output merger are shown in Listing 3.27.

```
void OMSetRenderTargets(
    UINT NumViews,
    ID3D11RenderTargetView **ppRenderTargetViews,
    ID3D11DepthStencilView *pDepthStencilView
);

void OMSetRenderTargetsAndUnorderedAccessViews(
    UINT NumViews,
    ID3D11RenderTargetView **ppRenderTargetViews,
    ID3D11DepthStencilView *ppDepthStencilView,
    UINT UAVStartSlot,
    UINT NumUAVs,
    ID3D11UnorderedAccessView **ppUnorderedAccessView,
    const UINT *pUAVInitialCounts
);
```

Listing 3.27. The device context methods for binding resources to the output merger stage.

Read-Only Depth Stencil Views

One other new feature in Direct3D 11 is the ability to use a depth stencil resource as a depth/stencil target in the output merger stage, and to also simultaneously view its contents through a shader resource view in one of the programmable shader stages. At first thought, this would seem to violate the rule that a resource cannot simultaneously be read from and written to from different resource, since the depth stencil resource would be written to in the depth test portion of the output merger.

However, if the depth stencil view is created with the appropriate read-only flags,[37] this ensures that the resource is not written to by the output merger stage. Instead it is only read by the output merger to determine if the depth test has passed or failed (if the depth test is currently enabled). This effectively makes both of the pipeline locations that the re-source is bound to read-only access points, which still follows the simultaneous read/write rules. This configuration is depicted in Figure 3.80. This is useful because a secondary rendering pass can use the depth buffer contents without having to copy them into another resource, while at the same time the resource is also being used for the depth test.

3.12.3 Output Merger Stage Processing

The output merger stage performs two types of operations—visibility tests and blending operations. The visibility tests provide configurable operations that are used to determine if each particular fragment should be blended into the output resources attached to the output merger. If the fragment passes both the depth test and the stencil test, it is passed to the blending function to be combined with the render targets. The blending function is also configurable, which allows for a wide variety of methods to apply pixel shader out-puts to the render target. In this section, we follow the path that a fragment would follow through the output merger stage. We first discuss each of the two visibility tests, to better understand how the tests operate and what modifications can be made to them. Next, we investigate how the blending function performs its duty and examine the available modes for blending.

Visibility Tests

Two visibility tests are performed simultaneously by the output merger stage for each frag-ment passed on to it: the depth test and the stencil test. Both of these tests use the depth stencil resource that is bound to the output merger stage with a depth stencil view (DSV), and require an appropriate data format to be selected for the depth stencil resource. The

---

[37] Two flags are available, D3D11_DSV_READ_ONLY_DEPTH and D3D11_DSV_READ_ONLY_STENCIL, which correspond to the depth and stencil portions of a depth/stencil target, respectively.

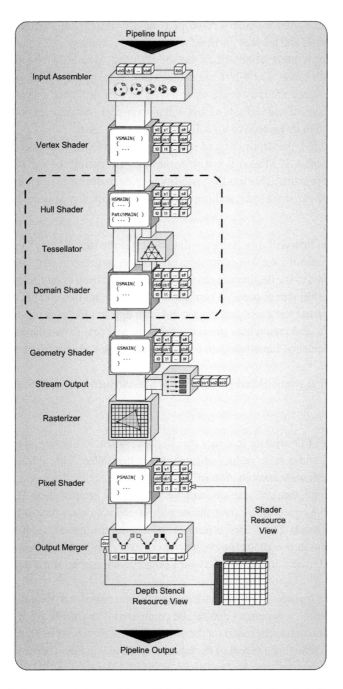

Figure 3.80. A single resource being used in multiple locations of the pipeline, including a depth stencil view in the output merger.

formats that can be used for such a resource typically contain one portion for storing the current depth value, and another portion for storing the current stencil value. One example of such a format is `DXGI_FORMAT_D24_UNORM_S8_UINT`, where 24 bits are dedicated to holding the depth value, and 8 bits are dedicated to holding the stencil value. There are some formats that contain only depth values, with no data dedicated as a stencil buffer. In these cases, the stencil test is disabled and will always pass. We will see how each of these quantities is used in its respective visibility determination tests.

When enabled, the depth test and stencil test are performed for every fragment sent to the output merger. If MSAA is enabled, the tests are performed for every subsample produced by the pipeline. This lets the visibility tests be performed with the same granularity that the rasterizer stage uses, and improves the image quality at geometry intersection areas.

Stencil test. The first visibility test we will consider is the stencil test. This test allows an application to perform a masking operation on a render target to control when a particular fragment is written to the destination render target. The test itself is quite configurable, with a relatively large number of possible configurations. There are two components to this test. The first is the actual test itself, and the second is an update mechanism used to update the stencil buffer. To understand how the test works, we will first define the equation that the stencil test implements. Equation (3.6) provides the pseudocode for the stencil test.

$$(\text{StencilRef} \ \& \ \text{StencilMask}) \ \text{CompFunc} \ (\text{StencilBufferValue} \ \& \ \text{StencilMask})$$

$$(3.6)$$

Equation (3.6) evaluates to either *true* or *false*, where a *true* result indicates that the test has passed, and a false result indicates that it has failed. The individual arguments of the stencil test are a combination of the configurable states discussed in the "Output Merger State Configuration" section, and we will review each of them here. Starting on the left side of the equation, we have the stencil reference value, which is bitwise-ANDed with the stencil mask. The stencil reference value is an unsigned integer value provided by the application in the `ID3D11DeviceContext::OMSetDepthStencilState()` method, while the stencil masks are set as the `StencilReadMask` member of the depth stencil state description.

If we take a moment to consider what this equation represents, we see an argument on the left side based on a parameter that can be configured with a device context method. On the right side is an argument based on the current stencil buffer value. Both arguments are masked to allow selecting a subset of the bits contained within their source variables. These two arguments are then compared with one of several comparison functions provided by the API. Thus, all three major portions of this equation are configurable, allowing for a wide variety of possible tests.

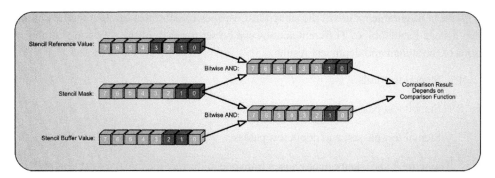

Figure 3.81. The bit registers for each of the arguments of the stencil test.

Because this test is highly configurable, it can be difficult to visualize how it operates. To help clarify this functionality, we will provide a small example scenario for a single fragment being passed through the stencil test. In this case, the stencil reference value will have bits 0, 1, and 3 set, while the stencil buffer has bits 1 and 2 set. The stencil mask has bits 0 and 1 set. These values are depicted in Figure 3.81.

Depending on which comparison function is chosen, a different result can be found for the stencil test. For example, if D3D11_COMPARISON_LESS is used, the stencil test would fail since the left side of the equation is greater than the right side. Of course, this function could be reversed if D3D11_COMPARISON_GREATER were used instead. The available comparison functions are shown in Listing 3.28.

```
enum D3D11_COMPARISON_FUNC {
    D3D11_COMPARISON_NEVER,
    D3D11_COMPARISON_LESS,
    D3D11_COMPARISON_EQUAL,
    D3D11_COMPARISON_LESS_EQUAL,
    D3D11_COMPARISON_GREATER,
    D3D11_COMPARISON_NOT_EQUAL,
    D3D11_COMPARISON_GREATER_EQUAL,
    D3D11_COMPARISON_ALWAYS
}
```

Listing 3.28. The D3D11_COMPARISON_FUNC enumeration.

Considering the ability to specify the stencil reference value, the stencil mask, and the comparison function, the developer has more or less complete control over how the stencil test will operate. After the test has completed, some action needs to be taken, depending on if the test has passed or failed. In fact, the action taken relies on both the stencil test result and the depth test result (which we will discuss in the next section). In keeping with the configurable nature of the stencil test, various possibilities can be specified in the depth stencil state object. Different actions can be set for each of the following combinations of the stencil and depth test results:

1. Stencil test fails.

2. Stencil test passes, but depth test fails.

3. Stencil test passes, and depth test passes.

The action to be taken can be chosen from one of the members of the `D3D11_STENCIL_OP` enumeration, which is shown in Listing 3.29.

```
enum D3D11_STENCIL_OP {
    D3D11_STENCIL_OP_KEEP,
    D3D11_STENCIL_OP_ZERO,
    D3D11_STENCIL_OP_REPLACE,
    D3D11_STENCIL_OP_INCR_SAT,
    D3D11_STENCIL_OP_DECR_SAT,
    D3D11_STENCIL_OP_INVERT,
    D3D11_STENCIL_OP_INCR,
    D3D11_STENCIL_OP_DECR
}
```

Listing 3.29. The D3D11_STENCIL_OP enumeration.

These options determine what is done with the stencil buffer value after the tests complete. There are options for keeping the existing stencil buffer value; clearing it to zero; replacing it with the stencil reference value; and incrementing, decrementing, or inverting it. Once again, these options provide significant freedom in using the stencil buffer for a variety of different algorithms, since different actions can be taken, depending on the test results. One final operation is performed before the stencil value is written to the stencil buffer. The value produced by the stencil operation is bitwise AND'ed with the `StencilWriteMask` before being stored.

In this section, we have seen the configuration of a complete stencil test. However, there are some additional configurations to discuss. All of the settings shown above can be individually configured for front faces and back faces with the `FrontFace` and `BackFace` members of the depth stencil state object. These members are actually structures that encapsulate one set of stencil test configurations (both the comparison function and the operations to perform in each test result case). This lets special tests to be implemented that perform different options for front and back faces. The most common example of using different settings in this manner is the shadow volumes algorithm.[38] In addition to the ability to use different configurations for front and back faces, the overall test can also be enabled or disabled with the StencilEnable Boolean parameter. If the stencil test is disabled, it is always considered to have passed, and the fragment will never be culled because of the stencil test.

Depth test. While the stencil test is being performed, the depth test is also carried out. This test essentially implements a classical Z-buffer algorithm (Williams, 1978) for performing a visibility test. The basic concept is to keep a second buffer that is the same size as the intended render target. However, when a primitive is rasterized, it stores the Z-component of the normalized device coordinates of each fragment, instead of storing the rasterized color values in the buffer. This produces a buffer that contains the post-divide Z-component of the fragment position with values in the range of $[0.0, 1.0]$. In the output merger stage, this Z-buffer is implemented as the depth portion of the depth stencil buffer, and is typically referred to as a *depth buffer*, since the Z-component represents a measure of the distance from the viewer.

When these depth values are stored, any additional primitives that are rasterized to create fragments can compare their own depth value with the one stored in the depth buffer. If the depth value stored in the buffer is smaller (closer to the viewer) than the value for the new fragment, then the new fragment can be discarded, since it is located behind another object in the scene. If the stored depth buffer value is larger than the new fragment (farther from the viewer), then the new fragment's color value is passed to the rest of the pipeline, and the Z-buffer value is updated to represent the new visibility information. In this way, the Z-buffer provides per-pixel (or per-sample if MSAA is used) visibility determination.

The output merger implements the Z-buffer algorithm with several additional configurations for preparing the depth value, as well as actually performing the depth comparison and writing the result to the Z-buffer. These configurations are primarily contained within the depth stencil state object, with the exception of the viewport. We will follow a fragment through the depth test in the same fashion as we have done for the stencil test, to gain a better understanding of how the process works.

The depth test can be enabled or disabled with the DepthEnable member of the depth stencil state object. This parameter only determines if the depth test is performed; it does not control if the depth writing functionality is enabled or disabled. The depth value is received either directly from the rasterizer stage, or if the pixel shader program modifies the depth value then it is received from the pixel shader stage. The value is then clamped to the min and max depth values specified in the viewport structure that was used to generate the fragment. This clamping is performed in a depth buffer format appropriate way. After the depth range is clamped, the depth value is read from the depth stencil buffer, and the two values are compared with a selectable depth-comparison function, selected

---

[38] This algorithm was first described in the "Rasterizer" section of this chapter.

with the `DepthFunc` member of the depth stencil state object. Listing 3.30 provides all of the available comparison functions.

```
enum D3D11_COMPARISON_FUNC {
    D3D11_COMPARISON_NEVER,
    D3D11_COMPARISON_LESS,
    D3D11_COMPARISON_EQUAL,
    D3D11_COMPARISON_LESS_EQUAL,
    D3D11_COMPARISON_GREATER,
    D3D11_COMPARISON_NOT_EQUAL,
    D3D11_COMPARISON_GREATER_EQUAL,
    D3D11_COMPARISON_ALWAYS
}
```

Listing 3.30. The D3D11_COMPARISON_FUNC enumeration.

These comparisons are performed with the fragment depth value on the left side, and the depth buffer value on the right side. If the comparison evaluates to *true*, then the depth test has passed, and the fragment continues on in the process. If the comparison evaluates to *false*, then the depth test is concluded, and the fragment is discarded. For example, the standard comparison is to use D3D11_COMPARISON_LESS, which provides functionality similar to the original *Z*-buffer algorithm. If the fragment depth is less than the depth buffer value, it will pass the depth buffer test. This indicates that the new fragment is closer to the viewer, and should be visible.

If the depth test fails, the fragment is discarded. If it passes, it will be passed to the blending functionality (which will be covered in the next section). The depth buffer may be updated, depending on a few conditions. If both the depth test and the stencil test have passed, the fate of the depth value depends on whether depth writes are enabled or not. This is specified with the DepthWriteMask member of the depth stencil state. If the value is set to D3D11_DEPTH_WRITE_MASK_ALL, then the depth value can be updated in the depth buffer; otherwise, the depth value is discarded.

It may seem counterintuitive that the depth buffer could be used for depth testing, but still have a configuration that allows it to not be written to. In fact, there are many situations where this is the preferred behavior in modern rendering schemes. It is quite common for a first rendering pass to be used to fill the depth buffer with values using the standard *Z*-buffer technique described above. Then, any subsequent passes would use the depth buffer as it is, since all of the geometry has been rasterized before. In effect, the visibility has already been determined, so there is no need to have depth writing enabled. A performance benefit is available when the depth writing is disabled, since the depth buffer would only need to be read from and never written to.

This "read-only depth buffer" concept can be taken a step further. You may recall that in Chapter 2 we saw that a depth stencil view could be created with flags to indicate that either the depth component or the stencil component, or both, can be created as read-only. This means that the view itself ensures that the resource can only be read from, which allows it to be used in more than one location in the pipeline. The resource can be bound for use in the depth test and simultaneously used as a shader resource. This can be very beneficial when the depth buffer is used as an input to later rendering passes, but still needs to be used for the depth test at the same time.

Blending

If a fragment has survived both the stencil test and the depth test, it is passed to the blending function. The blending function makes it possible to combine two selectable values prior to writing them to the output render target, and the function used to combine the values is also selectable. This functionality has traditionally been used to perform alpha-blending to implement partially transparent rendering materials. However, with the large number of possible blend sources, operations, and write options, there is plenty of additional functionality that can be used for other, less conventional techniques. The blending function is controlled by the configurations contained within the blend state object. Listing 3.31 provides the list members that can be manipulated in the blend state.

```
struct D3D11_BLEND_DESC {
    BOOL                          AlphaToCoverageEnable;
    BOOL                          IndependentBlendEnable;
    D3D11_RENDER_TARGET_BLEND_DESC RenderTarget[8];
}
```

Listing 3.31. The D3D11_BLEND_DESC structure.

Here we find two top level configurations, followed by an array of eight render target blend description structures. The first top-level member is the AlphaToCoverageEnable parameter, which we have already discussed in the "Pixel Shader" section of this chapter. The second top-level member is another Boolean value, which determines if all of the render targets bound for output in the output merger stage will be blended independently, or if they will all use the same blending configuration. As you may have guessed by now, the blending configurations are stored in the eight-element array member of the blend state, since there is a maximum of eight render target slots available. If individual blending is enabled by setting IndependentBlendEnable to *true*, each of these array elements defines the blending mode to use for the corresponding render target slot. If independent blending is disabled, then all render targets will use the blending configuration from index 0 of the array. If it is not needed, then independent blending should be disabled to allow for better performance.

To gain insight into how the blending configuration is used, we will use the same paradigm that we employed for the stencil and depth tests and follow the path of a fragment that is running through the blending functionality. Listing 3.32 shows the members of the D3D11_RENDER_TARGET_BLEND_DESC structure.

```
struct D3D11_RENDER_TARGET_BLEND_DESC {
    BOOL           BlendEnable;
    D3D11_BLEND    SrcBlend;
    D3D11_BLEND    DestBlend;
    D3D11_BLEND_OP BlendOp;
    D3D11_BLEND    SrcBlendAlpha;
    D3D11_BLEND    DestBlendAlpha;
    D3D11_BLEND_OP BlendOpAlpha;
    UINT8          RenderTargetWriteMask;
}
```

Listing 3.32. The D3D11_RENDER_TARGET_BLEND_DESC structure.

The first member, BlendEnable, performs exactly what its name states. It enables or disables the blending functionality. We will assume that blending is enabled for this example. Next, we see two sets of three states. The three members following BlendEnable are used to blend color values (the RGB components of the fragment color), and the next three are used for blending the alpha value (the A component of the fragment color). This separation of blending state allows completely different blending modes to be used for alpha and color, which can let them be used for separate purposes.

The three members that are supplied for both color and alpha are used to define the blending equation. The two data sources for the blending equation are selected by SrcBlend and DestBlend parameters for the color blending, and SrcBlendAlpha and DestBlendAlpha for alpha blending. These parameters define their data source as one of the members of the D3D11_BLEND enumeration, which is shown in Listing 3.33.

```
enum D3D11_BLEND {
    D3D11_BLEND_ZERO,
    D3D11_BLEND_ONE,
    D3D11_BLEND_SRC_COLOR,
    D3D11_BLEND_INV_SRC_COLOR,
    D3D11_BLEND_SRC_ALPHA,
    D3D11_BLEND_INV_SRC_ALPHA,
    D3D11_BLEND_DEST_ALPHA,
    D3D11_BLEND_INV_DEST_ALPHA,
    D3D11_BLEND_DEST_COLOR,
    D3D11_BLEND_INV_DEST_COLOR,
    D3D11_BLEND_SRC_ALPHA_SAT,
    D3D11_BLEND_BLEND_FACTOR,
    D3D11_BLEND_INV_BLEND_FACTOR,
```

```
        D3D11_BLEND_SRC1_COLOR,
        D3D11_BLEND_INV_SRC1_COLOR,
        D3D11_BLEND_SRC1_ALPHA,
        D3D11_BLEND_INV_SRC1_ALPHA
    }
```

Listing 3.33. The **D3D11_BLEND** enumeration.

As you can see, there are a substantial number of possible data sources. The general options in this list include data sources such constant values and color/alpha values, with several possible modifiers. The color and alpha sources all provide source and destination modifiers, where the source is the value from the current fragment and the destination is the value located in the "destination" render target. Each of these values also provides an "inverse" value, which is generated by using 1 minus the value. For example, the D3D11_BLEND_SRC_COLOR selects the current fragment color as its source. However, D3D11_BLEND_INV_DEST_COLOR selects the inverse of the destination render target value as its source. In addition to each of these modified color and alpha values, it is also possible to use one of several constant values. The constant values that can be chosen are either 0, 1, or the blend factor that is set by the application with the ID3D11DeviceContext::OMSet BlendState() method. The blend factor also provides an inverse value as well.

You may also notice that there are several parameters with the SRC1 modifier in their names, such as D3D11_BLEND_SRC1_COLOR and D3D11_BLEND_INV_SRC1_ALPHA. These values are actually taken from the second output register in the pixel shader, and can be used in blending operations at the same time as the first output register. This lets two source quantities be read from the current fragment, instead of using the contents of the destination buffer. Using both of the output register values is referred to as dual-source color blending, and can only be used with a single render target output (otherwise, the second color value would be passed to the second render target).

Once the data sources have been selected, the blending operation must be selected. This operation combines the two data source values into a single value to be written into the output render target destination. The available operations are provided in Listing 3.34. The available operations are fairly self-explanatory and implement what their names imply. There are three arithmetic operations (A+B, A-B, B-A), along with minimum and maximum selection options.

```
    enum D3D11_BLEND_OP {
        D3D11_BLEND_OP_ADD,
        D3D11_BLEND_OP_SUBTRACT,
        D3D11_BLEND_OP_REV_SUBTRACT,
        D3D11_BLEND_OP_MIN,
        D3D11_BLEND_OP_MAX
    }
```

Listing 3.34. The D3D11_BLEND_OP enumeration.

As previously mentioned, these blending operations can be performed separately for color values and alpha values. Once the blending operation has completed, a complete four-component RGBA color is available for writing. The final step in the blending process is to use the render target write mask to determine which channels of the output color will actually be written. This is specified in the `RenderTargetWriteMask` member of the blend state structure. The RGBA channels correspond to the bit positions 0, 1, 2, and 3 of the mask. When the bit is set, the channel is written to the render target.

3.12.4 Output Merger Pipeline Output

We have seen throughout this section what operations are performed on the pipeline data before it is written to the resources bound to the output merger stage. There are a total of three different types of resources that can be written to by a pipeline execution: render targets, unordered access views, and the depth stencil target.

The render targets receive the output from the blending functions. Each fragment written to a render target must have survived both the stencil test and the depth test before being blended and finally written to a render target. If MRTs are being used, then each render target will have a blended value written to it that originates from the pixel shader. If render target arrays are used, then the appropriate render target slice is determined back at the input to the rasterizer stage with the `SV_RenderTargetArrayIndex` system value semantic. Thus, by the time the data reaches the output of the blending function, the appropriate texture slice has already been selected.

In contrast, data written to a UAV is directly controlled by the pixel shader stage and does not need to pass any tests to successfully make it to the output resource. When the pixel shader modifies a resource, those changes are immediately submitted to the GPU and take effect as soon as the GPU's memory system writes the values to memory. There is some delay in this writing process, and hence, its results must not be assumed to be immediately available when designing an algorithm.

The depth stencil target is somewhat more complex. Output data must pass the depth test and the stencil test to be written to the depth portion of the output resource. In addition, depth writing must be enabled in the depth state object. Stencil data can be updated differently, depending on which combination of the depth and stencil tests pass. Since these are configurable settings, it is more or less up to the application to determine when and how the stencil data is updated by the data stream coming into the output merger stage.

After completion of the pipeline execution, the modified contents of each of these resources are available for use in further rendering passes and computation passes, or even for direct manipulation by the CPU. In the case of MSAA rendering, one further step is needed before using the resource. After the rendering is finished, it is typically desirable to resolve the MSAA render target to a non-MSAA texture. In Direct3D 11 this is done by calling the `ID3D11DeviceContext::ResolveSubresource()` method. Performing this resolve combines the values of the individual subsamples into a single value, resulting in a texture suitable for being displayed on the screen, or for being sampled using normal shader sampling methods. The filter applied when performing the resolve is specific to the hardware and driver, and to the quality level specified for the render target. However, in most cases, a box filter is used. For a box filter, all subsamples are weighted equally, making it a simple arithmetic average of the subsample values.

3.13 High-Level Pipeline Functions

After stepping through the entire pipeline, stage by stage, we have covered a significant amount of functionality in great detail. In fact, there is so much information presented in this chapter that it can be overwhelming to consider all of the different types of calculations that can be performed. Therefore, we will provide a few ways to consider the functions of each stage from a somewhat higher level, in an attempt to clarify how they are typically used. These groupings are generalized based on typical usage, but they don't necessarily reflect any type of requirement. After all, the pipeline's flexibility allows for very creative and unusual uses, so these are merely presented as a starting point for the reader to build from.

3.13.1 Vertex Manipulations

The early stages of the pipeline are generally used to build and manipulate the vertices that represent the geometric surfaces that will be rendered. This includes construction of vertices by the input assembler, as well as the individual per-vertex operations that are performed by the vertex shader. At the lowest level, these two stages perform the most basic operations on the input geometry that are seen throughout the pipeline. Because of this close proximity to the original input geometry, it is quite natural to make modifications to the geometry early in the pipeline. Traditionally, the transformation matrices are applied in the vertex shader, which is essentially manipulating the geometric properties of the model.

In general, since both of these stages operate at a per-vertex level, they will most likely be executed the least during a single pipeline execution. For each draw call, the input assembler will produce each vertex of the model, and will set up each primitive for further processing later in the pipeline. From these two streams, the vertex shader will operate only once per vertex. In contrast to the later stages in the pipeline, the input assembler and vertex shader perform their calculations at a relatively low frequency. While the input assembler is a fixed function stage, the vertex shader is programmable and can be used to implement a customized routine. This customizable processing can be used to our advantage by performing calculations in the vertex shader that can be passed on to the later stages, instead of calculating everything directly in the later stages.

It is also logical to perform any other calculations in the vertex shader that provide relatively low-frequency information to the surface of the geometry. For example, it is often adequate to calculate ambient lighting calculations in the vertex shader, since it doesn't change very quickly across the entire model. By performing the calculation at the vertex level, we relieve a small amount of calculation from each shader invocation later in the pipeline.

3.13.2 Tessellation Manipulations

The next set of pipeline stages implements the tessellation functionality of the rendering pipeline. The hull shader, tessellator, and domain shader are all used together to provide a very flexible and robust tessellation system. This can be seen as an extension of the previous vertex-based processes, since it ultimately produces vertices, as well. In many ways, the domain shader can be seen as a repeat of the vertex shader, with some additional responsibilities for creating the vertex before it is processed.

However, the tessellation stage is intended to amplify the available detail at a geometric level, which is a departure from what the input assembler and vertex shader can do. This means that the overall number of invocations of the domain shader will be significantly higher than it is for the vertex shader. Because of this, and because of the additional calculations that need to be performed in the domain shader, the vertex shader should be used as much as possible, instead of the domain shader. If all things are equal, the vertex shader should be chosen over the domain shader.

Even so, the tessellation system provides some unique opportunities for the developer. By amplifying how many vertices are available to represent the geometry, we can perform much higher frequency calculations much earlier in the pipeline than was possible in previous versions of Direct3D. In the extreme case, we can produce one vertex for every pixel, making the visual fidelity of the per-vertex data equal to that of the per-pixel data. We can use this to our advantage by moving calculations to earlier in the pipeline and away from the raster based stages, or we can simplify the later stages by performing more complex calculations at the vertex level.

3.13.3 Geometry Manipulations

The geometry shader and the stream output stages are very much specialized stages that sit between the vertex-based stages and the raster-based stages. As such, they are intended to perform higher level geometric operations than the vertex based stages do, before the geometry is rasterized. While this is extremely useful in some situations, it is less common to use the geometry shader in most algorithms. The majority of the algorithms that use geometry shaders use them because of its special features that aren't available in any other stage. The expansion of points into quads is a good example of functionality that can't be performed in any of the other stages.

This is probably at least partially due to the poor performance that the geometry shader became known for during its debut in Direct3D 10, and to a corresponding lack of development effort geared toward it. However, with the shared processor architectures that most current generation GPUs use, sufficient processing power is available to use the geometry shader. The biggest challenge is to ensure that the memory usage of the stage is appropriately balanced with the tasks that are being performed. For example, if each geometry shader invocation is used to produce 100 output triangles, the algorithm should likely be reevaluated to take advantage of the tessellation stages, instead of the geometry shader. However, for small-scale geometry manipulation, there is a much better balance of memory usage to computation. This makes point sprite expansion an attractive target—it performs some calculations, and a small amount of data amplification. At the very least, it will be interesting to see if more algorithms are developed to use the geometry shader in the coming years, or if it will remain as a specialty stage in the pipeline.

3.13.4 Raster-Based Manipulations

The final portion of the pipeline is the raster-based stages. These include the rasterizer, pixel shader, and output merger. These stages operate at a much higher frequency that the previous stages, due to the data amplification that is performed in the rasterizer. This allows for much higher frequency operations to be performed in these stages, which is why the pixel shader is so frequently used to add all of the detail to a rendering. The rasterizer is typically not a bottleneck in rendering algorithms, while the pixel shader can potentially be either a calculation or memory bandwidth bottleneck, due to the large number of invocations that are performed.

A somewhat less recognized performance issue can be introduced by the output merger. Since it reads and writes to the depth stencil buffer and can potentially read and write to the render targets when blending is enabled, the output merger can greatly increase bandwidth usage. If you don't need to use the blending function, make sure it is disabled! Likewise, if there is no reason to expect updating of the depth buffer (such as in a second or third rendering pass), ensure that the depth writing functionality is disabled.

In addition to these considerations, an entire class of algorithms is just now starting to be developed that use the unordered access views in the pixel shader. The ability to perform custom reading and writing of resources at arbitrary locations provides a totally new way to fill resources with data, from a location that was previously unavailable. Some totally new demos are being produced by the GPU manufacturers that efficiently use this new ability in very interesting ways. As we further explore this functionality, we will see how developers can manipulate it for either a performance or image quality improvement.

The Tessellation Pipeline

4.1 Introduction

Tessellation is one of the headline new features for Direct3D 11. While version 11 brings many improvements, only a few are wholly new compared with others that are incremental revisions on previous releases.

Tessellation not only introduces two new programmable units and one fixed-function, all of which require new knowledge and expertise from the developer; it also opens up new opportunities for artists and content creators. It can revolutionize how real-time computer graphics operates.

This chapter will take the concepts introduced earlier in Chapter 1 and Chapter 3, starting by providing a detailed discussion of the motivation and concepts behind the tessellation feature. The chapter concludes by covering numerous parameters and entry points exposed to the developer.

Chapter 9 will cover examples of using this technology.

4.1.1 What Is Tessellation?

The origins of the word *tessellate* date back to Latin, in the context of mosaics. The word *tessella* refers to the small fragments used to create a larger mosaic, while *tessellation* (or tiling) means to cover a surface, with no gaps between fragments, and no overlaps.

This can be extended to computer graphics in the general sense of using many smaller pieces of geometry to create, without gaps or overlaps, a larger complete surface. At its simplest, a common closed mesh of triangles satisfies this definition.

A number of other related terms are used within the field of computer graphics to formalize tessellation as a means of representing non-linear mathematical surfaces. Triangles are typically the preferred unit of raster-based computer graphics, due to their very useful properties of being convex and having coplanar vertices, but this convenience also limits them to representing flat geometric surfaces. Many small triangles may be used to approximate a smooth surface, but the basic unit of the triangle is still flat.

Two of the better known techniques for representing a curved surface are *non-uniform rational basis splines* (*NURBS*) and *SubDivision surfaces* (often abbreviated as *SubDs*). The former is a generalized mathematical form (Weisstein) for smooth, curved surfaces; it has been a staple feature in graphics software for decades. The latter is a general framework for mesh refinement, recursively adding triangles until it better represents the ideal surface. *Catmull-Clark subdivision surfaces* (Catmull & Clark, 1978) are a commonly used example. The crucial difference is that subdivision surfaces don't require a mathematical basis, whereas using NURBS does.

Mathematically defined surfaces are often referred to as *higher-order surfaces* because their underlying equations are defined in terms of their order—quadratic and cubic are most common—and because this order is above linear.

Various tessellation algorithms allow developers or artists to create idealized surface of curves and smooth surfaces to be mapped to conventional triangle-based raster hardware. The individual triangles are the small stone fragments of the final curved-surface mosaic.

4.1.2 Why Is Tessellation Useful?

As previously discussed, a large number of small triangles can be used to approximate a higher-order smooth surface. It is therefore useful to understand why we need additional complexity and hardware when we could simply stick with existing techniques.

This new technology available in Direct3D 11 solves real technical, artistic, and business problems in the domain of real-time computer graphics.

The demand for increasing image quality imposes a significant strain by greatly increasing the volume of data required to define high-resolution models. This increased volume of data requires a correspondingly large increase in on-disk and in-memory storage, I/O bandwidth, and number of calculations. A mathematical definition for a surface requires storage of only the coefficients or inputs into the appropriate function, much less than needed to store raw triangles. Being able to dynamically scale output based on fixed inputs also offers convenience for developers looking to scale image quality across multiple grades of hardware, thus reducing problems involved in targeting different hardware configurations.

Despite these immediately obvious savings in storage and bandwidth, it is important to note the trade-off. Functions that define higher-order surfaces must be sampled in real time and cannot be precalculated. These functions can be mathematically complex and require non-trivial processing time by the GPU. On the whole, GPU tessellation pays for itself, but it doesn't come for free.

Higher-order surfaces have been a common tool in content creation packages for decades and are well understand by any experienced artist. Previously, artists had to work only with triangle-based meshes, which complicated their workflow by requiring them to participate in the implementation details, as opposed to the purely artistic and conceptual.

Tessellation is therefore useful in both technical and artistic contexts. These advantages indirectly also benefit businesses, by reducing the amount of time and effort required.

A technical solution that uses resources more conservatively and scales well across different GPUs can allow for less complex code, having a single code path that can be applied to many configurations. One historical difficulty with real-time graphics has been the common need for many code paths to be implemented and fine tuned for each major hardware configuration, which quickly becomes a maintenance and development nightmare. Cleaner development can only be a good thing.

Enabling artists to be more productive greatly reduces the time needed to produce high-quality art assets. Regarding content creation, it has been noted that the drive for higher-quality real-time graphics has put more strain on art production than it has software. Direct3D 11 helps balance this.

4.1.3 History of Tessellation

Curved surfaces and tessellation algorithms are not new concepts. Traditionally, they have only been available to offline rendering, since their complexity was prohibitively high for real-time applications. A well-known example is films using computer-generated imagery (*CGI*) from Pixar (such as *Toy Story*) and DreamWorks (such as *Shrek*). Outside of media, design and engineering were common early applications of this style of mathematics and rendering. In 1962 Pierre Bézier popularized Paul de Casteljau's work on what has become known as a *Bézier curve* (or *Bézier surface*) while working on car designs for Renault. The advantages of using mathematical curves for design became immediately obvious, for many of the reasons discussed above.

Over the decades, various tools, often for computer-aided design (*CAD*), were developed to allow manipulation of higher-order surfaces. For higher-order surfaces to be combined with real-time rendering, as in the early days of commodity three-dimensional computer games, it was necessary to "bake" the mathematical surface into a fixed triangle equivalent that could be used by the rendering software and hardware of the day. This made it impossible to scale the quality and resolution of surfaces, since the required information had been lost.

In the context of computer games, it was ID Software's id Tech 3 graphics engine, used in *Quake 3 Arena* in December 1999, that made waves with regard to using curved surfaces for real-time graphics.

In late 2001, ATI Technologies introduced the Radeon 8500 GPU, which included its TruForm functionality (ATI, 2001). This was a first for commodity hardware, which, when combined with Direct3D 8.1, allowed for hardware-accelerated tessellation of higher-order surfaces.

The popularity and use of this pioneering new feature were short-lived, with the advent of the id Tech 4 engine (culminating in August, 2004's *Doom 3*) and real-time stencil shadow techniques. *Shadow volumes* were a geometric technique that required processing on the CPU for accurate results; this conflicted with TruForm, in which the finally rendered geometry was different from that available to the CPU. Visible differences between the silhouette of a model and its shadow made them incompatible. Ultimately, shadowing won the popularity contest.

Interestingly, around the same time, Valve Software were working on its Source Engine (culminating in *Counterstrike: Source* and *Half Life 2*, both released in summer, 2004). The Source Engine supported shadow mapping (Valve Software) as an equivalent feature to id Tech 4's stencil shadows; in particular, this technique was compatible with tessellated geometry.

Competition between hardware vendors and the two rival software engines was intense. Ultimately, TruForm lost out, due to the popularity of stencil shadows and a lack of universal support across all hardware vendors.

Following on from ID Software's id Tech 4, it became practical and desirable to simulate higher-resolution models by using low-density meshes and texture trickery. Tangent-space normal mapping and related environment mapping techniques can represent the lighting interactions of a geometrically complex surface, while still being texture-mapped to a far simpler planar model. Although this is generally considered to be a reasonable tradeoff between performance and image quality, it had a significant drawback, in that a model's silhouette was still low-detail, which showed up the trick for what it truly was.

4.2 Tessellation and the Direct3D Pipeline

At first glance, the new tessellation stages that have been added in Direct3D 11 don't appear overly complex. But on closer inspection, we can see that when one is designing and writing code, the flow of data and the responsibilities of each unit can quickly become confusing. This is compounded because the deeper pipeline is harder to visualize—a classic pipeline consisting of a vertex shader and pixel shader was relatively simple, and small

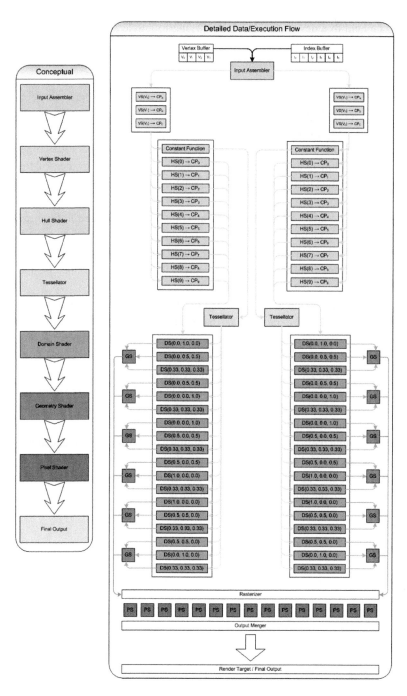

Figure 4.1. Pipeline flow with tessellation enabled.

enough to be held in a developers head. But trying to juggle all six programmable stages can be much more difficult!

The remainder of this chapter goes into more detail on the concepts and general pipeline issues introduced as part of Chapter 3. A knowledge of the higher-level concepts of the pipeline, resources, and shaders is assumed from here onward, so these concepts will not be explicitly described again.

Figure 4.1 can aid in visualizing what is happening at various stages of the new rendering pipeline, by mapping the logical pipeline states to the operations they perform. The precise algorithm is unimportant for the diagram, because the idea of tessellation, which will be described in more detail later, is relatively abstract from a pipeline execution perspective.

To help make Figure 4.1 clearer, the conceptual view discussed in the previous section has been included with a matching color scheme. Arrows show the flow of data and/or execution, and individual shader functions are denoted in the form $xx(yy) \rightarrow zz$, where xx is the abbreviation of the shader type, yy represents important parameters, and zz is the output.

Figure 4.1 assumes the naive approach of executing a stage as many times as there are outputs or inputs. It is expected that hardware can take advantage of commonality (using pre- or post-transform caches, for example) and reduce the number of invocations. The two best candidates are the vertex and domain shaders; in this example there are 6 VS invocations and 36 DS invocations, yet there are only 4 unique control points and 14 domain points. Specifically, the example used here would do 50% more vertex shading and 157% more domain shading. In a field where performance is crucial, it's easy to see why the hardware would want to be more efficient!

In this example there are four vertices defining a quad on the XZ plane and six indices defining two triangles. Unlike in many other tessellation algorithms, no adjacency information is required, so this doesn't appear to be any different from rendering a normal quad. Without tessellation, the output would like Figure 4.2.

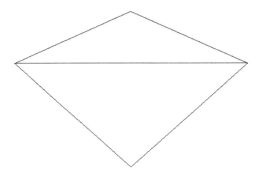

Figure 4.2. Quad rendered without tessellation.

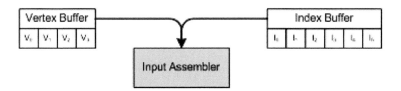

Figure 4.3. Input assembler stage.

4.2.1 Input Assembler

With Direct3D 11 it is now possible to create primitives with up to 32 vertices. But regardless of this, the input assembler functions exactly as it did in previous versions. It uses the vertex declaration (an ID3D11InputLayout created from an array of D3D11_INPUT_ELEMENT_DESCs), a vertex buffer (an ID3D11Buffer with binding of D3D11_BIND_VERTEX_BUFFER) and an index buffer (another ID3D11Buffer with a binding of D3D11_BIND_INDEX_BUFFER) (see Figure 4.3). The topology set on the device will be D3D11_PRIMITIVE_TOPOLOGY_n_CONTROL_POINT_PATCHLIST, where *n* is between 1 and 32. The input assembler will then read the index buffer in chunks of n and will pick out the appropriate vertices from the vertex buffer.

4.2.2 Vertex Shader

Chapter 3 introduced the subtle difference between *control points* and actual vertices; with tessellation, the vertex shader no longer has to output a clip-space vertex to SV_Position, as in Direct3D 10 (see Figure 4.4). (Technically, this could be done in a geometry shader in D3D10, but it was often more efficient to stick with the conventional vertex shader approach.) It is now completely free to operate on data in whatever form the application gives it (through the input assembler) and output that data in whatever coordinate system or format it chooses.

The common use-case for a vertex shader with D3D11 tessellation will be for animation—transforming a model according to the bones provided for a skeletal animation is a good example (and will be covered in more detail as part of Chapter 8). In this example, the vertex shader simply transforms the model-space vertex buffer data into world-space for the later stages. Note that once the vertex shader has

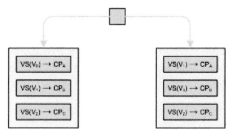

Figure 4.4. Vertex shader stage.

executed, the later stages cannot see any of the original data from the vertex buffer, so if any of this source data is useful, it should be passed down as part of an output.

4.2.3 Hull Shader

This is the first new programmable unit in the Direct3D 11 pipeline. It is made up of two developer-authored functions—the hull shader itself and a "constant function."

The constant function is executed once per patch. Its job is to compute any values that are shared by all control points and that don't logically belong as per-control-point attributes. As far as Direct3D 11 is concerned, the requirement is to output an array of SV_TessFactor and SV_InsideTessFactor values. The size of these arrays varies, depending on the primitive topology defined in the input assembler stage, details of which are discussed later in this chapter. The outputs from a constant function are limited to 128 scalars (32 float-4s), which provides ample room for any additional per-patch constants once the tessellation factors have been included.

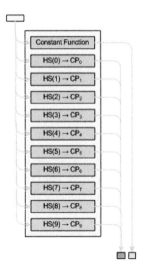

Figure 4.5. Hull shader stage.

An attribute on the main hull shader function declares how many output control points will be generated. This can be a maximum of 32 and does not have to match the topology set on the input assembler—it is perfectly legal for the hull shader to increase or decrease the number of control points. This attribute determines how many times the individual hull shader is executed, once for each of the declared output control points (the index is provided through SV_OutputControlPointID uint input). The quantity of data can be up to 32 float4sor 128 scalars, the same as for the per-patch constant function, but with one difference; the maximum output for all hull shader invocations triggered by a given input patch is 3,968 scalars. In practice, this means that if you're outputting 32 control points, you can only use 31 float4s, instead of 32. Putting these numbers together, the entire hull shader output size is clamped to 4 KB.

Both functions have full visibility of all vertices output by the vertex shader and deemed to be part of this primitive. This is represented in Figure 4.5 by the blue arrows leading from the Vertex Shading section to each of the hull shader and constant function invocations.

4.2.4 Fixed-Function Tessellator

The next stage of processing is entirely fixed function and operates as a black-box, except for the two inputs—the SV_TessFactor and SV_InsideTessFactor values output by the hull shader constant function (see Figure 4.6).

A very important point to note is that the control points output by the hull shader are not used by this stage. That is, it does all of its tessellation work based on the two aforementioned inputs. The control points are there for you as the developer to create your tessellation algorithm (they reappear again as an input to the domain shader), and the fixed-function stage doesn't pay them any attention. This is why there are no restrictions (other than storage) or requirements on the output from the individual invocations.

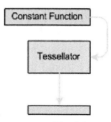

Figure 4.6. Fixed function stage.

The output from the tessellator is a set of weights corresponding to the primitive topology declared in the hull shader—line, triangle or quad. Each of these weights (presented as the SV_DomainLocation input) is fed into a separate domain shader invocation, which is discussed next. In addition to these newly created vertices, which the developer actually sees as part of the domain shading stage, the tessellator also handles the necessary winding and relations between domain samples, so that they form correct triangles that can later be rasterized.

4.2.5 Domain Shader

Domain shader invocations run in isolation, but as is necessary, they can see all the control points and per-patch constants output earlier in the pipeline by the hull shader stage. Simply put, the domain shader's job is to take the point on the line/triangle/quad domain provided by the tessellator and use the control points provided by the hull shader to create a complete new, renderable vertex (see Figure 4.7).

It is now the domain shader's responsibility to output a clip-space vertex coordinate to SV_Position. (Strictly speaking, the geometry shader could do this, but it will be less efficient.)

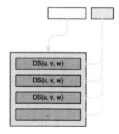

Figure 4.7. Domain shader stage.

4.2.6 Geometry Shader

With regard to tessellation, this stage remains unchanged from previous versions of Direct3D (see Chapter 3 for details on refinements affecting other aspects of Direct3D rendering) and effectively marks the end of any tessellation related programming. Unless the domain shader has passed along any of the control mesh

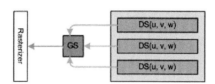

Figure 4.8. Geometry shader stage.

information as per-vertex attributes, the geometry shader has no knowledge of the tessellation work that preceded it.

Unlike in Direct3D 10, where the number of geometry shader invocations was directly linked to the parameters of a draw call by the number of primitives passed into the pipeline, the number of executions when tessellation is enabled is now linked to the SV_TessFactor and SV_InsideTessFactor values emitted from the hull shader constant function (see Figure 4.8). If these are constant and/or set by the application (such as through a constant buffer), you can derive the number of geometry shader invocations, but if a more intelligent LOD scheme is implemented, the number of invocations will be much more difficult to predict.

4.2.7 Rasterization and the Pixel Shaders

Tessellation is a geometry based operation. This means that the final rasterization stages remain completely unchanged and oblivious to anything that came before it.

4.2.8 Demonstration of Possible Outcomes

Earlier in this section, Figure 4.2 showed the "plain" 4-vertex quad of control points. This is the original data sent by the application. The result of the above flow of execution through the Direct3D 11 pipeline transforms Figure 4.9 into Figure 4.10.

In Figure 4.10, you can see the four triangles generated for each input triangle, and you can see that for a 2.0 integer partitioning, it simply splits each edge in half. The next section will detail the tessellation parameters.

Figure 4.11 demonstrates tessellation in a more complex setting. Taken from the DetailTessellation11 sample in the Microsoft DirectX SDK, it shows the result as it would be seen in the final rendered image, as well as the wireframe representation, which

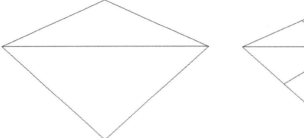

Figure 4.9. Without tessellation enabled.

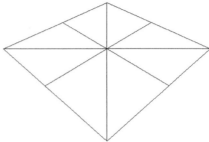

Figure 4.10. With tessellation enabled.

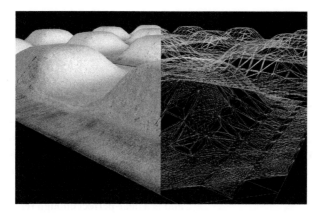

Figure 4.11. DetailTessellation11 from the Microsoft DirectX SDK.

depicts the various tessellation levels (contrast the foreground and background) defined by the underlying geometry.

Chapter 9 covers examples of tessellation algorithms, and the Microsoft DirectX SDK includes several others.

4.3 Parameters for Tessellation

In the preceding section, the detailed flow of execution and data revealed two stages where amplification could occur. First, the hull shader could alter the number of control points passed down the pipeline (3 vertices of a triangle become 10 control points on a cubic surface). Secondly, the fixed-function tessellator generated a number of new primitives, according to the output from the hull shader constant function (7 vertices defining 6 triangles).

As the author of Direct3D 11 shaders, you have full control over these amplification steps.

Generally speaking, there are two scales to consider when deciding the level of amplification—quality and performance. In most cases, these will be inversely proportional, such that higher quality implies greater amplification, which requires more processing and hence reduces performance.

4.3.1 A Simple Example

Figure 4.12 demonstrates the performance-versus-quality decision. The green curve behind the black arrows can be considered as the ideal surface—the one that an artist or

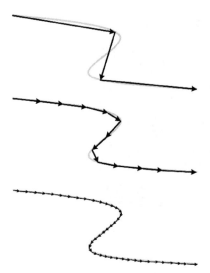

Figure 4.12. Approximations of a smooth curve.

designer would like to have displayed on screen; the black arrows are the line segments generated by the Direct3D 11 pipeline to try to approximate this surface.

The top example generates only 3 line segments. It is quite clear that the end result is not a great approximation. The middle example uses 12 segments and does a much better job of representing the ideal surface; in particular, it only really misses detail in the tight second turn of the surface. The bottom example uses 42 segments and is a near-perfect enough representation of the ideal surface.

Depending on requirements, the above example would ideally use somewhere between 10 and 40 lines to represent the surface. Less than 10, and the approximation would simply not be close enough; above 40 and there will be diminishing returns—extra processing work for little or no visible improvement.

A clever software algorithm would scale between 10 and 40, according to some heuristic of the contribution to the final image. When this particular curve is dominant (as in the center foreground) it could use the full 40 segments, whereas the opposite case (in the far distance off to one side of the image) it would use only 10. An equally clever algorithm may use a performance-based heuristic—if overall performance drops below a certain threshold, it will reduce the quality of the tessellation until performance increases again.

4.3.2 Defining the Tessellation Parameters

The fixed-function tessellation stage uses values from the hull shader's constant function to decide the level of tessellation for a primitive. The following is a simple example of such a function:

```
struct HS_PER_PATCH_OUTPUT
{
    float edgeTesselation[3]   : SV_TessFactor;
    float insideTesselation[1] : SV_InsideTessFactor;
};
HS_PER_PATCH_OUTPUT hsPerPatch
```

```
    (
        InputPatch<VS_OUTPUT, 3> ip
        , uint PatchID : SV_PrimitiveID
    )
{
    HS_PER_PATCH_OUTPUT o = (HS_PER_PATCH_OUTPUT)0;

    o.edgeTesselation[0]
        = o.edgeTesselation[1]
        = o.edgeTesselation[2]
        = 2.0f;

    o.insideTesselation[0] = 2.0f;
    return o;
}
```

Listing 4.1. A simple hull shader constant function.

The important detail is that the constant function *MUST* output `SV_TessFactor` and `SV_InsideTessFactor` values, which specify how finely to tessellate the current primitive. The tessellator can work on three fundamental types of primitives, and the corresponding output structure varies accordingly. Each of these primitive types, the number of components in their tessellation factors, and a brief sample declaration, are provided in Table 4.1.

Primitive Type	SV_TessFactor	SV_InsideTessFactor	Example

Line	2	0	`struct HS_PER_PATCH_OUTPUT`
			`{`
			`    float lenAndThick[2] : SV_TessFactor;`
			`    // ... other values`
			`};`

Triangle	3	1	`struct HS_PER_PATCH_OUTPUT`
			`{`
			`    float edges[3] : SV_TessFactor;`
			`    float inside[1] : SV_InsideTessFactor;`
			`    // ... other values`
			`};`

Quad	4	2	`struct HS_PER_PATCH_OUTPUT`
			`{`
			`    float edges[4] : SV_TessFactor;`
			`    float inside[2] : SV_InsideTessFactor;`
			`    // ... other values`
			`};`

Table 4.1. Primitive types, tessellation factors, and examples.

The HLSL compiler will consider it an error if the output structure does not match the declared primitive type.

The number of components in `SV_TessFactor` matches the number of edges the primitive has, except in the case of a line, where the first value is tessellation along the length of the line, and the second value is the thickness of the tessellation.

`SV_InsideTessFactor` makes more sense when seen visually and is covered later in this section.

The algorithms used to determine the tessellation parameters are subject to the common HLSL and shader programming rules. The code must reside in its own function and can use hard-coded constants, constant buffers, or regular buffers, as well as entirely from first principles without external inputs.

The constant function has visibility of all outputs from the vertex shader, but not the final control points as generated by the other part of the hull shader. The previously mentioned example of adaptive tessellation according to final image contribution could be performed with this information—a rough approximation of the final size within the rasterized image could be made, and the tessellation factors adjusted accordingly.

4.3.3 HLSL Helper Functions

For the most part, the selection of tessellation factors is left to the author of the hull shader's constant function. However Shader model 5.0 defines several utility functions that can be useful when determining intermediary values, as well as handling the `pow2` partitioning mode. This latter point is important, as the other three partitioning modes do not require these helper methods and are implemented "behind the scenes" by the compiler; `pow2` partitioning must be configured through these intrinsic functions. Table 4.2 provides some information.

The minimum, maximum, or average of the edge factors is then combined with the `InsideScale` parameter to determine the final internal tessellation factor(s) for the primitive being rendered.

In each of the methods listed above, the final three parameters are marked as outputs that can be retrieved and inspected by HLSL code within the hull shader. With knowledge of the partitioning function, it is easy to write a function of the raw (unrounded) and rounded tessellation factors to compute the current interpolation factor, which could be very important for the later domain shader stage. As explained in the next section, there are important cases where knowledge of neighboring tessellation factors is essential to ensure a continuous surface without gaps.

Function Prototype	Description
void ProcessQuadTessFactors**(float4 RawEdgeFactors, float InsideScale, float4 RoundedEdgeTessFactors, float2 RoundedInsideTessFactors, float2 UnroundedInsideTessFactors);	Applies to a quad where the two internal tessellation factors are independent
void Process2DQuadTessFactors**(float4 RawEdgeFactors, float2 InsideScale, float4 RoundedEdgeTessFactors, float2 RoundedInsideTessFactors, float2 UnroundedInsideTessFactors);	Applies to a quad where both internal factors are the same
void ProcessTriTessFactors**(float3 RawEdgeFactors, float InsideScale, float3 RoundedEdgeTessFactors, float RoundedInsideTessFactor, float UnroundedInsideTessFactor);	Applies to all triangles

Where ** refers to one of:

Min: takes the minimum of the provided edge factors
Max: takes the maximum of the provided edge factors
Avg: takes the average of the provided edge factors

Table 4.2. HLSL Helper Functions.

4.4 Effects of Parameters

Since the tessellation factors are simple scalar values that each apply to different portions of a primitive, it can be difficult to visualize what the factors ultimately influence in the output of the tessellation stages. The following section shows the tessellated output related to changes in the parameters provided by the hull shader constant function. For brevity, only the triangle and quad primitive types are covered, although line primitives share many of the same properties.

It is also important to note that any factors less than or equal to zero (or *NaN*) cull the primitive, which can be useful in more complex or efficient algorithms—why tessellate a back-facing primitive? Simply cull it in the hull shader! In most cases, it simply sets a minimum value of 1.0 for tessellation factors.

4.4.1 Edge Factors

We begin our discussion of the tessellation factors with *edge factors*. As the name suggests, the three or four values provided for triangle or quad domains correspond directly to the three or four edges that these two primitives have. The fixed function tessellator simply divides up each edge, according to the matching tessellation factor.

Given that a common use for this technology is to generate representations of a higher-order surface from a domain of points, it is natural to assume that the *surface area* (key word: *surface*) is of most importance. However, in a practical implementation it is likely that the edge factors will be the ones demanding attention.

The reasons for this are simply that individual pieces of geometry don't exist in isolation, and that many patches, if not all, will have adjacent patches. These patches join up at the edges; thus; for a continuous surface, you absolutely must tessellate the edges that match up. If you were to have two quads share the same edge but have different tessellation factors, you would likely see gaps in the geometry, as demonstrated in Figure 4.13 by the two orange arrows in gaps along a shared edge where tessellation levels clearly don't match.

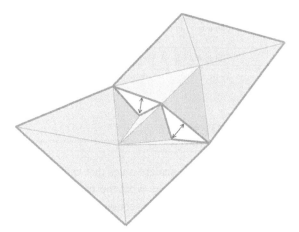

Figure 4.13. Gaps due to mis-matching edge factors.

The term *water tightness* expresses the concern about neighboring patches that share a common edge. Take, for example, a plastic bath toy—a rubber duck—that needs to float in water. If there are any gaps or holes in its surface, it would not be water tight and would sink. Similarly, if this rubber duck were a geometric model being rendered and had gaps between the patches that defined, it we could also declare it as not being water-tight.

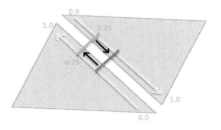

Figure 4.14. Mirrored values on neighboring edge.

There are guarantees from the fixed-function tessellator that assist in this job. First, for a given tessellation factor and partitioning method, it will generate consistent and repeatable distributions of samples—the precision and layout are tightly controlled by both the hardware and software specifications. Second, sample locations are symmetrical about the midpoint of any given edge.

Ultimately, it is responsibility of the author of the domain shader to ensure water tightness. Given *U*, *UV*, or *UVW* sample locations along a common edge, the domain shader code should generate the same output vertex.

Figure 4.14 demonstrates this mirrored distribution along an edge. Due to vertex ordering, or the geometric layout, it is possible that the *U*, *UV*, or *UVW* values are reversed in neighboring patches that share a common edge.

Samples are effectively mirrored around the midpoint (0.5 in Figure 4.14) such that a sample at *X* will be the same as the sample at *1.0–X* on an inverted edge.

A simple solution to this problem is to require neighboring patches to use the same tessellation factors along common edges. You then push the problem of blending between different tessellation factors to the inner surface area of the geometry.

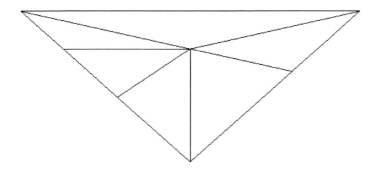

Figure 4.15. Independent edge factors.

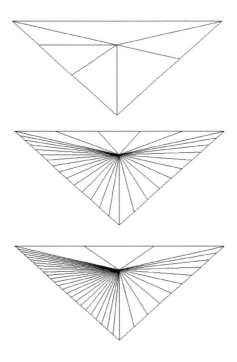

Figure 4.16. Examples of different edge factors.

Top: 1, 2, 3 (clockwise from top)
Middle: 5, 10, 15
Bottom: 3, 10, 25

In more complex settings it is necessary to *blend up* or *blend down* an edge's tessellation factor, based on knowledge of its neighboring patch. Exactly how and when this is done depends entirely on the algorithm being implemented.

Figure 4.15 demonstrates one key detail with edge factors—they are independent. The input mesh is a simple triangle, and the edge factors are 1.0, 2.0 and 3.0, clockwise from the top.

Individual edge factors are independent, and there is no technical requirement for water-tight geometry, a restriction that could be prohibitively complex to force upon artists. However, it is a very desirable property to achieve at the algorithmic level, and many implementations pay special attention to ensuring a smooth transition between neighboring primitives.

Figure 4.16 demonstrates several different edge tessellation factors. In all cases, the inside factor is set at 1.0.

4.4.2 Inside Factors

Figure 4.16 depicts varying independent edge factors which reveals one notable shortcoming with edge factors. The relatively small number of vertices generated in the middle of the primitive makes for a *focal point* on the patch. The edges can be highly tessellated, but the surface still always flows through one single central point, giving it a rather significant influence on the final shape and visual appearance of the surface.

The one (triangle) or two (quad) internal tessellation factors can be used help the developer mitigate this problem. Increasing the internal tessellation factors makes the inner area of the surface more densely tessellated.

In all of the cases shown in Figure 4.17, the four edge factors are 1.0, and the blue numbers 1, 3, 6, or 12 indicate the internal tessellation factor. For a quad, both a *U*- and *V*-axis of tessellation exist. In the diagram, it can be seen that they operate in essentially the

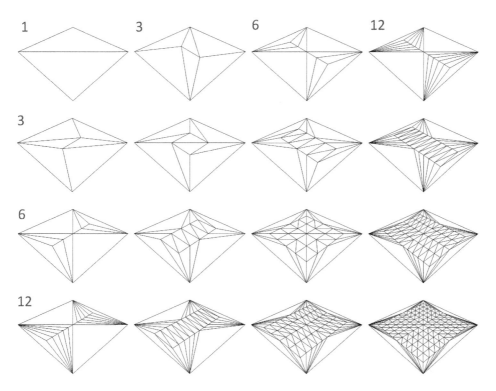

Figure 4.17. Varying internal factors for a quad.

same way, since the bottom-left and top-right halves of the diagram are effectively mirror copies.

Triangles have only one internal tessellation factor, and the surface appears to grow an island of triangles outward from the midpoint as demonstrated in Figure 4.18.

As can be seen in Figure 4.17 and Figure 4.18, having a high internal tessellation factor with a low edge tessellation factor produces the inverse issue that we saw earlier. In this case, the interiors of the primitives are highly tessellated, while they must meet up with very minimal geometry at the edges. Keeping all of the tessellation factors (both internal

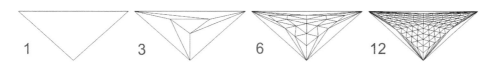

Figure 4.18. Varying internal factors for a triangle.

and external) in a tighter range will greatly improve the final results and avoid the case where factors at opposite extremes come into contact with each other. This will provide a more uniform sampling of the primitive, which in general produces a higher-quality output image.

4.4.3 Partitioning

A detail that has been omitted up to this point is the *partitioning method*, which is declared as an attribute on the main hull shader entry point. This can be `integer`, `fractional_even`, `fractional_odd`, or `pow2` and will affect how the fixed-function tessellator interprets the tessellation factors provided by the hull shader constant function.

Because this value is declared as a compile-time constant, it will be the same for all patches being rendered with this particular hull shader. This removes any potential hassle with trying to match patch edge factors between different partitioning functions—two quads sharing an edge where each quad defines a different edge factor is tricky enough without the two quads using different internal algorithms.

A key observation is that the factors are specified as *float* values by the constant function, yet it makes no real sense to have a fractional quantity as you either generate a new piece of geometry or don't—you can't create half of a triangle!

Some important considerations include:

- Integer partitioning rounds all floating-point values up to their nearest integer in the range 1 to 64.

- Using `fractional_even` rounds to the nearest even number in the range 2 to 64—basically, the series 2, 4, 6, 8, 10 . . . .

- Using `fractional_odd` rounds to the nearest odd number in the range 1 to 63—basically, the series 1, 3, 5, 7, 9, 11 ...

- Using `pow2` rounds according to the 2n series, where n is an integer between 0 and 7. Basically, the possible post-rounding values are 1, 2, 4, 8, 16, 32, and 64.

Figure 4.19 shows the four partitioning methods available—`integer` (left), `fractional_odd` (middle-left), `fractional_even` (middle-right) and `pow2` (right). The vertical axis represents the floating-point tessellation factor provided by the hull shader constant function—the same factor for all six tessellation factors expected for a quad (four edge factors and two internal).

The integer partitioning scheme is straightforward, and the output is as expected—a straight rounding up to the nearest integer. Although there are 9 distinct factors shown (one per row), there are only 3 distinct outputs due to rounding—3, 4, or 5. As soon as the

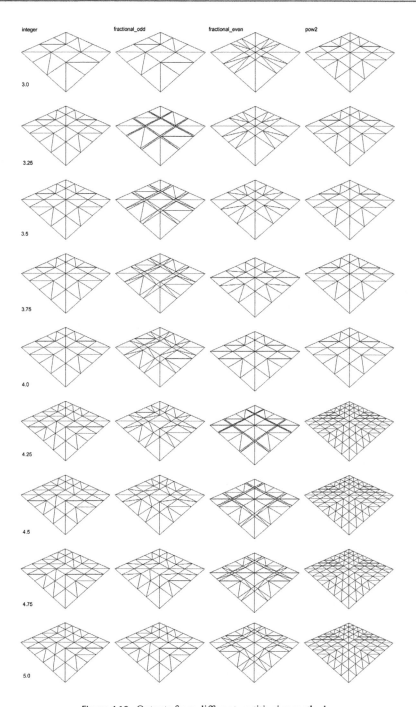

Figure 4.19. Outputs from different partitioning methods.

floating-point value tips over 3.0, the actual value used becomes 4.0. There is no smooth transition, and the change is binary—it's either 3 or 4, nothing else.

The `pow2` partitioning is similarly straightforward, in that it rounds completely and immediately up to the next 2^n integer value. With 9 distinct tessellation factors, there are only two distinct outputs—4 or 8—and as with integer partitioning, there is no interpolation between these two outputs.

Things get much more interesting with the fractional partitioning functions. A quick glance reveals that the size of some of the rows/columns in the grid of triangles changes with the tessellation factor. A closer look will show that they change smoothly in line with the floating-point factor provided by the hull shader.

As with integer partitioning, the two fractional methods will add new segments as soon as the floating point factor crosses one of the odd or even boundaries, but unlike in the integer mode, it will set two of these segments to be incredibly small and have them slowly grow to the desired size of the next factor up.

Expressing this transition as a percentage makes it easier to understand. For the middle column of Figure 4.19 the "raw" tessellation factors are 3.00, 3.25, 3.50, 3.75, 4.00, 4.25, 4.50, 4.75, and 5.00. But only the first and last images in that sequence have completely finished tessellation (only 3.00 and 5.00 are odd numbers!) so the sequence could be 3.00, 12.5%, 25%, 37.5%, 50.0%, 62.5%, 75.0%, 87.5%, and 4.00—the middle percentages are now more like the weighting used in the common linear interpolation equation. The `fractional_even`, right-hand column can be expressed similarly, except that only the 4.00 value in the middle is an even number, and thus of its all segments are equally sized. The right-hand column would become 75%, 87.5%, 4.00, 12.5%, 25%, 37.5%, 50%, and 62.5%.

An interesting observation is that for each integer increment of tessellation factor with `fractional_odd` or `fractional_even` (such as 2–4 or 3–5) will add two new segments. This makes the morphing pattern much more predictable, since the GPU implementation can place one smaller segment on either side of the reflection/mirror axis. A similar morphing effect with the integer partitioning method would be difficult for the same reasons; adding only one new segment per increment means, for even numbers, that it would have to choose one side of the axis and not be a pure reflection.

Transitioning between Levels of Detail (LODs) for integer partitioning can get tricky, but the relatively large jumps for `pow2` partitioning can be very convenient. First, the transition between segments matches the transition between texture mip-map levels (assuming that the convention of 2^n texture dimensions is being used), which may be helpful when the domain shader uses a texture to generate the final geometry. Second, each transition doubles the number of segments, which can also be interpreted as splitting each existing segment into two pieces. For custom morphing algorithms this can be a very useful piece of information, since every sample location from the lower/previous level is still present, and all new vertices are equidistant between previous positions, making it relatively simple to interpolate for the new values.

4.4.4 LOD Transitions and the Risk of Popping

Figure 4.19 in the previous section provides a clear view of one potentially big problem regarding the transition between different LODs. For both `integer` and `pow2` partitioning, there is a sudden jump where new geometry is added to the final output. In any animated rendering sequence where the LOD is dynamic, this sudden change can be very noticeable to the viewer and generally looks very bad—it is often referred to as *popping*.

Both `fractional_odd` and `fractional_even` have continuous morphing between levels, which greatly reduces this problem. The changing shape of the surface will still be visible, but it will be much less noticeable or irritating. However, it is still not completely immune to displaying the transition between tessellation levels. When the raw input factors are changing in a smooth and continuous manner, the partitioning will handle it well, but if the raw factors jump suddenly, the partitioning will similarly jump, and you'll get a "pop" in the final output.

Temporal considerations are important, here, as this visual artifact is almost always caused by inputs changing through time and across multiple frames in an animation. It is important to think about the selection of tessellation factors with regard to how they change through time, rather than as a single, isolated per-frame factor.

5 The Computation Pipeline

5.1 Introduction

Throughout the book up to this point, we have seen how the new features of Direct3D 11 can be used to perform extremely flexible rendering operations. However, there is yet another pipeline stage available for performing flexible computations that can be applied to a wider range of applications. This allows the GPU to be used in such diverse applications as ray tracing and physical simulations, and in some cases, it can even be used for artificial intelligence calculations. Of course, this pipeline stage is the *compute shader* stage, and it represents the implementation of a technology often referred to as *DirectCompute*. DirectCompute provides the first steps to adding a general-purpose processing paradigm to Direct3D, and supplies a more natural and flexible processing environment to the developer to harness the power of the GPU for non-rendering algorithms.

This chapter will explore this new computation pipeline, starting with the architectural details of how the compute shader pipeline stage operates. This includes details about its new threading model, memory model, synchronization system, and addressing scheme. With a clear understanding of how the compute shader functions, we will then turn our attention to how we can use it for various different computational tasks. We attempt to provide general guidelines to follow when designing an algorithm for DirectCompute. Since several thousand threads can be active on the GPU simultaneously, it is important to have a solid understanding of how to harness all of these threads to fully use the power of the GPU.

5.1.1 DirectCompute

DirectCompute introduces a new processing paradigm that attempts to make the massively parallel computational power of the GPU available for tasks outside of the normal raster-based rendering domain. Since the GPU is comprised of a large number of small processors working in parallel, it is especially well suited for computational tasks that allow for strong parallelization. In fact, this is not the first time that GPUs have been used for purposes other than rendering. An entire field of application development, named *general purpose GPU computations* (*GPGPU*), has sprung up, which uses existing rendering APIs to implement support for a desired algorithm.

This process has been effective, but it is not without its drawbacks. Although application developers started to use the 3D rendering API to perform calculations it wasn't originally designed for, the process of implementing an algorithm is significantly more difficult than it is when using a traditional programming language such as C++. In addition, there are certain operations which were not even possible to perform on the GPU through the rendering API, such as scattered writing to resources. The GPU manufacturers were well aware of this situation, and several new APIs have been developed to allow more flexible use of the GPU's power. APIs such as *CUDA* and *OpenCL* provide a more general processing environment than trying to use a 3D API. These solutions are an improvement over the do-it-yourself frameworks.

DirectCompute takes this improved flexibility a step further. DirectCompute is squarely embedded in the Direct3D 11 processing environment, and it shares much of the existing framework from the rendering portion of the API. Within this context, an application developer familiar with Direct3D 10 and/or Direct3D 11 already knows how to perform many of the normal interactions with the API code to get an application up and running. Furthermore, the resource models, execution paradigms, and general debugging process all leverage existing knowledge. These are all reasons that DirectCompute is an attractive technology to use. However, perhaps the single most important benefit that DirectCompute provides is that it is so close to the rest of the rendering pipeline and can easily be used directly to supply input to rendering operations. This makes the API instantly useable by a wide variety of Direct3D programmers, who can use it in their applications almost immediately. DirectCompute uses the exact same resources that are used in rendering, making interoperation between the two very simple. With general-purpose processing available with rendering, we have a very potent combination for use in a wide variety of applications.

Another great benefit of using DirectCompute is that the performance of a particular algorithm can easily scale with a user's hardware. If a user has a high-end gaming PC with two or more high-end GPUs, then an algorithm can easily provide additional complexity to a game without the need to rewrite any code. The threading model of the compute shader inherently supports parallel processing of resources, so adding additional work when more computational power is available is trivial. This is even more true for GPGPU applications, in which the user typically processes as much data as possible, as fast as possible. If an algorithm is implemented in the compute shader, it can easily scale to the current system's capabilities.

With such a wide variety of benefits, and tight integration with Direct3D 11, DirectCompute represents the first steps toward having a fully general-purpose coprocessor. Throughout the rest of this chapter, we will take a detailed look at how this technology works and how it can be used.

5.1.2 The Compute Shader Stage

The compute shader follows the same general usage concept as the other programmable shader stages. A shader program is compiled and then used to create a shader object through the device interface, which can then be loaded into the compute shader stage through the device context interface. The stage can take advantage of the same set of resources, constant buffers, and samplers that we have seen in Chapter 3 for the other rendering pipeline stages, with the additional capability to bind resources to the stage with unordered access views. However, the compute shader is fundamentally different than the other programmable pipeline stages, since it doesn't explicitly have an input or output that is passed from a previous stage or passed to the next stage. It can receive some system value semantics (which will be discussed shortly) as input arguments, but this is the only attribute -type data that can be use in a shader program. All of the remaining data input and output is performed through resources instead.

This arrangement indicates that the compute shader implements complete algorithms within a single program, as opposed to the rendering pipeline, which has the option to implement an algorithm over many different pipeline stages. Of course, the compute shader can be used to iteratively implement an algorithm in steps, but the choice of how best to design the algorithm is left to the developer. This is an interesting design, and it allows for algorithms to be composed in a different way than was possible prior to the introduction of the compute shader. We will further explore the details about how to use the compute shader from the API perspective later in the chapter.

5.2 DirectCompute Threading Model

We begin our journey through DirectCompute by examining its threading and execution model. We have already noted that the GPU is very good at processing parallel algorithms due to its large number of processing cores. With so many processors available to perform work, it is necessary to have some methodology that allows us to efficiently map a particular algorithm to run on many threads. The typical multithreading paradigm used in traditional CPU-based algorithms uses separate threads of execution, coupled with a shared memory space and manual synchronization. It is a fairly well known model and has been in use for many years on systems with multiple processors.

However, this model is not quite optimal when it is mapped onto a processing paradigm that requires thousands of threads operating simultaneously. DirectCompute uses a different type of threading and execution model, which attempts to provide a balance between generality and ease of use. As we will see later in this section, this model allows for easier mapping of threads to data elements. This provides a simple way to break a processing task down into smaller pieces and have it run on the GPU.

5.2.1 Kernel Processing

Also like the other programmable shader stages, the compute shader implements a kernel based processing system. The compute shader program itself is a function, which when executed can be considered to be a form of a processing kernel. This means that the shader program provides a kernel that will be used to process one unit of work. That unit of work will vary from algorithm to algorithm, but the currently loaded kernel is instantiated and applied to a single input set of data. In the case of the compute shader, the data set is provided through access to the appropriate resources bound to the compute shader stage.

This provides a very simple and intuitive way to program work for the thousands of threads that the GPU is capable of operating on. Each thread will be tasked with executing one individual invocation of the kernel on a particular data element. This simple concept reduces the complexity of designing algorithms for many threads by changing the algorithm design task from "What do I make each thread do?", to "Which piece of data does each thread process?" When the kernel is the same for all threads, the task becomes finding a data model that allows the desired data set to be broken into individual data elements that can be processed in isolation, instead of manually trying to synchronize the actions of all the threads. Instead of trying to orchestrate all the different responsibilities for each individual thread, the developer can instead focus on the best way to split a problem into a series of many instances of the same problem for every thread.

5.2.2 Dispatching Work

With an understanding of what the individual threads will be executing, we can continue on to consider how the developer actually executes a batch of work with the desired number of threads. This is performed with the device context ID3D11DeviceContext::Dispatch() method, or the ID3D11DeviceContext::DispatchIndirect() method,

which is similar to the indirect rendering methods we have already seen in Chapter 3. The
Dispatch method is analogous to the many draw call calls from the rendering pipeline.
It takes three unsigned integer parameters as input: x, y, and z. These three parameters
indicate how many *groups* of threads we would like to "dispatch" to execute the desired
processing kernel. These three parameters provide the dimensions of a three-dimensional
array of thread groups that will be instantiated, and the size of each parameter can range
from 1 to 64k. For example, if an application calls Dispatch(4,6,2), a total of 4*6*2 = 48
thread groups will be created. Each group of threads would be identified by a unique set of
indices within the specified dispatch arguments, ranging from 0 to size-1 in each of the
three dimensions.

Notice that the dispatch call defines how many *groups of threads* are instantiated,
and not how many *threads* are instantiated. The number of threads instantiated is defined
by specifying how many threads will be created for each thread group with a numthreads
HLSL function attribute preceding the compute shader program in its source code file. As
in the dispatch call, this numthreads statement also defines the size of a three dimensional array, except that this array is made up of threads instead of thread groups. The size
of each of these parameters is dependent on the shader model being used, but for cs_5_0
the x and y components must be greater than or equal to 1, the z component must be between 1 and 64, while the total number of threads ($x*y*z$) cannot exceed 1024. The statement shown in Listing 5.1 is an example of how this numthreads statement must appear in
the HLSL shader program.

```
[numthreads( 10, 10, 2 )]
// Shader kernel definition follows...
```

Listing 5.1. The number of threads per thread group declaration.

With this example, a total of 10*10*2 = 200 threads will be instantiated for each
thread group. Each of these threads can also be uniquely identified by its integer indices,
ranging from 0 to size-1 in each of the three dimensions. If we use the dispatch call
example from above, we would have a total of 48*200 = 6,400 threads instantiated for this
dispatch call.

However, with the three dimensional method of specifying and identifying these
threads, we can think of their organization in a geometric way. Consider the visualization
of a thread group shown in Figure 5.1.

Instead of thinking of the threads in our dispatch call as a linear list of 6,400 threads,
we can instead use the thread group size to provide a spatial orientation for each thread.
Since each thread has a unique identifier, it is easy to reference one particular thread within

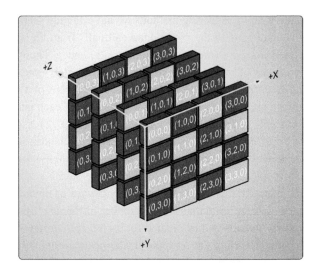

Figure 5.1. A visualization of the threads within a thread group.

the thread group. Similarly, we can visualize all of the thread groups defined by the dis-patch call as a three dimensional array, as shown in Figure 5.2.

If we combine the two previous images, we can get a sense for how each individual thread is located within the complete system of threads instantiated by the dispatch call. This is shown in Figure 5.3.

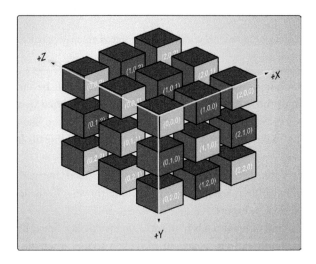

Figure 5.2. A visualization of the thread groups within a dispatch call.

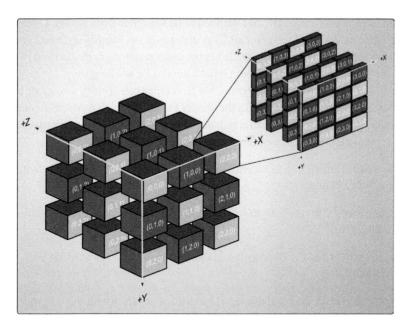

Figure 5.3. A visualization of individual threads within a dispatch call.

From this geometric interpretation of the thread locations, we can also consider an overall unique identifier for each of the threads from within the complete dispatch call. This identifier could essentially locate the thread with an *X*, *Y*, and *Z* coordinate within the three-dimensional grid shown in Figure 5.3. As described before, each of these threads will be used to execute one instance of our processing kernel.

5.2.3 Thread Addressing System

At this point, we have an understanding of how the compute shader instantiates threads, and subsequently, of how many instances of our processing kernel will be executed for a given dispatch size. From the example dispatch call in the previous section, we would have 6,400 threads/kernel instances being executed. But how do all of these individual threads know what data to operate on? If they are all instantiations of the same kernel program, then the desired data cannot be manually specified in the shader program. There must be a different mechanism for informing each thread about which portion of a data set it should be processing.

The geometric interpretation of the threading structure of a dispatch call holds the key to understanding how this information is provided. This geometric organization of the threading system was an intentional design decision, which allows for a very clear and concise thread addressing system. As we saw in the previous section, it is trivial to identify each of the threads of a dispatch call with a unique identifier from within a thread group, or from within the entire dispatch call. It is also trivial to identify a thread group from within a dispatch call. It turns out that this is the exact mechanism that is used in the computer shader to inform each thread about what it should be working on.

When authoring the shader program, the developer can specify a number of system value semantics as input parameters for the shader function, which provide an identifier for that particular invocation. The list of available system value semantics is provided below, with a short explanation of each of their meanings.

- SV_GroupID: Defines the 3D identifier (uint3) for which thread group within a dispatch a thread belongs to.

- SV_GroupThreadID: Defines the 3D identifier (uint3) for a thread within its own thread group.

- SV_DispatchThreadID: Defines the 3D identifier (uint3) for a thread within the entire dispatch.

- SV_GroupIndex: Defines a flattened 1D index (uint) for which thread group a thread belongs to.

Once each thread invocation has this identifier information, the developer can use that identifier to determine what portion of the input data set should be processed by each thread. To see the utility of such a processing scheme, it is beneficial to consider an example. One simple example would be a compute shader that doubles the value of every element within a Buffer<float> resource. If the buffer contains 2,000 elements, then we could define a thread group size of [20,1,1], which is dispatched with [100,1,1] thread groups. Listing 5.2 shows what the compute shader would look like for this example.

```
Buffer<float>    InputBuf : register( t0 );
RWBuffer<float>  OutputBuf : register( u0 );

// Group size
#define size_x 20
#define size_y 1

// Declare one thread for each texel of the input texture.
[numthreads(size_x, size_y, 1)]

void CSMAIN( uint3 DispatchThreadID : SV_DispatchThreadID )
{
    float Value = InputBuf.Load( DispatchThreadID.x );

    OutputBuf[DispatchThreadID.x] = 2.0f * Value;
}
```

Listing 5.2. A sample compute shader for doubling the contents of a buffer resource.

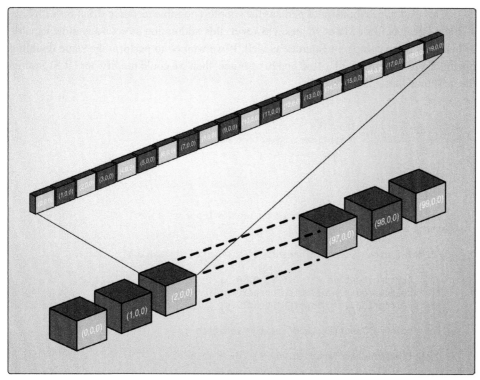

Figure 5.4. The threading structure for the sample compute shader program.

Notice that in the code listing, the size of the thread group is declared with the numthreads function attribute. This is followed by the function definition that represents the processing kernel for our compute shader. In the arguments to this function, we specify which system values we would like to receive to use for our addressing scheme. Since there is some redundancy in the system values (there is more than one way to identify each thread), we have some freedom in choosing which ones would be the most efficient or advantageous to use. In this case, we can simply use the SV_DispatchThreadID and use its X-component to provide a linear index into our buffer resource. This thread addressing structure can be visualized as shown in Figure 5.4.

With SV_DispatchThreadID, we can directly use its X-component as the index when reading from the buffer resource. The buffer can be accessed with the normal square bracket array syntax. We would bind the resource to the compute shader stage with an unordered access view (UAV), which allows each thread to read the float value, double it, and then write it back to the same resource. The dispatch thread ID provides a trivially simple way to have each thread access the elements of a resource, only leaving the developer to determine how large of a dispatch call to make.

This example demonstrates perhaps the simplest possible resource structure with only a single linear buffer of float values. However, this addressing structure is quite capable of handling more complex resources as well. If we wanted to perform the same doubling operation on all elements of a `Texture2D` resource, then we could modify our HLSL source file as shown in Listing 5.3.

```
Texture2D<float>    InputTex : register( t0 );
RWTexture2D<float>  OutputTex : register( u0 );

// Group size
#define size_x 20
#define size_y 20

// Declare one thread for each texel of the input texture.
[numthreads(size_x, size_y, 1)]

void CSMAIN( uint3 DispatchThreadID : SV_DispatchThreadID )
{
    int3 texturelocation = int3( 0, 0, 0 );
    texturelocation.x = DispatchThreadID.x;
    texturelocation.y = DispatchThreadID.y;

    float Value = InputTex.Load( texturelocation );

    OutputTex[DispatchThreadID.xy] = 2.0f * Value;
}
```

Listing 5.3. A sample compute shader for doubling the contents of a 2D texture resource.

Here we can see that we have changed the number of threads per thread group, as well as the type of resource that will serve as our input and output storage. There is no change in the system value semantics used for addressing remains, since the `dispatch` thread ID can also handle two-dimensional identifiers. If our input texture resource was 640×480, then to process it, we would bind it to the compute shader with a UAV and call `Dispatch` with a size of [32,24,1]. After the `dispatch` call was completed, the resource's data elements would all have doubled values. This same concept would also be simple to extend to a three-dimensional case, as well.

Up to this point, we have only considered direct resource mappings where the thread addressing structure maps directly to the shape of the resource we are working with. However, one of the benefits of allowing the developer to directly perform the resource accesses in the shader program is that it is also possible to create customized resource schemes. For example, if we wanted to store a four-dimensional data set in a resource, there are no native four dimensional resource types. However, we could easily use a linear buffer resource and then manually implement the data access patterns. If we wanted a [10,10,10,10] resource, we would create a buffer resource with 10*10*10*10 = 10,000 elements. Then we would modify our numthreads statement to use a size of [10,10,10], and use a dispatch call of size [10,1,1]. Listing 5.4 demonstrates how the shader would be modified to calculate the index to lookup in the buffer for each element.

```
Buffer<float>    InputBuf : register( t0 );
RWBuffer<float>  OutputBuf : register( u0 );

// Group size
#define size_x 10
#define size_y 10
#define size_z 10
#define size_w 10

// Declare one thread for each texel of the input texture.
[numthreads(size_x, size_y, size_z)]

void CSMAIN( uint3 DispatchThreadID : SV_DispatchThreadID, uint3 GroupID :
SV_GroupID )
{
    int index = DispatchThreadID.x +
                DipsatchThreadID.y * size_x +
                DipsatchThreadID.z * size_x * size_y +
                GroupID.x           * size_x * size_y * size_z +


    float Value = InputBuf.Load( index );

    OutputBuf[index] = 2.0f * Value;
}
```

Listing 5.4. A sample compute shader for doubling the contents of a custom 4D resource.

Here we simply change our resource access code to use each thread's SV_GroupID system value as the fourth-dimension index, and then calculate a linear address for that particular element's location in the buffer resource. The important consideration here is that how a resource is accessed does not need to be simple—it is quite possible to calculate arbitrary memory locations within a resource, which provides significant freedom for the developer to implement the desired access patterns.

5.2.4 Thread Execution Patterns

Throughout this section, we have seen how the threading model of the compute shader functions from the developer's perspective. However, it is important to understand that an implementation may or may not execute the the threading commands of the developer precisely how they are declared. For example, when a thread group is declared, it can have up to 1024 threads. From a programmatic point of view, all of these threads execute simultaneously. However, from a hardware perspective they may not all execute in parallel. Specifically, if a particular GPU doesn't have 1024 processing cores, then it is impossible for a complete thread group to be executed simultaneously.

Instead, the threads are executed in a manner that ensures that they behave as if they were operating at the same time. For example, whenever a point in the shader program requires a synchronization of all of the threads (synchronization is covered more later in this chapter), then each subgroup of threads will be executed to the synchronization point, and then are swapped out so that another subgroup can be executed to the same point. Only after all of the threads in a thread group have completed up to this synchronization point can they continue on. This method of operation can have some performance implications if there is excessive synchronization points in the compute shader, but how much of an impact will depend on the GPU hardware that it is executing on. As the number of processing cores continues to increase, this will be come less and less of a performance issue.

5.3 DirectCompute Memory Model

The overall compute shader execution model provides a great deal of flexibility for instantiating a suitable number of threads to execute the desired processing kernel on the elements of a resource. It is easy to map a complete resource to a given number of threads and perform some computation on each of its data elements. With this execution model in mind, we will now turn our attention to what can be done within the compute shader itself. We will investigate some of the unique features of the compute shader memory model that give developers even more flexibility in deciding how to implement an algorithm.

5.3.1 Register-Based Memory

The compute shader runs on the same programmable processing hardware as the other programmable shader stages. This means that it is based on the same general processing paradigm and also implements the common shader core. It uses a register-based processing concept similar to that for the other pipeline stages, with the exception that the computation pipeline only consists of a single stage. The set of registers that the computer shader supports is quite similar to that of the other programmable stages, and supports input attribute registers (v#), texture registers (t#), constant buffer registers (cb#), unordered registers (u#), and temporary registers (r#, x#). Since all shader programming is performed in HLSL, these registers are not directly used and are not directly visible to the developer. However, the shader programs are eventually compiled into an assembly form before they can be used to create shader objects for use in the pipeline. In addition, it is also possible to obtain an assembly listing of a compiled shader (using the fxc.exe tool), which does allow for inspection of the low-level details of a program.

Because of this, it is worth understanding some of the register functionality. The basic architecture of the common shader core has been covered in Chapter 3, so please review this section for reference during this discussion. In this section, we are the most interested in the temporary registers. These registers can be used to hold intermediate calculations during execution of a shader program. They are only accessible to the thread that is currently executing, and are typically extremely fast registers. Up to 4096 temporary registers (r# and x# combined) are specified in the common shader core, which seems like a fairly large amount for the small scale shader programs. However, while the hardware implementation must be able to handle 4096 registers, it doesn't necessarily need to provide each processing core its own set of registers. Many architectures share a pool of registers among many processing cores, meaning that the overall available number of registers can in reality be less than 4096. While this level of detail is hidden from the developer, the upper limit of 4096 registers is a good indication of how much data can be stored in these registers during the execution of a shader program.

The temporary registers used in this processor architecture can be considered the first class of memory that the compute shader can use. Due to its access speed, this storage memory mechanism is the first choice for data that must be kept available within the shader program. Selection and allocation of these registers are performed automatically by the compiler, so in general, once data has been loaded into the shader core, it will use temporary registers as much as possible.

5.3.2 Device Memory

While the register-based memory is very fast, it is finite in size. In addition, the desired data must be loaded into the shader core before the registers can be used, and after a thread has completed its shader program, the contents of these registers are reset for the next shader program. Of course, we need to have data that persists between executions of a shader program, and we also need much larger possible storage areas. This need is filled by the memory resources that we have already seen many times throughout the book. In the compute shader context, these resources are commonly referred to as *device memory resources* to distinguish them from the other available types of memory.

Direct3D 11 provides a significant array of resource types that can be used for read-only, write-only, or read/write access. The compute shader can use shader resource views and unordered access views to access device memory resources. These two resource views allow read and read/write access, respectively. In addition, the compute shader is also able to utilize constant buffers in the same manner we have seen in the rendering pipeline. Constant buffers provide read-only access to the data stored in them.

Among all the different types of resources that these access mechanisms can attach to, there are literally gigabytes of storage available for shader programs to use. However, to accommodate this large amount of memory, it must be stored in memory located outside of the GPU itself. It is currently not feasible to store such large amounts of data within a processor, so there is generally off-board memory modules used. This memory is normally accessed with a very high bandwidth connection, but there is also a relatively high latency between the time a value is requested and when it is returned. Because of this, device memory resources are considerably slower than register-based memory. While an unordered access view can be used to implement the same operations in device memory as in register-based memory, there would be significant performance penalties when frequent read and write operations are performed.

Another consideration for device memory resources is that access to these resources is provided to all threads that are executing the current shader program. This means that if there are 6400 threads (as we saw in our initial `dispatch` example), each of those threads can read or write to any location within a resource through an unordered access view. Naturally, this requires manual synchronization of access to the resource, either by using atomic operations or by defining an access paradigm that can adequately ensure that threads will not overwrite each other's desired data ranges.

5.3.3 Group Shared Memory

As discussed in Chapter 3, "The Rendering Pipeline," all of the programmable shader stages in the rendering pipeline are kernel based. Each instance of the rendering pipeline kernels is performed in complete isolation from one another. The compute shader breaks free from this paradigm and allows the use of a shared memory area that can be accessed simultaneously by more than one thread. In fact, every thread in a complete thread group is allowed to access the same memory area. This gives the shared memory its name—*group shared memory* (*GSM*). The group shared memory is limited to 32 KB for each thread group, but it is accessible to all of the threads in the thread group. It is intended to reside on the GPU processor die, which allows for much faster access than the device memory resources.

The group shared memory is declared in the global scope of the compute shader with a special storage scope identifier called groupshared. This memory area can be declared either as an array of basic data types, or as structures of more complex data arrangements. Once a thread group is instantiated, the group shared memory is available to all of its threads simultaneously. Since the entire group shared memory is available to all threads in the group, the compute shader program must determine how the threads will interact with and use the memory, and hence it also must synchronize access to that memory. This will depend on the algorithm being implemented, but it typically involves using the thread addressing mechanisms described in the previous sections to ensure that access to the GSM is performed safely, without the need for atomic functions.

While group shared memory provides a fast and efficient means for threads within a thread group to share information, it does come with some limitations. Since it is limited to 32 KB for each thread group, in any situation where a larger shared pool of memory is needed it is not sufficient. In addition, sharing information is confined to within a single thread group. If an algorithm calls for a large number of threads to all have access to the shared memory pool, group shared memory is not an option.

The group shared memory joins the register-based memory of the programmable shader cores and the larger resource-based memory that can be bound to the pipeline. These three types of memory provide a variety of access speeds and available sizes, which can be used in different situations that match their abilities. Register-based memory is the fastest to access, but it has the smallest amount of memory available. Device memory resources provide gigantic available memory sizes, but also exhibit the slowest access times. Group shared memory strikes a balance between the other two. It is faster to access than resource memory and provides a larger available memory size than register-based memory.

This is a major increase in flexibility for the compute shader. With a memory area that is accessible to multiple threads, it is possible for threads to share information with one another. It also increases the potential for improved efficiency of a thread group as a whole. For example, texture accesses needed by more than one thread could be loaded by one thread and then shared with all other threads in the thread group. This would effectively lower the overall number of texture accesses for the thread group, and would thus improve the overall efficiency of the algorithm. In this way GSM can be used as a customized memory cache that can be directly controlled by the shader program. Sharing between threads is not limited to simple texture accesses, either. It is also possible to share intermediate calculations between threads, which would otherwise need to be performed individually. We will see examples of both of these optimizations in the second half of this book.

5.4 Thread Synchronization

With a large number of threads operating simultaneously, and with the ability for threads to interact with one another through either the group shared memory or through unordered access views of resources, there is clearly a need to be able to synchronize memory access between threads. As with traditional multithreaded programming, where many threads can read and write the same memory locations, there is a potential for memory corruption due to read-after-write hazards (additional details about multithreading programming on the

CPU can be found in Chapter 7, "Multithreaded Rendering"). How can such a massive number of threads be efficiently synchronized without losing the performance that the GPU's parallelism provides? Fortunately, several different mechanisms are available for synchronizing the threads of a thread group. We will explore each of these possibilities in the following sections.

5.4.1 Memory Barriers

We will first look at the highest-level synchronization techniques, referred to as *memory barriers*. HLSL provides a number of intrinsic functions that can be used to synchronize memory accesses across all threads in a thread group. It is important to note that this is an access mechanism that synchronizes only the threads within a thread group, and not across an entire dispatch. These functions have two properties that differentiate them from one another. The first is the class of memory that the threads are synchronizing across when the function is called. It is possible to synchronize access to the group shared memory, device memory, or both. The second property specifies whether all of the threads in a given thread group are synchronized to the same point within their execution. These two properties provide a range of different synchronization behaviors for the developer to choose from. The different versions of these intrinsic functions are listed in Table 5.1 below.

Without Group Synchronization	With Group Synchronization
GroupMemoryBarrier()	GroupMemoryBarrierWithGroupSync()
DeviceMemoryBarrier()	DeviceMemoryBarrierWithGroupSync()
AllMemoryBarrier()	AllMemoryBarrierWithGroupSync()

Table 5.1. Intrinsic Functions: without and with group synchronization.

Each of these functions will block a thread from continuing until that function's particular conditions have been met. The first function, GroupMemoryBarrier(), blocks a thread's execution until all writes to the group shared memory from all threads in a thread group have been completed. This is used to ensure that when threads share data with one another in the group shared memory that the desired values have had a chance to be written into the group shared memory before being read by other threads. There is an important distinction here between the shader core executing a write instruction, and that instruction actually being carried out by the GPU's memory system and being written to memory, where it would then be available again to other threads. Depending on the hardware implementation, there can be a variable amount of time between writing a value and when it actually ends up at its destination. By performing a blocking operation until these writes

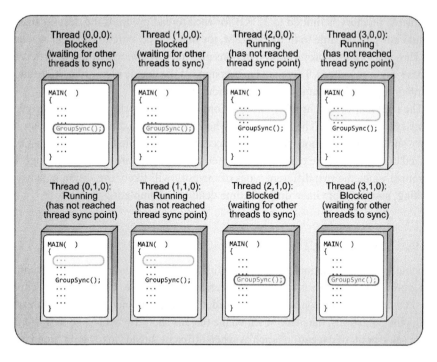

Figure 5.5. The synchronization of multiple threads within a thread group with a memory barrier with group synch.

are guaranteed to have completed, the developer can be certain there won't be any issues with read-after-write errors.

The sister function for `GroupMemoryBarrier()` is `GroupMemoryBarrierWithGroup Sync()`. This function blocks a thread from continuing until all group shared memory writes are completed, but it also blocks execution until all of the threads in the group have advanced to this function. This allows the developer to be certain that all threads have had a chance to execute to this function call before other threads advance farther. This is a useful mechanism to utilize when each thread in a thread group is loading a set of data into the group shared memory for the other threads to use. It is clear that we wouldn't want any thread to advance until all of the group shared memory has been loaded, making this the perfect synchronization method. Figure 5.5 demonstrates how this thread management is performed.

The second pair of synchronization functions performs similar operations, except that they operate over the *Device* memory pool. This means that all pending memory write operations that write to a resource through an unordered access view can be synchronized before continuing execution of the shader program. This can be very useful for synchronizing larger amounts of memory, which would require the use of device resources since the group shared memory is limited to 32 KB for each thread group. If the desired size of the shared memory is too large to fit in the GSM, the data can be stored in a larger resource instead. These *Device* synchronization functions provide a synching method when a program uses these larger resources. There are two versions of the device memory barrier functions, one with group synchronization and one without it.

The final pair of synchronization functions essentially performs both of the previous types of synchronization. They are useful when there is a mixture of both group shared memory and resource memory that will be accessed by multiple threads and must be synchronized before use. These types of mixed memory scenarios are particularly important to synchronize, since the two different types of memory will likely use different subsystems for writing, and therefore have different time delays for completing the writes.

5.4.2 Atomic Functions

The memory barrier functions are very useful for synchronizing all of the threads in a thread group. However, this is not always necessary or desirable. There are many situations when smaller scale synchronizations are needed, perhaps among only a few threads at a time. In other situations, the place at which the threads should synchronize may or may not be at the same point of execution (such as when different threads in a thread group perform heterogeneous tasks). In these cases, the memory barrier functions are not an appropriate synchronization method.

Instead, Shader model 5 has introduced a number of new atomic functions that can provide more fine-grained synchronization between threads. These functions are guaranteed to be executed in the order that they are programmed, and they can thus be used from one thread, and the result of the function will be propagated to any other threads trying to access the same destination. The following list specifies all of the available functions.

- `InterlockedAdd()`

- `InterlockedMin()`

- `InterlockedMax()`

- `InterlockedAnd()`

- `InterlockedOr()`

- `InterlockedXor()`

- `InterlockedCompareStore()`

- `InterlockedCompareExchange()`

- `InterlockedExchange()`

Like the memory barrier functions, these atomic functions can be used on group shared memory as well as resource memory, which allows for a wide range of potential uses. Each function performs an operation that can be used to turn the contents of either a group shared memory location or a device resource location into a synchronization primitive. For example, if a compute shader program wants to keep a count of the number of threads that encounter a particular data value, then the total count can be initialized to zero, and each thread can perform an `InterlockedAdd()` function on either a GSM location (for the fastest access speed) or a resource (which persists between `dispatch` calls). These atomic-style functions ensure that the total count will be incremented properly without any overwriting of the intermediate values by different threads.

Since each of these functions provides a different type of operation, developers have a significant amount of freedom to implement a desired type of synchronization. For example, the `InterlockedCompareExchange()` function can be used to compare the value of a destination to a reference value, and if the two match, a third argument could be written to the destination. This functionality allows for the implementation of data that can be "checked-out" by a thread and later checked back in for use by another thread. Since these functions are very low level, they can be applied to a particular situation very flexibly. Each function has its unique input requirements and operations, so the Direct3D 11 documentation should be referenced when selecting an appropriate function. These functions are also available to the pixel shader stage, allowing it to also synchronize across resources (since it doesn't have a group shared memory, it can only use the device resources for thread-to-thread communication).

5.4.3 Implicit Synchronization

The final form of synchronization that we will discuss is actually not an explicit part of the compute shader functionality. We will refer to this as *implicit synchronization*, which occurs when an algorithm is designed to access its data set in such a way that there are no potential interactions between threads. This is the preferred method of synchronization—when no synchronization is needed! Each of the other two methods mentioned above is effective and useful, but both come at some cost to performance. If an algorithm can access a memory resource in an explicit and orchestrated manner, then no additional functions are needed, and no extra thread context switching is needed either.

For example, in our simplified example program from earlier in the chapter, we started by reading a value from a resource, then doubled it, and then stored it back to the resource. In this case, no communication was needed from thread to thread, and hence no synchronization was needed. Each thread can operate completely independently from one another. Another example of this type of algorithm is the creation of a particle system, where the particle state is stored in a structured buffer and accessed with an append/

consume structured buffer resource object. In this case, each thread would read one particle's data using the consume intrinsic function. Since the thread doesn't know which particle it is getting, it is completely independent of any other particle's data and can hence execute independently of the other threads. After the particle is updated, it can be added back into an append structured buffer with the append intrinsic function. There is no need to synchronize between threads, and hence the individual GPU processing elements can execute without managing any extra interthread communication. This type of particle system is explored further in Chapter 12, "Simulations."

5.5 Algorithm Design

Throughout this chapter, we have learned about the various capabilities of the compute shader. Some of the concepts we have seen are quite similar to the other programmable shader stages, but some are quite different from what we have seen before. Indeed, the concept of having a raw-computation-based shader stage is a completely new idea that has been added in Direct3D 11. With the introduction of so much new functionality, it can be somewhat difficult to approach a completely new algorithm and decide what tools to use to implement it. This section aims to provide some general design guidelines that can be applied when developing an algorithm. Of course, there is no perfect methodology to design an algorithm, so these guidelines should be taken as suggested starting points that can be built on for a particular scenario.

5.5.1 Parallelism

The first area we will consider is how to maximize the parallelism of an algorithm. The whole reason that the compute shader has been added to Direct3D 11 is to allow developers to harness all of the GPU's available parallel processing power. We have alluded to this throughout the chapter, but it should be an explicit design goal when developing an algorithm to run in the compute shader. The data to be processed should be organized in such a way that it can be processed with a minimal amount of memory access and computation, which will result in a generally faster algorithm. If the problem can be broken down into smaller, coherent parts, then the compute shader should be a good candidate for performing the calculations.

Minimize Synchronization

As discussed in the previous section, there are many ways to synchronize data between threads. Group shared memory, device resources, atomic functions, and memory barriers all provide different varieties of synchronization techniques. However, these synchronization techniques all introduce some overhead during the execution of the compute shader. If it is possible to perform the same calculations without synchronization, the algorithm will run faster. It is often easier to design an algorithm with synchronization than without, but unless the synchronization methods are used to increase efficiency, their use may be detrimental to performance.

Sharing Between Threads

The next point may seem contradictory to our previous comments, but in some cases, explicitly designing synchronization into an algorithm can result in improved performance. When memory bandwidth or the computational load can be reduced by sharing data between threads, it is certainly possible to speed up an algorithm's execution time by synchronizing data across multiple threads. The key is to determine when it is appropriate to do so.

Share loaded memory. One of the primary uses for compute shader programs is in image processing algorithms. Because image-like resources are accessed, it is very natural to map the compute shader onto an image-processing basis. This also happens to be one of the areas that can take advantage of the group shared memory to improve performance. Depending on the algorithm being implemented, image processing is typically bound by the memory accesses it performs. However, if multiple pixels within a thread group can use the same sampled values, then it is quite possible to load a small number of values into the GSM in each thread, followed by using a memory barrier with group synchronization that can be accessed by all of the threads from that point on. This effectively reduces the number of device memory accesses that each thread needs to perform and moves the desired data into the GSM, which is in general faster to access.

However, care must be taken with this approach as well. There is some latency involved with writing to the GSM and then reading from it. If the reduced device memory bandwidth does not offset the costs of accessing the GSM, then this technique could actually hurt performance. This topic becomes even less clear when texture caches are taken into consideration. It is quite possible that the built in texture caches are already performing sufficient data caching, making it difficult to predict which technique will be faster. This can also vary by GPU manufacturer, complicating matters even further. A good suggestion is to write your algorithms in a way that lets you quickly test them in both scenarios, and the higher-performing method can be chosen appropriately. An even better approach is to allow your algorithms to test the current platform and dynamically decide which technique to use.

Share long calculation results. Just as loaded device memory contents can be cached, threads can also share calculated values in the GSM. This is also difficult to predict if it is faster to share calculations, or to just perform them independently in each thread. Modern

GPUs are typically able to perform many arithmetic logic unit (ALU) operations in the time it takes to fetch content from device memory, and there is a somewhat smaller difference when reading from the GSM. Once again, a best practice would be to develop an algorithm so it can either independently calculate the desired values, or share them through the GSM, and then either profile the performance or dynamically decide which version to use.

5.5.2 Choose Appropriate Resource Types

Another very important consideration is the selection of the resource type that will be used. Depending on the data being processed, one resource type may provide a better overall algorithm design than others. The resource type will dictate how the resource is accessed by a thread group, and will also dictate the actual thread group shape and sizes.

Memory Access Patterns

In some cases, the best resource type to choose is fairly obvious. For example, image processing algorithms typically work with a two-dimensional texture, since that is the format that images are stored and used in. However, other algorithms may not be as clearly defined. For example, when implementing a GPGPU algorithm, the data set can usually be manipulated into whichever resource type makes the most sense. One of the most important considerations in this regard is *how* the data must be accessed. Earlier, we mentioned a particle system using append/consume buffers. Since the algorithm doesn't care about the order the particles are processed in, it can use a buffer resource that allows the append/consume functionality. Alternatively, if there are data sets that are not directly accessed, but are rather spatially sampled, texture resources would be a much better choice. There are also intrinsic functions, such as the `gather` function, that can return multiple values from texture resources, but that can't be used on buffer resources. The resource type should be chosen so that the algorithm can take advantage of all of the available built-in hardware and software functionality.

Thread Group and Dispatch Size

Once a resource type has been selected, an appropriate threading pattern needs to be chosen. In reality, this will probably be decided in conjunction with the resource type to select the best method of accessing the needed data. The dimensions of a thread group will dictate the thread addressing system that can be used to access a resource, both for reading input data and for eventually writing output data. In addition, the total size of the resource must be covered by the dispatch dimensions, which instantiates thread groups. Therefore, these two dimensions can be chosen simultaneously to provide the simplest and most efficient method of accessing a resource.

This can be a tricky decision, and it can take some trial and error to find the best technique for a given situation. In general, these two sets of dimensions should be chosen so that a thread group contains threads that will be accessing coherent memory locations and/ or can share intermediate calculations with each other. Then the dispatch size can be used to ensure that the complete resource is processed accordingly.

Mixing Computation with Rendering

The final design consideration we will look at regards the use of a resource for performing some rendering. If the compute shader is used to process a data set, and the end results of the processing will be used in a rendering pass, the output resource must be chosen such that it will allow for the most efficient possible access. This will be a balancing act between choosing a resource type and layout that allow for efficient calculations in the compute shader, while also allowing for efficient use of the output data in rendering operations.

6

High Level
Shading Language

6.1 Introduction

The core of the Direct3D 11 pipeline is its various programmable shader stages. These stages are where the vast majority of rendering tasks occur, and are also the means by which a GPU's power and flexibility are made available to the user. While various API functions can manipulate the state in which a shader program will be executed, the actual authoring of shader functionality and behavior is not done with these APIs. Instead, shader programs (also referred to as simply *shaders*) are authored, using a specially designed programming language known as *High Level Shading Language*.

High Level Shading Language (commonly abbreviated as *HLSL*), is primarily a C/C++-derived language with a simplified feature set. Things like semicolons for line termination, braces for statement blocks, and C-style function and struct declarations make the language immediately familiar to a C or C++ programmer. The major exceptions are that pointer types and operations are not supported, as well as C++-style templates. In addition, the language does not support a dynamic memory allocation model. As in C or C++, programs are written by authoring a function that serves as the entry point. This code is executed at runtime and runs until the end of the function is reached. It should be noted that in D3D11, the language is always compiled in advance, before shaders are bound to the pipeline, as opposed to being dynamically interpreted, like a scripting language. To support static compilation, the compiler features a C-style preprocessor. The preprocessor supports common uses such as macro definitions, conditional compilation, and include statements.

One of the key features of HLSL in D3D11 is that the same language is used to author shader programs for all shader stages. This provides a consistent set of standard features for all stages, allowing common code to be shared among shader programs written for each stage of the pipeline. Authoring in a high level language also allows many of the low-level details of the underlying graphics hardware to be abstracted, allowing shader programs to extract performance from a wide range of graphics hardware.

6.2 Usage Process

Figure 6.1 shows the typical sequence for authoring and running an HLSL program.

6.2.1 Authoring and Compilation

The usage process begins by authoring the shader program in HLSL. Typically, this is done by creating a standard ASCII text file containing the code for one or more shader programs. This can be done with any standard text editor or development environment. Once the code for the program is completed, it is compiled by the D3D shader compiler. The compiler is accessed through the D3DCompile function exported by the D3DCompiler DLL, or through the command line by using the standalone fxc.exe tool. There are also D3DX helper functions (D3DXCompileFromFile\D3DXCompileFromResource) that simplify the process of compiling directly from a file or from an embedded resource. In all cases, the compiler accepts configuration parameters, in addition to code itself. These parameters indicate the entry point function, the shader profile to use for compilation, a list of preprocessor macro definitions, a list of additional files to be included by the preprocessor, and a list of compilation flags (including optimization and debug options). If compilation fails or warnings are produced, the D3DCompile and D3DX functions store the warning or error messages in a string embedded in an ID3D10Blob object. As with C++ compilation, these messages include an informative message and a line number, so that the errors or warnings can be easily found and corrected. If fxc.exe is used, the messages will simply be output to the command line.

For debug builds of applications, it is typically desirable to compile shader programs with the D3D10_SHADER_DEBUG (/Zi switch in fxc) flag enabled. This causes the compiler to embed debug information into the bytecode stream, which can be used to map instructions and constants back to their respective lines in the HLSL source code. This enables source-level debugging of shader programs, which can be much more convenient than directly debugging the assembly. Note that it may also be desirable to disable optimizations by using D3D10_SHADER_SKIP_OPTIMIZATIONS (/Od switch in fxc), as this will prevent instruction reordering and folding, making it easier to debug the programs.

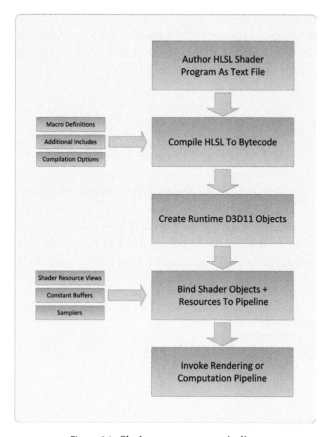

Figure 6.1. Shader program usage pipeline.

Shader programs can be debugged using PIX, a debugging and profiling utility included with the DirectX SDK. Consult the SDK documentation for additional information and usage instructions.

6.2.2 Bytecode

Once compilation successfully completes, the compiler outputs to an intermediate format known as *bytecode*. This is an opaque binary stream containing all of the assembly instructions for the shader, as well as embedded reflection and debugging metadata. If a programmer wants to inspect the resulting assembly instructions for verification, it can be obtained by using the D3DDisassemble function exported by the D3DCompiler DLL. The command-line tool fxc.exe will also output the assembly to the command line when compilation completes, or it can be configured to output the assembly listing to a file. In addition, PIX can be used to inspect D3D11 shader objects and view the assembly code. For more information on shader assembly, see the HLSL reference section in the DirectX SDK documentation.

6.2.3 Runtime Shader Objects

Once the bytecode is produced for a shader program, it is ready to be used to create a D3D11 shader object. The ID3D11Device interface provides six methods for creating shader objects, one for each programmable stage of the pipeline. These methods are CreateVertexShader, CreateHullShader, CreateDomainShader, CreateGeometryShader, CreatePixelShader, and CreateComputeShader. Each of these accepts a bytecode stream and produces an ID3D11*Shader interface that represents the runtime shader object that can be bound to the pipeline. See Listing 6.1 for a simple example of compiling a vertex shader and creating a runtime shader object.

```
ID3D10Blob* compiledShader;
ID3D10Blob* errorMessages;
HRESULT hr = D3DX11CompileFromFile( filePath, NULL, NULL, "VSMain",
                                    "vs_5_0", 0, 0, NULL, &compiledShader,
                                    &errorMessages, NULL );
if ( FAILED( hr ) )
{
    if ( errorMessages )
    {

        const char* msg = (char*)( errorMessages->GetBufferPointer() );
        Log::Get().Write( msg );
    }
    else
    {
        Log::Get().Write( "D3DX11CompileFromFile failed" );
    }
}
else
{
    ID3D11VertexShader* vertexShader = NULL;
    device->CreateVertexShader( compiledShader->GetBufferPointer(),
                                compiledShader->GetBufferSize(),
                                NULL,
                                &vertexShader );
}
```

Listing 6.1. Compiling a vertex shader from a file.

6.2.4 Binding to the Pipeline

Shader objects are bound to the pipeline by calling the various *SetShader methods available on an ID3D11DeviceContext. Once all required shader objects are bound to their corresponding stages, a Draw call can be issued, and the bound shader programs will be used to render the geometry. Or in the case of compute shaders, a Dispatch call can be issued, instead, to use the compute pipeline. If the shader program uses shader resources, samplers, or constant buffers, these must be bound before issuing the Draw or Dispatch call. This is done by calling the appropriate *SetShaderResources, *SetConstantBuffers, and *SetSamplers methods on the device context. Unordered access views are bound to the compute shader stage using CSSetUnorderedAccessViews, and are bound to the pixel shader stage using OMSetRenderTargetsAndUnorderedAccessViews. See Chapter 3, "The Rendering Pipeline," and Chapter 5, "The Computation Pipeline," for more information on configuring shader stages.

6.3 Language Basics

6.3.1 Primitive Types

High Level Shader Language features several primitive types that are common across all shader profiles. Like the primitive types in C++, HLSL primitives consist of various integral and floating-point numerical types. These types are listed in Table 6.1.

Type	Description
bool	32-bit integer that can contain the logical values true and false
int	32-bit signed integer
uint	32-bit unsigned integer
half	16-bit floating point value (provided only for backward compatibility; D3D11 shader profiles will map all half values to float)
float	32-bit floating point value
double	64-bit floating point value

Table 6.1. HLSL primitive types.

As in C or C++, these types can be used to declare both scalar and array values. Scalar variables support a standard set of mathematical operators, including addition, subtraction, negation, multiplication, division, and modulus. Scalars also support a standard set of logical and comparison operators, using the same syntax as C/C++. Array variables must have a fixed, statically declared size determined at the time of compilation, because HLSL does not support dynamic memory allocation.

6.3.2 Vectors

HLSL also supports declaring vector and matrix variables. Vector types support 1-4 components, with each component using the storage format specified by the primitive type. Vector variables are declared with the syntax shown in Listing 6.2.

```
vector<float, 4> floatVector;    // 4-component vector with float components
vector<int, 2> intVector;        // 2-component vector with int components
```

Listing 6.2. Vector declaration syntax.

For convenience, HLSL predefines shorthand-type definitions for all combinations of primitive types and components. Listing 6.3 demonstrates the declaration of vector variables using the shorthand notation.

```
float4 floatVector;    // 4-component vector with float components
int2 intVector;        // 2-component vector with int components
```

Listing 6.3. Vector shorthand declaration syntax.

Vector types can be initialized when declared using array initializer syntax, with the number of values matching the number of components. Vector types also support constructor syntax, which allows passing scalar values or even other vectors. In addition, vectors can be initialized using a single scalar value, in which case the scalar value is replicated to all components. Listing 6.4 demonstrates these initialization patterns.

```
float2 vec0 = { 0.0f, 1.0f };
float3 vec1 = float3( 0.0f, 0.1f, 0.2f );
float4 vec2 = float4( vec1, 1.0f );
float4 vec3 = 1.0f;
```

Listing 6.4. Vector initialization syntax.

Individual components of a vector can be accessed using notation similar to accessing structures in C++. The members *x*, *y*, *z*, and *w* each correspond to the first, second, third, and fourth components of the vector, respectively. Alternatively the members *r*, *g*, *b*, and a can be used in the same manner. Vector components also support being accessed using array index syntax, which can be useful for looping through components. Listing 6.5 demonstrates usage of these accessors.

```
float4 floatVector = 1.0f;
float  firstComponent = floatVector.x;
float  secondComponent = floatVector.g;
float  thirdComponent = floatVector[2];
```

Listing 6.5. Vector component access syntax.

In addition to supporting single-component access, vectors also support accessing multiple components simultaneously through the use of *swizzles*. A swizzle is an operation that returns one or more components of a vector in an arbitrary ordering. Swizzles can also replicate a component more than once, to create a new vector value with a greater number of components than in the source vector. While there are no restrictions on the ordering of components in swizzles, the notation used must be consistent. In other words, the *xyzw* notation cannot be mixed with the *rgba* notation within a single swizzle. Listing 6.6 contains sample code that uses swizzles in various ways.

```
float4 vec0 = float4(0.0f, 1.0f, 2.0f, 3.0f);
float3 vec1 = vec0.xyz;
float2 vec2 = vec1.rg;
float3 vec3 = vec0.zxy;
float4 vec4 = vec2.xyxy;
float4 vec5;
vec5.zyxw = vec0.xyzw;
```

Listing 6.6. Vector swizzle syntax.

If a vector value is assigned to a vector variable with a fewer number of components, the value is implicitly truncated when the assignment is performed. Programmers should always be careful not to inadvertently cause such a truncation, since values will be discarded. To make the truncation explicit, a C++-style cast or a swizzle can be used. As a convenience, the HLSL compiler will emit a warning when an implicit truncation occurs.

Vectors support the same set of mathematical, logical, and comparison operators supported by scalar variables. When these operators are used on vector types, the operation is performed on a per-component basis and produces a vector result value with a number of components equal to that of the operands.

6.3.3 Matrices

Matrix variables are declared and used in a manner similar to vector variables. Matrix declarations specify a base primitive type, in addition to the number of rows and columns. The number of rows and columns are listed to 4 each, making for a maximum of 16 individual components. Component access can be done with two-dimensional array syntax, where the first index specifies the row and the second specifies the column. Additionally, when only a single array index is specified, it returns the corresponding row of the matrix as a vector type. Matrices also support their own syntax for member access, which differs from the format used for vectors. These formats are shown in Listing 6.7.

```
float4x4 worldMatrix = float4x4( float4( 1.0f, 0.0f, 0.0f, 0.0f ),
                                 float4( 0.0f, 1.0f, 0.0f, 0.0f ),
                                 float4( 0.0f, 0.0f, 1.0f, 0.0f ),
                                 float4( 0.0f, 0.0f, 0.0f, 1.0f ) );
float matVal0 = worldMatrix._m00;      // Value from first row, first column
float matVal1 = worldMatrix._12;       // Value from first row, second column
float matVal2 = worldMatrix[0][1];     // Value from first row, second column
float2 matVal3 = worldMatrix._11_12;   // Swizzles
```

Listing 6.7. Matrix component access syntax.

As with vectors, operators used with matrix types are performed per-component. Consequently the multiplication operator should not be used to perform matrix multiplications and transformations. Instead the mul intrinsic function is provided for matrix/matrix and matrix/vector multiplications. See the "Intrinsic Functions" section in this chapter or the SDK documentation for more details.

Typically, shader programs initialize matrices in a row-major format and handle all vector/matrix transformations accordingly. However, in many cases the compiler will optimize the assembly such that the matrices contain column-major data, because that format allows for a more efficient expression in assembly using four dot products. In addition, the compiler defaults to treating all matrices declared in constant buffers as if they contained column-major data, even when row-major transformation is performed with the mul intrinsic. Consequently, matrices may need to be transposed by the host application before being set into the constant buffer, or the matrix would have to be transposed in the shader program. This behavior can be changed by passing D3D10_SHADER_PACK_MATRIX_ROW_MAJOR to shader compilation functions, or by declaring matrices with the row_major modifier.

6.3.4 Structures

HLSL allows for custom structure declarations, with rules very similar to C++. Structures can contain an arbitrary number of members with scalar, vector, or matrix primitive types.

They can also contain members with array types, or with other custom structure types. Listing 6.8 demonstrates a simple structure declaration.

```
struct MyStructure
{
    float4 Vec;
    int Scalar;
    float4x4 Mat;
    float Array[8];
};
```

Listing 6.8. Structure declaration syntax.

6.3.5 Functions

HLSL allows for declaration of free functions in a manner similar to C/C++. Functions can have any return type (including void), and can accept an arbitrary number of parameters. Since HLSL does not support reference or pointer types, it has its own syntax for providing input/output semantics for parameters. This syntax includes four modifiers for parameters, which are listed in Table 6.2.

Modifier	Description
in	Parameter is an input to the function. Any changes to the parameter value are temporary, and are not reflected in the calling code. Similar to pass-by-value in C++. This is the default modifier.
out	Parameter is an output of the function. The value of the parameter must be set by the function, and this change is reflected in the calling code.
inout	Parameter is both an input and an output to the function. The function can access the value of the parameter set by the calling code, and changes that occur in the function are reflected in the calling code. Similar to pass-by-reference in C++.
uniform	Parameter is constant for the duration of execution. For functions that aren't entry points, this modifier is equivalent to in. For entry point functions, the parameter is declared as part of the $Param default constant buffer.

Table 6.2. Function parameter modifiers.

When using functions, it is important to keep in mind that shader programs do not use a traditional stack as in C++. Consequently, it is not possible to recursively call functions. Instead dynamic loop constructs must be used to implement recursive algorithms.

6.3.6 Interfaces

Interfaces in HLSL are similar to abstract virtual classes in C++, and are primarily used to enable dynamic shader linkage. Interfaces can only contain member functions, and not member variables. Listing 6.9 contains a declaration of a simple interface type.

```
interface MyInterface
{
    float3 Method1( );
    float2 Method2( float2 param );
};
```

Listing 6.9. Interface declaration syntax.

The methods declared in an interface are always considered to be pure virtual functions, and thus no virtual keyword is necessary. See the "Dynamic Shader Linkage" section for more information on using interfaces in HLSL.

6.3.7 Classes

HLSL classes, like classes in C++, can contain member variables, as well as member functions. They can also inherit from a single base class, and can inherit from multiple interfaces. However if a class inherits from an interface, it must fully implement all methods declared in that interface. Listing 6.10 contains a simple class declaration that implements an interface.

```
interface MyInterface
{
    float3 Method1( );
    float2 Method2( float2 param );
};

class MyClass : MyInterface
{
    float3 Member1;
```

```
    float3 Method1( )
    {
        return Member1;
    }
};

float2 MyClass::Method2( float2 param )
{
    return Member1.xy + param.xy;
}
```

Listing 6.10. Class declaration syntax.

While instances of classes can be declared and have methods called on them in normal shader programs, they are primarily used to enable dynamic shader linkage. See the "Dynamic Shader Linkage" section for more information on using classes in HLSL.

6.3.8 Conditionals

Conditional code execution is supported through the `if` and `case` statements, which work in the same manner as in C/C++. All `if` statements operate on a single Boolean value, which can be created through the use of logical and comparison operators. It should be noted that the Boolean result of vector operations cannot be directly used, because such operations produce a vector result, rather than a single Boolean value. By extension this same principle applies to `switch` statements.

It should be noted that conditional branching based on values that are only known during the execution of a program can be expressed by the compiled shader assembly in one of two ways: *predication* or *dynamic branching*. When predication is used, the compiler will emit code that evaluates the expressions on both sides of the branch. A compare instruction is then used to "select" the correct value, based on the result of a comparison. Dynamic branching, on the other hand, emits branching instructions that actually control the flow of execution in the shader program. Thus, it can be used to *potentially* skip unneeded calculations and memory accesses.

Whether or not instructions in a branch are skipped is based on the `coherency` of the branching. In other words, simultaneous executions of a shader program must all choose the same branch, to prevent the hardware from executing both sides of the branch. Typically, the hardware will simultaneously execute a number of consecutive vertices in the vertex shader, consecutive primitives in the geometry shader, and consecutive grids of pixels in the pixel shader. For compute shaders, the threads are explicitly mapped to thread groups, requiring coherency within each group. In addition, dynamic branching can impose a constant performance penalty, due to the overhead incurred from executing the actual branch instructions. The basic texture `Sample` method also cannot be used within a dynamic branch in a pixel shader, since the partial derivatives may not be available, because pixels in a quad may potentially take different branches. Consequently, it is important to weigh the additional overhead of dynamic branching against the number of potentially skipped instructions, while also considering the coherency of the branch.

By default, the compiler will automatically choose between predication and dynamic branching, using heuristics. However, it is possible to override the compiler's decision-using attributes. See the "Attributes" section in this chapter for more information.

6.3.9 Loops

HLSL supports `for`, `while`, and `do while` looping constructs. Like the conditional statements, their syntax and usage is identical to their C/C++ equivalents. They are also similar to conditionals, in that they can potentially produce dynamic flow control instructions when looping based on runtime values, and they thus have the same performance characteristics with regard to coherency.

6.3.10 Semantics

In HLSL, semantics are string metadata attached to variables. They serve three primary purposes when used in shader programs:

1. To specify the binding of variables used for passing values between shader stages

2. To allow shader programs to accept special system values generated by the pipeline

3. To allow shader programs to pass special system values interpreted by the pipeline

The code in listing 6.11 demonstrates these three use cases.

```
cbuffer PSConstants
{
    float4x4 WVPMatrix;
}
```

```
void VSMain(  in float4 VPos : POSITION          // Binds variable to Input
                                                 // Assembler
              in uint VID : SV_VertexID          // Binds variable to a
                                                 // system generated value

              out float4 OPos : SV_Position      // Tells the pipeline to
                                                 // interpret the value as the
                                                 // output vertex position

            )
{
    OutPos = mul( VPos, WVPMatrix );
}
```

Listing 6.11. Using semantics in a vertex shader.

In the example given in Listing 6.11, applying the POSITION semantic to the input allows the shader to specify which element from a vertex it requires. Thus, the input assembler will use the input layout to determine the proper byte offset within a vertex that is required to read in the appropriate element. In the same way, values passed from one shader stage to another can be identified with semantics. This allows the runtime to ensure that the proper output value is matched to a given input parameter.

System-value semantics, denoted with an SV_ prefix, are instead used to accept a value from the runtime, or to pass a value to the runtime. In the example shader, SV_VertexID specifies that the parameter should be set to a value that indicates the index of the vertex within the vertex buffer. The SV_Position semantic indicates that the value being output is the final transformed vertex position that should be used for rasterization. For a full listing of supported system-value semantics, consult the HLSL reference section of the SDK documentation. You can also refer to Chapter 3 for more information on the meaning and usage of each of these system value semantics.

6.3.11 Attributes

Attributes are special tags that can be inserted into HLSL code to modify the assembly code emitted by the compiler, or to control how the shader is executed by the pipeline. Some are purely optional, while others are required for certain shader types. In all cases, they are used by being declared immediately before the function, branch, loop, or statement they are affecting. Table 6.3 contains a list of all attributes, with a short description.

Attribute	Description
[branch]	Causes the compiler to emit dynamic branching instructions to express a conditional statement.
[flatten]	Causes the compiler to "flatten" a branch by emitting compare instructions to select the appropriate values.
[loop]	Causes the compiler to emit dynamic looping instructions to express a loop construct.
[unroll]	Causes the compiler to "unroll" a loop by replicating the instructions inside the loop. Can optionally take a parameter specifying the maximum number of iterations.
[maxvertexcount]	Required attribute for geometry shaders; specifies the maximum number of vertices emitted by the shader.
[domain]	Specifies the patch type (*tri*, *quad*, or *isoline*) used in the hull shader.
[earlydepthstencil]	Used on a pixel shader to force the depth-stencil test to be performed before the shader is executed.
[instance]	Specifies that multiple instances of a geometry shader should be run.
[maxtessfactor]	Specifies the tessellation factor returned by the hull shader.
[numthreads]	Specifies the number of threads dispatched in a compute shader thread group in three dimensions.
[outputcontrolpoints]	Specifies the number of control points emitted by the hull shader.
[outputtopology]	Specifies the primitive topology (`line`, `triangle_cw`, or `triangle_ccw`) to be used by the tessellator.
[partitioning]	Specifies the tessellation scheme (`integer`, `fractional_even`, `fractional_odd`, `pow2`) used in the hull shader.
[patchconstantfunc]	Specifies that the tagged function is the patch constant function for the hull shader.

Table 6.3. HLSL attributes.

6.4 Constant Buffers

As mentioned in the previous chapter regarding D3D11 resources, constant buffers are a type of resource designed for storing small amounts of data that is constant throughout the duration of a `Draw` or `Dispatch` call. To access data from a constant buffer in a shader program, the layout of the constant buffer must be declared in HLSL. Defining this layout is done with syntax very similar to declaring a structure, where multiple members are declared with both a name and a type. Once a member of a constant buffer is declared, it is globally accessible throughout the shader program by simply using the name of the member. This makes constant buffers very convenient, as no special code is required to access the resource (unlike other resource types, such as buffers and textures). Listing 6.12 contains a very basic constant buffer definition, as well as its usage in a simple vertex shader program.

```
cbuffer VSConstants
{
    float4x4 WorldMatrix;
    float4x4 ViewProjMatrix;
    float3 Color;
    uint EnableFog;
}
float4 VSMain( in float4 VtxPosition : POSITION ) : SV_Position
{
    float4 worldPos = mul( VtxPosition, WorldMatrix );
    return mul( worldPos, ViewProjMatrix );
}
```

Listing 6.12. HLSL Constant buffer declaration and usage

When the programmer is mapping a constant buffer in the host application and setting constant values, the offset from the start of the buffer must match the offset specified by the HLSL declaration of the constant buffer. The offset of any individual constant can be queried using the reflection APIs, which can then be used when setting the constant data. Or alternatively, a C or C++ structure can be declared in the host application with a data layout that exactly matches the HLSL layout. With this approach, the entire contents of a constant buffer can be set with a single `memcpy`, without requiring the reflection APIs. However, special attention must be paid to HLSL's alignment and packing rules, which differ from those used by most C and C++ compilers. By default, the HLSL compiler will attempt to align constants such that they don't span multiple float4 registers. Consequently, certain members of a C/C++ `struct` must be declared with 16-byte alignment to match the HLSL layout. Alternatively, padding can be inserted into the C/C++ structure to produce a matching alignment. Some compilers also support pragmas for specifying the packing alignment. The packing for an HLSL constant buffer can also be manually specified though the `packoffset` keyword. Listing 6.13 demonstrates usage of `packoffset` to declare a tightly packed constant buffer with four-byte alignment.

```
cbuffer VSConstants
{
    float4x4 WorldMatrix : packoffset(c0);
    float4x4 ViewProjMatrix : packoffset(c4);
    float3 Color : packoffset(c8);
    uint EnableFog : packoffset(c8.w);
    float2 ViewportXY : packoffset(c9);
    float2 ViewportWH : packoffset(c9.z);
}
```

Listing 6.13. Specifying constant buffer packing.

When a constant buffer is declared in HLSL, the compiler automatically maps it to one of 15 constant buffer registers for the corresponding stage of the pipeline. These registers are named `cb0` through `cb14`, and the index of the register directly corresponds to the slot passed to `*SSetConstantBuffers` when binding a constant buffer to a shader stage. The register index can be queried for a shader program using the reflection APIs, so that the host application knows which slot to specify. Alternatively, the register index can be manually specified in the HLSL declaration, using the `register` keyword. Listing 6.14 demonstrates the usage of this keyword.

```
cbuffer VSConstants : register(cb0)
{
    float4x4 WorldMatrix : packoffset(c0);
    float4x4 ViewProjMatrix : packoffset(c4);
    float3 Color : packoffset(c8);
    uint EnableFog : packoffset(c8.w);
    float2 ViewportXY : packoffset(c9);
    float2 ViewportWH : packoffset(c9.z);
}
```

Listing 6.14. Specifying the constant buffer register index.

6.4.1 Default Constant Buffers

Any global variables declared without the `static const` modifiers will be treated by the compiler as constants inside a default constant buffer named `$Globals`. Similarly, parameters to a shader entry point function marked as `uniform` will be placed inside another default constant buffer, named `$Params`.

6.4.2 Texture Buffers

Since constant buffers are optimized for small sizes and uniform access patterns, they can have undesirable performance characteristics in certain situations. One common situation is an array of bone matrices used for skinning. In this scenario, each vertex contains one or more indices indicating which bone in the array should be used to transform the position and normal. On many hardware types, this access pattern will cause the threads executing the shader program to serialize their access to the bone data, stalling their execution.

To rectify this issue, D3D11 allows a special type of constant buffer known as a *texture buffer*. A texture buffer uses the same syntax for declaring constants and accessing them in a shader program. However, under the hood, memory access will be done through the texture fetching pipeline. Thus it will have the same cached, asynchronous access pattern as a texture sample. Listing 6.15 demonstrates declaration and usage of a texture buffer in a vertex shader program.

```
cbuffer VSConstants : register(cb0)
{
    float4x4 WVPMatrix;
}

tbuffer Bones : register(t0)
{
    float4x4 BoneMatrices[256];
}
float4 VSMain( in float4 VtxPosition : POSITION,
              in uint BoneIndex : BONEINDEX ) : SV_Position
{
    float4x4 boneMatrix = BoneMatrices[BoneIndex];
    float4 skinnedPos = mul( VtxPosition, boneMatrix );
    return mul( skinnedPos, WVPMatrix );
}
```

Listing 6.15. Using a texture buffer in a vertex shader program.

One important point to keep in mind when declaring a texture buffer is that it must be mapped to a texture register, rather than a constant buffer register. Also, the host application must create a shader resource view for the texture buffer and bind it through `*SSetShaderResources`, as opposed to calling `*SSetConstantBuffers`. The texture buffer resource itself must also be created as a texture resource, rather than a buffer resource.

6.5 Resource Objects

HLSL provides several types of resource objects that allow HLSL shader programs to interact with the various resource types available in D3D11. Each object type exposes the functionality of the resource through a set of methods that can be called on a declared object of that type. Most of these methods provide a means of reading data from the resource when given an address or index, while some also provide information about the resource itself. In the case of read/write resources, methods are also exposed for writing to the resource data.

All of the read-only resource objects correspond to a type of shader resource view, while the read/write resource objects correspond to a type of unordered access view. As with constant buffers, each declaration of a resource object is mapped by the compiler to a register index that corresponds to the slot passed to *SSetShaderResources, OMSetRenderTargetsAndUnorderedAccessViews, or CSSetUnorderedAccessViews. Like constant buffers, the register can be manually specified when declaring the resource object, using the register keyword. For shader resource views the registers are t0 through t127, and for unordered access views the registers are u0 through u7. Refer to Chapter 2 for more information regarding resources, and how they are bound to the pipeline.

6.5.1 Buffers

The buffer resource objects correspond to a shader resource view created for an ID3D11Buffer resource. For the most part, they have a very simple interface that consists of methods for obtaining the size of the resource, a Load method for reading a value when given an address, and an array operator for reading a value when given an index. In cases where a buffer can contain different types, the data type returned by the Load method or array operator is specified using a template-like syntax. Listing 6.16 demonstrates declaration of a standard Buffer object with float4 return type, corresponding to DXGI_FORMAT_R32G32B32A32_FLOAT.

```
Buffer<float4> Float4Buffer : register(t0);
```

Listing 6.16. Declaring a float4 buffer.

Buffer

The Buffer object is the simplest resource object type in HLSL. A Buffer provides read-only access to its data through the Load method and an array operator, both of which return the data at the specified index. It also provides the GetDimensions method for querying the size of the buffer in bytes.

ByteAddressBuffer

The ByteAddressBuffer type allows access to a buffer using a byte offset, rather than an index. This functionality is exposed through the Load, Load2, Load3, and Load4 methods, which return 1, 2, 3, or 4 uint values, respectively. Since those methods only return type uint, buffers containing other types of data must have the return value manually converted using one of the conversion/casting intrinsics.

StructuredBuffer

A StructuredBuffer provides read-only access to a buffer containing elements defined by a structure, as opposed to the regular Buffer type, which can only access types corresponding to DXGI_FORMAT values. The structure must be declared in HLSL as a struct, and that type must be specified as the template argument when declaring the StructuredBuffer object.

Read/Write Buffers

Read/Write buffers allow random-access reads and writes to buffer resources. Since the reads and writes are not automatically synchronized by the device, synchronization intrinsics or atomics must be used to control memory access across threads. By default, a synchronization intrinsic will only cause a write to flush across that thread group. In order for the write to be flushed across the entire GPU, the buffer declaration must use the globallycoherent prefix.

The three types of read/write buffers are RWBuffer, RWByteAddressBuffer, and RWStructuredBuffer. For reads, they behave exactly like their read-only equivalents, and use the same methods. For writing to memory, the array operator can be used for RWBuffer and RWStructuredBuffer, while RWByteAddressBuffer provides the Store, Store2, Store3, and Store4 methods. RWByteAddressBuffer also provides a set of interlocked methods that allow atomic operations to be performed on the contents of the resource. For a RWStructuredBuffer that contains a hidden counter (which is the case for structured buffer resources created with the D3D11_BUFFER_UAV_FLAG_COUNTER flag), it can be incremented or decremented with IncrementCounter and DecrementCounter. See Chapter 2 for more details on structured buffer resources.

6.5.2 Append/Consume Buffers

Append and consume buffers allow adding and removing of values from a resource in a pixel or compute shader. The addition and removal operations are done at the end of a buffer, causing it to behave somewhat like a stack. However unlike traditional stacks, the ordering of additions and removals is not guaranteed, because the resource is accessed by many threads simultaneously. Addition of values is done with the `Append` method of the `AppendStructuredBuffer` type, while removal is done with the `Consume` method of the `ConsumeStructuredBuffer` type. As their names suggest, these types are both structured buffer resources.

6.5.3 Stream-Output Buffer

The stream-output buffer is a simple resource object used by the geometry shader to emit primitive vertex data. Unlike other resource types, it is not declared globally and mapped to a register. Instead, it is taken as an `inout` parameter to the geometry shader entry point function. Three types of stream-output buffers can be declared, each corresponding to a different primitive type. They are `PointStream`, `LineStream`, and `TriangleStream`, which correspond to a point strip, a line strip, and a triangle strip, respectively. The parameter declaration should also include a type as the template argument, which indicates the data format of the vertices. Typically, this type is a structure containing all vertex data that needs to be passed to the pixel shader, or that will be output by the stream output stage. Adding a vertex to the strip is done with the `Append` method, while `RestartStrip` cuts the current strip and begins a new strip. If a strip is cut before the minimum number of vertices is appended, the incomplete primitive is discarded.[1]

6.5.4 Input and Output Patches

HLSL includes two patch types, which are used for accessing an array of control point data in the hull and domain shaders, and are also available as input to the geometry shader.[2] The `InputPatch` contains an array of points that can be declared as an input attribute for the hull shader, the patch constant function, and the geometry shader. The `OutputPatch` contains an array of points that is declared as an input to the domain shader. Both of types are accessed using the array operator, and expose a Length property for determining the number of elements they contain.

---

[1] Further details on the declaration and usage of stream output buffers can be found in the "Geometry Shader" and "Stream Output" sections of Chapter 3.

[2] The geometry shader is typically not thought of as being able to receive an *InputPatch* as its input, but it is valid to declare and use this in the same manner as seen with the hull shader. This can be used to perform higher-order surface operations in the geometry shader if the control patches are not consumed by the tessellation stages.

6.5.5 Textures

Textures are among the most commonly-used resources in the rendering pipeline, and make use of specialized texture sampling units in graphics hardware. Consequently HLSL exposes a large interface for sampling texture data, to allow shader programs to make the most use of the hardware's capabilities. HLSL includes read-only texture resource object types, corresponding to the various texture resource types in D3D11. A full listing of these types is given in Table 6.4.

Intrinsic	Description
Texture1D	One-dimensional texture
Texture1DArray	Array of one-dimensional textures
Texture2D	Two-dimensional texture
Texture2DArray	Array of two-dimensional textures
Texture2DMS	Two-dimensional texture with multisampling
Texture2DMSArray	Array of two-dimensional textures with multisampling
Texture3D	Three-dimensional texture
Texture3DArray	Array of three-dimensional textures
TextureCube	Array of six 2D textures, representing faces of a cube
TextureCubeArray	Array of cube textures

Table 6.4. Texture resource objects.

The texture resource object types each support a subset of all of the texture operations, and consequently, each object type has its own interface. Consult the HLSL documentation to see if a method is supported for a given texture object type. The following sections describe how each of the methods can be used for accessing a texture object.

Sample Methods

Traditional texture sampling operations are performed with the `Sample` family of methods available on the texture objects. These methods each take a set of floating-point texture coordinates representing the memory location to sample, where each component is normalized to the range [0,1]. The number of components depends on the texture type. For instance, a `Texture1D` only accepts a single float, while a `Texture2D` accepts two floats, and a `Texture3D` accepts three. If a texture array is used, an additional floating-point component is passed, indicating the index of the array to use.

The `Sample` method allows for hardware texture filtering (minification, magnification, and mip-map interpolation) to be performed, according to a given sampler state. The sampler state must be declared as a `SamplerState` object in HLSL, which is mapped to a sampler register (`s0` through `s15`) in the same way constant buffers and resource objects are mapped to their register types. The index of the register corresponds to the slot passed to the `*SSetSamplers` methods available on the device context. A declared `SamplerState` object can then be passed to the `Sample` method, which will in turn use the specified sampler states for filtering. Listing 6.17 contains a simple pixel shader that demonstrates calling `Sample` on a `Texture2D` using a sampler state.

```
SamplerState LinearSampler : register(s0);
Texture2D ColorTexture : register(t0);

float4 PSMain( in float2 TexCoord : TEXCOORD ) : SV_Target
{
    return ColorTexture.Sample( LinearSampler, TexCoord );
}
```

Listing 6.17. Sampling a texture in a pixel shader.

When `Sample` is called on a texture that contains multiple mip-map levels, the mip-map level is automatically selected based on the screen-space derivatives (also known as `gradients`) of the texture coordinates. This is why `Sample` is only available in pixel shaders, and only outside of dynamic loop or branch constructs. To use texture sampling in other shader stages, or inside a dynamic branch/loop in a pixel shader, other variants of the `Sample` method are available, which allow the gradients or the mip level to be explicitly specified. These are `SampleGrad` and `SampleLevel`, respectively. A third method, `SampleBias`, works like `SampleLevel` with the addition of a bias value.

The return type of a sample method depends on the `DXGI_FORMAT` specified when creating the shader resource view bound for the texture object. Thus, `DXGI_FORMAT_R32G32B32A32_FLOAT` will return a `float4`, `DXGI_FORMAT_R32G32_UINT` will return a `uint2`, and `DXGI_FORMAT_R8G8B8A8_UNORM` will return a `float4`.

SampleCmp Methods

Rather than directly returning a texture value, `SampleCmp` returns the result of comparing the sampled value against a value passed to the method as the `CompareValue` parameter. The returned value is equal to 1.0 if the comparison passes, and 0.0 if the comparison fails. This makes the method quite natural for implementing shadow mapping techniques.

The inequality used in the comparison is specified in the `ComparisonFunc` member of the `D3D11_SAMPLER_DESC` structure used to create the sampler state passed to the method. Note that the filter specified for the sampler state must be one of the `COMPARISON` values of the `D3D11_FILTER` enumeration. Also, the sampler state declared in HLSL must have the `SamplerComparisonState` type if it is to be passed to `SampleCmp`. If a linear filtering mode is specified for the sampler state, the texture will be sampled multiple times and compared against the comparison value. The final return value is then the filtered result of all comparisons. This can be used to efficiently implement percentage closer filtering for shadow maps.

Gather Methods

The `Gather` family of methods returns four values when given a single texture coordinate. The values come from a 2×2 quad of texels, using the same location that would be used for bilinear filtering. `GatherRed` returns the four red components of the texels, `GatherGreen` returns the four green components, `GatherBlue` returns the four blue components, and `GatherAlpha` returns the four alpha components. `Gather` returns the four red components, making it functionally equivalent to `GatherRed`. Since the method returns a 2×2 quad, it can only be used on `Texture2D` and `Texture2DArray` resource objects. `GatherCmp` methods are also available, which work similarly to `SampleCmp`.

One of the more useful cases for the `Gather` methods is in pixel shaders that perform image processing. Normally gathering rgb values for a 2×2 quad of texels requires four sample operations, with one for each texel. This can be made more efficient by calling `GatherRed`, `GatherGreen`, and `GatherBlue` retrieve the same data in only three instructions. Another useful case for `Gather` instructions is implementing custom shadow map filtering kernels, which can be done efficiently with `GatherCmp`.

Load Methods

The `Load` method provides direct, unfiltered, read-only access to texture data in a manner similar to what we have seen with the buffer types. Like the `Buffer Load` method, it takes an `int` index parameter, rather than a [0,1] texture address. Since it is the most basic means of accessing texture data, it is available on all texture resource types and in all stages. The array operator is also available, which provides similar functionality. For multisampled textures, `Load` provides access to the individual subsamples of a pixel. This functionality can be used to implement custom MSAA resolves, or to integrate MSAA into a deferred rendering pipeline. Like the `Load` method for a `Buffer` object, the return type is determined by the type that was used as a template parameter when declaring the texture object.

Get Methods

Texture resource objects implement several methods for querying information about the underlying resource. All texture resource object types implement `GetDimensions`, which returns the size, number of mip levels, number of samples (for multisampled textures), and number of elements (for texture arrays). `Texture2DMS` objects also support the `GetSamplePosition` method, which returns the position of a MSAA sample point within the pixel given a sample index.

Cube Textures

Cube textures are a special case of texture arrays meant for representing the six sides of a cube, and are often used for reflection or environment maps. While they can be accessed like a normal texture array through the `Texture2DArray` type, they can also be declared as a `TextureCube` type. When calling a `Sample` method on `TextureCube`, the texture coordinate passed in is a normalized three-component direction vector. The texture is then sampled by choosing the texel pointed to by the direction vector. This usage is further described in the cube texture section of Chapter 2.

Read/Write Textures

Read/write textures function in a similar manner to read/write buffer resources. Random-access reads and writes are supported through the array operator, and `GetDimensions` can be called to query the size of the resource. The read/write resource object types supported in HLSL are `RWTexture1D`, `RWTexture1DArray`, `RWTexture2D`, `RWTexture2DArray`, and `RWTexture3D`.

6.6 Dynamic Shader Linkage

Since programmable graphics hardware has become mainstream, applications employing three-dimensional graphics have developed increasingly large and complex sets of shader code for implementing their graphical features. However prior to Direct3D 11, HLSL provided no built-in means of enabling dynamic dispatch in shader programs. Shader Model 3.0 introduced the ability for shaders to dynamically branch on values. However, the heavy performance penalties of doing so made it prohibitively expensive as a means for implementing dynamic dispatch. Static branching has also been supported since Shader Model 2.0, but its limitations on the number of branches, as well as performance implications, have made it similarly inadequate. Consequently many applications resorted to statically compiling all required permutations of a shader program, either by using macros and conditional compilation or by piecing together code from smaller fragments. While this approach has the advantages of not requiring dynamic branching, and provides the optimizer with full access to a complete program, the number of required permutations grows exponentially as new options are added. This is often referred to as the *shader combinatorial explosion*. Thus, programmers have often been faced with a difficult choice between decreasing shader performance, increasing shader compilation times, or making their shader build pipeline more complex.

To rectify this situation, Direct3D 11 has introduced a feature known as *dynamic shader linkage*. It essentially allows applications to dynamically choose from multiple implementations of an HLSL code path when binding a shader program to the pipeline, effectively allowing dynamic dispatch at the level of a Draw or Dispatch call. We will explore this capability further in the following sections to understand how it can be used in a real-time rendering application.

6.6.1 Authoring Shaders for Dynamic Linkage

Shader programs that will make use of dynamic linkage must be authored using HLSL interfaces and classes. Essentially, the procedure is similar to virtual dispatch in C++: interfaces are declared with a set of methods, and the shader code calls methods on those interfaces. Then at runtime, the host application assigns an instance of a class that implements the interface to be used for the duration of a Draw or Dispatch call. Shader programs using interfaces can declare a global instance of an interface, just like any other variable type. That instance then functions like a polymorphic pointer in C++, and forwards any methods called on it to the class instance specified by the host application. Listing 6.18 demonstrates the syntax for declaring an interface instance and calling one of its methods.

```
interface Light
{
    float3 GetLighting( float3 Position, float3 Normal );
};

Light LightInstance;

float4 PSMain( in float3 Position : POSITION,
               in float3 Normal : NORMAL ) : SV_Target
{
    float3 lightColor = LightInstance.GetLighting( Position, Normal );
    return float4( lightColor, 1.0f );
}
```

Listing 6.18. Calling a method on an interface.

Any classes that can possibly be used to implement an interface must also be declared in the HLSL shader program, or included through the use of a #include pragma. If the class has member variables, an instance of the class must be declared in a constant buffer. This allows the host application to specify values for those members at runtime. Listing 6.19 demonstrates the syntax for declaring a class that implements an interface, and declaring an instance in a constant buffer.

```
class DirectionalLight
{
    float3 Color;
    float3 Direction;
    float3 GetLighting( float3 Position, float3 Normal )
    {
        return saturate( dot( Normal, Direction) ) * Color;
    }
};

cbuffer ClassInstances : register( cb0 )
{
    DirectionalLight DLightInstance;
}
```

Listing 6.19. Declaring a class instance.

6.6.2 Linking Classes to Interfaces

At runtime, applications employing dynamic shader linkage must specify which class will be used to implement an interface used by a shader program. The first step is to create an ID3D11ClassLinkage interface, which is done by calling ID3D11Device::CreateClass Linkage. Once the class linkage interface is created, it can be bound to an instance of a shader program. This is done by passing the ID3D11ClassLinkage interface as the pClassLinkage parameter of CreateVertexShader, CreateHullShader, CreateDomainShader, CreateGeometryShader, CreatePixelShader, or CreateComputeShader. Listing 6.20 demonstrates this process for a pixel shader.

```
ID3D10Blob* compiledShader;
ID3D10Blob* errorMessages;
HRESULT hr = D3DX11CompileFromFile( filePath, NULL, NULL, "PSMain",
                                    "ps_5_0", 0, 0, NULL, &compiledShader,
                                    &errorMessages, NULL );

ID3D11ClassLinkage* classLinkage = NULL;
if ( SUCCEEDED( hr ) )
{
    device->CreateClassLinkage( &classLinkage );
    device->CreatePixelShader(  compiledShader->GetBufferPointer(),
                                compiledShader->GetBufferSize(),
                                classLinkage,
                                &pixelShader );
}
```

Listing 6.20. Creating and binding a class linkage.

Once the class linkage is bound to a shader, an ID3D11ClassInstance interface can be retrieved. A class instance represents one of the classes declared in the shader that can be used to implement an interface. The simplest means to acquire one is to call GetClassInstance on the ID3D11ClassLinkage interface, and pass the name of a class instance declared in the shader program. CreateClassInstance can also be used for shader programs where a class instance wasn't declared in a constant buffer, which is useful for classes that don't contain member variables. Listing 6.21 demonstrates the process of acquiring a class instance interface.

```
ID3D11ClassInstance* dLightInstance = NULL;
classLinkage->GetClassLinkage( L"DLightInstance", 0, &dLightInstance );
```

Listing 6.21. Acquiring a class linkage.

The final step for using dynamic linkage is to specify an array of class instances when binding a shader program to the pipeline. The ID3D11DeviceContext::*SSet Shader methods all have a ppClassInstances parameter that accepts an array of ID3D11ClassInstance interfaces. This array must contain one class instance for each the interfaces used in the shader program. Listing 6.22 demonstrates initializing and passing such an array for a pixel shader.

```
ID3D11ClassInstance* classInstances[1];
classInstances[0] = dLightInstance;
deviceContext->PSSetShader( pixelShader, classInstances, 1 );
```

Listing 6.22. Specifying class instances when binding a shader.

Each interface used in a shader program has a unique index, which corresponds to an index of the array passed to *SSetShader containing the class instances. This index can be retrieved using the ID3D11ShaderReflectionVariable interface, which is part of the shader reflection APIs. See the "Shader Reflection" section for more details.

6.7 Intrinsic Functions

HLSL provides a built-in set of global functions known as *intrinsics*. Like intrinsics in C or C++, HLSL intrinsics often directly map to specific shader assembly instructions. This provides HLSL shader programs with full access to the functionality provided by the assembly instruction set. In a few cases, intrinsic functions also provide an optimized set of common mathematical operations, which spares the programmer from implementing them manually. The following sections contain tables listing intrinsic functions grouped by category, with a brief description for each. For a full description (including return types and parameters), consult the SDK documentation

6.7.1 Mathematical Functions

Vector/Matrix Operations

Table 6.5 contains a list of all vector and matrix math intrinsics supported by HLSL. These intrinsics implement mathematical operations that can be performed on vector and matrix data types.

Intrinsic	Description
mul	Matrix/Matrix, Matrix/Vector, or Vector/Vector multiplication
dot	Vector dot product
cross	Vector cross product
transpose	Matrix transpose
determinant	Matrix determinant
length	Vector length/magnitude
normalize	Geometric vector normalize
distance	Computes the distance between two points
faceforward	Flips a surface normal direction such that it points in a direction opposite to an incident vector
reflect	Computes a reflection vector, given normal and incident vectors
refract	Computes a refraction vector, given an entering direction, normal vector, and refraction index

Table 6.5. Vector/matrix math intrinsics.

General Math Functions

Table 6.6 contains a listing of HLSL's general math intrinsics. These instrinsics are mostly scalar math operations, similar to those found in the C Standard Library.

Intrinsic	Description
cos	Cosine function
sin	Sine function
tan	Tangent function
acos	Arcosine function
asin	Arcsine function
atan	Arctangent function
atan2	Signed arctangent function
sincos	Performs sine and cosine simultaneously
cosh	Hyberbolic cosine
sinh	Hyperbolic sine
tanh	Hyperbolic tangent
log	Natural logarithm (base e)
log2	Logarithm (base2)
log10	Logarithm (base10)
exp	Exponential (base e)
exp2	Exponential (base 2)
pow	Raises a number to a power
sqrt	Square root
abs	Absolute value
trunk	Floating point truncation
floor	Return largest integer less than a value
ceil	Returns smallest integer greater than a value
round	Rounds to nearest integer value
frac	Returns the fractional part of a value
fmod	Floating point remainder
modf	Returns integer and fractional parts of a value
countbits	Number of storage bits for an integer
sign	Returns the sign of a value
all	Returns true if all components of a value are non-zero
any	Returns true if any components of a value are non-zero
clamp	Restricts a value to a specified minimum and maximum
degrees	Converts a value from radians to degrees
firstbithigh	Gets the first set bit of an integer, starting at the highest-order bit and moving downward
firstbitlow	Gets the first set bit of an integer, starting at the lowest-order bit and moving upward

frexp	Returns the mantissa and exponent of a floating-point value
isfinite	Returns true if a floating point value is not infinite
isinf	Returns true if a floating point value is equal to -INF or +INF
isnan	Returns true if a floating point value is equal to NAN or QNAN
ldexp	Multiplies a value to another value equal to 2 raised to a specified exponent
lerp	Linearly interpolates between two values, using a specified scale value
lit	Computes ambient, diffuse, and specular values for a Blinn-Phong BRDF
mad	Multiplies two values and sums the product with a third value
max	Returns the maximum of two values
min	Returns the minimum of two values
modf	Splits a floating-point value into integral and fractional parts
radians	Converts a value from degrees to radians
rcp	Calculates a fast, approximate reciprocal of a value
reversebits	Reverses the order of the bits in an integer value
rsqrt	Calculates the reciprocal of the square root of a value
saturate	Clamps a value to the [0, 1] range
smoothstep	Uses a Hermite interpolation to calculate a value between 0 and 1, using a specified value, minimum, and maximum
step	Returns 1 if one specified value is greater than another, and 0 otherwise

Table 6.6. General math intrinsics.

6.7.2 Casting/Conversion Functions

In some cases, a resource will contain values in a particular data format that cannot be directly read in HLSL. Consequently, HLSL provides casting and conversion intrinsics so that shader programs can cast or convert a resource value to an appropriate data type. These intrinsics are listed in Table 6.7.

Intrinsic	Description
asfloat	Reinterprets a value as a float
asdouble	Reinterprets two 32-bit values as a double
asint	Reinterprets a value as an int
asuint	Reinterprets a value as a uint
f16to32	Converts a 16-bit floating-point value to a 32-bit floating-point value
f32to16	Converts a 32-bit floating-point value to a 16-bit floating-point value

Table 6.7. Casting/Conversion intrinsics.

6.7.3 Tessellation Functions

These functions can be used in a hull shader to generate corrected tessellation factors for a patch. A listing of all tessellation intrinsics can be found in Table 6.8

Intrinsic	Description
Process2DQuadTessFactorsAvg	Computes tessellation factors for a 2D quad patch using the average of the edge tessellation factors
Process2DQuadTessFactorsMax	Computes tessellation factors for a 2D quad patch using the maximum of the edge tessellation factors
Process2DQuadTessFactorsMin	Computes tessellation factors for a 2D quad patch using the minimum of the edge tessellation factors
ProcessIsolineTessFactors	Computes rounded tessellation factors for an isoline
ProcessQuadTessFactorsAvg	Computes tessellation factors for a quad patch using the average of the edge tessellation factors
ProcessQuadTessFactorsMax	Computes tessellation factors for a quad patch using the maximum of the edge tessellation factors
ProcessQuadTessFactorsMin	Computes tessellation factors for a quad patch using the minimum of the edge tessellation factors
ProcessTriTessFactorsAvg	Computes tessellation factors for a triangle patch using the average of the edge tessellation factors
ProcessTriTessFactorsMax	Computes tessellation factors for a triangle patch using the maximum of the edge tessellation factors
ProcessTriTessFactorsMin	Computes tessellation factors for a triangle patch using the minimum of the edge tessellation factors

Table 6.8. Tessellation intrinsics.

6.7.4 Pixel Shader Functions

Several intrinsics are specific to pixel shader programs. These intrinsics are listed in Table 6.9.

Intrinsic	Description
clip	Discards the pixel shader result if the specified value is less than 0. Can be used to implement clipping planes.
ddx	Returns the partial derivative of a value with respect to the screen space X-coordinate of the pixel being shaded.

ddx_coarse	Returns a low precision partial derivative of a value with respect to the screen space *X*-coordinate of the pixel being shaded.
ddx_fine	Returns a high precision partial derivative of a value with respect to the screen space *X*-coordinate of the pixel being shaded.
ddy	Returns the partial derivative of a value with respect to the screen space *Y*-coordinate of the pixel being shaded.
ddy_coarse	Returns a low precision partial derivative of a value with respect to the screen space *Y*-coordinate of the pixel being shaded.
ddy_fine	Returns a high-precision partial derivative of a value with respect to the screen space *Y*-coordinate of the pixel being shaded.
fwidth	Returns the absolute value of the sum of ddx and ddy for a value.
EvaluateAttributeAtCentroid	Interpolates a pixel shader input using the centroid of all covered sample points, as if the attribute were marked with the centroid modifier.
EvaluateAttributeAtSample	Interpolates a pixel shader input using the sample point indicated by the specified index, as if the attribute were marked with the sample modifier.
EvaluateAttributeSnapped	Interpolates a pixel shader input using the centroid position, with an offset.
GetRenderTargetSampleCount	Retrieves the number of samples in the current render targets.
GetRenderTargetSamplePosition	Retrieves the *XY* position of a sample point for the given sample index.

Table 6.9. Pixel shader intrinsics.

6.7.5 Synchronization Functions

The synchronization intrinsics listed in Table 6.10 are only available for compute shaders, and are generally used to synchronize access to shared memory or resource objects. These functions and their uses are described in detail in Chapter 5.

Intrinsic	Description
AllMemoryBarrier	Ensures that all pending memory accesses have completed for all threads in the thread group

AllMemoryBarrierWithGroupSync	Ensures that all pending memory accesses have completed for all threads in the thread group, and blocks until all threads reach the `sync` instruction
DeviceMemoryBarrier	Ensures that all pending device memory accesses (reads and writes to texture and buffer resources) have completed for all threads in the thread group
DeviceMemoryBarrierWithGroupSync	Ensures that all pending device memory accesses (reads and writes to texture and buffer resources) have completed for all threads in the thread group, and blocks until all threads reach the `sync` instruction
GroupMemoryBarrier	Ensures that all pending shared memory accesses (reads and writes to groupshared variables) have completed for all threads in the thread group
GroupMemoryBarrierWithGroupSync	Ensures that all pending shared memory accesses (reads and writes to groupshared variables) have completed for all threads in the thread group, and blocks until all threads reach the sync instruction

Table 6.10. Synchronization intrinsics.

6.7.6 Atomic Functions

These intrinsics, listed in Table 6.11, perform a guaranteed atomic operation on a local variable, a shared memory variable, or a resource variable. They can only be used on `int` or `uint` variables, and only in a pixel or compute shader. Most atomic intrinsics can optionally return the original value of the variable, at the expense of additional runtime cost.

Intrinsic	Description
InterlockedAdd	Atomic add operation, optionally returning the previous value
InterlockedAnd	Atomic AND operation
InterlockedCompareExchange	Compares the variable to a comparison value, and exchanges it with another value
InterlockedCompareStore	Compares the variable to a comparison value
InterlockedExchange	Exchanges a variable with another value
InterlockedMax	Atomic max operation
InterlockedMin	Atomic min operation
InterlockedOr	Atomic OR operation
InterlockedXor	Atomic XOR operation

Table 6.11. Atomic intrinsics.

6.7.7 Debugging Functions

HLSL debugging instrinsics, listed in Table 6.12, allow shader programs to output debugging messages to the information queue.

Intrinsic	Description
abort	Outputs a debug message, and aborts the current draw or dispatch
errorf	Outputs an error message to the information queue
printf	Outputs a debug message to the information queue

Table 6.12. Debugging intrinsics.

6.7.8 Format Conversion Functions

The DirectX SDK includes a file named D3DX_DXGIFormatConvert.inl, which contains a variety of inline format conversion functions. These functions are designed to be used in a pixel shader or compute shader to convert from the raw integer representation of a DXGI format to a vector floating point or integer format that can be used in standard shader arithmetic. For normal read-only textures, this conversion is normally done in the hardware's texture units as part of the Sample or Load operation. However, for byte-address or structured buffer resources accessed through unordered access views, the hardware texture units are not used, and thus the conversion needs to be performed manually in the shader. These inline functions spare the programmer from needing to write these conversion functions manually. For a full listing of all available conversion functions, consult the SDK documentation.

6.8 Shader Reflection

Direct3D 11 provides a rich, full-featured set of APIs for programmatically querying detailed information about a compiled shader program. Effective use of these APIs can enable applications to develop sophisticated content pipelines that determine the requirements of shader programs and prepare runtime data in a custom format. Or alternatively, it can allow applications to dynamically set up an appropriate runtime environment for a generic shader program.

6.8.1 Shader Program Information

General information and statistics regarding a single shader program can be queried through the ID3D11ShaderReflection interface. This interface also provides the means for accessing the interfaces used for querying constant buffer, variable, or type information. To obtain the ID3D11ShaderReflection interface for a compiled shader program, an application must call the D3D11Reflect function exported by the D3DCompiler DLL. Note that this function is actually just a wrapper for D3DReflect, which it simply calls with the interface ID of ID3D11ShaderReflection. Listing 6.23 demonstrates a simple case of compiling a vertex shader program, and obtaining its reflection interface.

```
ID3D10Blob* compiledShader;
ID3D10Blob* errorMessages;
HRESULT hr = D3DX11CompileFromFile( filePath, NULL, NULL, "VSMain",
                                   "vs_5_0", 0, 0, NULL, &compiledShader,
                                   &errorMessages, NULL );

ID3D11ShaderReflection* reflection = NULL;
if ( SUCCEEDED( hr ) )
    D3D11Reflect( compiledShader->GetBufferPointer(),
                  compiledShader->GetBufferSize(),
                  &reflection );
```

Listing 6.23. Obtaining the reflection interface for a shader program.

Once the reflection interface is obtained, it can be queried for information by calling its various methods. These include methods for retrieving the thread group size, the minimum device feature level required to run the shader, and the frequency of execution for a pixel shader to name just a few. See the SDK documentation for a full list of available methods.

Most of the information provided by the ID3D11ShaderReflection interface is available by calling the GetDesc method. This method returns a D3D11_SHADER_DESC structure containing a wealth of information. Listing 6.24 contains the declaration of this structure.

```
struct D3D11_SHADER_DESC {
    UINT                        Version;
    LPCSTR                      Creator;
    UINT                        Flags;
    UINT                        ConstantBuffers;
    UINT                        BoundResources;
    UINT                        InputParameters;
    UINT                        OutputParameters;
    UINT                        InstructionCount;
```

```
    UINT                                    TempRegisterCount;
    UINT                                    TempArrayCount;
    UINT                                    DefCount;
    UINT                                    DclCount;
    UINT                                    TextureNormalInstructions;
    UINT                                    TextureLoadInstructions;
    UINT                                    TextureCompInstructions;
    UINT                                    TextureBiasInstructions;
    UINT                                    TextureGradientInstructions;
    UINT                                    FloatInstructionCount;
    UINT                                    IntInstructionCount;
    UINT                                    UintInstructionCount;
    UINT                                    StaticFlowControlCount;
    UINT                                    DynamicFlowControlCount;
    UINT                                    MacroInstructionCount;
    UINT                                    ArrayInstructionCount;
    UINT                                    CutInstructionCount;
    UINT                                    EmitInstructionCount;
    D3D10_PRIMITIVE_TOPOLOGY                GSOutputTopology;
    UINT                                    GSMaxOutputVertexCount;
    D3D11_PRIMITIVE                         InputPrimitive;
    UINT                                    PatchConstantParameters;
    UINT                                    cGSInstanceCount;
    UINT                                    cControlPoints;
    D3D11_TESSELLATOR_OUTPUT_PRIMITIVE      HSOutputPrimitive;
    D3D11_TESSELLATOR_PARTITIONING          HSPartitioning;
    D3D11_TESSELLATOR_DOMAIN                TessellatorDomain;
    UINT                                    cBarrierInstructions;
    UINT                                    cInterlockedInstructions;
    UINT                                    cTextureStoreInstructions;
}
```

Listing 6.24. The D3D11_SHADER_DESC structure.

6.8.2 Constant Buffer Information

As previously mentioned, the ID3D11ShaderReflection interface can be used to obtain an ID3D11ShaderReflectionConstantBuffer interface containing information for a constant buffer used in a shader program. This is done by calling either GetConstantBufferByName, or GetConstantBufferByIndex. The former accepts a string parameter, and is useful if an application has prior knowledge of a shader program and simply needs to query a few specific details. The latter is more useful when dealing with an unknown shader program, as it allows enumeration of all constant buffers. The number of constant buffers in a program is provided by the D3D11_SHADER_DESC structure returned by GetDesc. Listing 6.25 demonstrates the common pattern of retrieving the number of constant buffers, and iterating through the ID3D11ShaderReflectionConstantBuffer interfaces.

```
D3D11_SHADER_DESC shaderDesc;
shaderReflection->GetDesc( &shaderDesc );
const UINT numCBuffers = shaderDesc.ConstantBuffers;
for ( UINT i = 0; i < numCBuffers; i++ )
{
    ID3D11ShaderReflectionConstantBuffer* cbReflection = NULL;
    cbReflection = shaderReflection->GetConstantBufferByIndex( i );

    // Query constant buffer interface for information
}
```

Listing 6.25. Retrieving constant buffer reflection interfaces.

Like the ID3D11ShaderReflection interface, the ID3D11ShaderReflectionConstant Buffer interface provides a GetDesc method that returns a structure containing general information. In this case, GetDesc returns a D3D11_SHADER_BUFFER_DESC structure, whose declaration is shown in Listing 6.26.

```
struct D3D11_SHADER_BUFFER_DESC {
    LPCSTR              Name;
    D3D11_CBUFFER_TYPE  Type;
    UINT                Variables;
    UINT                Size;
    UINT                uFlags;
}
```

Listing 6.26. The D3D11_SHADER_BUFFER_DESC structure.

The main members of interest in the D3D11_SHADER_BUFFER_DESC structure are Size and Variables. The Size member indicates the total size of all constants in the constant buffer, and can be used to initialize an ID3D11Buffer containing the actual constant data bound to the pipeline at runtime. The Variables member indicates the number of constants within the constant buffer, and can be used in conjunction with GetVariableByIndex to enumerate all individual constants within the constant buffer.

6.8.3 Variable Information

Information for individual constants in a constant buffer is available through the ID3D11 ShaderReflectionVariable interface, which can be obtained by calling the Get VariableByIndex or GetVariableByName methods on an ID3D11ShaderReflection ConstantBuffer interface. It can also be obtained for any global variable in a shader program by calling `GetVariableByName` on an `ID3D11ShaderReflectionInterface`. Global variables include not just constants, but also resource objects, samplers, and interface instances. The `GetDesc` method of the `ID3D11ShaderReflectionVariable` interface returns a `D3D11_SHADER_VARIABLE_DESC` structure, which contains general information about the variable. The declaration of this structure is listed in Listing 6.27.

```
struct D3D11_SHADER_VARIABLE_DESC {
    LPCSTR Name;
    UINT   StartOffset;
    UINT   Size;
    UINT   uFlags;
    LPVOID DefaultValue;
}
```

Listing 6.27. The `D3D11_SHADER_VARIABLE_DESC` structure.

The `Name` member contains a string indicating the name of the variable in the HLSL source code. This value is useful for creating a system of allowing constant values to be set at runtime via by a string containing the name of the value. `StartOffset` and `Size` contain the offset in bytes from the start of the constant buffer, and the size of the value in bytes. These can be used to `memcpy` a value intended for specific constant into the correct portion of a mapped buffer. The `DefaultValue` member contains the value that a constant was initialized to in the shader source code, if one was provided. It is typically used to initialize the contents of a constant buffer when it is first created.

If the `ID3D11ShaderReflectionVariable` represents an interface instance, the `GetInterfaceSlot` method can be called to query the interface slot for that variable. This value indicates the array index that should be used for this particular interface when passing the array of class instances to a `*SSetShader` method.

6.8.4 Type Information

Type information for a variable can be queried by obtaining an `ID3D11ShaderReflectionType` interface, which is obtained by calling `GetType` on the `ID3D11ShaderReflectionVariable` interface for a variable. Calling `GetDesc` on this interface returns a `D3D11_SHADER_TYPE_DESC` structure, which contains general information regarding the type of the variable. The declaration of this structure is available in Listing 6.28.

```
struct D3D11_SHADER_TYPE_DESC {
    D3D10_SHADER_VARIABLE_CLASS  Class;
    D3D10_SHADER_VARIABLE_TYPE   Type;
    UINT                         Rows;
    UINT                         Columns;
    UINT                         Elements;
    UINT                         Members;
    UINT                         Offset;
}
```

Listing 6.28. The D3D11_SHADER_TYPE_DESC structure.

The `Class` member of the `D3D11_SHADER_TYPE_DESC` structure indicates whether the variable is a scalar, a vector, a matrix, a resource object, a structure, a class, or an interface pointer. The `Type` member indicates the primitive type of a scalar, vector, or matrix variable (float, uint, double, etc.). For resource objects it indicates the resource type, such as `Texture2D`, `Buffer`, `AppendStructuredBuffer`, or others. The `Rows` and `Columns` members indicates the number of rows and columns for a matrix variable, or are set to 1 for other numeric types. `Elements` contains the number of elements in array types, while `Members` contains the number of members for structure or class types. `Offset` contains the number of bytes from a parent structure.

The `ID3D11ShaderReflectionType` interface also contains a variety of methods that can be used to determine the full inheritance tree for interface or class types. These include a method for querying the base class type, a method for querying all supported interfaces, and a method for querying whether a class implements an interface. The interface also contains `GetMemberTypeByName` and `GetMemberTypeByIndex` methods for obtaining the `ID3D11ShaderReflectionType` interfaces for all members in a structure or class.

6.8.5 Input/Output Signature

The `ID3D11ShaderReflection` interface provides a means of querying the full input and output signatures for a shader program. Such information can be important when matching a shader program from one pipeline stage to a shader program for another stage, or when matching a vertex buffer to a vertex shader. Obtaining the input signature is done by calling `GetInputParameterDesc` for each input parameter (the number of input and output parameters is available as part of the `D3D11_SHADER_DESC` structure), and calling `GetOutputParameterDesc` for each output parameter. Both methods return a `D3D11_SIGNATURE_PARAMETER_DESC` structure, which contains information about the parameter. The declaration for this structure is available in Listing 6.29.

```
struct D3D11_SIGNATURE_PARAMETER_DESC {
    LPCSTR                          SemanticName;
    UINT                            SemanticIndex;
    UINT                            Register;
    D3D10_NAME                      SystemValueType;
    D3D10_REGISTER_COMPONENT_TYPE   ComponentType;
    BYTE                            Mask;
    BYTE                            ReadWriteMask;
    UINT                            Stream;
}
```

Listing 6.29. The D3D11_SIGNATURE_PARAMETER_DESC structure.

The SemanticName and SemanticIndex members indicate which semantic is bound to the parameter, whether a user-defined semantic or a system-value semantic. The Register indicates which register (*v0* through *v31*) the parameter was mapped to by the compiler. For parameters mapped to a system-value semantic, the SystemValueType member will contain a value indicating which semantic was used. ComponentType indicates whether the parameter is a float, int, or uint, while mask indicates which components of the register are used to store the parameter value. For input parameters the ReadWriteMask member indicates which components are read by the shader program, while for output parameters it indicates which components were written. Finally, the Stream member indicates which stream a geometry shader is using for the parameter.

6.8.6 Resource Bindings

While the ID3D11ShaderReflectionVariable and ID3D11ShaderReflectionType interfaces can be used to obtain information about the resources required by a shader program, they are not very convenient for determining how resources need to be bound to the pipeline to provide a suitable execution environment for a shader. In particular, they don't provide one key piece of information, which is the slot index that a resource is bound to. Instead, this information can be obtained by calling GetResourceBindingDesc or GetResourceBindingDescByName on an ID3D11ShaderReflection interface. These methods both return a D3D11_SHADER_INPUT_BIND_DESC structure, whose declaration is in Listing 6.30.

```
struct D3D11_SHADER_INPUT_BIND_DESC {
    LPCSTR                   Name;
    D3D10_SHADER_INPUT_TYPE  Type;
    UINT                     BindPoint;
    UINT                     BindCount;
```

```
    UINT                      uFlags;
    D3D11_RESOURCE_RETURN_TYPE ReturnType;
    D3D10_SRV_DIMENSION       Dimension;
    UINT                      NumSamples;
}
```

Listing 6.30. The D3D11_SHADER_INPUT_BIND_DESC structure.

Basic pieces of information about the resource binding, such as the name and type of the resource, are available through the Name and Type members ,respectively. BindPoint indicates the slot index for singular resources, while for arrays of resources, it indicates the start slot. BindCount indicates the number of resources in resource arrays. The uFlags member can be used to determine whether a resource was manually mapped to its slot through the register statement, and whether a sampler is a comparison sampler. ReturnType indicates the return type that was specified as the template parameter of a resource object, while Dimension indicates the required type of shader resource view that can be bound for the resource. Finally, NumSamples indicates the number of samples specified as the template parameter of a Texture2DMS or Texture2DMSArray resource object.

6.9 Using fxc.exe

The DirectX SDK includes a command-line utility named fxc.exe, which can compile shaders and programs and effects using the same compiler provided by the D3DCompiler DLL. While it can be used to precompile a shader program to bytecode and save it to a file for quick loading, it can also produce some helpful diagnostic information. The simplest way to obtain this information is to run fxc and only specify the shader profile, the entry point, and the file name. This will cause the resulting information to be output to the command line. Listing 6.31 shows the output from fxc when compiling a pixel shader from the LightPrepass sample project.

```
Microsoft (R) Direct3D Shader Compiler 9.29.952.3111
Copyright (C) Microsoft Corporation 2002-2009. All rights reserved.
//
// Generated by Microsoft (R) HLSL Shader Compiler 9.29.952.3111
//
//
//   fxc /T ps_5_0 /E PSMainPerSample LightsLP.hlsl
//
//
// Buffer Definitions:
//
```

```
// cbuffer CameraParams
// {
//
//   float4x4 ViewMatrix;               // Offset:    0 Size:    64 [unused]
//   float4x4 ProjMatrix;               // Offset:   64 Size:    64
//   float4x4 InvProjMatrix;            // Offset:  128 Size:    64 [unused]
//   float2 ClipPlanes;                 // Offset:  192 Size:     8 [unused]
//
// }
//
//
// Resource Bindings:
//
// Name                              Type  Format       Dim Slot Elements
// -----------------------------  ---------- ------- ----------- ---- --------
// GBufferTexture                    texture float4       2dMS 0           1
// DepthTexture                      texture float        2dMS 1           1
// CameraParams                      cbuffer NA           NA   0           1
//
//
//
// Input signature:
//
// Name                 Index   Mask Register SysValue Format   Used
// -------------------- ----- ------ -------- -------- ------ ------
// SV_Position              0   xyzw        0      POS float   xy
// VIEWRAY                  0   xyz         1     NONE float   xyz
// RANGE                    0     w         1     NONE float     w
// POSITION                 0   xyz         2     NONE float   xyz
// COLOR                    0   xyz         3     NONE float   xyz
// SV_SampleIndex           0     x         4   SAMPLE  uint   x
//
//
// Output signature:
//
// Name                 Index   Mask Register SysValue Format   Used
// -------------------- ----- ------ -------- -------- ------ ------
// SV_Target                0   xyzw        0   TARGET float   xyzw
//
// Pixel Shader runs at sample frequency
//
ps_5_0
dcl_globalFlags refactoringAllowed
dcl_constantbuffer cb0[7], immediateIndexed
dcl_resource_texture2dms(0) (float,float,float,float) t0
dcl_resource_texture2dms(0) (float,float,float,float) t1
dcl_input_ps_siv linear noperspective v0.xy, position
dcl_input_ps linear v1.xyz
dcl_input_ps linear v1.w
dcl_input_ps linear v2.xyz
dcl_input_ps linear v3.xyz
dcl_input_ps_sgv v4.x, sampleIndex
dcl_output o0.xyzw
dcl_temps 4
ftoi r0.xy, v0.xyxx
mov r0.zw, l(0,0,0,0)
```

```
ldms_indexable(texture2dms)(float,float,float,float) r0.z, r0.xyzw, t1.yzxw, v4.x
ldms_indexable(texture2dms)(float,float,float,float) r0.xyw, r0.xyww, t0.xywz,
v4.x
add r0.z, r0.z, -cb0[6].z
div r0.z, cb0[6].w, r0.z
mad r1.xyz, -v1.xyzx, r0.zzzz, v2.xyzx
dp3 r1.w, r1.xyzx, r1.xyzx
sqrt r1.w, r1.w
div r1.xyz, r1.xyzx, r1.wwww
div r1.w, r1.w, v1.w
add r1.w, -r1.w, 1(1.000000)
max r1.w, r1.w, 1(0.000000)
mad r2.xyz, -v1.xyzx, r0.zzzz, r1.xyzx
dp3 r0.z, r2.xyzx, r2.xyzx
rsq r0.z, r0.z
mul r2.xyz, r0.zzzz, r2.xyzx
mov r3.zw, 1(0,0,-1.000000,1.000000)
mov r3.xy, r0.xyxx
dp3 r0.z, r3.xywx, -r3.xyzx
sqrt r2.w, r0.z
mul r0.xy, r2.wwww, r3.xyxx
mad r0.xyz, r0.xyzx, 1(2.000000, 2.000000, 2.000000, 0.000000), 1(0.000000,
0.000000, -1.000000, 0.000000)
dp3_sat r2.x, r0.xyzx, r2.xyzx
dp3_sat r0.x, r0.xyzx, r1.xyzx
log r0.y, r2.x
mul r0.y, r0.y, r0.w
add r0.z, r0.w, 1(8.000000)
mul r0.z, r0.z, 1(0.039789)
exp r0.y, r0.y
mul r0.y, r0.z, r0.y
mul r0.y, r0.x, r0.y
mul r0.xzw, r0.xxxx, v3.xxyz
mul o0.xyzw, r1.wwww, r0.xzwy
ret
// Approximately 35 instruction slots used
```

Listing 6.31. Diagnostic output from fxc.exe.

The first section ("Buffer Definitions:") is a listing of all constant buffers used in the shader program. This includes the offset and size of each individual constant, as well as whether or not the constant is actually used in the shader program. The next section ("Resource Bindings:") contains the name, type, and slot of all resource objects and constant buffers used in the shader. The third section ("Input signature:") contains a listing of all inputs required by the shader program. The input signature can be very important when matching shaders for one stage with shaders from another stage, since the earlier shader must produce enough outputs to satisfy the input signature. This is also true of vertex shaders, except now the bound vertex buffer and corresponding input layout must provide the elements required by the input signature. The final section ("Output signature:") simply lists all values returned by the shader. It also indicates if the pixel shader runs at per-sample frequency, which is due to taking `SV_SampleIndex` as an input.

After the diagnostic information, the output also contains the fully compiled shader program in assembly. While it's not typically useful to examine the generated assembly, it can be helpful for verification purposes during performance analysis. In particular, it is common to look for dynamic branching or looping constructs, since these can have a drastic effect on performance. Dynamic branches can be spotted by looking for an `if_`"comp" instruction, where `comp` is a two-letter abbreviation of a comparison. For instance, `lt` will be used for a less-than comparison, and `ge` for a greater-than-or-equal comparison. Dynamic loops will begin with a `rep` instruction. The output also contains an instruction count at the end, which is simply the number of assembly instructions in the program.

While it is possible to get a *very* rough estimate of the relative performance cost of a shader through this number, in general, it is not a reliable figure. This is because shader assembly is merely an intermediate format that is further compiled by the driver into a microcode format that can be executed by the hardware. Thus, the final program could have a very different number of instructions, and the number of cycles required to execute those instructions could also vary, depending on the hardware. More importantly, even the actual microcode instructions will not properly reflect the larger-scale performance characteristics caused by memory accesses and multithreaded execution. To obtain more accurate performance statistics regarding a shader program, specialized analysis and profiling tools are available from the major graphics hardware vendors.

7 Multithreaded Rendering

7.1 Introduction

As we have discussed throughout the first half of this book, Direct3D 11 has introduced some very interesting and powerful new concepts and abilities to its arsenal. However, some of the most important new features of the API revolve around *multithreading*. The average number of CPU cores in a typical user's PC has been steadily increasing, while the frequencies at which those cores operate have plateaued, and this trend is expected to continue for the foreseeable future (Sutter). With the availability of these additional processing cores, the developer must find techniques to use them for tasks that have traditionally been performed in single threads of execution. It is in the developer's interest to convert as many tasks as possible to use parallel processing.

Direct3D 11 has been specifically designed to allow an application's rendering system to take advantage of multiple cores, and hence, of multiple threads of execution. The core interfaces that an application uses to interact with Direct3D 11 have been carefully designed to exhibit well-defined behaviors in multithreaded programs. This promotes the use of multithreaded programming, while at the same time not explicitly requiring it either. All prior versions of Direct3D had either poor or non-existent multithreading support, making D3D11 a fundamentally different type of API. This also requires a fundamental review of existing software designs to take advantage of these new abilities. As we will see later in this chapter, these multithreading capabilities have been designed to provide the potential for increased rendering performance by spreading the work required to render a frame across multiple CPU cores when they are available. When used properly, these designs can provide flexibility and the ability for your program's performance to scale with an increasing number of CPU cores in a user's system. This is referred to as the *scalability* of the software, and it will become increasingly more important as the average number of CPU cores continues to increase.

This chapter introduces and discusses the tools that have been added to the Direct3D 11 API to facilitate multithreaded rendering operations. This is distinct from, and should not be confused with, *general multithreaded computation*.[1] Here, we are concerned with using multiple threads of execution to render a frame as quickly as possible, without considering the other processing tasks that need to be performed within the application. However, if the time required to perform a frame's rendering requests is minimized by the use of multiple threads, the overall performance of the application improves, leaving more processing time for other tasks. These new Direct3D 11 tools allow the developer to use multiple threads for two types of operations—resource creation, and *draw submission* sequences. These are the two primary tasks that any Direct3D 11 application performs, creating the objects it needs, and then using them to feed the GPU with work to keep it busy. Efficiently performing these two key tasks lets a program use multithreaded code for the majority of the interactions with the D3D11 API, allowing its performance to scale with additional CPU cores.

7.2 Motivations for Multithreaded Rendering

Before we dive into the details of the Direct3D 11 threading model, it is important to understand what we want to be able to achieve with by introducing multithreading to rendering-related operations. After all, if there is no benefit, why should we introduce the complexity of multithreading into our rendering code? The truth is that there are very tangible reasons to want to break the rendering code path out of serial execution. The areas that we will focus on in this section are *resource creation* and *pipeline manipulation*.

During these discussions, it is useful to understand how other recent iterations of Direct3D were used. In both D3D9 and D3D10, it was possible to create a device with a *multithreaded* flag that would make the methods of the rendering device thread-safe. However, this thread-safe property was achieved by using coarse synchronization primitives to effectively serialize access to the code within the methods. Therefore, if two threads simultaneously called the same device method, one thread would have to wait until the other method invocation was completed. This hinders any performance gains from using

---

[1] Many resources are available to begin exploring this vast topic. The reader is encouraged to view the following threading library resources as a place to start (Microsoft Corporation) (OpenMP Architecture Review Board).

multiple threads, because of the frequent synchronization forced through the API. For this reason, the general advice when using these APIs was to only interact with the API from a single thread, often referred to as the *rendering thread*. This lets a program use general multithreading for other areas of the program, but restricts all rendering operations to being performed from a single serial thread of execution.

7.2.1 Resource Creation

The topic of resource creation spreads across a wide variety of tasks, all of which must interact with the Direct3D 11 API to create API resources. This includes the simple object creation tasks such as shader objects and pipeline state objects, and also extends to device memory based resources such as buffers and textures. To gain an insight into how multithreading can help with this type of operation, we can consider a few use cases from traditional serial execution applications.

A simple application structure can create all of the API based resources that it needs at startup, and then simply use the loaded resources during the execution of the application. Since we are considering serial execution of these tasks, the application must create all of its resources before proceeding into its normal operation. Depending on the number of resources to create, and how complex they are to load, this can lead to a significant time period during which the user is waiting to use the program. Resources such as textures require reading texture data from the hard disk, which can introduce delays for each read request. Other types of resources, such as shader objects, require some processing to be performed on them before they are ready to use. In the case of a shader, the source code for the shader must be compiled before the shader object is created by the API. When there are many of these tasks to perform, the net effect can produce a significant overall delay.

In more complex application structures, it is common for the application to require more resources than can fit into memory. A flight simulator provides a typical scenario, in which there is simply too much data to naively create resources for the entire run of the program. In this scenario, the terrain and world object data must be dynamically loaded into memory depending on where the viewer currently is within the scene. This means reading the data from hard disk (or perhaps over a network connection) during the execution phase of the program. Depending on how large the amount of data to be loaded is, this can cause a short-term reduction in the frame rate of an application, which is very objectionable for the user.

If a user has a multicore CPU, the impact of these two issues can be somewhat reduced in some cases. As described above, in recent modern iterations of Direct3D, the APIs can be used in a multithreaded environment even though they don't directly support multithreading themselves. This means that in situations where a delay is caused by reading from the hard disk, a program could create and use a secondary thread to load the desired data into memory, and could then allow the main rendering thread to use the data and create the API resource. This technique provides some relief from the delays described above, but unfortunately, it is not available or applicable in all situations. For example, when creating shader objects out of the compiled byte code, multiple threads are not helpful, since the device is only accessible from one thread.

In older Direct3D editions, the general use of multiple threads for resource creation is limited to trying to find ways around the limitations imposed by the fact that the API was not designed for use in a multithreaded environment. This adds complexity to the overall application structure, and it requires additional development time to ensure that all of the interactions between secondary threads and the rendering thread are properly implemented.

7.2.2 Draw Submission Sequences

The second area of rendering that can benefit from multiple threads of execution is draw submission sequences. These consist of all interactions that the application has with the pipeline, including all pipeline state settings, and any pipeline executions done with draw calls or dispatch calls. Once an application moves out of the startup phase and into the normal execution mode, this is the primary work that gets performed to render a frame. This type of work is depicted in Figure 7.1, with several state changes followed by draw calls.

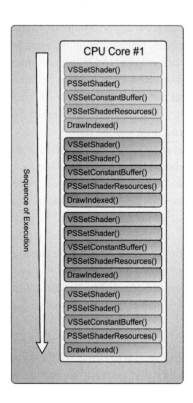

As you can see in Figure 7.1, the stream of API calls are grouped together to render one object after another, with each object identified by a different color. You can also see that all of these API calls are performed in a sequential manner, because rendering operations are traditionally restricted to a single thread. Due to the high number of these API calls that are commonly used to render a single frame, the CPU may become a bottleneck to the application's overall performance. A large portion of the driver model changes made between D3D9 and D3D10 were intended to reduce the amount of overhead for each of these API calls, and hence, to increase the maximum number of calls programmers could make for a given frame while still achieving a target frame rate. In fact, many optimization techniques for the

Figure 7.1. A series of API calls required to perform a rendering operation.

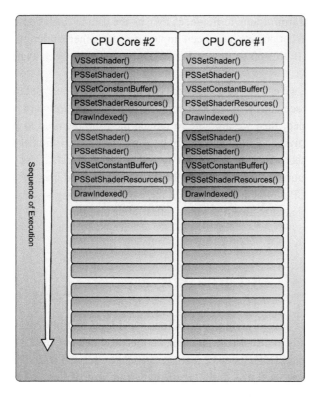

Figure 7.2. Parallelizing the API calls required for a rendering operation.

older APIs are geared toward increasing *batch sizes*. A *batch* is a group of objects that can be rendered either together, in the same draw call, or one after another, without other state changes—all aimed at reducing the number of API calls that need to be executed to render a frame.

This is clearly a task that could benefit from parallelization—the operations being performed are already neatly grouped and can be executed independent of one another. The colored groups could easily be split up and processed by other threads, effectively dividing the time used to submit the API calls by the number of available processing cores. That is, if the rendering API allowed the calls to be submitted on multiple threads, which up until Direct3D 11 they could not be! Such a parallel execution system is shown in Figure 7.2.

This can also be taken a step further, with the concept of using objects to represent a series of API calls. These objects, which are commonly referred to as *display lists* or *command lists*, could be generated once by recording a set of all of the desired API functions. Then the list could be executed quickly and efficiently by the driver, with minimal CPU time required to submit all of the API calls (since they aren't necessary anymore). The list can also be reused from frame to frame, as long as the data in them doesn't change. These types of command list objects have been implemented in other rendering systems (such as OpenGL or the XBOX 360 variant of Direct3D) but have not been available in D3D9 or D3D10.

In summary, we can say that the overall desire regarding draw submission sequences is to increase the scalability of the rendering software. If the software can correctly parallelize its workload, then we are able to reduce the time required to perform the CPU work for a given frame, and automatically become even more efficient as the number of CPU cores increases. This includes both spreading the API calls over multiple threads, and reducing the number of serialized API calls with command lists. If we can reduce the potential for the CPU to be a bottleneck in the overall rendering system, we increase the likelihood that we can use the GPU to the fullest of its abilities.

7.3 Direct3D 11 Threading Model

Now that we have several strong motivations for adding multithreading to a rendering system, we are ready to discuss the threading model provided in Direct3D 11. As described earlier, the D3D9 and D3D10 device interfaces could be created with a flag to specify that the device would be used in a multithreaded environment. If this flag was not used, the devices were considered to be *thread-unsafe* and therefore should only be accessed from a single thread. When devices are created with the multithreading flag, they are created with coarse synchronization primitives to ensure that the device methods are only used by one thread at a time. This is referred to as *thread-safe*, which means that multiple threads can call the same method of a device, while synchronization ensures that there is no interference between the calls.

Direct3D 11 was designed with multithreading as one of the primary requirements. Devices, in comparison to how they were handled in D3D9/10, have been split into two different interfaces: the *device* and the *device context*. We will discuss each of these interfaces, and how they have been implemented for multithreading, in the following sections.

7.3.1 The Device

Before Direct3D 11, the responsibilities of the device interface had remained more or less unchanged for the most recent generations of Direct3D. It represented the entire GPU, including all draw submission sequences, as well as all resource creation activities. This made the device the primary interface that was used in Direct3D. Now, Direct3D 11 has taken the device and split up its responsibilities, leaving only a subset of its original functionality in the new version of the device. This new, leaner device retains resource creation responsibilities, but it has given up the direct management of the pipeline to the device context. This change in responsibilities is described in great detail in Chapters 1–3.

However, this was not just a splitting of responsibilities. The split was made along the lines where there are different multithreaded requirements for the functionality. The resource creation methods have been implemented as *free-threaded*, which means that similar to thread-safe, they can be called simultaneously from multiple threads, but an important difference is that that the methods now don't use the heavyweight synchronization primitives. Instead, they are designed to be reentrant, so that there are no dependencies between multiple invocations of a given method. This lets the device perform object creation on multiple threads without the artificial serialization that was used to provide thread safety in previous versions of Direct3D.

With resource creation now available in a true multithreaded environment, and with minimal overhead introduced by the implementation technique, Direct3D 11 has provided facilities to allow parallel loading of resources with very straightforward implementations. Now, the same thread that has loaded a resource from disk can directly create a texture or buffer resource with the loaded data. This minimizes the amount of communication needed between threads and reduces the complexity of these types of operations.

7.3.2 The Device Context

The other half of functionality that has been split off from the device has been put into its own interface, called the *device context*. The device context comes in two flavors: the *immediate context* and the *deferred context*. We will discuss both of these objects, which implement the same interface, in the following sections.

The Immediate Context

The immediate context represents more or less the pipeline draw submission abilities from previous iterations of Direct3D's device interface—it provides the gateway for an application to directly interface with the GPU. The pipeline state setting methods are sent to the driver as soon as they are called, and are executed more or less "immediately" (in reality, this may not be immediate, because of driver implementations that use a buffer to queue up operations, but it is as close to immediate as the application can get).

This new interface has been implemented to be *thread-unsafe*, meaning that it doesn't use synchronization primitives, and that it hasn't been designed to intrinsically allow multiple threads to simultaneously use it. This is essentially the opposite of the multithreading behavior exhibited by the device, which is free-threaded. Why would this be implemented in such a thread-unfriendly way? The answer is that this is actually a very thread-friendly design—the immediate context should be used only from within one thread, but the other threads in a multithreaded program can use a different type of context—the *deferred context*.

The Deferred Context

The deferred context provides much of the same functionality as the immediate context, and it also serves as an interface to the pipeline. This includes the same multithreading behavior, meaning that the deferred context should be used from a single thread. However, as its name implies, all state changes and pipeline execution requests are deferred until a later point in time. Instead of immediately executing each state change and each draw and dispatch call, these are instead queued into what is called a *command list*. The command list can then be executed either by the immediate context, which would apply the list to the GPU immediately, or by another deferred context, which would insert the command list into that context's current command list. This functionality provides a fairly simple and intuitive introduction to generating a command list, since the deferred context is used in the same way that the immediate context is used. The only difference is that the immediate context applies the calls immediately, and the deferred context applies the calls to a command list. You can consider the deferred context to be a per-thread, per-core command list generator.

There are a few other functionalities that don't apply to the deferred context that can be used on the immediate context. The deferred context does not support directly performing queries—they must be performed on the immediate context instead. In addition, the deferred context cannot read data back from resources. This makes sense when you consider that the deferred context is creating a list of its instructions, instead of immediately doing the work. However, the deferred context is allowed to write to the contents of resources, as long as the resources are mapped with the D3D11_MAP_WRITE_DISCARD flag. This also makes sense, since this allows the command list to contain requests to write the entire contents of the resource, further enforcing the concept that the deferred context may not access the contents of a resource.

7.3.3 Command Lists

Since command lists play such an important role in the overall multithreading implementation of Direct3D 11, it is worth understanding precisely how they work. Command lists are generated by calling the deferred context's FinishCommandList() method. This creates a command list that includes all of the state change and draw/dispatch calls since the most recent previous call to FinishCommandList(). Once a command list has been created, the sequence of operations it contains cannot be changed—the object itself is immutable. The command list is submitted for replaying by calling `ExecuteCommandList()` on the imme-diate context or another deferred context. So the general paradigm for using command lists is that they are generated by a deferred context, then are either consumed by the immedi-ate context or applied to another deferred context, which effectively inserts that list into another list's stream of operations.

After a command list has been used, it is disposed of by calling its `Release()` method. However, determining when a command list should be released is up to the developer. Even though the command list itself is immutable, there is no limit to the number of times it can be executed. This introduces some interesting possibilities with respect to caching particular subsets of calls into command lists and permanently reusing them throughout the duration of the application. Later in this chapter, we will see several different levels of granularity for command lists, which can take advantage of this property.

This discussion also requires us to provide a distinction between *static* and *dynamic* *states* for a given object rendering sequence. A *static state* is any pipeline configuration that doesn't change from frame to frame, while *dynamic states* do change in some way over time. Examples of static states could be something like the blend state, a vertex shader, or a texture. Examples of dynamic states would be any transformation matrices that change from frame to frame, and any other state that doesn't remain the same over time. Typically, every object rendering requires some combination of both static and dynamic states. At first, this concept of reusing command lists may seem somewhat unusable, since you would normally want to render an object in a different location in each frame, because the camera, or the object itself, is moving. This would require new transformation matrices to be gener-ated and uploaded for use with the vertex shader, which would seem to be incompatible with reusing command lists, which are immutable.

However, a command list records the states that are set on the deferred context. To use the example of transformation matrices from above, the application would bind a constant buffer (constant buffer *A*) that contains the transformation matrices to the vertex shader. Once the command list has been generated with the `FinishCommandList()` method, the specific constant buffer bound to the vertex shader cannot be changed in the command list. This means that there is no way to bind a different constant buffer (constant buffer *B*) in its place. However, you can modify the *contents* of constant buffer *A* to contain the desired transformation matrices, without requiring a new command list. This is the same concept as passing function arguments by value or by reference in C/C++. The command list essentially records a pointer to a particular resource, while the value that it points to can be changed independently of the command list. This lets you inject dynamic state into command lists, even though the command list itself cannot change. To summarize this idea, the command list cements the `resources` that are bound in its API call sequence—but it does not cement `the contents of those resources`. This concept is depicted graphically in Figure 7.3.

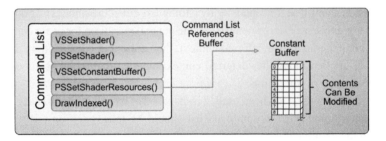

Figure 7.3. A depiction of binding resources in a deferred context, while the contents of the resource can be changed at a later time.

This gives us two general ways to use command list objects. The first is to generate and use a command list within every frame. This allows for simple dynamic updates to any state that is used in the rendering of an object, and the command list is released after the rendering is complete. This usage is intended to allow spreading the costs of API calls across multiple threads. The second way to use a command list is to generate it once, and then update its `resources` from frame to frame, and reuse it many times. This reduces the cost of generating the command list in every frame, at the expense of some additional complexity to ensure that only resources contain the dynamic state for the object rendering.

7.3.4 Using the Device and Context Interfaces

With this understanding of how the deferred context works, now would be a good opportunity to consider how a rendering system could be implemented to use deferred contexts. The general advice regarding deferred contexts is to use at most one context/thread per core in the user system's CPU.2 Then the application would have each thread perform some rendering work that can be queued into a command list. Once the command lists have all been generated, the immediate context would iterate through them, executing them one by one in the proper order. This basic setup is depicted below in Figure 7.4 for a quad core CPU.

As you can see from the figure, this new setup significantly reduces the amount of time needed to submit the `draw` state change calls to the API. In addition to reducing the overall time to submit the rendering requests, command lists are created in a format that lets the driver quickly and efficiently play back the sequence, which will further reduce the

---

[2] There are several ways to determine the number of CPU cores on a user's computer. The *GetLogicalProcessorInformation()* function has been in existence since Windows XP, but it includes hyper-threaded CPUs as a separate core. While this is partially correct, we are more interested in the actual number of CPU cores. Windows 7 has introduced the *GetLogicalProcessorInformationEx()* function, with extra flags to better specify precisely which count to return. Since Direct3D 11 runs on both Windows Vista and Windows 7, it may be necessary to use both functions, depending on which operating system is in use.

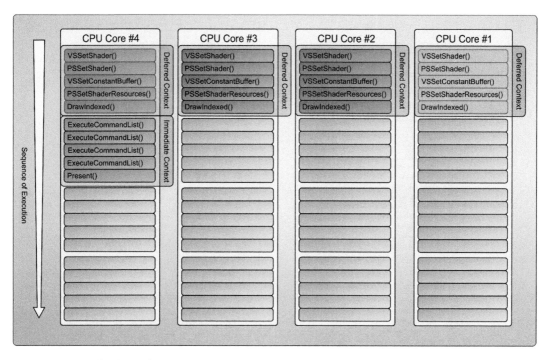

Figure 7.4. Using four deferred contexts to generate command lists to be executed on the immediate context.

total time required to finish rendering the final output on the GPU side of the application. The modified design of the Direct3D 11 interfaces provides the potential for a significant speed increase, without requiring too many changes in the application itself.

7.4 Context Pipeline State Propagation

With the structural details explained, we need to consider another topic concerning how command lists are generated and executed. Specifically, we need to understand how pipeline state is preserved, restored, and eliminated during the various operations that are performed when using command lists.

7.4.1 The Immediate Context State

We will start by investigating the behavior of the immediate context's pipeline state throughout a command list execution phase. Figure 7.5 shows the immediate context and

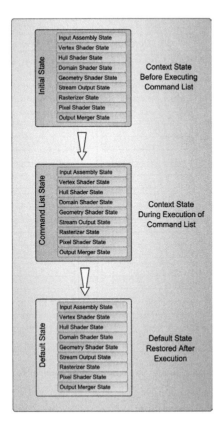 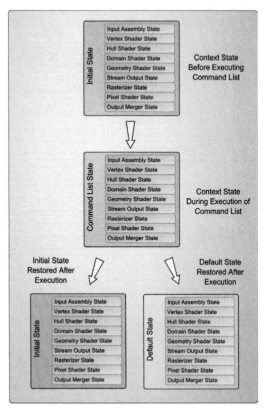

Figure 7.5. Immediate context state before, during, and after the execution of a command list.

Figure 7.6. The two possible ways to handle the immediate context pipeline state after executing a command list.

its pipeline state prior to executing a command list. It's current configuration reflects any state change requests that were performed directly on the immediate context with its SetXXX() methods. When a command list is executed on the immediate context, its pipeline state is replaced with the default pipeline state. This default state is the same state in which the device was initially created.[3]

There are two possible outcomes for the existing pipeline state that was originally in the immediate context. Which one is chosen depends on a Boolean parameter passed to the ExecuteCommandList() method. If the passed parameter is true, the original pipeline state is stored until the command list has been executed, and is then restored to the immediate context. If the passed parameter is set to false, the original state is simply deleted, and the immediate context is reset to the default state after the command list has been executed.

---

[3] The individual values that comprise the default pipeline state are further described in the DXSDK documentation.

In both cases, the existing state of the pipeline is never shared with the command list, and the state introduced by the command list execution is never left active on the immediate context. These two possible execution paths are depicted below in Figure 7.6.

7.4.2 The Deferred Context State

The deferred context also contains the same pipeline state structure that the immediate context possesses. When the deferred context is created, it starts out with a default pipeline state. As the pipeline state is modified through use of the deferred context, the state-changing calls are accumulated until the application calls `FinishCommandList()` to generate a command list. Similar to the immediate context state when calling `ExecuteCommandList()`, the deferred context pipeline state will be handled by a Boolean parameter passed to the `FinishCommandList()` method. If the passed parameter is set to `false`, the deferred

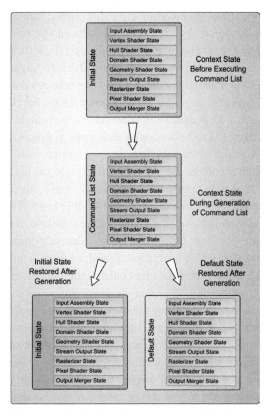

Figure 7.7. The two possible ways to handle the deferred context pipeline state after generating a command list.

context pipeline state is reset to the default pipeline state after the command list is created. If the passed parameter is set to `true`, the deferred context pipeline state is preserved beyond the `FinishCommandList()` invocation. This is depicted in Figure 7.7.

The differences between these two options determine the pipeline state with which the deferred context's *next* command list starts out with. This allows the application to control the starting state of the command list, and potentially shares the cost of setting that state initially across several command lists if the same settings are always used.

7.4.3 Performance Considerations

With some level of control over whether a context either retains or resets its pipeline state, it is helpful to consider a few situations in which it would be potentially advantageous to use these different behaviors. The primary considerations depend on the size of the command lists being used, as well as on the command-list execution strategy that the application will use. If the application is generating many small command lists within a single frame, and those command lists all share some common pipeline states, then a net reduction in the number of calls to set the deferred context pipeline state can be achieved by maintaining the state between calls to `FinishCommandList()`. Conversely, if the subsequent command lists that are being generated are very long or don't share common states, then there is likely little or no benefit to maintaining the context state between command lists.

The immediate context state propagation depends somewhat more on the frequency of command list execution. In general, it is faster for the immediate context to return to a default pipeline state than it is to both save its state prior to execution and then restore it afterwards. With this in mind, we can see that if many command lists will be executed in a row, there is no reason to propagate the state since the command list is executed in a fresh default pipeline state. It wouldn't make any sense to save and restore the original state if it isn't used in between command list executions. On the other hand, if the immediate context is used for normal rendering in between command list executions, then it may make sense to restore its state after executions. Of course, this also depends on how many of the states are common before and after the command list is used, so there is no general rule to determine which technique is better suited for a particular rendering sequence. This is a decision best left up to profiling and testing later on in the development cycle.

One final point to consider regarding the use of a previously set state relates to debugging of rendering operations. In this case, if an application inadvertently relies on a state that is set in some other portion of the rendering code then it is quite possible that a change to the rendering code of one object could affect the state of other objects that occur later in the rendered frame. In general, it is good practice to minimize these types of dependencies, especially if the rendering code is data-driven instead of hard coded. At the very least, it is a good idea to occasionally run the rendering code with no state propagation active to determine if these dependencies exist.

Determining Driver Support

The two main multithreading features provided by Direct3D 11, resource creation and draw submission, are guaranteed to be safe to use in a multithreaded development environment. However, they also rely on the GPU driver to operate optimally. If the driver cannot provide fully free-threaded resource creation, then the runtime will default to using its own lightweight synchronization. Similarly, if the driver does not provide a native implementation for multithreaded command lists, then the runtime will emulate the feature itself and provide a similar functionality. In both of these cases, the performance of the individual implementations may vary slightly, but having an emulation mode still allows the multithreading features to be used, even when the driver doesn't support them. The benefits of using multiple threads should far outweigh any discrepancy in performance between native implementations and emulated ones.

The `ID3D11Device` interface provides a facility that indicates the level of support for these two multithreading features in the user's video card driver. The process for determining multithreading support requires an initialized device interface that can then be queried for its level of support for multithreading. (The process for creating a device is described in detail in Chapter 1.) When an application intends to use multithreading, it must initialize the device without the `D3D11_CREATE_DEVICE_SINGLETHREADED` device creation flag. Once the device has been successfully initialized, it can be queried for multithreading support using the `ID3D11Device::CheckFeatureSupport()` method. Listing 7.1 demonstrates how to do this. Once this method returns successfully, the passed-in `D3D11_FEATURE_DATA_THREADING` structure contains two Boolean variables: `DriverConcurrentCreates` and `DriverCommandLists`. These variables indicate the driver level support for resource creation and command lists, respectively.

```
D3D11_FEATURE_DATA_THREADING ThreadingOptions;
m_pDevice->CheckFeatureSupport( D3D11_FEATURE_THREADING,
                        &ThreadingOptions, sizeof( ThreadingOptions ) );
```

Listing 7.1. Checking for driver support of multithreading features.

While it is not mandatory to check these flags to determine the level of multithreading support (since emulation is available), doing so provides nice side benefit for previous-generation GPU hardware. Since the D3D11 runtime can be used with down-level hardware, it is possible for a GPU manufacturer to release updated driver implementations for D3D10 hardware. This provides a potentially significant performance improvement by allowing true multithreading on older hardware simply by having an updated driver and using D3D11. In this case, having the ability to poll the driver for support can be extremely beneficial for the end user.

7.5 Potential Usage Scenarios

The discussion of the multithreading behavior of the Direct3D 11 interfaces up to this point has focused on what they are capable of from a technical standpoint, and on how they function. Now we can take a look at a few situations where these tools could be used in a rendering application. We will look at potential cases where the use of multithreading can improve the performance of a given task, and will touch on several design considerations for designing a new rendering system.

7.5.1 Multithreaded Terrain Paging

We will start with how to apply Direct3D 11's multithreading capabilities to a *terrain paging system*. A *page* is a term for a memory management technique that uses more memory than is physically available on a computer. As data is needed, it can be *paged in* to a system memory block, replacing whatever page is already there. When paging is used to manage the data is needed at any given time, more data can be used than fits into the physical system memory. When designing a terrain paging system, the focus is typically on providing your scene with a very large space that can be explored. Because the scene data is often quite large, it is normally not possible to directly create all of the resources needed to hold all of the terrain data. So, that data must be paged into and out of a resource at runtime, piece by piece, based on the current location of the viewer. This provides a good use case for the multithreaded systems that have been discussed, since it can use both parallel resource creation and as parallel draw submission.

Such a system could use *worker threads* during startup and runtime. Each of these threads would be assigned a deferred context to use for the duration of the application, and would also have a reference to the device. Since we would likely have more terrain pages than CPU cores, the threads would have to be assigned in some repeating pattern to manage a given terrain page. At startup, each worker thread could start loading one terrain page from disk. It could then use the device's free threaded buffer resource creation method to allocate a vertex buffer and initialize it with the loaded terrain data. Since terrain loading is happening in parallel, the startup speed would likely be limited only by the available I/O bandwidth, which would vary, depending on what media is being used to store the data. For example, the hard disk, DVD drive, or network storage might all be used, each with different access characteristics. In addition, since resources can be created in parallel to rendering operations, the application can simultaneously start rendering a startup screen or debriefing message while the terrain pages are being loaded and created.

During runtime, each worker thread could generate a command list that configures the rendering pipeline with the appropriate shaders, resources (such as textures or constant buffers), and states needed to render that terrain page. For each frame to be rendered, the command list of every visible terrain pages would be executed on the immediate context. These command lists could potentially be reused, since the view matrix required for rendering (or some combination of matrices including the view matrix) would be updated in a constant buffer, whose contents are not included in the command list. Only the reference to the constant buffer would be included in the list, not its contents.

As the viewer moves around in the scene, the worker threads could dynamically load new terrain pages from disk, as needed. Using their deferred contexts, they could add the Map/UnMap sequences for updating the terrain page's vertex buffers with the new data into their command list sequences. Then, the next time that particular terrain page is rendered, the vertex buffer resource could be updated and made available for rendering. This provides a simple approach to updating the terrain pages, with minimal synchronization required between the main rendering thread and the worker threads.

7.5.2 Multithreaded Shader Creation

The terrain paging system discussed in the previous section would certainly benefit from being run on a multithreaded system. However, not all applications need to use a data set that is so large that it must be paged into memory dynamically. However, there are some tasks that most, if not all, applications need to implement, and that could also benefit from multithreaded resource creation. Our next example fits into this broader context—creating shader objects at startup to be used during rendering. This seems like a somewhat trivial task, but depending on the number of shader programs your application will use, as well as their complexity, and the situations that they need to be used in, a "combinatorial explosion" can produce a very large number of required shader objects. The time required to create all of these objects can easily become unmanageable.

Creating a shader program requires two steps. First, the shader source code must be compiled to a byte-code format. Then, that byte code is submitted to the free-threaded device methods to create a shader program for a given programmable pipeline stage (additional details about shader creation are available in Chapter 6). Since shader compilation is relatively CPU intensive, it presents a good opportunity for parallelization when more than one shader must be compiled on a system with more than one CPU core.

In the second step, the free-threaded methods of the device allow multiple threads to simultaneously create shader objects after the compilation step. This eliminates the need to synchronize multiple threads to create the shader objects, which would have been required in older versions of Direct3D. A simple implementation to allow parallel loading of shader programs would create one worker thread for each available processing core. This arrangement is essentially a *thread pool*, a concept that may be familiar to the reader from standard multithreading designs. Then the list of shader programs to be loaded, and their respective types and required shader models, would be distributed among the worker threads. Each thread would compile and create its list of shader objects in complete isolation from the others, eliminating possible synchronization issues. Then the main application thread would simply wait for all of the worker threads to complete, and would then read the resulting objects and store them centrally for use later on.

7.5.3 Multithreaded Submissions

The previous two examples are interesting, and they provide some insight into how resource-intensive tasks can use some of the Direct3D 11 multithreading tools to gain some performance advantage. However, the largest performance potential lies with the ability to split the rendering work for a frame among several CPU cores. As described above, this allows the overall CPU/driver costs for rendering a scene to be amortized over several threads simultaneously, reducing the time spent to send work to the GPU. Of course, there are many variations of how to implement this concept, and some may fit a particular application better than others. We will discuss a few possibilities here, and try to provide some context about why each of them would be a good choice in a particular situation.

The general scenario for the following discussion is the following. You have one main thread that can house the immediate context, and several worker threads (one for each CPU core in the system), each with a deferred context. When a frame needs to be rendered, its total workload is split up in some manner among the worker threads to generate a command list. When all of the command lists are executed in the proper order on the immediate context the final rendered image is produced. The order of execution is only important when the contents of one resource must be modified before the resource is used. For example, if one command list generates a shadow map, and then a second command list uses the shadow map to render a scene, the shadow map must be generated before being used. There are other cases where the order will not matter, such as per-object command lists (discussed below).

Deciding how to split up the work among the threads will likely depend on the types of scenes being rendered, as well as their contents. Let's look a little closer at three potential techniques for splitting up the scene's rendering workload. Please note that these are only three possible techniques that could be used, and that many other variations may or may not perform better in a given situation.

Per-View Command Lists

The first method of dividing a scene for rendering is perhaps the least intrusive of the three. The general idea is to generate a command list for each *view* of a scene. In this sense, a view can be considered one complete rendering pass in which the complete scene is rendered.

For example, the generation of a shadow map would be one view of a scene. The generation of an environment map would be another, and the regular perspective rendering of a scene would also be considered a view. If you consider the normal single-threaded series of rendering commands to be one continuous list, then a general rule of thumb for splitting the list into views would be to break them whenever the render target is changed or cleared (or in the case of computation pipeline execution, when the UAVs are changed). This splitting is shown in Figure 7.8.

With this partitioning, we would break up the rendering workload into these discrete chunks and hand them off to the worker threads to be processed into command lists. Each scene view would proceed with the rendering code it normally uses, except that it would be working with a deferred context, instead of an immediate context. Implementing this segmentation of the work may be somewhat difficult, depending on how a rendering pass is currently implemented in a rendering framework. One topic that must be considered is that of data synchronization. As an example, let's assume that an application uses a scene graph to contain its scene elements. If the graph is traversed once for each of our rendering views, and these transversals are carried out on different threads, we must ensure that the state of the scene graph itself is not modified in any way by any of the threads—otherwise, we risk (and almost certainly induce) data corruption between rendering threads. We also would like to avoid using manual synchronization code, since it will detract from any performance benefit we get from multithreading. In general, all updating for a given frame must be done prior to the rendering pass, or else any data modification must be very carefully planned out!

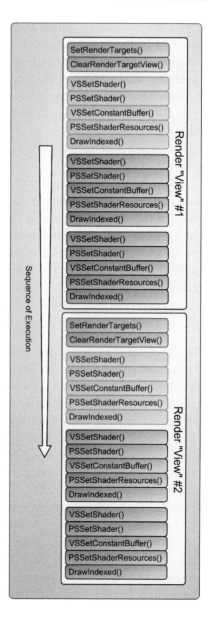

Figure 7.8. High-level rendering operations can be used to split a rendering sequence into "views."

Our view-level multithreading granularity provides good potential for reduced CPU overhead. Some rendering passes use very similar rendering effects for most, if not all of the scene objects. Consider a shadow map generation pass—most objects will use the exact same pixel shader to output the appropriate depth value, and most will also use a similar transformation shader setup (vertex and/or tessellation based shaders) with some variations for static versus dynamic geometry. These types of rendering passes are typically presorted at the view level, so all of the operations that a deferred context executes to set up a rendering pass (such as setting render targets or stencil setup) will be amortized over many draw calls.

However, due to the generally larger command lists, this view-based processing doesn't have much of a chance to reuse command lists from frame to frame. As discussed above, it is possible to update the dynamic state of objects by modifying the contents of the resources used. However, this does not allow the application to change which objects are rendered and which are culled in a given frame. This should not be seen as a critical problem, however, since there will not be very many view-sized command lists being generated for each frame.

Per-Object Command Lists

The next level of granularity we could use is to render the scene objects at the individual object level. In this scheme, the worker threads would generate one command list for each object that will be rendered. This introduces a much finer level of processing and consequently increases the number of command lists that must be generated and executed. Because of larger number of command lists, it is probably advantageous to use deferred context state propagation between command list generations. This would alleviate the need to make more frequent calls to higher-level rendering setup functions, such as setting render targets, because so many command lists are used. The additional command lists would also require a higher number of `FinishCommandList()` and `ExecuteCommandList()` calls, which may or may not impact performance. Since the command lists are generated in isolation from the rest of the scene, it also reduces the amount of batching that can be performed.

However, this paradigm also has some interesting side effects that could prove to be beneficial. Once a command list is generated for a particular object for a particular rendering pass, there is likely no reason to release and recreate the command list for every frame. Since any per-frame dynamic rendering data, such as view or skinning matrices, is provided to the shader programs in constant buffers, the command list will not change from frame to frame as long as the same constant buffers are updated and used for every frame. With no overhead for generating the command list, any additional costs discussed above could largely be overcome. In addition, the simplicity of such a scheme would be attractive as well—each object would simply receive its own command list and use it as necessary.

Another potential optimization for this technique is provided by the fact that deferred contexts can also consume command lists. This means that it is possible to build larger command lists from a group of smaller ones to minimize the number of command lists needed to be executed on the immediate context. Depending on the implementation of this feature in drivers, this could also be used to reduce the overall cost associated with rendering a frame with low-level command lists.

Per-Material Command Lists

The final solution that we will consider is to process a scene in per-material chunks. When it is time to render a frame, some preprocessing of the scene is required to group objects that share common materials, so that they can be processed together as a group by a single worker thread. Then, each newly generated command list is passed on to the main rendering thread for processing by the immediate context. This type of object sorting was a fairly standard practice in D3D9 programs, due to the relatively high CPU cost for each API call. By sorting objects, the total number of state changes could be reduced by finding a more optimal sequence to render them in. Thus, this technique could draw upon some of these existing routines for grouping suitable objects together.

This model strikes somewhat of a balance between the two prior methods. The size of the groups of objects that are rendered depends solely on how common a given material is, so this method provides the potential to handle the largest number of objects within a single rendering batch. Since the per-material command lists encompass more objects with a relatively small number of state changes between them, they provide better performance than the per-object command lists since there would be fewer lists to submit to the immediate context. However, there is also the additional cost of having to sort the objects by material before creating the command list, so the benefits of this technique will vary with the scene complexity and variations in material types. For scenes that contain relatively homogeneous rendering materials, this technique can provide a good way to efficiently process and submit them to the GPU.

7.6 Practical Considerations and Tips

We will conclude this chapter with a section devoted to providing some general design advice that should be useful when you start to build an application or framework that will use the multithreading features of Direct3D 11. Due to the typical complexity of multithreaded programming, it is best to minimize potential issues from the beginning to avoid trying to find (and fix) them later on.

7.6.1 Things to Do

We begin this section with a discussion of some design points that should be incorporated if possible into your initial design. These are suggested practices that are not required by Direct3D, but that have been found to be helpful in the authors' experiences with the API.

Variable Levels of Threading

The first topic is to ensure that your application/rendering framework is not locked into a particular number of threads. This is important for several reasons. First, there is no way to know how many CPU cores an end user's system will have, so it makes sense to dynamically decide the number of threads to create and use. If possible, this should be extended to allow the number of threads to be selectable at runtime, to allow for future possible optimizations for switching additional threads on or off as the rendering workload needs them.

The second reason for using a variable number of threads is related to debugging your programs. If you are not receiving the final rendered frame that you want, this could be due to the multithreaded rendering itself (or more precisely, to how it is being used). To help minimize debugging time, it is extremely advantageous to be able to switch among the following setups:

- Multi-threaded, deferred context rendering: Multiple threads are used with deferred contexts to generate command lists, which are then rendered on the immediate context.

- Single-threaded, deferred context rendering: A single thread is used with a deferred context to generate command lists, which are then rendered on the immediate context.

- Full single-threaded rendering: A single thread is used with only the immediate context to directly render without generating command lists.

These various mixtures allow for isolating rendering issues between a regular rendering error, an incorrect state setting due to the multiple contexts being used, and data contention between multiple threads of execution. This type of flexibility is assisted by the fact that the immediate and deferred contexts outwardly appear identical. Therefore, if the rendering system batches are incorporated into objects that can be passed to a thread for processing, then each thread would require one context to do the processing. The rendering framework could then decide dynamically at runtime how many threads to use, and which types, and then pass the rendering work to the corresponding objects.

Using PIX

Some additional considerations are needed when using PIX for debugging and analyzing your application. PIX is a very useful tool that can let you inspect what your program has

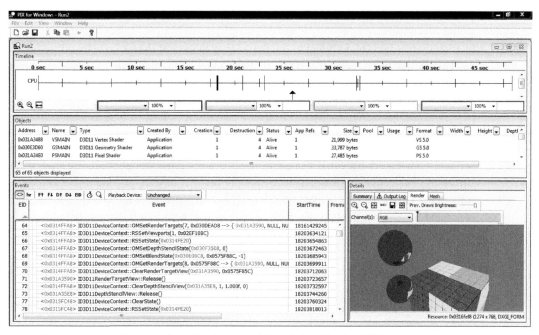

Figure 7.9. A sample screen shot taken from a multithreaded application PIX frame capture.

done with the D3D11 API for a given frame. When using multiple contexts and multiple threads to generate and consume command lists during a frame, you can still perform a frame capture and obtain significant information about how the frame was constructed. Specifically, you can see each of the API calls that are made from each context, with the calling context identified by its address. Figure 7.9 shows a snapshot of the information available from a frame capture.

Unfortunately, at the time of this writing some of the visualization tools in PIX are not operational when used in conjunction with deferred contexts/command lists. In the single context frame capture, you can step through each API call and visualize surface contents and the geometry being rendered, and you can inspect the contents of memory resources. But when deferred contexts are used, the API calls are wrapped into a command list, and the visualizations are not possible since the entire command list is executed at the same time, which is much later than when the command is actually sent to the deferred context. Even so, the time and order of the calls are still available and can help you understand if there is a mistake somewhere in the sequencing of the API calls.

This also provides another reason to be able to switch from using deferred contexts to only using the immediate context—you can use the immediate context mode to provide a complete PIX frame capture with visualizations to check for errors, and then switch back to deferred rendering once you are sure there aren't any errors in the rendering code.

7.6.2 Things to Avoid

In addition to advice about what to do, there are also several things you should try not to do. This section briefly describes a few of these. Once again, these are simply suggestions based on past experiences.

Pipeline Stalls

Although using deferred contexts and command lists lets you use multiple CPU cores to perform your rendering work, the same hazards and advice about rendering in general still apply. The command lists efficiently submit many state changes and draw/dispatch calls to the Direct3D 11 runtime and driver. However, if there is an option to submit the command lists in a variable order, then be sure to put them into the order that will allow

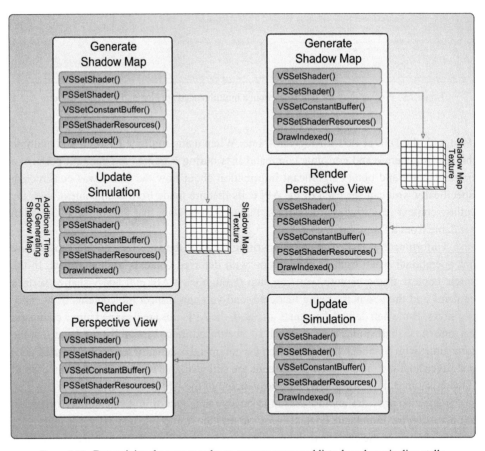

Figure 7.10. Determining the proper order to execute command lists, based on pipeline stalls.

overall execution to continue without pipeline stalls. For example, consider if we have one command list that generates a shadow map, another one that updates a simulation with the compute shader, and a third one that performs the final scene rendering. If the simulation contains two buffer resources for containing its state, then we can render the current frame with the current buffer and use the GPU resources to update the buffer for the next frame. With this set of command lists, we can execute them in the following order: the shadow map first, then the simulation, and finally, the perspective rendering. The reason behind this is depicted in Figure 7.10.

When the command lists are submitted in this order, execution of the final rendering pass doesn't start until the shadow map has already been completed. This means that the GPU will remain busy the entire time, without waiting for one pass to complete before continuing with the final pass.

Switching Modes

Another general piece of advice regarding the ordering of command lists is to minimize the number of times an application switches between using the GPU for *rendering* and using it for *computation*. If a full-frame rendering consists of several computation workloads (i.e. using `Dispatch` calls) as well as several rendering workloads (such as using `Draw` calls) then they should be ordered such that the computation command lists are executed together (as much as possible) to minimize how many times the GPU switches modes. This is general advice based on current GPU implementations, which may or may not continue to be the case in future generations of hardware.

Context State Assumptions

If you are porting an existing rendering system to Direct3D 11, it is quite possible that your previous rendering system makes some optimizations about setting pipeline states to eliminate redundant state changes and minimize the number of API calls in every frame. A typical example is to keep a reference to the current render target, and ensuring that when your application tries to set a render target, it checks to see if the current target already matches the desired one.

In previous versions of Direct3D, this would reduce the number of API calls and relieve the application code from having to manage the state itself. However, due to the way that context state is implemented in Direct3D 11, there are significant differences to how any such state-caching system will need to operate. These systems can still be used, but they must be aware that the complete pipeline state may or may not be reset based on the calls to `FinishCommandList` and `ExecuteCommandList`, as detailed above. State caching can still reduce unnecessary API calls within a command list generation pass, but the caching system needs to be reset at the appropriate points, depending on how the command lists are generated and executed.

7.7 Conclusion

With an API architecture specifically designed to allow multithreaded programming, there is a very rich set of possible performance-improving techniques available to the developer. In addition to simplifying some portions of an application, such as resource loading during startup, there is also great potential for parallelizing the rendering sequence API calls to minimize the cost of submitting work to the GPU. In addition, as the number of available CPU cores continues to increase, it is quite likely that the rest of an application framework will become multithreaded. This trend applies pressure on the rendering framework to fit into such a processing system, which provides even further incentive to build support for multithreading into a Direct3D 11-based renderer.

We can also consider these additional features in a higher-level view, as well. By directly supporting multithreading, Direct3D 11 essentially allows developers to break down the sequential rendering processes that have been used for many generations of D3D. From a design perspective, you can now logically think of a frame as being a collection of tasks and dependencies, rather than as a monolithic sequence of events.

8   Mesh Rendering

Our first sample chapter will cover the fairly broad topic of mesh rendering. As we have seen in Chapter 3, there are many different ways to submit work for the GPU to perform. To introduce the beginner to a basic example of how this can be achieved, we begin with a discussion of transformation matrices and how they are used in the context of the rendering pipeline to achieve basic mesh placement within a scene. After the basics of animating an object have been covered, we extend the simple transformation matrices to allow for animation to take place within the mesh itself. A technique for vertex skinning is presented, which allows a number of transformation matrices to be used to deform the basic shape of a model for more flexibility in how a mesh can be manipulated. Finally, we discuss adding displacement mapping to our vertex skinning sample. Displacement mapping allows the higher-frequency details of a model to be stored in a texture instead of within the mesh itself, potentially allowing for improvements on the standard vertex skinning technique.

8.1   Mesh Transformations

This section will cover some of the basic rendering operations found in a typical real-time rendering application. The rendering of a triangle mesh at a desired location, orientation, and scale within a scene can be considered the absolute minimum operation that a rendering framework must be able to perform. This chapter serves as an introduction to using the Direct3D 11 rendering pipeline, and also provides an example of the methodology that will be followed throughout the remainder of the book to design and implement rendering techniques.

8.1.1 Theory

In a typical 3D application, it is common for the 3D model data that will be rendered to be generated in an external modeling package for later use. This can be an artist's modeling program, a computer-aided engineering program, or any of a host of other software packages. These programs are often referred to as *digital content creation* tools, or *DCCs* for short. When the real-time rendering program is started, it loads the 3D model data from disk, processes it into an appropriate format, and then renders the model in a given scene, as needed.

To properly render the model within the application's scene, we must specify the location, orientation, and scale that we want it to appear with. Since the model is defined in object space, we need to manipulate the vertex positions prior to rasterizing the model's triangles, so that they appear where we want them to. This manipulation is executed by performing matrix multiplication on the vertex positions with specially created transformation matrices.

We will provide a brief discussion on transformation matrices here, but will not review the mathematical foundations behind their functionality. Our intention in this book is to provide the best possible resource to Direct3D 11, which means that we could not possibly provide a complete introduction to computer graphics. The formulas described below are taken from the DirectX SDK and are suitable for our uses in the current example. Since we are only performing a basic-level matrix manipulation, we can use the matrices without a complete understanding of their details. If readers are interested in further information, we refer them to (Eberly, 2007) and (Akenine-Moeller, 2002) for a more detailed discussion of transformation matrices. Unless otherwise stated, all transformation matrices are defined such that a position is represented with a 1×4 vector that is right-multiplied by a 4×4 transform matrix to produce another 1×4 row vector.

World Space

To render a triangle mesh, the first model manipulation is to determine the spatial state of the model within the scene. More specifically, we need to specify the desired scale, orientation, and position of the model. Each of these desired properties can be expressed in a 4×4 matrix form, which is commonly referred to as a *homogenous matrix representation*. As we will see later in this section, using homogenous matrices for our object manipulations allows for easier manipulations and combination of the properties they represent.

Scale matrices. The first property of the mesh that we will examine is its *scale*. This is commonly used to convert the model from its original size (as created in the artist's DCC tool) to the desired size within the rendered scene. In addition, the scale of an object can be modified to achieve some simple effects, such as shrinking, expanding, or oscillating the size of an object. The size of the model is manipulated with a scale matrix, which is created

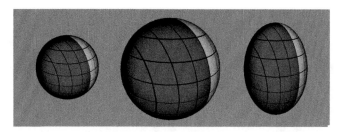

Figure 8.1. An example model (left) scaled with a uniform scaling matrix (middle) and a non-uniform scaling matrix (right).

in the form shown in Equation (8.1):

$$S = \begin{bmatrix} sx & 0 & 0 & 0 \\ 0 & sy & 0 & 0 \\ 0 & 0 & sz & 0 \\ 0 & 0 & 0 & 1 \end{bmatrix}. \tag{8.1}$$

In this equation, we can see that a scaling value can be applied to each of the three coordinates individually. When all three of these values are the same, the matrix is called a *uniform scaling matrix*. Uniform scaling effectively only changes the size of the object being transformed, while *non-uniform scaling* changes both its size and shape. Figure 8.1 demonstrates various scaling matrices being applied to a model, in comparison with the original model.

Rotation matrices. The next property of our model to manipulate is the **orientation**. Typically, the orientation of an object is modified by applying a rotation around one of the three principal axes at a time. The angles of rotation are referred to as *Euler angles*, since they were first described by Leonhard Euler. Each of these rotations can be applied with an individual rotation matrix, and the complete orientation of an object is represented by three rotation matrices—one for rotation about each of the three principal axes. Each rotation is performed about its respective axis, meaning that the overall rotation can be considered to occur around the origin of the model's frame of reference.

When dealing with rotations, care must be taken to ensure that the rotations are all applied in a consistent order, since matrix multiplication is not commutative. A rotation about each axis can be created as shown in Equation (8.2):

$$R_X = \begin{bmatrix} 1 & 0 & 0 & 0 \\ 0 & \cos\theta & \sin\theta & 0 \\ 0 & -\sin\theta & \cos\theta & 0 \\ 0 & 0 & 0 & 1 \end{bmatrix}, \quad R_Y = \begin{bmatrix} \cos\theta & 0 & -\sin\theta & 0 \\ 0 & 1 & 0 & 0 \\ \sin\theta & 0 & \cos\theta & 0 \\ 0 & 0 & 0 & 1 \end{bmatrix}, \quad R_Z = \begin{bmatrix} \cos\theta & \sin\theta & 0 & 0 \\ -\sin\theta & \cos\theta & 0 & 0 \\ 0 & 0 & 1 & 0 \\ 0 & 0 & 0 & 1 \end{bmatrix}.$$

$$\tag{8.2}$$

Figure 8.2. Several example model rotations are shown. Many combinations are possible, such as rotation about the *x*-axis (left), the *y*-axis (middle), or both (right). Model courtesy of Radioactive Software.

In some cases, you would restrict one or more of the potential rotations, such as the camera orientation in a *first person shooter* (*FPS*) game that doesn't allow rotation about the Z axis. However, it is generally possible for the objects in a scene will allow rotations about all three axes. Several examples of rotations being applied to a model are shown in Figure 8.2.

Translation matrices. Finally, the position of the model within our scene can be manipulated with a *translation matrix*. This matrix is used to move an object from one location to another, without modifying the other properties discussed above. The method for creating a translation matrix is shown in Equation (8.3):

$$T = \begin{bmatrix} 1 & 0 & 0 & 0 \\ 0 & 1 & 0 & 0 \\ 0 & 0 & 1 & 0 \\ t_x & t_y & t_z & 1 \end{bmatrix}. \tag{8.3}$$

The translation matrix is arguably the simplest of the three operations, and it is very intuitive to consider the results of a translation operation. Several examples are provided in Figure 8.3.

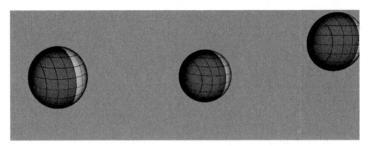

Figure 8.3. Several model translations, including movement along the *x*-axis (middle) and the *x*- and *y*-axes (right).

Matrix concatenation. When taken together, these three types of transformations can be concatenated into a single matrix. These transforms are applied in the order shown in Equation (8.4) when multiplying the vertex positions as a row vector on the left side of the matrices. The rotation transformation is assumed to be the concatenation of the three individual rotations, which together determine the orientation of the model. The order that these transformations are performed in is scaling, then rotation, then translation:

$$W = S * R * T. \tag{8.4}$$

By concatenating all of the matrices into a single one, it can be applied to the model in a single matrix multiplication, instead of in several individual ones. Since this combined transform converts the model from its native *object space* to the scene's *world space*, it is commonly referred to as the *world matrix*. Further consideration must be made for transforming the normal vectors of the model. Since the normal vectors are vectors and not positions, it does not necessarily produce the correct results if transformation with a scaling component is used. For a transformation matrix containing a uniform scaling, normalizing the result of the transformation is sufficient to produce the correct normal vector. However if the transformation contains a non-uniform scaling, the normal vector must be transformed with the transpose of the inverse of the transformation used for the vertex positions. This is shown in Equation (8.5):

$$N = (W^{-1})^{T}. \tag{8.5}$$

View Space

Once the model's vertices have been positioned within the scene's world space, we need to again reposition them, according to where they are being viewed from. This is typically performed by applying a *view space transformation matrix,* which represents a translation and rotation of the world space coordinates to place them into a frame of reference relative to the virtual camera that the scene is being rendered from. To construct a typical view matrix, we can use the formula shown in Equation (8.6):

$$V = T * R_{z} * R_{y} * R_{x}. \tag{8.6}$$

In this case, the translation actually is the negated position of the camera. This is because we are moving the world with respect to the camera instead of moving an object with respect to the world, as we did in the world space section. Likewise for the rotations, each of them represents the opposite of the rotation amount that would be applied to the camera if it were being rendered. Direct3D 11 provides the `D3DXMatrixLookAtLH()` C++ function in the D3DX library for constructing a view matrix based on more intuitive parameters, such as the location of the camera and the point it is looking at. Figure 8.4 shows where a

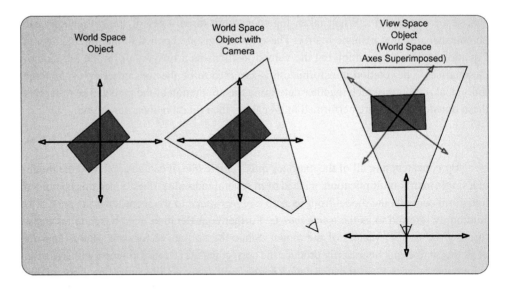

Figure 8.4. A demonstration of how a view matrix is used to transform a scene into view space.

camera is located in world space, and the subsequent movement of the other objects when they are converted to view space.

Projection Space

After obtaining the model data in view space, we need to apply a projection. It is quite common in games and many visualization applications to use a perspective projection that attempts to mimic the real-world perspective. This is not the only type of projection though—many CAD packages also support orthographic projections as well. Since the focus of this book is on real-time rendering, we will focus on the perspective projection. While the view matrix defines the location and orientation of the camera viewing the scene, the projection matrix can be considered to define the camera's other characteristics, such as the *field of view* (*FOV*), the aspect ratio, and the near and far clipping planes to be used. These parameters define a viewing frustum, which indicates which objects are visible for a given view position and set of projection characteristics.

This projection matrix will transform our vertices into what is referred to as *clip space*. Clip space has been discussed in some detail in Chapter 3, but we can summarize the concept here. In general, clip space is the frame of reference immediately after the projection matrix has been applied. If the positions of each vertex are divided by their *w*-component, they will be contained in a space that represents the viewable area of a scene within the *unit cube*. This unit cube extends from the origin of this space to +1 and -1 in

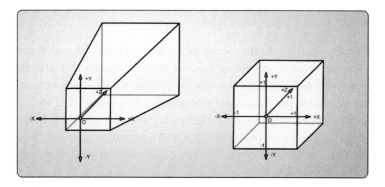

Figure 8.5. The view frustum as seen in world space, and how it maps to the unit cube.

the x- and y-directions, and from 0 to 1 in the z-direction. In essence, the eight corners of the viewing frustum are mapped to the eight corners of the unit cube. This means that if an object is within the viewing frustum in view space, it will end up inside the unit cube after projection and w-divide. Figure 8.5 demonstrates a view frustum being converted to clip space.

The projection process is accomplished through a single projection matrix. A typical projection matrix formula is shown in Equation (8.7), which is also implemented in the `D3DXMatrixPerspectiveFovLH()` C++ function:

$$P = \begin{bmatrix} xScale & 0 & 0 & 0 \\ 0 & yScale & 0 & 0 \\ 0 & 0 & z_f/(z_f - z_n) & 1 \\ 0 & 0 & -z_n * z_f/(z_f - z_n) & 0 \end{bmatrix} \tag{8.7}$$

$$yScale = \cot\left(\frac{fovY}{2}\right)$$

$$xScale = yScale/aspect.$$

The three different transformation matrices we have discussed up to this point—the world, view, and projection matrices—can also be concatenated into a single complete transform matrix. When they are used to render a large number of vertices (as we will be doing), applying a single matrix instead of individually multiplying each matrix on its own will provide a significant performance benefit by reducing the number of calculations performed, while still producing the same result.

8.1.2 Implementation Design

With a clear understanding of what we want to accomplish with our transformation matrices, we can move on to designing how we will implement this algorithm in the Direct3D 11 rendering pipeline. The first step is to define what our input model data will look like. We can assume that the model was generated by some DCC tool in a format that can be read by our sample program. The model will be a static model, with vertex positions, normal vectors generated by the tool, and a texture map applied to the model. This implies that we will have per-vertex texture coordinates as well. The specific vertex format is shown in Figure 8.6.

Now that we know what information we will be storing, we must make a decision about how to structure the resources that will hold our model data. These resources will be bound to the pipeline as inputs to the input assembler stage. It is common to store our model data in one buffer resource to hold the per-vertex data (referred to as a *vertex buffer*), and one buffer resource to hold our primitive specification data (referred to as an *index buffer*). Let's consider the vertex buffer first. As shown in Figure 8.6, in this sample application we will use a per-vertex position, normal vector, and texture coordinate for our vertex format. The vertex buffer will simply contain an array of these structures to represent each of the vertices of the input model.

Next, we will look at our index buffer. There are many primitive types available for use in Direct3D 11, especially when you add in the control patch formats used with the tessellation system. There are also different index ordering semantics involved with each of the primitive types. For example, a triangle list uses three indices for each triangle primitive, while a triangle strip only uses two indices plus one additional index per primitive. For our first implementation, we will start with a basic triangle list primitive type, for ease in specifying and inspecting data. Thus, the index buffer will provide a list of indices, which point into the vertex buffer. The total number of indices is indicated by multiplying the number of triangles by three.

With the input buffers defined, we can consider how the output of the pipeline will be configured. The normal output configuration is to have both a render target and a depth stencil buffer bound to the pipeline in the output merger stage. The render target is typically acquired from the swap buffer that is being used to present the rendered output to a window. Unless otherwise noted, this is the configuration that will be used in the subsequent examples.

If we consider our current pipeline design, we can see that the input resources and output resources have been identified. Now we need to make a decision regarding which portions of the pipeline we will use to implement this algorithm. As noted before, we will be applying our transformation

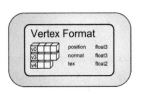

Figure 8.6. A visualization of the vertex format that we will be using.

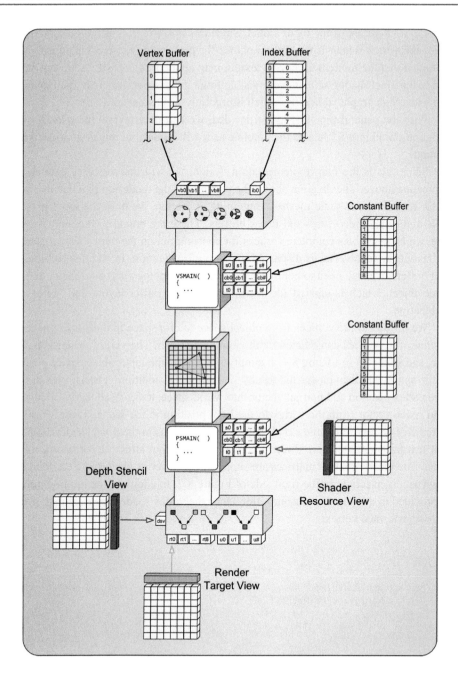

Figure 8.7. The pipeline configuration for rendering a static triangle mesh in our current rendering scheme.

matrix to each vertex of the input model. Since this operation is carried out once on each vertex, the vertex shader is the natural choice for performing these calculations. We also mentioned that the model will have a texture map applied to it. Textures are typically applied in the pixel shader, so that every visible fragment can perform a texture lookup. This results in a high graphical fidelity, which is precisely what we want.

With the general pipeline layout decided, we can now review the overall pipeline configuration. Figure 8.7 shows the pipeline as it will be used to render a model with this technique.

Since this is the first implementation design, we will discuss each state shown in the pipeline above. The diagram shows the portions of the rendering pipeline that we will use for rendering our static meshes. Beginning at the top, we have the input data to the whole algorithm—the vertex and index buffers. These are bound to the input assembler stage, which assembles complete vertices for consumption in the vertex shader stage. The data layout for each vertex is displayed, for easy visualization. In addition to binding the input buffers, we also need to configure the primitive topology and bind the current input layout object, which is created for this specific vertex buffer layout and vertex shader combination.

We transform the vertices from object space to clip space in the vertex shader, so it consumes the vertices constructed by the input assembler. The vertex shader is relatively basic, and is shown in Listing 8.1. It simply transforms the input object space position to the clip space, and then passes the result to its output. In addition, it transforms the object space vertex position and normal vector into world space, and subsequently calculates the world space vector from the vertex to the light position in the scene. These world space vectors will be used for some simple lighting calculations in the pixel shader. Finally, the input texture coordinates are directly copied to the output structure for use in the pixel shader. The transformation matrices are supplied to the vertex shader in a constant buffer, shown as a connection on the right side of Figure 8.7. In addition, the lighting properties are provided in a separate constant buffer. Once the vertex shader has executed, it emits a single transformed vertex.

```
cbuffer StaticMeshTransforms
{
    matrix WorldMatrix;
    matrix WorldViewProjMatrix;
};

cbuffer LightParameters
{
    float3 LightPositionWS;
    float4 LightColor;
};
```

```
Texture2D        ColorTexture : register( t0 );
SamplerState     LinearSampler : register( s0 );


struct VS_INPUT
{
    float3 position : POSITION;
    float2 tex      : TEXCOORDS0;
    float3 normal   : NORMAL;
};

struct VS_OUTPUT
{
    float4 position : SV_Position;
    float2 tex      : TEXCOORD0;
    float3 normal   : NORMAL;
    float3 light    : LIGHT;
};

VS_OUTPUT VSMAIN( in VS_INPUT input )
{
    VS_OUTPUT output;

    // Generate the clip space position for feeding to the rasterizer
    output.position = mul( float4( input.position, 1.0f ), WorldViewProjMatrix );

    // Generate the world space normal vector
    output.normal = mul( input.normal, (float3x3)WorldMatrix );

    // Find the world space position of the vertex
    float3 PositionWS = mul( float4( input.position, 1.0f ), WorldMatrix ).xyz;

    // Calculate the world space light vector
    output.light = LightPositionWS - PositionWS;

    // Pass through the texture coordinates
    output.tex = input.tex;

    return output;
}
```

Listing 8.1. The vertex shader for static mesh rendering.

Since no hardware tessellation is needed in this technique, we can skip the hull, tessellator, and domain shader stages. We also don't need per-primitive level manipulations of the model data, so the geometry shader stage is also not used. This means that the vertices emitted from the vertex shader are directly consumed by the rasterizer stage. Once the rasterizer receives a primitive (consisting of three vertices for our triangle lists) it executes, and produces fragments to be consumed by the pixel shader. Each fragment uses a format similar to that of

Figure 8.8. The final results of performing the rendering of a static mesh. Model courtesy of Radioactive Software.

the vertices consumed by the rasterizer, except that the position information is no longer needed and hence doesn't appear in the output format of the rasterizer.

The pixel shader receives these fragments from the rasterizer and must determine the color of the model at each fragment. The pixel shader is shown in Listing 8.2. We want the pixel shader to sample a texture map to determine the surface properties of the mesh. As described in Chapter 2, the Texture2D resource requires a shader resource view to bind it to the pixel shader stage for read-only access. The pixel shader samples the texture and then calculates the amount of illumination that is available at this point on the surface with the dot product of the normal and light vectors. Finally, the surface color is modulated by the illumination value, and the result is emitted as output color information to the output merger stage. Finally, the output merger writes the output color data to the attached render target, as well as the depth stencil resource, after performing the depth test (the stencil test and blending functionality are not used).

```
float4 PSMAIN( in VS_OUTPUT input ) : SV_Target
{
        // Normalize the world space normal and light vectors
        float3 n = normalize( input.normal );
        float3 l = normalize( input.light );
        // Calculate the amount of light reaching this fragment
        float4 Illumination = max(dot(n,l),0) + 0.2f;

        // Determine the color properties of the surface from a texture
        float4 SurfaceColor = ColorTexture.Sample( LinearSampler, input.tex );

        // Return the surface color modulated by the illumination
        return( SurfaceColor * Illumination );
}
```

Listing 8.2. The pixel shader for static mesh rendering.

Figure 8.8 shows the results of our rendering as we have configured it for several orientations of the model and camera.

8.1.3 Conclusion

Rendering a static mesh is somewhat of a first point of entry into developing real-time rendering algorithms. This technique allows for an object to become animated by changing some of its properties over time. For example, if the object were spinning in a circle, this could easily be accomplished by animating the rotation about the y-axis. The operation described above would be executed once for each object that needs to be rendered in a given scene. Since we are using a transformation matrix and a texture map that will be unique for each object, each object must be rendered with its own draw calls after the associated resources have been updated and bound to the pipeline.

8.2 Vertex Skinning

In the previous section, we saw how to perform the transformation of an object to control its spatial properties within a scene. This provides a surprisingly large amount of functionality. For instance, it could be used to implement a chess game where each piece is represented by a static mesh that only moves within the scene. However, many other types of objects would require a different type of animation to appear correct. For example, an object composed of multiple pieces that move with respect to one another (like a robotic arm) can't be accurately portrayed with a static mesh. We could use multiple individual meshes and manipulate their transform properties to provide a more convincing result, but this would not be applicable to any type of organic mesh, such as an animal or a humanoid figure. Static meshes with transformations are simply not capable of properly rendering this class of objects.

Instead, another technique must be used: *vertex skinning*. This algorithm lets individual portions of a triangle mesh be animated with their own object space, prior to the transformation matrix being applied. This allows each of the scenarios described above to be more convincingly rendered, and provides a flexible method for incorporating animation data directly into the model, instead of having to manually control every movement of the object at a higher level. In this section, we will introduce the concept of vertex skinning, and the theory behind its operation. We will develop an implementation of this algorithm for Direct3D 11, and will finally consider the performance implications of such an approach.

8.2.1 Theory

The key concept behind vertex skinning is the ability to use transformation hierarchies. In the previous section, we learned how an object can be placed throughout a scene in various orientations or scales. This rendering technique allows each object to be positioned with respect to the world space origin. However, if we consider the robotic arm example from above, if there are multiple meshes that need to be positioned relative to another object in the scene instead of the origin, it can quickly become difficult to manage two different world-space transforms that depend on one another. This would be even more difficult if there are more than two joints in the arm. Figure 8.9 shows such an arm, where the components of the arm that are connected together can only move with respect to the joint that they are attached to.

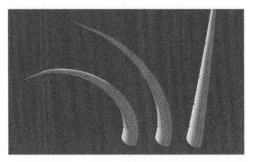

Figure 8.9. An arm that demonstrates the concept of multiple mesh sections constrained to one another.

Transformation Hierarchies

The type of movement constraint that we need can be accomplished with *transformation hierarchies*. If one object's transformation matrix is generated such that it is applying the position and orientation of its mesh with respect to the mesh that it is attached to, it can be rendered by concatenating its own world space matrix with that of its "parent" object. Returning to our robotic arm example, if each of these arm components is connected to a parent object, it only needs to manage the transformation state with respect to its parent. When it is time to render each component, we would begin with the component that has no parent and calculate its transformation matrix. This would be followed by the first child calculating its transformation matrix and then concatenating this value with the parent's transformation matrix. This process of concatenating matrices essentially moves the child object from a coordinate space defined by its parent to the world space in which its parent is located. This is repeated until all of the components have been updated, at which point we could render them individually with their newly calculated world matrices.

This concept was used in computer graphics long before programmable GPUs were developed. However, this is still an interesting topic to consider, which can be applied to the more difficult problem of rendering organic objects. Using transformation hierarchies, we can render individual portions of the robotic arm, since they actually are individual pieces. However, a human arm not only has bones that are separate pieces, but also has a muscular system and skin that are continuous across the arm. This doesn't lend itself to rendering an arm in pieces, since the object should appear continuous, as it does in the real world.

Transformation hierarchies still hold the key to finding a solution to this problem. Instead of trying to split the arm into multiple discrete pieces, it would be better to be able to say how much each vertex is influenced by each the objects it is in contact with. For example, the skin on an elbow moves partially with the upper arm and partially with the lower arm. If we develop a system that allows us to assign a weighting for each vertex to each specify how much it is influenced by a particular component, we could calculate the vertex position as if it were attached to each component and then interpolate its final position based on the weights applied to each component.

Skin and Bones

This is the primary idea behind vertex skinning. The vertices of a mesh define the *skin* of a model, and each component included in the transformation hierarchy is referred to as a *bone*. The complete group of bones is referred to as the *skeleton* of the model. These names provide a simple visualization of what the individual pieces of this algorithm represent. There are many different variations of this algorithm that can be more or less suitable for a given situation, but the algorithm provided here is a fairly general version. We will allow each vertex to be associated with up to four bones each, which will allow for a significant amount of flexibility in making an animation look correct.

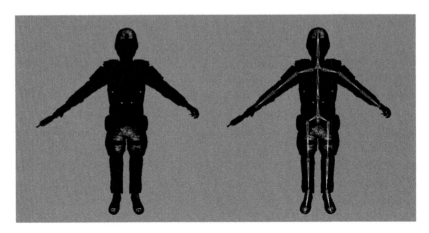

Figure 8.10. A sample mesh shown with and without its skeleton. Model courtesy of Radioactive Software.

To develop a model for rendering, we need to determine the number of bones that a model will use, and how they will be connected. This is typically determined in the design stage and can be chosen to suit the needs of a particular application. It is quite common for humanoid models to define their bones in similar locations to where the human body has joints between bones. The root of the bone hierarchy is typically chosen somewhere in the lower back, where it can be used to position the entire object within world space. The current bone transformation for each bone is the calculated by multiplying the transformation matrix from the root bone with each successive bone's transformation matrix, until you reach the current joint. An example mesh is shown with and without its skeleton in Figure 8.10.

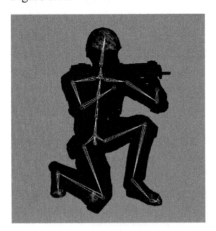

Figure 8.11. A sample model shown in a crouching pose. Model courtesy of Radioactive Software.

In this case, we see the skeleton oriented in what is referred to as its *bind pose*. This is the base orientation of all of the bones that a model will be created in, and all of the vertices will have their individual bones selected, and weights applied, in this base orientation. After the model has been created in this manner, the individual animations can be created using the base information. Instead of trying to manually place each vertex, the animator can simply manipulate the position and orientation of the bones to create the desired pose of the model. As an example, our sample model can be seen in a crouching pose in Figure 8.11.

Mathematics of Skinning

Bind poses are constructed primarily to allow the content creator to have a simple, easy-to-understand method of creating the model. However, the use of bind poses has some implications for the mathematics of our solution. When the vertices are created and placed around a model, the positions are defined relative to the coordinate space of the entire object—not with respect to its bone. This means that before we can apply the hierarchical bone transformation matrix to a vertex, it must be converted from an object space position into a *bone space position*. However, since there are multiple bones allowed per vertex, this can't be done in the modeling package. Instead, this must be performed during runtime, when each bone transformation matrix is created.

Fortunately, matrix transformations can easily perform this kind of operation. Essentially, we are trying to "undo" the bone transformations of the bind pose, and then reapply the bone transformations of the currently desired pose. This is done by creating a transformation matrix that applies the inverse of the bind pose transformation before applying the desired pose transformation matrices. This is shown in Equation (8.8), where the final transformation matrices for each bone are the product of the inverse of the bind pose and the current bone transformation matrix:

$$\boldsymbol{B}[\boldsymbol{n}]_{\text{final}} = \boldsymbol{B}[\boldsymbol{n}]_{\text{bind}}^{-1} * \boldsymbol{B}[\boldsymbol{n}]_{\text{curr}}. \tag{8.8}$$

We can visualize what this series of transforms does by considering the vertices associated with a hand bone. In the bind pose, the vertex positions are specified in the object space of the model. When the inverse of the bind pose is applied to the vertices, the hand bones moves to the origin of the coordinate space, so that it would be sticking out of the origin at the wrist. Then the current bone pose transformation moves the hand bone to the desired location. Since the bind pose is always the same throughout the lifetime of an application, it can be calculated once at startup and then simply applied to the hierarchical bone transformation matrices that are calculated for each frame of animation.

8.2.2 Implementation Design

With a clear concept in mind about how vertex skinning works, we can now design an implementation to be used with Direct3D 11. We can use a similar rendering configuration to what we saw in the static mesh rendering section, with two exceptions: we need to incorporate the new per-vertex data into the vertex structure, and then the portion of the vertex shader that transforms the vertices must be updated to use an array of bone matrices and perform the bone-weighting interpolation. We will take a closer look at each of these steps. The updated algorithm layout is shown in Figure 8.12 for reference during the discussion.

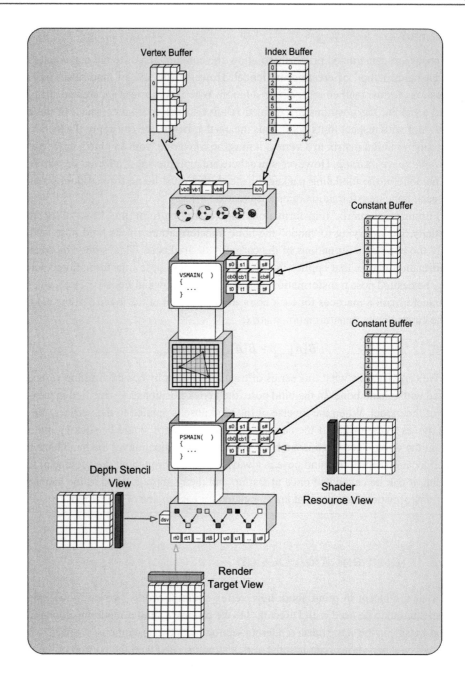

Figure 8.12. The vertex-skinning pipeline configuration.

To support the vertex skinning algorithm, our model im-
porter must support a different type of per-vertex information.
The new vertex format must include the information that specifies which bone it is associated with, as well as what weighting
amount each bone should have on the outcome of the interpolation. To accomplish this, each bone must be assigned an integer
index that can be used as an index into an array of matrices. We
will allow up to four bones to influence each vertex, which indicates that we will need a four-component set of bone indices, as well

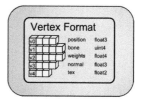

Figure 8.13. The vertex layout for our vertex-skinning
implementation.

as a four-component set of bone weights. The updated vertex structure is shown in Figure 8.13. Of
course, with a new vertex structure a new input layout object must be generated as well.

To use the new vertex data, we must modify our vertex shader according to the algorithm
we specified in the theory section. The first step is to gain access to the transformation matrices
that the application provides. The matrices are made available to the vertex shader through a
constant buffer, and essentially replace the world transformation matrix portion of the concatenated transform from our previous example. This means that the will calculate the world-space
position of each vertex within the vertex shader, and then transform this resulting position by
the view and projection matrices that will be concatenated into a single matrix. In addition to the
bone transformation matrices, we also include the normal vector bone transformation matrices,
to produce a correct world-space skinned vertex-normal vector as well. The modified vertex
shader is shown in Listing 8.3.

```
cbuffer SkinningTransforms
{
    matrix WorldMatrix;
    matrix ViewProjMatrix;
    matrix SkinMatrices[6];
    matrix SkinNormalMatrices[6];
};

cbuffer LightParameters
{
    float3 LightPositionWS;
    float4 LightColor;
};

Texture2D       ColorTexture : register( t0 );
SamplerState    LinearSampler : register( s0 );

struct VS_INPUT
{
    float3  position  : POSITION;
    int4    bone      : BONEIDS;
    float4  weights   : BONEWEIGHTS;
    float3  normal    : NORMAL;
    float2  tex       : TEXCOORDS;
};
```

```
struct VS_OUTPUT
{
    float4 position    : SV_Position;
    float3 normal      : NORMAL;
    float3 light       : LIGHT;
    float2 tex         : TEXCOORDS;
};

VS_OUTPUT VSMAIN( in VS_INPUT input )
{
    VS_OUTPUT output;

    // Calculate the output position of the vertex
    output.position  = (mul( float4( input.position, 1.0f ),
      SkinMatrices[input.bone.x] )
                      * input.weights.x);
    output.position += (mul( float4( input.position, 1.0f ),
      SkinMatrices[input.bone.y] )
                      * input.weights.y);
    output.position += (mul( float4( input.position, 1.0f ),
      SkinMatrices[input.bone.z] )
                      * input.weights.z);
    output.position += (mul( float4( input.position, 1.0f ),
      SkinMatrices[input.bone.w] )
                      * input.weights.w);

    // Transform world position with viewprojection matrix
    output.position = mul( output.position, ViewProjMatrix );

    // Calculate the world space normal vector
    output.normal  = (mul( input.normal, (float3x3)SkinNormalMatrices[input.
      bone.x] )
                      * input.weights.x).xyz;
    output.normal += (mul( input.normal, (float3x3)SkinNormalMatrices[input.
      bone.y] )
                      * input.weights.y).xyz;
    output.normal += (mul( input.normal, (float3x3)SkinNormalMatrices[input.
      bone.z] )
                      * input.weights.z).xyz;
    output.normal += (mul( input.normal, (float3x3)SkinNormalMatrices[input.
      bone.w] )
                      * input.weights.w).xyz;

    // Calculate the world space light vector
    output.light = LightPositionWS - output.position.xyz;

    // Pass the texture coordinates through
    output.tex = input.tex;

    return output;
}
```

Listing 8.3. The vertex shader used in the vertex skinning operation.

As we can see in Listing 8.3, the input position is calculated by transforming the object space position for each of the bones specified in the vertex's bone ID attribute, and the resulting position is scaled by the bone weighting value from the vertex's bone weight attribute. The normal vector is transformed in a similar manner, except with the inverse transpose of the transformation matrix. Finally, the four weighted results are added together to find the final world space position and normal vector. The world space position can then be transformed into clip space with the `ViewProjMatrix` parameter, and the remainder of the vertex shader is the same as in the static mesh rendering case.

This implementation provides a few additional indicators of how the various parts of the rendering pipeline are typically applied to particular aspects of a model's appearance. For example, we modified how the geometry of a model is calculated, which required only a change to the vertex shader. The remainder of the pipeline configuration (other than the input layout) is the same. This is often the case when the geometric properties of a model are determined in the earlier stages and the surface properties of the geometry are determined in the later stages. With this separation of duties, we can use individual shader programs for more than one pipeline configuration, when it is appropriate. This reduces the number of different shader programs that need to be written. It is a good idea to keep this in mind when developing a new algorithm, since a shader program that is already written is much faster to develop than one written from scratch!

8.2.3 Conclusion

If we consider how vertex skinning operates, we can gain some insight into how the performance of the algorithm will vary with different input models. If we take the static mesh algorithm as our baseline for performance, then add vertex skinning, we must upload a larger constant buffer (to hold the bone matrices), and must perform some additional math on the vertex before it is passed out of the vertex shader. Since we are allowing four bones per-vertex, our vertex shader is written to access and calculate one transformed position for each bone, which is then weighted and summed to find the final location. Even if fewer than four bones are used in each vertex (if some of the weights are equal to 0), they must all still be calculated, regardless of if the resulting weighted position produces an all-zero position that will not contribute to the output position. Therefore, the cost associated with performing vertex skinning should be constant from vertex to vertex.

This also means that the algorithm's performance will scale with the number of vertices included in the model. If a highly detailed geometry model is rendered, its corresponding rendering cost will be proportionally higher than that of a less-detailed geometric model. Depending on the overall pipeline configuration and the hardware that an application is running on, this can limit the available model detail that can be used for a given scene. In many cases this may not make any difference, but if geometry processing is the bottleneck of the algorithm, vertex skinning may introduce a performance limitation. Luckily, there are techniques available that can reduce this dependency on vertex counts.

8.3 Vertex Skinning with Displacement Mapping

By using vertex skinning, we have significantly expanded the number of objects we can represent with our rendering algorithms. We have techniques for both static and dynamic meshes, and can perform animation with both of these mesh types. However, as mentioned in the last section, using vertex skinning can introduce a performance bottleneck, due to the additional mathematic operations performed on each vertex. It would be better to diminish the link between geometric complexity and the number of vertices that must be processed. Fortunately, some of the new abilities in the Direct3D 11 rendering pipeline can help us achieve this decoupling.

Ideally, we would like to perform the skinning operation on a low-resolution model, but we still want the detailed look of a high-resolution model. The tessellation stages of the rendering pipeline allow us to perform this geometry detail injection. There are many available techniques, but we will focus on one of the more versatile solutions—*displacement mapping*. Displacement mapping essentially applies a height-map to every triangle in a triangle mesh. By using the tessellation stages, we can tessellate a low-resolution mesh to generate a higher number of vertices, and then sample the displacement map and modify the position of each of the vertex with respect to the original lower-resolution triangles.

This effectively moves the highly detailed geometry out of the static vertex information and into a texture instead. This has many advantages. The first is that we can perform skinning on the lower-resolution geometry, but there are other benefits. Since a displacement map can be mip-mapped, it allows for a simple level of detail (LOD) system to be implemented automatically. In the following section, we will see how this technique works, and how to implement it in the context of the Direct3D 11 rendering pipeline.

8.3.1 Theory

To displace the surface of the low-resolution triangle mesh, we must perform two operations. The first is to determine how much detail to add into the model, and the second is to understand more precisely how we can perform this displacement operation. We will consider these two items before moving on to the implementation details of this algorithm.

Dynamic Tessellation

Since we will be dynamically determining how much detail to add into the lower-resolution mesh, we must consider the appropriate measures to select how much geometry is

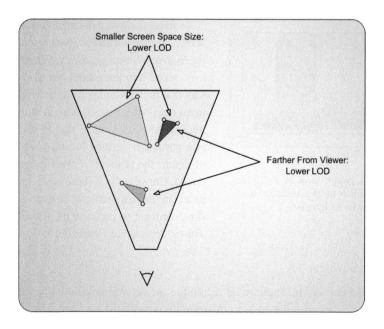

Figure 8.14. A number of different triangle properties used for determining the appropriate tessellation level.

needed for a given triangle. There are many metrics that we could use, so we will consider what situations would require the highest level of detail.

Clearly, when a triangle face points away from the viewer, we don't want to increase its complexity at all. In fact, if possible, we want to eliminate it from further processing completely, since it won't contribute to the rendered output image. At the transition point from a back-facing triangle to a front-facing triangle, we have silhouette edges. These edges have very high importance, because the user will be able to fairly clearly differentiate the outline of the object from its background. If there is a low-resolution silhouette, the user will spot it quite easily. After the silhouette edge, we have the triangles that are potentially visible to the user as long as they are not occluded by other geometry within the scene. If a triangle is potentially visible, we want to increase its level of detail to the highest degree that will be visible under the current viewing conditions. This particular type of situation provides a number of potential metrics.

The simplest property is the distance from the camera. If a triangle is very close to the camera, it should use the highest level of detail possible. If it is very far away from the camera, it should use a much lower detail. This calculation depends on the average size of a triangle, as well as the angle the triangle is being viewed with. When combined, these parameters describe the effective screen-space size of the triangle. Figure 8.14 demonstrates the various situations we have described up to this point.

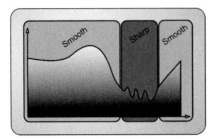

Figure 8.15. A profile view of a displacement map, and how much tessellation would be appropriate for each region.

In addition to a distance-based metric or a screen-space-size metric, we must also consider the high resolution geometry. When geometry is introduced into the model, we also want to ensure that there is adequate detail in the displacement map before increasing the number of vertices to be processed. If the displacement map at a particular surface location has a low level of detail, the amount of tessellation can be reduced. Conversely, if there is a large amount of detail in the displacement map at a given location, we want to increase the tessellation level. This is depicted in Figure 8.15.

Displacing Surfaces

Once we have chosen the appropriate technique for determining the required amount of tessellation, we can consider how to displace the tessellated vertices. The displacement function actually doesn't need to know the method that was used to determine the tessellation level; it is only concerned with taking a vertex location and finding the required displacement to apply. This operation is slightly more complicated than it may initially seem, because we are applying a flat texture to a semi-flat triangle mesh. Since neighboring triangle faces are typically not coplanar, there is at least some change in orientation between them. This makes it impossible to perform a simple displacement away from the surface due to either a gap or an overlap at the transition between faces. This is shown in Figure 8.16.

In some respects, this is why normal vectors are specified for each vertex, instead of for each triangle. The triangle mesh is an approximation of a smooth surface. Thus, the vertex normal vectors are used to make surface lighting appear smoother than it really is by

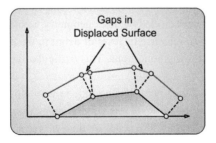

Figure 8.16. A profile of a triangle mesh, showing that non-coplanar triangles can't use simple surface offsets.

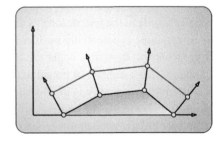

Figure 8.17. Using vertex normal vectors for displacing tessellation-generated vertices.

providing a vector that is a combination of the triangle normal vectors that it is touching. Displacement mapping can follow the same methodology by using the vertex normal vector. For each point on a triangle mesh being displaced, we read the displacement amount from the displacement map and then shift the new vertex along its interpolated vertex normal vector. This allows the displacement to be smoothly transitioned between triangle faces. This is updated scheme shown in Figure 8.17.

8.3.2 Implementation Design

Once again, we will be able to use our previous implementation's pipeline configuration, with just a few changes. Even though we will be passing lower-resolution model data into the pipeline, the vertex skinning process is oblivious to this, except that the final projection to clip space must be moved to later in the pipeline. Since we are still using skinning in the same general way on our input vertices, the vertex layout and the vertex shader will remain mostly the same. However, since we are using the tessellation stages, we must change our input topology setting to indicate that we are using control patches instead of triangles. In effect, there is no difference for our input buffers, since we will be using a three-point control patch list, which is topologically the same as a triangle. However, the runtime will require a patch list type as input to the hull shader, so we will make the change. Since we will be using the tessellation stages to increase the resolution of our geometry, the hull shader, tessellator, and domain shader all must be used, since they are used in conjunction. Finally, the pixel shader will remain the same, to apply the color texture to the geometry that is produced in the tessellation stages. The updated pipeline configuration is shown in Figure 8.18.

With the vertex shader remaining the same except for the removal of the clip space projection, the first new step to implementing our displacement-mapping addition is to configure the hull shader to set up the tessellation process. The hull shader will receive skinned vertices from the vertex shader and will interpret them as three point control patches. The hull shader requires two different functions—its main function, as well as a patch constant function. Since the patch constant function is responsible for determining the tessellation factors, we don't need to do anything in the hull shader. This lets us simply pass the hull shader input control points as the output control points, which will also remain as three-point control patches.

We will use the patch constant function to determine how finely we should split up the input control patches, by specifying one edge-tessellation factor for each edge of the triangle patch, plus a single tessellation factor for the interior portion of the triangle. Each of the metrics discussed in our theory section could be used to determine the necessary tessellation level, but we will restrict our patch constant function to using a fixed value for the amount of tessellation. This fixed value is passed to the output of the patch constant

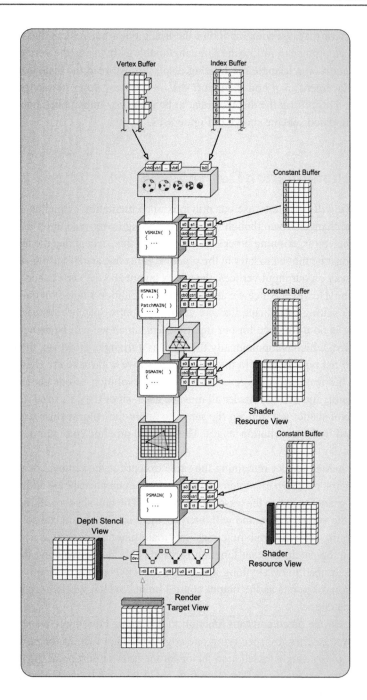

Figure 8.18. The pipeline configuration for vertex skinning with displacement mapping.

function, including each of the tessellation factor system value semantics. Both of the functions for the hull shader are shown in Listing 8.4.

```
cbuffer SkinningTransforms
{
    matrix WorldMatrix;
    matrix ViewProjMatrix;
    matrix SkinMatrices[6];
    matrix SkinNormalMatrices[6];
};

cbuffer LightParameters
{
    float3 LightPositionWS;
    float4 LightColor;
};

Texture2D    ColorTexture : register( t0 );
Texture2D    HeightTexture : register( t1 );
SamplerState LinearSampler : register( s0 );

struct VS_INPUT
{
    float3 position : POSITION;
    int4   bone     : BONEIDS;
    float4 weights  : BONEWEIGHTS;
    float3 normal   : NORMAL;
    float2 tex      : TEXCOORDS;
};
//-----------------------------------------------------------------
struct VS_OUTPUT
{
    float4 position : SV_Position;
    float3 normal   : NORMAL;
    float3 light    : LIGHT;
    float2 tex      : TEXCOORDS;
};
//-----------------------------------------------------------------
struct HS_POINT_OUTPUT
{
    float4 position : SV_Position;
    float3 normal   : NORMAL;
    float3 light    : LIGHT;
    float2 tex      : TEXCOORDS;
};
//-----------------------------------------------------------------
struct HS_PATCH_OUTPUT

{
    float Edges[3]  : SV_TessFactor;
    float Inside    : SV_InsideTessFactor;
};
```

```
//-----------------------------------------------------------------------------
struct DS_OUTPUT
{
    float4 position : SV_Position;
    float3 normal   : NORMAL;
    float3 light    : LIGHT;
    float2 tex      : TEXCOORDS;
};
//-----------------------------------------------------------------------------
VS_OUTPUT VSMAIN( in VS_INPUT input )
{
    VS_OUTPUT output;

    // Calculate the output position of the vertex
    output.position  = (mul( float4( input.position, 1.0f ),
      SkinMatrices[input.bone.x] )
                        * input.weights.x);

    output.position += (mul( float4( input.position, 1.0f ),
      SkinMatrices[input.bone.y] )
                        * input.weights.y);
    output.position += (mul( float4( input.position, 1.0f ),
      SkinMatrices[input.bone.z] )
                        * input.weights.z);
    output.position += (mul( float4( input.position, 1.0f ),
      SkinMatrices[input.bone.w] )
                        * input.weights.w);

    // Calculate the world space normal vector
    output.normal  = (mul( input.normal, (float3x3)SkinNormalMatrices[input.
      bone.x] )
                        * input.weights.x).xyz;
    output.normal += (mul( input.normal, (float3x3)SkinNormalMatrices[input.
      bone.y] )
                        * input.weights.y).xyz;
    output.normal += (mul( input.normal, (float3x3)SkinNormalMatrices[input.
      bone.z] )
                        * input.weights.z).xyz;
    output.normal += (mul( input.normal, (float3x3)SkinNormalMatrices[input.
      bone.w] )
                        * input.weights.w).xyz;

    // Calculate the world space light vector
    output.light = LightPositionWS - output.position.xyz;

    // Pass the texture coordinates through
    output.tex = input.tex;

    return output;
}
//-----------------------------------------------------------------------------
HS_PATCH_OUTPUT HSPATCH( InputPatch<VS_OUTPUT, 3> ip, uint PatchID : SV_PrimitiveID )
{
    HS_PATCH_OUTPUT output;
```

```
        const float factor = 16.0f;

        output.Edges[0] = factor;
        output.Edges[1] = factor;
        output.Edges[2] = factor;

        output.Inside = factor;

        return output;
}
//----------------------------------------------------------------
[domain("tri")]
[partitioning("fractional_even")]
[outputtopology("triangle_cw")]
[outputcontrolpoints(3)]
[patchconstantfunc("HSPATCH")]
HS_POINT_OUTPUT HSMAIN( InputPatch<VS_OUTPUT, 3> ip,
                        uint i : SV_OutputControlPointID,
                        uint PatchID : SV_PrimitiveID )
{
    HS_POINT_OUTPUT output;

    // Insert code to compute Output here.
    output.position = ip[i].position;
    output.normal = ip[i].normal;
    output.light = ip[i].light;
    output.tex = ip[i].tex;

    return output;
}
```

Listing 8.4. The vertex and hull shader programs and the patch constant function for setting up the tessellation process.

After the hull shader, the tessellator stage consumes the tessellation factors and generates a set of barycentric coordinates that will be passed to the domain shader. The domain shader also receives the control points that were passed from the hull shader main function, and must convert the barycentric coordinates into clip-space vertices. This is done by first calculating the world-space position of the barycentric point by interpolating the input control point positions. These are the world-space skinned-vertex positions that we calculated in the vertex shader, meaning that if we passed these vertices out with these interpolated positions, we would effectively generate a triangle mesh that appears the same as our normal skinned mesh. In addition to interpolating the position, we also interpolate the per-vertex normal vector, as well as the texture coordinates of the vertices, and renormalize the normal vector afterward to ensure that it remains a unit length vector. This process can be seen in Listing 8.5 which shows the complete domain shader program.

```
[domain("tri")]
DS_OUTPUT DSMAIN( const OutputPatch<HS_POINT_OUTPUT, 3> TriPatch,
                  float3 Coords : SV_DomainLocation,
                  HS_PATCH_OUTPUT input )
{
    DS_OUTPUT output;

    // Interpolate world space position
    float4 vWorldPos = Coords.x * TriPatch[0].position
                     + Coords.y * TriPatch[1].position
                     + Coords.z * TriPatch[2].position;

    // Calculate the interpolated normal vector
    output.normal = Coords.x * TriPatch[0].normal
                  + Coords.y * TriPatch[1].normal
                  + Coords.z * TriPatch[2].normal;

    // Normalize the vector length for use in displacement
    output.normal = normalize( output.normal );

    // Interpolate the texture coordinates
    output.tex = Coords.x * TriPatch[0].tex
               + Coords.y * TriPatch[1].tex
               + Coords.z * TriPatch[2].tex;

    // Calculate the interpolated world space light vector.
    output.light  = Coords.x * TriPatch[0].light
                  + Coords.y * TriPatch[1].light
                  + Coords.z * TriPatch[2].light;

    // Calculate MIP level to fetch normal from
    float fHeightMapMIPLevel =
      clamp( ( distance( vWorldPos.xyz, vEye.xyz ) - 100.0f ) / 100.0f,
             0.0f,
             3.0f);

    // Sample the height map to know how much to displace the surface by
    float4 texHeight =
           HeightTexture.SampleLevel( LinearSampler, output.tex,
fHeightMapMIPLevel );

    // Perform the displacement. The 'fScale' parameter determines the maximum
    // world space offset that can be applied to the surface. The displacement
    // is performed along the interpolated vertex normal vector.
    const float fScale = 0.5f;
    vWorldPos.xyz = vWorldPos.xyz + output.normal * texHeight.r * fScale;

    // Transform world position with viewprojection matrix
    output.position = mul( vWorldPos, ViewProjMatrix );

    return output;
}
```

Listing 8.5. The domain shader program for implementing displacement mapping.

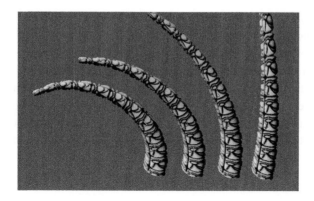

Figure 8.19. The results of using displacement mapping for mesh rendering.

With the interpolated world-space position, vertex normal vector, and texture coordinate, we can move ahead with the displacement mapping process. We first sample the displacement map (`HeightTexture`) to determine the desired magnitude of the displacement. This sampling process must use the `SampleLevel` function, since the domain shader stage can't automatically determine what mip-map level to use. Here we perform a simple calculation based on the distance from the viewer, then clamp the result between levels 0 and 3. The sampled height value is scaled by a constant after being read to allow for adjustments of the overall range of displacement values, if desired. The displacement magnitude is then used to scale the normal vector, and the result is added to the interpolated world-space position. The result of this calculation is the final output world-space position of the tessellated point, which represents a piece of our injected higher-resolution geometry.

Finally, the domain shader transforms the new vertex to clip space with the `ViewProjMatrix` transform and passes the texture coordinate to the pixel shader. As stated earlier, the pixel shader remains the same as in our earlier examples. From the point of view of the pixel shader, the geometry could have come from either the input vertex buffer or from the tessellation system—both would appear identical to the pixel shader. The resulting rendering contains higher-detail geometry where the user can see it, and can reduce the level of detail of the geometry where the user can't see it. Several sample renderings using the new algorithm are shown in Figure 8.19.

8.3.3 Conclusion

To review the algorithm we have implemented, we can now perform detailed animation of complex objects, while also reducing the amount of geometry that must be processed to obtain an equal amount of surface detail. Displacement mapping allows the expensive skinning calculations to be done at a lower detail level, but it still introduces high detail back in after skinning. Using the tessellation stages lets us ramp up the detail according to whatever metric we choose, which also provides a simple way to scale performance according to the hardware currently being used. Using a displacement map lets us obtain our additional geometric complexity relatively quickly with a single texture sample, and it also requires a relatively small amount of memory in comparison to an equivalent number of vertices. Finally, using a displacement map also allows for very simple use of LOD techniques to restrict the detail level returned by the sampled displacement map.

9

Dynamic Tessellation

Earlier in the book, the new tessellation-specific stages of the pipeline were introduced (Chapter 3) and the key parameters and implementation details were discussed in depth (Chapter 4). However, neither of these sections provided real-world demonstrations of this key new Direct3D 11 technology.

In the introduction to Chapter 4, two common approaches to tessellation were introduced, namely *subdivision* and *higher-order surfaces*. This chapter offers demonstrations of both, to allow comparison between the two methods, and to demonstrate how the various parameters and new shader units work together to produce a final output.

Subdivision is a form of tessellation based on refinement, in which we progressively add more detail (in the form of additional triangles) until the final mesh better represents an ideal shape. The first section of this chapter, "Terrain Tessellation," follows this approach.

Thinking about higher order is closer to the understanding that tessellation is about representing curved or smooth surfaces. The ideal mesh will usually be described in terms of a parametric mathematical equation, typically quadratic or cubic; the second section of this chapter demonstrates this.

Broadly speaking, using subdivision is more algorithmic, while using higher-order surfaces is more mathematical. We will explore the differences between these two approaches throughout this chapter, and will also build on the earlier introduction to the tessellation stages to provide a well-rounded understanding of the technology.

9.1 Terrain Tessellation

Terrain rendering is probably one of the most common forms of procedural or run-time generated graphics in the real-time rendering space. Unlike the majority of other art assets that will be rendered to make the final image, the terrain is often created by software algorithms, rather than hand-sculpted by an artist (however, a combination of both is common). A typical input is a *height map*—a monochrome texture where each pixel represents the elevation of the terrain at that particular point on the grid.

Using terrain rendering can easily require large data sets, both in terms of textures and geometry. It is not uncommon to have huge draw distances, stretching from the immediate foreground where the camera is located, out to the distant horizon. Many algorithms have been invented and refined over the years to efficiently solve this problem and, for the most part, the problem is well understood. As hardware performance and features have improved (such as the move from CPU to GPU processing) these algorithms have been adapted to take advantage.

Direct3D 11's tessellation capabilities may not revolutionize this problem space, but they do offer some interesting alternatives. Terrain rendering stands as a good example of the new functionality. Most current algorithms require at least some intervention and processing by the CPU before the GPU does the actual rendering; but with Direct3D 11 it is now possible to offload all processing to the GPU. Both existing "classic" algorithms can be adapted to suit this new hardware and the reverse is also true—new hardware opens up avenues for wholly new algorithms and rendering approaches.

9.1.1 GPU-Accelerated Interlocking Tiles Algorithm

Figure 9.1 shows a naïve rendering of a traditional heightmap-based terrain. The geometry is uniformly distributed across the entire terrain, with no regard to the characteristics of the terrain or viewer. Figure 9.2 shows the same inputs rendered using the interlocking terrain tiles algorithm algorithm introduced in this section—detail is now applied only where it is

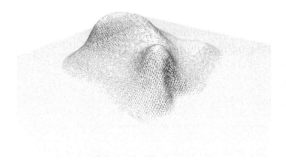

Figure 9.1. Naïve rendering.

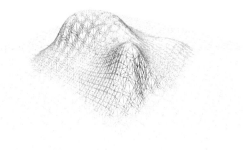

Figure 9.2. Interlocking Terrain Tiles Algorithm.

legitimately required, and performance is no
longer wasted by extra geometry that has no
impact on the final image.

In "Simplified terrain using interlocking tiles," (Snook, 2001), Greg Snook introduced an algorithm that mapped very nicely
to early GPU architectures—around the time
when hardware transform and lighting were
possible, but slightly predating programmable shader units.

The algorithm was particularly useful
because it didn't require modifying the vertex or index buffers; instead, is used multiple
index buffers to provide a LOD mechanism.
This removed any need for the CPU to modify resources used for rendering, an operation
that both then and now can be prohibitively
expensive, but which is important for lever-

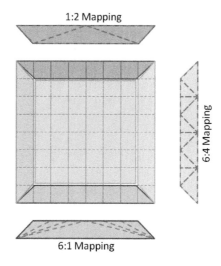

Figure 9.3. Index buffer layout for a single tile.

aging hardware transform and lighting capabilities. By simply rendering with a different
index buffer, Snook's algorithm could alter the LOD and balance performance and image
quality.

The algorithm breaks up the 2D height-map texture into a number of smaller tiles,
with each tile representing a 2^n+1-dimension area of vertices (9×9 was used in the original
implementation). This vertex data was fixed at load-time and simply represented the height
at each corresponding point on the height map, which also corresponds to the highest possible detail that can be rendered.

For each level of detail, there are 5 index buffers. For n=3, there are 4 LODs (n+1) in
a 9-dimensional area of vertices ($2^3+1=9$). The layout of the regions that these index buffers represent is depicted in Figure 9.3.

The five index buffers represent the central area of the tile, plus one index buffer for
each of the four edges shown in Figure 9.3. These are each assigned a different color. The
break-out edge diagrams show examples of mappings between different LODs, shown in
inner:outer ratios. An observant reader might notice that this is suspiciously similar to the
quad tessellation introduced in Chapter 4!

The index buffer for the middle area represents all but the outer ring of vertices, so
for the 9×9 grid used in our example, it will always represent 7×7 vertices, regardless of
the chosen level of detail. The four "skirt" index buffers essentially provide a mapping
between neighboring tiles. It is possible to match either the higher or lower LOD, but it's
more common to blend down toward the lower one as shown in Figure 9.4. Technically, a
jump between any two levels of detail is possible, but as mentioned in Chapter 4, this may

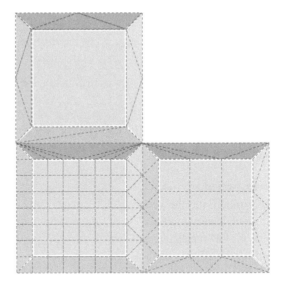

Figure 9.4. Neighboring tiles blending down.

cause visual *popping* to become noticeable, detracting from the final result. The simplest way to avoid this popping is to only transition between neighboring LOD levels, or in severe cases, to consider vertex blending or other morphing enhancements.

This algorithm was published around the time that Direct3D 8 was current, and could only use a single pair of index and vertex buffers. Consequently, it became a useful trick to use DrawIndexedPrimitive() parameters to offset the base vertex for rendering each patch, to avoid unnecessary state changes (an expensive operation with previous versions of Direct3D). In addition, due to the small storage size required for an index, it was possible to store all unique combinations and simply tweak the draw call parameters to change the LOD for each patch.

This means that there can be a great disadvantage, because of the potentially large number of draw calls. The GPUs of the time could not dynamically select which index buffer to use for each segment, so this had to be set up per-tile by the CPU, requiring a draw call for each patch. Clearly, this is not good, given that developers spend significant effort to reduce the number of these calls. The often-cited target of 500 or fewer draw calls per frame thus effectively limits rendering to 484 tiles per frame in a 22×22 grid ($\sqrt{500} = 22.36$, which when rounded down to 22×22 is 484). The use of quad trees[1] and frustum culling[2]

---

[1] A two-dimensional terrain height-map would be subdivided in half along both the X and Y axis to produce smaller tiles; each subdivision would then result in 4 new tiles. Being hierarchical allows for fast rejection or acceptance of large areas of terrain.

[2] Culling is a mathematical test to determine if geometry is visible given the current view properties.

usually meant this wasn't a huge problem, but it definitely one that had to be handled. Additionally, this handling stole CPU cycles that could have been used for better things and further coupled the CPU and GPU.

One could also reduce `draw` calls by increasing the tile size—going from 9×9 to 17×17 or even 33×33 was perfectly legitimate. However, this made the algorithms other draw-back, popping, even more noticeable. As discussed in Chapter 4 and earlier in this section, popping can be a very unsightly and irritating artifact. And in the context of terrain rendering, it can be even more irritating for the viewer, because of the viewing angles used. The primary axis of change (vertical) when moving between levels of detail also tends to be the primary vertical axis (Y) on the final image; hence, problems are more noticeable than if the axis of change were lined up with the view direction (such as into and out from the final image).

Typically, the biggest and most noticeable pops are transitions at the lower end of the LOD scale, such as from 0 to 1, where the geometric difference between the two states is highest. Transitioning at the top-end, such as from 4 to 5, won't be as noticeable, since the mesh is already quite detailed, and the triangles are relatively small, making the geometric difference small. Large changes to the silhouette of the terrain (such as mountains on the horizon) tend to be particularly noticeable due to their prominence in the final image, and should ideally be avoided. Unfortunately, the simplest metrics for deciding the level of detail will typically set the geometry farthest from the camera to the lowest level of detail. However, this can be mitigated with a more intelligent metric.

In such cases, working around this problem becomes non-trivial, because we do not want to dynamically change the underlying buffers (as stated earlier, this is typically an expensive operation, negating many benefits of using this algorithm). Also, in the days of Direct3D 8 and Direct3D 9, vertex and geometry processing on the GPU was still rela-tively simplistic. The first good solution to the problem was using more elaborate schemes for LOD selection—weighting by screen contribution (projecting the bounding box to the screen and using the 2D area as the LOD scalar) could handle the case where distant ob-jects on the horizon popped unnecessarily. Vertex blending[3] was trickier to implement, but was also an option that allowed for interpolation between detail levels.

Ultimately, the high number of `draw` calls and the popping artifacts were the major problems in an otherwise GPU-friendly algorithm, and this approach was not used exten-sively as it possibly merited. However, the Direct3D 11 implementation suffers from nei-ther of these issues—a single `draw` call can render the entire landscape, and the tessellator can handle smooth transitions between levels of detail.

The Implementation

As a high-level overview, this demonstration of Direct3D 11 tessellation focuses on the input assembler, hull shader, and domain shader stages. The sample application will build

---

[3] A method commonly used in animation to interpolate (blend) between two vertex positions based on weights associated with dynamic transformation matrices.

vertex and index buffers and texture resources representing the inputs to the pipeline, and all remaining functionality will be implemented in HLSL shaders.

In this case, the vertex shader will be responsible for transforming model-space geometry into world-space and nothing else. The majority of the work will be done in the hull shader constant function, which determines the LOD for the current patch, using the patch being rendered, as well as its four immediate neighbors. The main hull shader stage will then act as a simple pass-through of the four corner vertices for the current patch. The domain shader will then take the tessellated points and will implement displacement mapping to generate the actual terrain geometry that will finally be rasterized.

Here are the steps for implementing the interlocking terrain tiles algorithm.

Step 1: Creating the Input Data

Three sources of input data need to be created for this algorithm to work: a vertex buffer, and index buffer, and heightmap texture.

Listing 9.1 creates the vertex buffer. It is very straightforward, as it is simply a grid of points that bounds new geometry generated in the space between these points.

```
// Set up the actual resource
SAFE_RELEASE( m_pTerrainGeometry );
m_pTerrainGeometry = new GeometryDX11( );

// Create the vertex data
VertexElementDX11 *pPositions
    = new VertexElementDX11( 3, (TERRAIN_X_LEN + 1) * (TERRAIN_Z_LEN + 1) );
    pPositions->m_SemanticName = "CONTROL_POINT_POSITION";
    pPositions->m_uiSemanticIndex = 0;
    pPositions->m_Format = DXGI_FORMAT_R32G32B32_FLOAT;
    pPositions->m_uiInputSlot = 0;
    pPositions->m_uiAlignedByteOffset = 0;
    pPositions->m_InputSlotClass = D3D11_INPUT_PER_VERTEX_DATA;
    pPositions->m_uiInstanceDataStepRate = 0;

VertexElementDX11 *pTexCoords
    = new VertexElementDX11( 2, (TERRAIN_X_LEN + 1) * (TERRAIN_Z_LEN + 1) );
    pTexCoords->m_SemanticName = "CONTROL_POINT_TEXCOORD";
    pTexCoords->m_uiSemanticIndex = 0;
    pTexCoords->m_Format = DXGI_FORMAT_R32G32_FLOAT;
    pTexCoords->m_uiInputSlot = 0;
    pTexCoords->m_uiAlignedByteOffset = D3D11_APPEND_ALIGNED_ELEMENT;
    pTexCoords->m_InputSlotClass = D3D11_INPUT_PER_VERTEX_DATA;
    pTexCoords->m_uiInstanceDataStepRate = 0;

Vector3f *pPosData = pPositions->Get3f( 0 );
Vector2f *pTCData = pTexCoords->Get2f( 0 );

float fWidth = static_cast< float >( TERRAIN_X_LEN );
```

```
float fHeight = static_cast< float >( TERRAIN_Z_LEN );

for( int x = 0; x < TERRAIN_X_LEN + 1; ++x )
{
    for( int z = 0; z < TERRAIN_Z_LEN + 1; ++z )
    {
        float fX = static_cast<float>(x) / fWidth - 0.5f;
        float fZ = static_cast<float>(z) / fHeight - 0.5f;
        pPosData[ x + z * (TERRAIN_X_LEN + 1) ] = Vector3f( fX, 0.0f, fZ );
        pTCData[ x + z * (TERRAIN_X_LEN + 1) ] = Vector2f( fX + 0.5f, fZ + 0.5f );
    }
}

m_pTerrainGeometry->AddElement( pPositions );
m_pTerrainGeometry->AddElement( pTexCoords );
```

Listing 9.1. Creating the vertex buffer.

The TERRAIN_X_LEN and TERRAIN_Z_LEN constants are defined in the context of how many tiles should be generated. Therefore, 1 is added to each of these constants in various places—for example, a 3×3 grid of tiles will require a 4×4 grid of vertices.

It is also worth noting that the raw geometry is defined as being a flat grid on the *XZ* plane in 3D space, and that the height of the terrain is defined entirely along the *Y* axis. Listing 9.1 initializes all height points to 0.0 at this stage.

Listing 9.2, shown below, builds up the index buffer data that defines each patch. This is by far the most complex and most important resource that this algorithm must initialize. The input assembler will use this data to index into the vertex buffer and map all the necessary control points that the hull shader and domain shader need to do their job.

```
// Code below makes reference to the following macro:
#define clamp(value,minimum,maximum) (max(min((value),(maximum)),(minimum)))
for( int x = 0; x < TERRAIN_X_LEN; ++x )
{
    for( int z = 0; z < TERRAIN_Z_LEN; ++z )
    {
        // Define 12 control points per terrain quad

        // 0-3 are the actual quad vertices
        m_pTerrainGeometry->AddIndex( (z + 0) + (x + 0) * (TERRAIN_X_LEN + 1) );
        m_pTerrainGeometry->AddIndex( (z + 1) + (x + 0) * (TERRAIN_X_LEN + 1) );
        m_pTerrainGeometry->AddIndex( (z + 0) + (x + 1) * (TERRAIN_X_LEN + 1) );
        m_pTerrainGeometry->AddIndex( (z + 1) + (x + 1) * (TERRAIN_X_LEN + 1) );

        // 4-5 are +x
        m_pTerrainGeometry->AddIndex
        (
            clamp(z + 0, 0, TERRAIN_Z_LEN)
```

```
                + clamp(x + 2, 0, TERRAIN_X_LEN) * (TERRAIN_X_LEN + 1)
        );
        m_pTerrainGeometry->AddIndex
        (
            clamp(z + 1, 0, TERRAIN_Z_LEN)
            + clamp(x + 2, 0, TERRAIN_X_LEN) * (TERRAIN_X_LEN + 1)
        );

        // 6-7 are +z
        m_pTerrainGeometry->AddIndex
        (
            clamp(z + 2, 0, TERRAIN_Z_LEN)
            + clamp(x + 0, 0, TERRAIN_X_LEN) * (TERRAIN_X_LEN + 1)
        );
        m_pTerrainGeometry->AddIndex
        (
            clamp(z + 2, 0, TERRAIN_Z_LEN)
            + clamp(x + 1, 0, TERRAIN_X_LEN) * (TERRAIN_X_LEN + 1)
        );

        // 8-9 are -x
        m_pTerrainGeometry->AddIndex
        (
            clamp(z + 0, 0, TERRAIN_Z_LEN)
            + clamp(x - 1, 0, TERRAIN_X_LEN) * (TERRAIN_X_LEN + 1)
        );
        m_pTerrainGeometry->AddIndex
        (
            clamp(z + 1, 0, TERRAIN_Z_LEN)
            + clamp(x - 1, 0, TERRAIN_X_LEN) * (TERRAIN_X_LEN + 1)
        );

        // 10-11 are -z
        m_pTerrainGeometry->AddIndex
        (
            clamp(z - 1, 0, TERRAIN_Z_LEN)
            + clamp(x + 0, 0, TERRAIN_X_LEN) * (TERRAIN_X_LEN + 1)
        );
        m_pTerrainGeometry->AddIndex
        (
            clamp(z - 1, 0, TERRAIN_Z_LEN)
            + clamp(x + 1, 0, TERRAIN_X_LEN) * (TERRAIN_X_LEN + 1)
        );
    }
}

// Move the in-memory geometry to be
// an actual renderable resource
m_pTerrainGeometry->LoadToBuffers();
m_pTerrainGeometry->SetPrimitiveType( D3D11_PRIMITIVE_TOPOLOGY_12_CONTROL_
POINT_PATCHLIST );
```

Listing 9.2. Creating the index buffer.

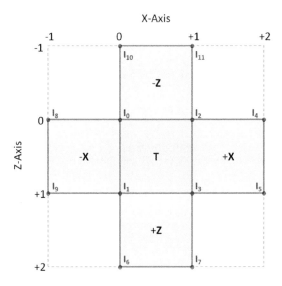

Figure 9.5. Input assembler layout for a single tile.

As shown in Figure 9.5, the first 4 control points (I_0–I_3 surrounding the red square) define the area that will have geometry generated and rendered, while the 8 control points defining the neighboring cells (I_4–I_{11}) are purely informational and allow the hull shader to correctly assign the tessellation factors. Note that the code makes use of the `clamp()` macro, which literally does what it suggests—clamps the first parameter to be in the range defined by the second and third parameters. The effects of this will be better demonstrated later, but for patches around the edge of the terrain, this sets the midpoint of the neighboring patch to be the midpoint of the edge, instead. This gives good results, is significantly better than any sort of modulus/wrapping/border operator, and is more analogous to texture gutters.

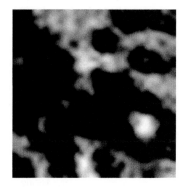

Figure 9.6. Example heightmap.

The final input required is the heightmap itself. The code in Listing 9.3 is a bit of an anticlimax compared with the previous two—it simply loads Figure 9.6 into memory as a texture resource.

```
// Load the texture
m_pHeightMapTexture = m_pRenderer11->LoadTexture
                ( std::wstring( L"../Data/Textures/TerrainHeightMap.png" ) );
```

```
// Store the height/width to the param manager
D3D11_TEXTURE2D_DESC d = m_pHeightMapTexture->m_pTexture2dConfig->GetTextureDesc();
Vector4f vTexDim = Vector4f(
                              static_cast<float>(d.Width)
                            , static_cast<float>(d.Height)
                            , static_cast<float>(TERRAIN_X_LEN)
                            , static_cast<float>(TERRAIN_Z_LEN)
                          );
m_pRenderer11->m_pParamMgr->SetVectorParameter( L"heightMapDimensions", &vTexDim );

// Create the SRV
ShaderResourceParameterDX11* pHeightMapTexParam = new ShaderResourceParameterDX11();
pHeightMapTexParam->SetParameterData( &m_pHeightMapTexture->m_iResourceSRV );
pHeightMapTexParam->SetName( std::wstring( L"texHeightMap" ) );

// Map it to the param manager
m_pRenderer11->m_pParamMgr->SetShaderResourceParameter
                              ( L"texHeightMap", m_pHeightMapTexture );

// Create a sampler
D3D11_SAMPLER_DESC sampDesc;
sampDesc.AddressU = D3D11_TEXTURE_ADDRESS_CLAMP;
sampDesc.AddressV = D3D11_TEXTURE_ADDRESS_CLAMP;
sampDesc.AddressW = D3D11_TEXTURE_ADDRESS_CLAMP;
sampDesc.BorderColor[0] =
    sampDesc.BorderColor[1] =
    sampDesc.BorderColor[2] =
    sampDesc.BorderColor[3] = 0;
sampDesc.ComparisonFunc = D3D11_COMPARISON_ALWAYS;
sampDesc.Filter = D3D11_FILTER_MIN_MAG_MIP_LINEAR;
sampDesc.MaxAnisotropy = 16;
sampDesc.MaxLOD = D3D11_FLOAT32_MAX;
sampDesc.MinLOD = 0.0f;
sampDesc.MipLODBias = 0.0f;
int samplerState = m_pRenderer11->CreateSamplerState( &sampDesc );

// Set it to the param manager
m_pRenderer11->m_pParamMgr->SetSamplerParameter( L"smpHeightMap", &samplerState );
```

Listing 9.3. Loading the heightmap as a texture.

Care needs to be taken when matching the size of the heightmap to the geometry being created. Keeping the heightmap 4–8 times larger than the underlying geometry typically works well. If it is less than this, the tessellation introduces very little new detail, and if it is much higher, this can lead to quite substantial differences between neighboring LODs. This consideration about sampling rate is a general one for tessellation—if extra generated geometry has no higher-frequency data to represent, there is little if any benefit in adding it. The example code for this uses a 512×512 monochrome texture for rendering a grid of 32×32 tiles.

Assuming successful execution of the previously described code, the application has now created all of the resources required for rendering.

Step 2: The Hull Shader

The previous step was all about generating the source data and binding it to the pipeline, in preparation for this step and the one after it. It is here that the first code is written to directly implement the algorithm described.

The introduction to Chapter 4 described the hull shader as being comprised of two pieces of HLSL—the patch constant function and the control point shader. Listing 9.4 implements our hull shader.

```
struct VS_OUTPUTww
{
    float3 position : WORLD_SPACE_CONTROL_POINT_POSITION;
    float2 texCoord : CONTROL_POINT_TEXCOORD;
};

struct HS_OUTPUT
{
    float3 position : CONTROL_POINT_POSITION;
    float2 texCoord : CONTROL_POINT_TEXCOORD;
};

[domain("quad")]
[partitioning("fractional_odd")]
[outputtopology("triangle_cw")]
[outputcontrolpoints(4)]
[patchconstantfunc("hsPerPatch")]
HS_OUTPUT hsSimple
            (
                InputPatch<VS_OUTPUT, 12> p,
                uint i : SV_OutputControlPointID
            )
{
    HS_OUTPUT o = (HS_OUTPUT)0;

    o.position = p[i].position;
    o.texCoord = p[i].texCoord;

    return o;
}
```

Listing 9.4. The hull shader control point function.

Listing 9.4 is the control point shader, which due to the attributes applied is also the main entry point as far as Direct3D is concerned. At first glance, the body of this shader appears to be a simple pass-through/identity operation. While this is partially, true a subtle detail exists in the definition of the `InputPatch`, which has 12 control points, and in the definition of `outputcontrolpoints`, which has 4.

As shown in the next section, the domain shader only needs to know about the 4 control points that define the boundary of the tile being currently rendered. Only the hull shader constant function needs the 8 neighboring points to generate tessellation factors. Consequently, there is no need to send the data further down the pipeline. This is good, because unless the GPU driver is extremely clever and automatically removes this data, it, would serve only to increase resource usage unnecessarily.

The `patchconstantfunc` attribute in Listing 9.4 points to the `hsPerPatch` entry point in Listing 9.5. This is where Greg Snook's algorithm is really implemented.

```
struct HS_PER_PATCH_OUTPUT
{
    float edgeTesselation[4]   : SV_TessFactor;
    float insideTesselation[2] : SV_InsideTessFactor;
};

HS_PER_PATCH_OUTPUT hsPerPatch
                (
                        InputPatch<VS_OUTPUT, 12> ip
                      , uint PatchID : SV_PrimitiveID
                )
{
    HS_PER_PATCH_OUTPUT o = (HS_PER_PATCH_OUTPUT)0;

    // Determine the midpoint of this patch
    float3 midPoints[] =
    {
        // Main quad
        ComputePatchMidPoint(ip[0].position,ip[1].position,ip[2].
            position,ip[3].position)

        // +x neighbor
        , ComputePatchMidPoint(ip[2].position,ip[3].position,ip[4].
            position,ip[5].position)

        // +z neighbor
        , ComputePatchMidPoint(ip[1].position,ip[3].position,ip[6].
            position,ip[7].position)

        // -x neighbor
        , ComputePatchMidPoint(ip[0].position,ip[1].position,ip[8].
            position,ip[9].position)

        // -z neighbor
        , ComputePatchMidPoint(ip[0].position,ip[2].position,ip[10].
            position,ip[11].position)
    };
```

```
// Determine the appropriate LOD for this patch
float dist[] =
{
    // Main quad
    ComputePatchLOD( midPoints[0] )

    // +x neighbor
    , ComputePatchLOD( midPoints[1] )

    // +z neighbor
    , ComputePatchLOD( midPoints[2] )

    // -x neighbor
    , ComputePatchLOD( midPoints[3] )

    // -z neighbor
    , ComputePatchLOD( midPoints[4] )
};

// Set it up so that this patch always has an interior matching
// the patch LOD.
o.insideTesselation[0] =
    o.insideTesselation[1] = dist[0];

// For the edges its more complex as we have to match
// the neighboring patches. The rule in this case is:
//
// - If the neighbor patch is of a lower LOD we
//   pick that LOD as the edge for this patch.
//
// - If the neighbor patch is a higher LOD then
//   we stick with our LOD and expect them to blend down
//   towards us

o.edgeTesselation[0] = min( dist[0], dist[4] );
o.edgeTesselation[1] = min( dist[0], dist[3] );
o.edgeTesselation[2] = min( dist[0], dist[2] );
o.edgeTesselation[3] = min( dist[0], dist[1] );

return o;
}
```

Listing 9.5. Hull shader constant function

The code in Listing 9.5 can be divided into three stages of execution. First, the tile being rendered and its 4 neighbors have their midpoints computed. Second, the distance from each midpoint to the camera is used to generate a "raw" level of detail for each of the 5 patches. And finally, these LOD values are assigned to the 6 output values that Direct3D expects (2 inner factors, SV_InsideTessFactor and 4 edge factors, SV_TessFactor).

The control points input into this stage are all in world space, which is the only transformation that the vertex shader performs. Thus, ComputePatchMidPoint() (see

Listing 9.6) can be done as a simple mean average of the four corners. Two things to note about this section are that for the edge patches where the application will have clamped the inputs, the midpoint is actually the midpoint of the edge, and not the tile; second the indexing into the ip[] array matches Figure 9.5, which was part of step 1.

```
float3 ComputePatchMidPoint(float3 cp0, float3 cp1, float3 cp2, float3 cp3)
{
    return (cp0 + cp1 + cp2 + cp3) / 4.0f;
}
```

Listing 9.6. Definition of ComputePatchMidPoint()

With an array of five midpoints generated, these can be compared against the camera position to generate five individual scalars representing the raw LOD for each of the patches. Of particular interest is that the raw values are continuous in nature (as continuous as IEEE-754 floating point values can be!) and there is only clamping between the minLOD and maxLOD values provided by the application. This, when combined with the partitioning attribute set to fractional_odd is all that is necessary to ensure a smooth transition between levels of detail on a frame-by-frame basis.

The ComputePatchLOD() function, shown in Listing 9.7, is very naïve and could be greatly improved, but for the purpose of an example it works well. The helper function simply scales linearly between application-provided minimum and maximum levels according to the distance from the camera—the further away, the lower the detail.

```
float ComputeScaledDistance(float3 from, float3 to)
{
    // Compute the raw distance from the camera to the midpoint of this patch
    float d = distance( from, to );

    // Scale this to be 0.0 (at the min dist) and 1.0 (at the max dist)
    return (d - minMaxDistance.x) / (minMaxDistance.y - minMaxDistance.x);
}

float ComputePatchLOD(float3 midPoint)
{
    // Compute the scaled distance
    float d = ComputeScaledDistance( cameraPosition.xyz, midPoint );

    // Transform this 0.0-1.0 distance scale into the desired LOD's
    // note: invert the distance so that close = high detail, far = low detail
    return lerp( minMaxLOD.x, minMaxLOD.y, 1.0f - d );
}
```

Listing 9.7. Definition of ComputePatchLOD().

More intelligent implementations could implement the idea of a near plane as well as a far plane, consider a nonlinear scale (similar to depth buffers) to apply more detail nearer the viewer, and consider using entirely different metrics. One more expensive possibility mentioned earlier is to use screen-space contribution as the LOD scalar; simply project the bounding coordinates to projection space and compute the area. A patch that contributes more pixels to the screen should have a higher LOD.

The final section of Listing 9.8 assigns the raw values to the outputs expected by Direct3D, which hides a couple of subtle details. First, as commented in the code, the choice of interpolating up or down the LOD scale—either is acceptable, but the choice will influence how much new geometry is generated—thus becomes a decision on quality versus performance. Second, and much less obvious, is the effect of the ordering of tessellation factors in the output array.

The Direct3D 11 specification defines, for a quad, that the 0th element is $U = 0.0$, the 1st is $V = 0.0$, the 2nd is $U = 1.0$, and the 3rd is $V = 1.0$. These U and V coordinates come into play downstream in the domain shader, but their life begins here. Incorrect ordering is a very easy bug to introduce and can generate unexpected results. Figure 9.7 shows this pattern in the context of this implementation.

In this case, the pattern is transposed from the more conventional texture coordinate system, but only for convenience and consistency throughout the implementation—there is no particular reason why it must be this way.

Note that the hull shader constant function has visibility of all upstream data provided by the vertex shader, and that this code, unlike the control point shader function, only produces additional intermediary values and does not remove any existing data.

These six output tessellation factors, stored in the HS_PER_PATCH_OUTPUT struct, are forwarded to the fixed-function tessellator stage, which is the next pipeline stage after the hull shader.

Step 3: The Domain Shader

By the time execution reaches the domain shader, we are only concerned with the patch being rendered. This is confirmed by the fact that it only has visibility of the four control points (CP_0-CP_3 in Figure 9.7) making up the patch itself, and has no knowledge of the four neighboring patches. The fixed-function tessellator has now generated a number of new vertices to match the primitives required to render this tessellated patch.

It is now the job of the domain shader to take individual UV coordinates and turn them into

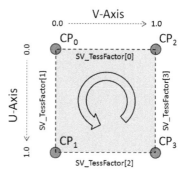

Figure 9.7. Output element ordering for a single tile.

projection space outputs that the rasterizer can work with. In particular, all geometry up until this point has been in world space, but upon exit from the domain shader, it must be in clip space—the only exception to this is if a geometry shader is bound to the pipeline, in which case you can defer this projection until then.

```
struct DS_OUTPUT
{
    float4 position : SV_Position;
    float3 colour : COLOUR;
};

float SampleHeightMap(float2 uv)
{
    // - Must use SampleLevel() so we can specify the mip-map. The DS has no
    //   gradient information it can use to derive this detail...

    // arbitrary bias to make the output more aesthetically pleasing...
    const float SCALE = 3.0f;
    return SCALE * texHeightMap.SampleLevel( smpHeightMap, uv, 0.0f ).r;
}

[domain("quad")]
DS_OUTPUT dsMain( HS_PER_PATCH_OUTPUT input,
                      float2 uv : SV_DomainLocation,
                      const OutputPatch<HS_OUTPUT, 4> patch )
{
    DS_OUTPUT o = (DS_OUTPUT)0;

    // We need to take the three world space
    // coordinates in patch[] and the interpolation
    // values in uvw (barycentric coords) and determine
    // the appropriate interpolated position.
    float3 finalVertexCoord = float3( 0.0f, 0.0f, 0.0f );

    // u,v
    // 0,0 is patch[0].position
    // 1,0 is patch[1].position
    // 0,1 is patch[2].position
    // 1,1 is patch[3].position

    /*
    0--1
     /
     /
    2--3
    */

    finalVertexCoord.xz = patch[0].position.xz * (1.0f-uv.x) * (1.0f-uv.y)
                        + patch[1].position.xz * uv.x * (1.0f-uv.y)
                        + patch[2].position.xz * (1.0f-uv.x) * uv.y
                        + patch[3].position.xz * uv.x * uv.y
                        ;
```

```
        float2 texcoord     = patch[0].texCoord * (1.0f-uv.x) * (1.0f-uv.y)
                            + patch[1].texCoord * uv.x * (1.0f-uv.y)
                            + patch[2].texCoord * (1.0f-uv.x) * uv.y
                            + patch[3].texCoord * uv.x * uv.y
                            ;

        // Determine the height from the texture
        finalVertexCoord.y = SampleHeightMap(texcoord);

        // We then need to transform the world-space
        // coord to be a proper projection space output
        // that the rasterizer can deal with. Could delegate
        // to the GS, but no need this time!
        o.position = mul( float4( finalVertexCoord, 1.0f ), mViewProj );

        // Perform a sobel filter on the heightmap to determine an appropriate
        // normal vector
        float3 normal = Sobel( texcoord );
        normal = normalize( mul( float4(normal, 1.0f), mInvTposeWorld ).xyz );
        o.colour = min(0.75f, max(0.0f, dot( normal, float3( 0.0f, 1.0f, 0.0f ) ) ) );

        return o;
}
```

Listing 9.8. The domain shader.

Listing 9.8 is all that is necessary to build the final terrain geometry that will be rendered. The domain shader shown above does not really implement a specific part of Greg Snook's original algorithm; rather, it implements a simple form of displacement mapping.

The output of the hull shader and fixed-function tessellator stages essentially just define a pattern on the *XZ* plane for where the heightmap texture should be sampled. Unlike more trivial terrain renderers, which generate a uniform grid of sample locations, this implementation generates an increasing density of sample locations, according to the LOD scheme in use. The domain shader simply takes these new points as being the correct ones to generate more detail for—the above code does not actively decide or influence the distribution of geometric detail.

The domain shader presented above can be broken down into three fundamental sections. First, the *UV* coordinate is decoded into a final world-space position on the *XZ* plane. This is a straightforward interpolation that follows the diagram in the preceding section. The same set of equations is used to determine the texture coordinate for this new vertex, which is then used to look up a value from the heightmap texture. These values are put together into the final vertex position and then transformed into projection space for the rasterizer.

The final section is a trivial lighting model to give the rendered terrain a more aesthetically pleasing appearance. Naturally, a more robust implementation would implement proper texturing and a more realistic lighting model. The above code makes references to the Sobel() function, which is shown in Listing 9.9.

```
cbuffer sampleparams
{
    // xy = pixel dimensions
    // zw = geometry dimensions
    float4 heightMapDimensions;
}

float3 Sobel( float2 tc )
{
    // Useful aliases
    float2 pxSz = float2( 1.0f / heightMapDimensions.x, 1.0f
                          / heightMapDimensions.y );

    // Compute the necessary offsets:
    float2 o00 = tc + float2( -pxSz.x, -pxSz.y );
    float2 o10 = tc + float2(    0.0f, -pxSz.y );
    float2 o20 = tc + float2(  pxSz.x, -pxSz.y );

    float2 o01 = tc + float2( -pxSz.x, 0.0f   );
    float2 o21 = tc + float2(  pxSz.x, 0.0f   );

    float2 o02 = tc + float2( -pxSz.x,  pxSz.y );
    float2 o12 = tc + float2(    0.0f,  pxSz.y );
    float2 o22 = tc + float2(  pxSz.x,  pxSz.y );

    // Use of the sobel filter requires the eight samples
    // surrounding the current pixel:
    float h00 = SampleHeightMap(o00); // NB: Definition provided in listing 9.8
    float h10 = SampleHeightMap(o10);
    float h20 = SampleHeightMap(o20);

    float h01 = SampleHeightMap(o01);
    float h21 = SampleHeightMap(o21);

    float h02 = SampleHeightMap(o02);
    float h12 = SampleHeightMap(o12);
    float h22 = SampleHeightMap(o22);
    // Evaluate the Sobel filters
    float Gx = h00 - h20 + 2.0f * h01 - 2.0f * h21 + h02 - h22;
    float Gy = h00 + 2.0f * h10 + h20 - h02 - 2.0f * h12 - h22;

    // Generate the missing Z
    float Gz = 0.01f * sqrt( max(0.0f, 1.0f - Gx * Gx - Gy * Gy ) );

    // Make sure the returned normal is of unit length
    return normalize( float3( 2.0f * Gx, Gz, 2.0f * Gy ) );
}
```

Listing 9.9. Definition of the Sobel() function.

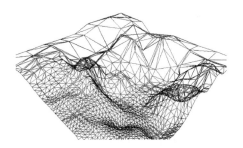

Figure 9.8. Wireframe.

Figure 9.9. Solid.

This is just a simple Sobel operator on the heightmap, a useful trick that efficiently generates acceptable normal vectors from a heightmap by using gradient detection. The important detail to note is that this technique lets the domain shader generate normal vectors in isolation, so that it does not need to be aware of any of its neighboring triangles or the underlying surface. For a displacement-mapping algorithm, this is very useful and greatly simplifies the implementation. Without this trick, the normal vectors would need to be read from a matching texture (which could increase quality), or simple "face normals" would need to be generated by the geometry shader.

The Result

The three steps shown above are all that is necessary to entirely implement Greg Snook's interlocking tiles algorithm on a Direct3D 11 GPU. Figures 9.8 and 9.9 show examples of the output, firstly in wireframe and secondly as a solid-shaded. The following is an analysis of the results:

- Resources:
    - Vertex Buffer: 81 vertices, forming a 9×9 grid, requiring 1,620 bytes
    - Index Buffer: 768 indices, with 12 for each of the 64 patches, requiring 1,536 bytes
    - Texture: 64×64 32-bit texture, requiring 16,384 bytes
    - Total storage: 19.1 KB
- Rendering:
    - 1 DrawIndexed() call
    - 768 vertex shader invocations

- 64 hull shader constant function invocations

- 256 hull shader control point function invocations

- 3,596 domain shader invocations

- 2,872 triangles generated

Regarding the storage requirements, it is worth noting that a CPU implementation of the same geometry pattern could require as much as 70 KB just for the vertex data, and this would vary according to the LODs used for any given frame. For less than one third of this storage, the inputs remain totally fixed and constant, and the underlying runtime and drivers have to handle the varying output sizes internally.

9.1.2 Extending the Interlocking Tiles Implementation

It was noted in the previous section that the LOD calculation in Listing 9.7 was very naïve. This aspect proves very useful for exploring one of the best enhancements that can be made to this implementation: the LOD equation.

Better results can be achieved on both the quality and performance axes. With a given performance budget (for example, being able to render 1 million triangles per second), for the sake of quality it is preferable to focus tessellation on areas where it is most noticeable—why waste triangles on pieces of terrain that won't benefit from them? An alternative view on the performance axis is that for a given terrain surface, there is no point in over-tessellating and creating more work than necessary; if 650 triangles can represent a given area, why generate 1000?

Figure 9.11 takes the inside tessellation factor for a patch and converts it into a shade of red (high detail), green (mid detail) and blue (low detail). It is immediately obvious that there is little red in the image, just the corners of three patches that are visible along the bottom edge. It is also noticeable that a large part of the final image is comprised of tiles that are shaded blue; in particular, many of the tiles shaded in blue are the main geographical features of the terrain being represented!

For reference, Figure 9.10 shows the heightmap that was used to generate Figure 9.11; areas of significant change (from dark to light or vice-versa) do not correlate directly with the shading in Figure 9.11. This is a good indication that the algorithm is currently naïve in its distribution of detail.

The above image clearly demonstrates that a naïve LOD algorithm like the one originally presented will neither generate high quality, nor high-performance results. The areas of the image that need the most detail are being given the least, and the areas with the most detail are being clipped by the rasterizer and not even shown.

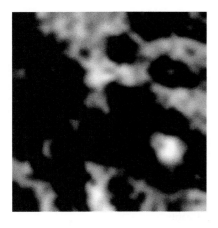 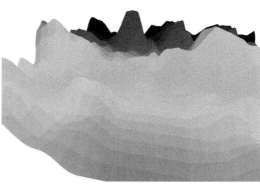

Figure 9.10. Original height map.   Figure 9.11. Debug rendering using original algorithm.

One simple change to the algorithm can help reduce this problem and also exercise another new feature in the Direct3D 11 pipeline—the compute shader.

The preceding diagrams show a case where a 256×256 heightmap is represented by a 16×16 grid. This means that for each patch rendered, the domain shader has a 16×16 grid of pixels to draw samples from. However, the hull shader, which is responsible for selecting the LOD, has no knowledge of this information. If it were able to analyze the appropriate grid of pixels later used by the domain shader, it could quite easily make a much more informed decision about the tessellation factors.

While a hull shader can read pixels from a texture bound to one of its samplers, it would be unnecessarily inefficient to have the constant function make 256 (16×16) texture samples on each invocation. However, if it had a lookup table that stored precomputed values, it could simply pull in the necessary values with a single sample operation.

Pre-Processing the Height Map

Although it is a pre-processing step in this context, the approach that will be taken is very similar to the concept of post-processing, which has been common in real-time graphics for several years.

The input heightmap will be divided up into kernels; each kernel area will have its pixels read in, and four values will be generated from the raw data, which can then be stored in a 2D output texture. This texture will then be indexed by the hull shader to enable it to have the previously described context when making LOD computations.

The key design decision is how to map the 16×16 height map samples down to a single per-patch output value.

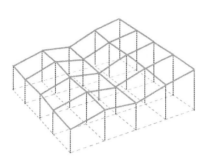
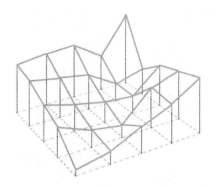

Figure 9.12. Examples of coplanarity.

It is relatively straightforward to compute a variety of statistics from the source data, but really, all that the hull shader cares about is having a measure of how much detail the patch requires. Is this piece of terrain flat? If yes, generate less detail. Alternatively, is this piece of terrain very bumpy and noisy? If yes, generate more detail.

A good objective for this pre-pass is to find a statistical measure of *coplanarity*—to what extent do the 256 samples lie on the same plane in 3D space?

This section covers a good solution to this question, one that maps cleanly to the GPU. However, there is scope for potentially higher-quality results using Fourier or Harmonic Analysis techniques familiar to mathematicians and physicists for deriving LOD metrics. This is left as an exercise to the more adventurous reader!

Consider Figure 9.12. The left-hand side shows a relatively uniform slope, possibly the side of a hill or valley. However, the right-hand side originates from a more complex section of terrain and is consequently much more erratic and noisy. Ideally, the hull shader would give the right-hand side a much higher level of detail, because, quite simply, that

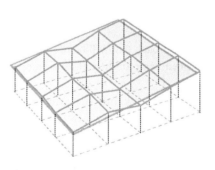
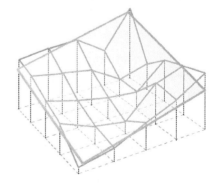

Figure 9.13. Planes fit over examples of different coplanarity.

side requires more triangles to accurately represent it. Consider Figure 9.13, which shows the same two examples as in Figure 9.12, but with a plane inserted into the dataset.

Although Direct3D cannot render quads natively,[4] the plane would be the best possible surface if there were no tessellation involved and only a single primitive were used to represent this piece of landscape. Notice that the plane in the left-hand side is a much closer match to the surface than in the right side of the figure.

Figure 9.14 shows a side-on view of a terrain segment, with plane and lines indicating how far each sample is from the plane. It is from this basis that we can measure coplanarity—the shorter the lines are between the samples and the plane, the more coplanar the data is.

Picking the plane to base these calculations off requires a "best fit" approach. It needs to be representative of the overall shape of the patch, yet it is unlikely that any plane generated will be a perfect match to the real data.

Figure 9.15 demonstrates one computationally efficient method of getting an acceptable "best fit" plane. On the left is the original patch geometry introduced earlier, and on the right is the same geometry, but with only the four corners joined together. Although this simplified primitive appears coplanar in this case, there is no guarantee that this will always be true.

For each of the 4 corners, that corner's 2 adjacent neighbors are also known, and from here it is trivial to generate the pairs of vectors denoted in red. The cross-product of each pair of vectors results in a normal vector for that corner, which is denoted in blue. Adding and

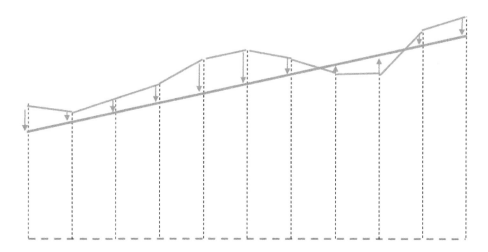

Figure 9.14. Side-on view of a terrain segment.

---

[4] It is true that geometry with four or more vertices can be sent into the pipeline, but these must be converted (by the tessellation stages) into triangles that are ultimately rasterized into the final image.

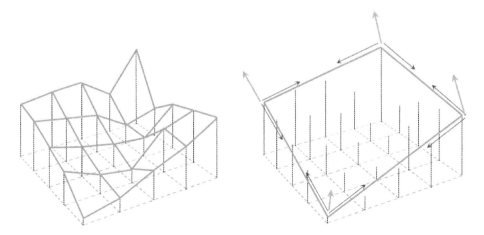

Figure 9.15. A "best fit" plane.

normalizing these 4 raw normal vectors results in a single unit-length normal vector for the patch, one that is generally representative of the underlying surface.[5] By taking any of the four corner positions, it is possible to derive a standard plane equation, as shown in Equation (9.1):

$$Ax + By + Cz + D = 0,$$
$$A = N_x,$$
$$B = N_y, \tag{9.1}$$
$$C = N_z,$$
$$D = -(N \bullet P).$$

With this plane equation known, the compute shader can evaluate each height map sample for the distance between it and the plane.

Implementing with a Compute Shader

Notation and indexes in the compute shader are not immediately obvious; Figure 9.16 introduces two of the key variables in the context of a terrain rendering pre-pass.[6] The core HLSL shader has an entry point, shown in Listing 9.10, with the [numthreads(x, y, z)] attribute attached to it.

---

[5] The process of calculating a normal vector is demonstrated in the vertex shader and rasterizer sections of Chapter 3.

[6] Compute shaders are discussed in detail in Chapter 5.

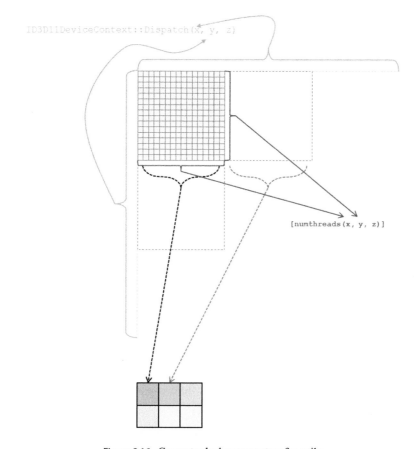

Figure 9.16. Compute shader parameters for a tile.

```
[numthreads(16, 16, 1)]
void csMain
    (
        uint3 Gid : SV_GroupID
        , uint3 DTid : SV_DispatchThreadID
        , uint3 GTid : SV_GroupThreadID
        , uint GI : SV_GroupIndex
    )
{
    /* Shader Code Here */
}
```

Listing 9.10. Compute shader entry point.

This attribute defines a thread group, also known as a *kernel*. In Listing 9.10 it is defining a 16×16×1[7] array of threads per group. The body of the csMain method is executed for a single thread, but through system generated values, it is able to identify which of these 256 (16×16×1) threads this actually is. Because it is know which thread this is, the code can be written to ensure that each thread reads from and writes to the correct location.

In Figure 9.16 the Dispatch(x, y, z) call is also introduced. This is made by the application and is analogous to a draw call as it begins execution of the compute shader. At this level, the parameters indicate how many groups of 16×16×1 thread groups to create. For this particular algorithm, the application simply divides the input height map texture dimensions by 16 and uses this as the number of kernels.[8]

For example, for a 1024×1024 height map, there will be 64×64 kernels, each kernel being 16×16×1 threads. Conceptually, this would imply a very large number of threads, one per pixel in this case, but it is up to the implementation how these tasks will be scheduled on the GPU and how many actually execute concurrently.

A key detail that has been omitted until now is how an invocation can identify itself relative to its group, as well as the entire dispatch call. Direct3D defines four system-generated values for this purpose:

1. SV_GroupID
   This uint3 returns indexes into the parameters provided by ID3D11Device Context::Dispatch(). It allows this invocation to know which group this is, relative to all others being executed. In this algorithm, it is the index into the output texture where the results for the whole group are written.

2. SV_GroupThreadID
   This uint3 returns indexes local to the current thread group—the parameters provided at compile-time as part of the [numthreads()] attribute. In this algorithm, it is used to know which threads represent corner pixels for the current 16×16 area.

3. SV_DispatchThreadID
   This uint3 is a combination of the previous two. Whereas they index relative to only one set of input parameters (either ::Dispatch() or [numthreads()]), this is a global index, essentially the two axes multiplied together. For a 64×64×1 dispatch of 16×16×1 threads, this system value will vary between 0 and 1023 in both axes (64*16=1024). Thus, for this algorithm, it provides the thread with the address of the source pixel to read from.

---

[7] The API expects *(X,Y,Z)* thread group notation, which is maintained in this text (and introduced in Chapter 5)—even if *Z*=1 is not actually significant to this implementation.

[8] Multiples of 16 were chosen for convenience in the sample code; this can be changed, as appropriate, for different-sized terrains.

4. `SV_GroupIndex`
    This `uint` gives the flattened index into the current group. For a 16×16 area, this
    value will be between 0 and 255. For the purpose of this algorithm, it is essen-
    tially the thread ID, used only to coordinate work across the group.

The final piece in the puzzle is the ability for threads to communicate with each
other. This is done through a 4-KB chunk of group shared memory, and synchronization
intrinsics. Variables defined at the global scope, such as those shown in Listing 9.11 with
the `groupshared` prefix, can be both read from and written to by all threads in the current
group.

```
groupshared float       groupResults[16 * 16];
groupshared float4      plane;
groupshared float3      rawNormals[2][2];
groupshared float3      corners[2][2];
```

Listing 9.11. Compute shader state declarations.

Synchronization is done through a choice of six barrier functions. The code can be
authored with either a `*MemoryBarrier()` or `*MemoryBarrierWithGroupSync()` call.
The former blocks until memory operations have finished, but progress can continue
before remaining ALU instructions complete. The latter blocks until all threads in the
group have reached the specified point—both memory and arithmetic instructions must
be complete. The barrier can either be `All`, `Device`, or `Group`—, with decreasing scope
at each level. Thus, an `AllMemoryBarrierWithGroupSync()` is the heaviest intrinsic to
employ, whereas `GroupMemoryBarrier()` is more lightweight. In this algorithm, only
`GroupMemoryBarrierWithGroupSync()` is used. Figure 9.17 shows the first phase of the
algorithm.

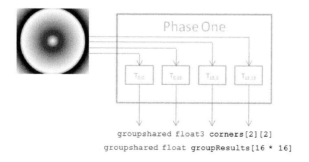

Figure 9.17. Compute shader phase one.

Notice that the first phase uses four threads, one for each corner of the 16×16 pixel group. Each of the four threads reads in a single sample. It then stores the height in group `Results[]` and stores a 3D position in `corners[][]`. All other threads are idle at this point. Listing 9.12 shows the code for this.

```
if(
    ((GTid.x ==  0) && (GTid.y ==  0))
    ||
    ((GTid.x == 15) && (GTid.y ==  0))
    ||
    ((GTid.x ==  0) && (GTid.y == 15))
    ||
    ((GTid.x == 15) && (GTid.y == 15))
  )
{
    // This is a corner thread, so we want it to load
    // its value first
    groupResults[GI] = texHeightMap.Load( uint3( DTid.xy, 0 ) ).r;

    corners[GTid.x / 15][GTid.y / 15]
        = float3(GTid.x / 15, groupResults[GI], GTid.y / 15);
    // The above will unfairly bias based on the height ranges
    corners[GTid.x / 15][GTid.y / 15].x /= 64.0f;
    corners[GTid.x / 15][GTid.y / 15].z /= 64.0f;
}

// Block until all threads have finished reading
GroupMemoryBarrierWithGroupSync();
```

Listing 9.12. Phase one of the compute shader.

Figure 9.18 depicts the next phase, where the same four threads continue to process the corner points. In this instance, they need to know about their neighboring corners, so that they can generate the cross-product and hence a normal vector for each corner—entirely ALU work. Concurrently, the other 252 threads can be reading in the remaining height map samples. Listing 9.13 shows this in action.

```
if((GTid.x ==  0) && (GTid.y ==  0))
{
    rawNormals[0][0] = normalize(cross
                        (
                            corners[0][1] - corners[0][0],
                            corners[1][0] - corners[0][0]
                        ));
}
else if((GTid.x == 15) && (GTid.y ==  0))
```

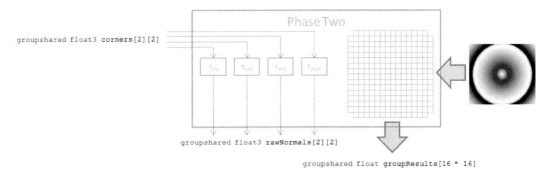

Figure 9.18. Compute shader phase two.

```
{
    rawNormals[1][0] = normalize(cross
                        (
                            corners[0][0] - corners[1][0],
                            corners[1][1] - corners[1][0]
                        ));
}
else if((GTid.x ==  0) && (GTid.y ==  15))
{
    rawNormals[0][1] = normalize(cross
                        (
                            corners[1][1] - corners[0][1],
                            corners[0][0] - corners[0][1]
                        ));
}
else if((GTid.x ==  15) && (GTid.y ==  15))
{
    rawNormals[1][1] = normalize(cross
                        (
                            corners[1][0] - corners[1][1],
                            corners[0][1] - corners[1][1]
                        ));
}
else
{
    // This is just one of the other threads, so let it
    // load in its sample into shared memory
    groupResults[GI] = texHeightMap.Load( uint3( DTid.xy, 0 ) ).r;
}

// Block until all the data is ready
GroupMemoryBarrierWithGroupSync();
```

Listing 9.13. Phase two of the compute shader.

Phase four, shown in Figure 9.20, is where the next big chunk of work takes place, but before this, the group must have a plane from which to measure offsets, as shown in Figure 9.19. This only requires a single thread and simply implements the plane-from-point-and-normal equations, as shown in Listing 9.14.

```
// The following fragment of the compute shader uses
// this utility function:
float4 CreatePlaneFromPointAndNormal(float3 n, float3 p)
{
    return float4(n,(-n.x*p.x - n.y*p.y - n.z*p.z));
}

// Phase 3 of the CS starts here:
if(GI == 0)
{
    // Let the first thread only determine the plane coefficients

    // First, decide on the average normal vector
    float3 n = normalize
                (
                    rawNormals[0][0]
                    + rawNormals[0][1]
                    + rawNormals[1][0]
                    + rawNormals[1][1]
                );

    // Second, decide the lowest point on which to base it
    float3 p = float3(0.0f,1e9f,0.0f);
    for(int i = 0; i < 2; ++i)
        for(int j = 0; j < 2; ++j)
            if(corners[i][j].y < p.y)
                p = corners[i][j];

    // Third, derive the plane from point+normal
    plane = CreatePlaneFromPointAndNormal(n,p);
}

GroupMemoryBarrierWithGroupSync();
```

Listing 9.14. Phase three of the compute shader.

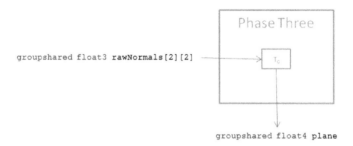

Figure 9.19. Compute shader phase three.

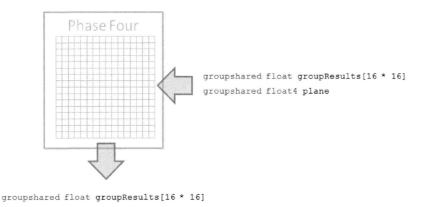

Figure 9.20. Compute shader phase four.

As mentioned earlier, Figure 9.20 shows phase four. With a plane available, it is necessary to process each of the raw heights as originally loaded from the height map. Each thread takes a single height. It then computes the distance between this sample and the previously-computed plane, replaces the original raw height value.

```
// Following fragment of CS relies on this
// utility function:
float ComputeDistanceFromPlane(float4 plane, float3 position)
{
    return dot(plane.xyz,position) - plane.w;
}

// Begin phase 5 of the compute shader:
groupResults[GI] = ComputeDistanceFromPlane
                    (
                        plane
                        , float3
                        (
                            (float)GTid.x / 15.0f
                            , groupResults[GI]
                            , (float)GTid.y / 15.0f
                        )
                    );

GroupMemoryBarrierWithGroupSync();
```

Listing 9.15. Phase four of the compute shader.

Figure 9.21 shows the final phase of the algorithm. This phase takes all of the height values and computes the standard deviation from the surface of the plane. This single value

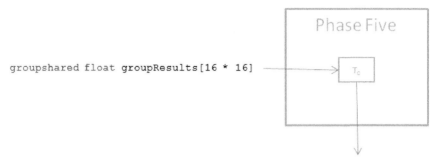

```
groupshared float groupResults[16 * 16]
```

Figure 9.21. Computer shader phase five.

is a good metric of how coplanar the 256 individual height samples are—lower values imply a flatter surface, and higher values imply a noisier and more varying patch. This single value and the plane's normal vector are written out as a float4 in the output texture, so 256 height map samples have been reduced to four numbers. Listing 9.16 shows the last phase of the computer shader.

```
if(GI == 0)
{
    // Let the first thread compute the standard deviation for
    // this patch. The 'average' is really just going to be 0.0
    // as we want the deviation from the plane and any point on the
    // plane now has a 'height' of zero.
    float stddev = 0.0f;
        for(int i = 0; i < 16*16; ++i)
        stddev += pow(groupResults[i],2);
    stddev /= ((16.0f * 16.0f) - 1.0f);

    stddev = sqrt(stddev);

    // Then write the normal vector and standard deviation
    // to the output buffer for use by the Domain and Hull Shaders
    bufferResults[uint2(Gid.x, Gid.y)] = float4(plane.xyz, stddev);
}
```

Listing 9.16. Phase five of the compute shader.

Integrating the Compute Shader

The previous section details the actual compute shader that implements our algorithm, but the application must still coordinate this work.

First, the output texture needs to be created. This will be bound as an output to the compute shader, but used later as an input to the hull shader. The underlying resource type is a regular 2D texture, with the important detail of having a D3D11_BIND_UNORDERED_ACCESS as one of its bind flags, as shown in Listing 9.17.

```
// Assert on the input texture dimensions
D3D11_TEXTURE2D_DESC d = m_pHeightMapTexture->m_pTexture2dConfig->GetTextureDesc();
_ASSERT( 0 == (d.Width % 16) );
_ASSERT( 0 == (d.Height % 16) );

// Create the output texture
Texture2dConfigDX11 LookupTextureConfig;
LookupTextureConfig.SetFormat( DXGI_FORMAT_R32G32B32A32_FLOAT );
LookupTextureConfig.SetColorBuffer( TERRAIN_X_LEN, TERRAIN_Z_LEN );
LookupTextureConfig.SetBindFlags( D3D11_BIND_UNORDERED_ACCESS
                                | D3D11_BIND_SHADER_RESOURCE );

m_pLodLookupTexture = m_pRenderer11->CreateTexture2D( &LookupTextureConfig, 0 );

// Create the effect
SAFE_DELETE( m_pComputeShaderEffect );
m_pComputeShaderEffect = new RenderEffectDX11( );

// Compile the compute shader
m_pComputeShaderEffect->m_iComputeShader =
    m_pRenderer11->LoadShader( COMPUTE_SHADER,
        std::wstring( L"../Data/Shaders/InterlockingTerrainTilesComputeShader.
                        hlsl" ),
        std::wstring( L"csMain" ),
        std::wstring( L"cs_5_0" ) );
_ASSERT( -1 != m_pComputeShaderEffect->m_iComputeShader );
```

Listing 9.17. Creating the compute shader.

At this point, the necessary resources have been created, so they simply need to be bound to the pipeline. Also, the compute shader must be initiated, as shown in Listing 9.18.

```
// Bind the resources
m_pRenderer11->m_pParamMgr->SetUnorderedAccessParameter
                            ( L"bufferResults", m_pLodLookupTexture );
m_pRenderer11->m_pParamMgr->SetShaderResourceParameter
                            ( L"texHeightMap", m_pHeightMapTexture );

// Determine number of threads
D3D11_TEXTURE2D_DESC d = m_pHeightMapTexture->m_pTexture2dConfig->GetTextureDesc();

// Run the compute shader
m_pRenderer11->pImmPipeline->Dispatch
```

```
                              (
                                *m_pComputeShaderEffect
                                , d.Width / 16
                                , d.Height / 16
                                , 1
                                , m_pRenderer11->m_pParamMgr
                              );
// Bind the output to the hull shader
m_pRenderer11->m_pParamMgr->SetShaderResourceParameter
                              ( L"texLODLookup", m_pLodLookupTexture );
```

Listing 9.18. Executing the compute shader.

The code after the `Dispatch()` call in Listing 9.18 is particularly important. Without this being executed, the *UAV* will still be bound to the pipeline referencing the 2D output texture; Direct3D will then keep it from also being bound as an input to the hull shader, since it is illegal to have a resource set to be both an input and output at the same time!

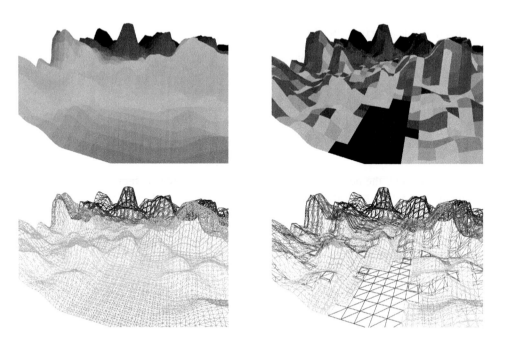

Figure 9.22. Naïve Distance Based LOD 32,500 Triangles generated, 76% rasterized.

Figure 9.23. Compute Shader Based LOD 22,232 triangles generated, 77% rasterized.

Results

Figure 9.22 and Figure 9.23 demonstrate the difference made by the new algorithm.

Although Figure 9.22 may appear more aesthetically pleasing due to the smooth gradients, the more chaotically shaded Figure 9.23 is by far the better image, from a geometric perspective. In both images, the patch detail is translated into a color—red for high detail, green for mid detail, and blue for low detail. Black is the lowest detail.

The lower half of each image is the corresponding wireframe representation, which clearly and obviously shows the approximately 40% reduction in triangle count.

Consider the bottom center area of both image. In Figure 9.22 it is mostly shaded in orange, whereas in Figure 9.22, it is predominantly black. It is also important to notice that the surface being represented is flat for this section of terrain. The naïve distance-based calculation assigns this geometrically simple piece of terrain a high number of triangles, although it simply doesn't need it. Why waste processing power and memory bandwidth generating and rasterizing extra triangles that add nothing to the final image?

Examining Figures 9.22 and 9.23 initially suggests that the computer shader approach has greatly reduced the LOD across the whole image. This is based on the observation that most of the terrain tiles are shades of blue, while the majority are green and yellow in the naïve implementation. Figure 9.24 makes it easy to compare the naïve approach (on the left) and the computer shader approach (on the right).

When rendering is performed using a simple N•L directional lighting shader with solid shading, we get an image much closer to what would be rendered in a real application. The only major omission is the lack of textures (such as grass and rock). In addition naïve and computer shader approaches shown, the middle of Figure 9.24 shows a simple image-based difference of the two results on either side. Here, black means no difference, and white means that they are completely different. Comparing the rasterized image is preferable to comparing the underlying geometry because ultimately, it is the

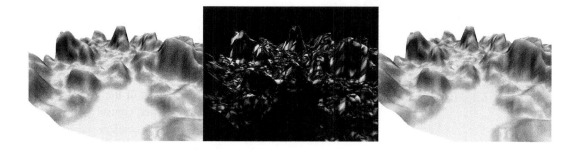

Figure 9.24. Comparison of naïve and compute shader approaches.

image displayed on the screen that is most important, and geometry is just a means by which to construct it.

On close inspection it isn't too hard to see the differences between the left and right images in Figure 9.24. However, the differences are actually quite minimal. Apart from the silhouette (discussed next) the difference is very rarely above 15%, which, for a reduction in geometry of 40%, is a good tradeoff. The exact weightings and biases can be easily tweaked in the appropriate HLSL code to achieve an aesthetically pleasing result.

Further Extensions

By weighting the level of detail by the actual complexity of the landscape, the available processing time can be more accurately shifted to areas that genuinely benefit from it. Despite being a worthwhile improvement over the original, there are still further extensions that could be implemented.

The discussion so far has calculated LODs as a pre-pass before any rendering. However, there is no reason that dynamic terrain (where the shape changes on a regular basis) can't be used, given that a simple rerun of the compute shader will update all the necessary inputs for the actual rendering.

One particular limitation of this implementation is that it still retains an element of distance in the LOD calculation. Geometry further from the camera is always assumed to be less important, and to require less detail.

Consider the horizon shown in Figure 9.25. By definition, a horizon is a long way from the viewer; in the current implementation, that works to reduce the final level of detail, even though the horizon is a significant part of the image as a whole.

It is worth extending the terrain algorithm to weight features on the horizon higher than distant objects that are less pronounced. The silhouette of a model is one of the biggest visual clues of its true level of detail, and is one of the more obvious places for the human eye to pick up on approximations or other visual artifacts.

Another possibility is to use a secondary height-map. We could use the approach already discussed to get the basic shape of the landscape for the hull shader control points, and could use a secondary height map to add further detail to the vertices generated by the domain shader. This has the neat potential of allowing

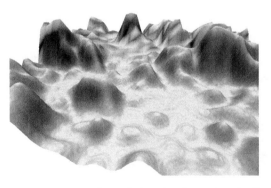

Figure 9.25. Example rendering showing a low-detail horizon.

overhangs or caves in the terrain, a feature that is typically missing from most height-map-based terrain renderers.

One significant problem that hasn't been mentioned thus far, and which is not really in the scope of this chapter, is that of interacting with the terrain.

In particular, the core application running on the CPU has no knowledge of the final geometry, which is generated entirely on the GPU. This is problematic for physics-oriented or user-oriented algorithms, such as picking (clicking to select geometric objects and features) or having objects interact (as in collision detection) or having objects track the landscape (such as having a car follow the terrain of a racing track).

Two ways to mitigate this problem are for the CPU to always use the highest level of detail, or to move picking or ray-casting to the GPU. In the former case, and for an adaptive GPU tessellation like the one described here, the rendered terrain should be very close to the highest LOD in all cases where it would be evident to a user. In the latter case, it is possible to use the geometry shader, combined stream output,[9] to output a candidate set of geometry that intersect points of interest. This is quite complicated and requires careful work to ensure that CPU read-back doesn't stall the GPU; but does present a robust solution.

9.2 Higher-Order Surfaces

This category of tessellation is the more commonly known approach. It covers the algorithms and equations used by the majority of the art and design software used over the last few decades, and is being closer to the assumption that tessellation is about *curved* or *smooth* surfaces.

Consider Equation (9.2), shown below. In particular, note that it has two basic constituents—constants (a, b, c, and d) and variables (only x in this case). In the context of tessellation, the constants would be control points output by the hull shader stage, and the variables would be the sample locations input into the domain shader stage from the fixed-function tessellator. The domain shader is responsible for taking both of these and evaluating the final result of the equation:

$$f(x) = a + bx + cx^2 + dx^3. \qquad (9.2)$$

The exact form of an equation used in this style of tessellation will vary, and the properties (such as complexity or classes of shape or the surface it can replicate) will be key to

---

[9] Geometry shaders and stream output capabilities are discussed in detail in Chapter 3.

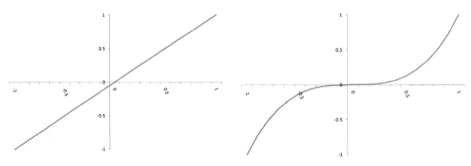

Figure 9.26. Linear.   Figure 9.27. Quadratic.

evaluating the usefulness of it in any tessellation context. This can be seen in Figures 9.26 through Figure 9.28.

In most cases, the mathematical functions employed for higher-order surfaces will be quadratic (Figure 9.27) or cubic (Figure 9.28); Figure 9.26 is included for reference with regards to planar triangle/line segments used in conventional rendering.

Individual quadratic or cubic sections are limited in the number of shapes they can represent, such that any real-world application of higher-order surfaces will want to compose many of these smaller curves into a larger model representing far more complex surfaces. How these segments are joined together has strongly impacts the overall appearance of the surface and is described as *geometric continuity*. The following figures introduce four levels of continuity.

Initially, G_2 continuity appears to be the most desirable. While aesthetically it does result in "perfect" curves, some characteristics of the other continuity levels can be beneficial. The primary motivation for non-G_2 continuity is that of sharp edges, which are typically required for non-organic or otherwise artificial objects. For example, aspects of a

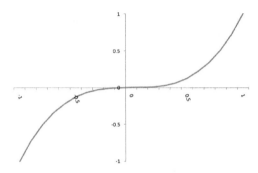

Figure 9.28. Cubic.

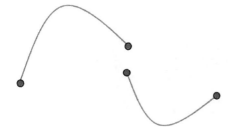

Figure 9.29. No continuity.

Figure 9.30. G_0 - start and end points match.

Figure 9.31. G_1 - incoming and outgoing tangents are parallel.

Figure 9.32. G_2 - Incoming and outgoing tangents are parallel and rate of change is same.

boat might be smooth and curved, but a boat still has a lot of edge detail that is not smooth or curved.

Utilizing G_0 and G_1 continuity can allow for sharper and more precise geometric shapes, but it is worth noting that this may require adding additional geometry around key details to ensure an accurate representation. Figures 9.34 and 9.35 demonstrate this particular case.

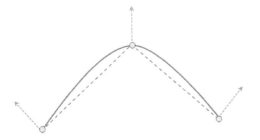

Figure 9.33. Without extra geometry.

Figure 9.34. With extra geometry.

Figure 9.35. Example of silhouette enhancement.

9.2.1 Curved Point Normal Triangles

Chapter 4 introduced the history of tessellation in modern computer graphics. In particular, it mentioned the TruForm feature in ATI Technology's Radeon 8500 GPU and the fact that this was an early attempt at consumer hardware tessellation. Despite its various merits, the technology never attained critical mass at the time, due in part to fierce competition, as well as to software advances.

The obvious benefit of this algorithm both, originally and now, is the improved visual aesthetics. However, it isn't just the continuity and smoothness of the surface itself that is significant. It is also worth noting the silhouette. Take Figure 9.35 as a demonstration of the silhouette enhancements; the key is that the right side does not present an obviously artificial outline comprised of triangle edges. Advances in recent years with per-pixel lighting and bump mapping can greatly improve the perceived quality of a model's interior, but they still leave a reminder that we are merely seeing a handful of triangles, something the human brain is prone to pick up on and smash the illusion.

Although the Direct3D 8 API didn't explicitly expose details about the algorithm it used, it did resolve down to what was discussed in the "Curved PN Triangles" paper (Vlachos, Peters, Boyd, & Mitchell, 2001) which ATI Technologies released with its own hardware implementation. Direct3D 11 removes the main barriers to the acceptance of Vlachos et. al's Curved PN Triangles algorithm from first time around, and the remainder of this chapter will discuss an implementation of the original approach, along with several potential improvements.

Algorithm Overview

A key advantage of the algorithm is that it requires no additional data to be provided by the application. This was crucial for commercial success and ease of use when applied to consumer hardware of the era (even though that success didn't happen then!). The intention was that developers could flip a switch in the form of an API render state, and the hardware would transparently apply the algorithm, producing smooth surfaces without any substantial code changes. It also didn't require different geometry to be created by artists. This would have been prohibitively difficult, due to targeting multiple types of hardware, and to not suiting the commonly used tools of the period.

The mathematics of the algorithm therefore only require a conventional triangle comprised of three positions and three normal vectors. Even with Direct3D 11, where adjacency (up to 32 control points per primitive) is relatively easy to provide, this simple property can be quite convenient. Most importantly, allows for an "upgrade" path of sorts—existing art assets can be rendered via Direct3D 11 using this algorithm without any changes, while developers still have the option to use more complex modeling formats and algorithms when the resources support it.

Curved Point Normal Triangles also neatly map on to the Direct3D 11 tessellation pipeline. The first stage is to take the incoming triangle and generate a more detailed *control mesh*—a hull shader program that amplifies. After sample locations are generated (by the fixed-function tessellation unit) the mathematical surface is evaluated using the detailed control mesh—a classic Direct3D 11 Domain Shader program.

Geometric Components

Tessellation of the geometry that actually makes up the final rasterized surface is done on a cubic basis, requiring an additional seven control points.

Looking back at Figure 9.27 and Figure 9.28, it becomes clear that a quadratic function cannot represent a full range of surface variations. However, a cubic function can capture surface inflections, resulting in a more accurate and aesthetically pleasing representation.

Figure 9.36 shows the locations of the additional seven control points (green) as well as the original three vertex positions (blue), as provided by the application. These seven control points, two per side and one in the middle, are evenly distributed across the surface of the triangle.

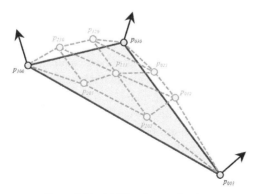

Figure 9.36. Additional control points.

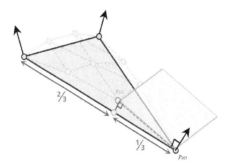

Figure 9.37. Control point equation shown visually.

The six control points along the edge are computed using the tangent plane defined by each control point's nearest vertex. Identifying the location of the edge mid-point is a trivial linear interpolation along the edge based on the corresponding vertices, as shown in Equation (9.3). Defining the tangent plane using the vertex and its normal, Equation (9.4), is a stock formula used throughout computer graphics. Combining these two elements into Equation (9.5) will generate the appropriate control point. Figure 9.37 shows how these three equations work together to generate the control points introduced in Figure 9.36:

$$Q = p_i + \tau \ (\ p_j - p_i); \tag{9.3}$$

$$
\begin{aligned}
P &= (P_x, P_y, P_z), \\
N &= (N_x, N_y, N_z), \\
0 &= xN_x + yN_y + zN_z + (- N \bullet P);
\end{aligned}
\tag{9.4}
$$

$$Q' = Q - N \times ((Q - P) \bullet N). \tag{9.5}$$

This still leaves the center control point to be generated. Computing the six control points along the edges actually forms a ring around this remaining location, a property that can be used to our advantage. Raising the midpoint from its position on the original triangle (Equation (9.6)) relative to this ring of surrounding control points (Equation (9.7)) using simple linear interpolation yields a desirable result. Choosing the halfway point between these keeps the final tessellated surface close to the original triangle geometry, leading to more predictable outputs with respect to their original coarse triangulations. Equation (9.8) shows the final position for this control point:

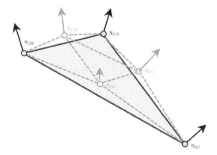

Figure 9.38. Additional normal vector control points.

$$V = \frac{P_{300} + P_{030} + P_{003}}{3},$$ (9.6)

$$E = \frac{P_{120} + P_{210} + P_{021} + P_{012} + P_{201} + P_{102}}{6},$$ (9.7)

$$P_{111} = E + \frac{E - V}{2}.$$ (9.8)

Normal Vectors

Tessellation of normal vectors is done on a quadratic basis, requiring an additional four control points. Initially, this seems unexpected, because it differs from the positional tessellation, which is cubic.

The original research paper (Vlachos, Peters, Boyd, & Mitchell, 2001) makes a particular case for this decision, based on simplicity. The added complexity of calculations to generate matching cubic normals for a generalized case is prohibitively difficult without producing an equivalent improvement in final image quality. While modern GPUs are orders of magnitude faster than those that introduced TruForm, this argument is still valid—why waste valuable processing time for no good reason?

Figure 9.38 shows the additional control points (green) required for quadratic normal interpolation: it is simply a case of putting a midpoint on each edge of the triangle, along with the original per-vertex normal vectors (blue) provided by the application.

To generate the control vector for the midpoint of each edge, we use reflection about a plane perpendicular to the edge currently being processed. This is necessary to allow the quadratic basis to approximate the geometric inflections that are possible with the cubic position tessellation.

Consider Figure 9.39, which shows the midpoint normal vector (orange) as a simple average of the vector at either end of the edge (dark green). Figure 9.40 shows the same

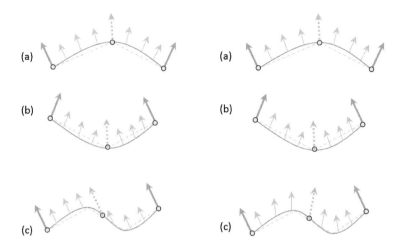

Figure 9.39. Simple averaging.   Figure 9.40. Averaging reflected.

calculation, but reflected about a plane perpendicular to the edge. Both figures show the three primary cases for normal vectors—(a) both pointing outwards, (b) both pointing inwards and (c) both pointing in the same direction. In all cases, it is assumed that the geometric surface is subject to a cubic tessellation function.

Equation (9.9) demonstrates how to generate a reflected normal vector in the general case (where n is the vector to reflect, and b is the unnormalized vector for the plane), while Equation (9.10) applies this to the context of quadratic normal vector interpolation:

$$n' = n - 2 \times \left(\frac{b \cdot n}{b \cdot b} \right) \times b; \tag{9.9}$$

$$
\begin{aligned}
N_1 &= n_{200}, \\
N_2 &= n_{020}, \\
N_3 &= n_{002}, \\
P_1 &= p_{300}, \\
P_2 &= p_{030}, \\
P_3 &= p_{003}, \\
v_{ij} &= 2 \times \frac{(P_j - P_i) \cdot (N_i - N_j)}{(P_j - P_i) \cdot (P_j - P_i)}, \\
n_{110} &= N_1 + N_2 - v_{12} \times (P_2 - P_1), \\
n_{011} &= N_2 + N_3 - v_{23} \times (P_3 - P_2), \\
n_{101} &= N_3 + N_1 - v_{31} \times (P_1 - P_3).
\end{aligned}
\tag{9.10}
$$

Generating Normals and Positions in the Hull Shader

With the equations for calculating the control mesh understood, it is now necessary to fit these into the Direct3D pipeline. This naturally aligns with the hull shader's responsibilities. It also demonstrates a characteristic introduced as part of Chapter 4, in that the hull shader stage can amplify or reduce the amount of geometry passed down the pipeline—in this case it is necessary to amplify. From the original three vertices, the hull shader must output ten positions and six normals.

There is also an interesting design decision to be made when implementing this part of the algorithm with regard to whether the control mesh is generated by the constant or by the per-element function. Logically, it fits in the per-element definition, but this results in code with more branches and conditionals (to determine which equation to apply for a given input value of `SV_OutputControlPointID`), which may not be as efficient to execute on a GPU. And not only may it be less efficient to execute a shader program with branches, but this branched code will be executed many times, once for each output control point. To avoid branching, it becomes necessary to place the code in the constant function, which ensures only a single evaluation.

However, it is important to note that this choice only exists for tessellation algorithms with small amounts of output data from the hull shader. The constant function is limited to 128 scalars of output (32 `float4`s), which must include the tessellation factors. Using Curved Point Normal Triangles requires at least 48 scalars of output (10 positions and 6 normals, each `float3` in size) before considering additional properties such as texture coordinates, color, or tangent vectors.

Listing 9.19 shows how to implement this using the logical approach of evaluating each output control point with branching. Readers can refer to the `PNTriangles11` (AMD & Microsoft) sample in the DirectX SDK for an implementation that uses the constant function approach.

```
cbuffer TessellationParameters
{
    float4 EdgeFactors;
};

struct VS_OUTPUT
{
    float3 position        : WORLD_SPACE_CONTROL_POINT_POSITION;
    float3 normal          : WORLD_SPACE_CONTROL_POINT_NORMAL;
};

struct HS_OUTPUT
{
    float3 position        : WORLD_SPACE_CONTROL_POINT_POSITION;
    float3 normal          : WORLD_SPACE_CONTROL_POINT_NORMAL;
};
struct HS_CONSTANT_DATA_OUTPUT
```

```
{
    float Edges[3]        : SV_TessFactor;
    float Inside          : SV_InsideTessFactor;
};

HS_CONSTANT_DATA_OUTPUT hsConstantFunc( InputPatch<VS_OUTPUT, 3> ip, uint
PatchID : SV_PrimitiveID )
{
    HS_CONSTANT_DATA_OUTPUT output;

    output.Edges[0] = EdgeFactors.x;
    output.Edges[1] = EdgeFactors.y;
    output.Edges[2] = EdgeFactors.z;

    output.Inside = EdgeFactors.w;

    return output;
}

[domain("tri")]
[partitioning("fractional_even")]
[outputtopology("triangle_cw")]
[outputcontrolpoints(13)]
[patchconstantfunc("hsConstantFunc")]
HS_OUTPUT hsDefault( InputPatch<VS_OUTPUT, 3> ip, uint i : SV_
OutputControlPointID, uint PatchID : SV_PrimitiveID )
{
    HS_OUTPUT output;

    // Must provide a default definition just in
    // case we don't match any branch below
    output.position = float3(0.0f, 0.0f, 0.0f);
    output.normal = float3(0.0f, 0.0f, 0.0f);
        switch(i)
    {
        // Three actual vertices:

        // b(300)
        case 0:
        // b(030)
        case 1:
        // b(003)
        case 2:

            output.position = ip[i].position;
            output.normal = ip[i].normal;
            break;

        // Edge between v0 and v1

        // b(210)
        case 3:
            output.position = ComputeEdgePosition(ip, 0, 1);
            break;
```

```
            // b(120)
            case 4:
                output.position = ComputeEdgePosition(ip, 1, 0);
                break;

            // Edge between v1 and v2

            // b(021)
            case 5:
                output.position = ComputeEdgePosition(ip, 1, 2);
                break;
            // b(012)
            case 6:
                output.position = ComputeEdgePosition(ip, 2, 1);
                break;

            // Edge between v2 and v0

            // b(102)
            case 7:
                output.position = ComputeEdgePosition(ip, 2, 0);
                break;
            // b(201)
            case 8:
                output.position = ComputeEdgePosition(ip, 0, 2);
                break;

            // Middle of triangle

            // b(111)
            case 9:
                float3 E =
                        (
                            ComputeEdgePosition(ip, 0, 1) +
                            ComputeEdgePosition(ip, 1, 0)
                            +
                            ComputeEdgePosition(ip, 1, 2) +
                            ComputeEdgePosition(ip, 2, 1)
                            +
                            ComputeEdgePosition(ip, 2, 0) +
                            ComputeEdgePosition(ip, 0, 2)
                        ) / 6.0f;
                float3 V = (ip[0].position + ip[1].position + ip[2].position)
                        / 3.0f;

                output.position = E + ( (E - V) / 2.0f );

                break;

            // Normals

            // n(110) - between v0 and v1
            case 10:
                output.normal = ComputeEdgeNormal(ip, 0, 1);
                break;
```

```
        // n(011) - between v1 and v2
        case 11:
            output.normal = ComputeEdgeNormal(ip, 1, 2);
            break;

        // n(101) - between v2 and v0
        case 12:
            output.normal = ComputeEdgeNormal(ip, 2, 0);
            break;
    }

    return output;
}
```

Listing 9.19. Hull shader for curved point normal triangles.

Equations (9.61) and (9.70) allow for aspects of each evaluation to be broken out into separate functions. Listing 9.20 covers these definitions.

```
float ComputeWeight(InputPatch<VS_OUTPUT, 3> inPatch, int i, int j)
{
    return dot(inPatch[j].position - inPatch[i].position, inPatch[i].normal);
}

float3 ComputeEdgePosition(InputPatch<VS_OUTPUT, 3> inPatch, int i, int j)
{
    return (
            (2.0f * inPatch[i].position) + inPatch[j].position
            - (ComputeWeight(inPatch, i, j) * inPatch[i].normal)
           ) / 3.0f;
}
float3 ComputeEdgeNormal(InputPatch<VS_OUTPUT, 3> inPatch, int i, int j)
{
    float t = dot
            (
                inPatch[j].position - inPatch[i].position
                , inPatch[i].normal + inPatch[j].normal
            );

    float b = dot
            (
                inPatch[j].position - inPatch[i].position
                , inPatch[j].position - inPatch[i].position
            );

    float v = 2.0f * (t / b);
        return normalize
            (
                inPatch[i].normal + inPatch[j].normal
                - v * (inPatch[j].position - inPatch[i].position)
            );
}
```

Listing 9.20. Hull shader utility functions.

The Domain Shader

So far in this section, the discussion has centered on construction of the control mesh from the input positions and normal vectors. As discussed earlier, these are analogous to the coefficients of the equation that represents the ideal curved surface.

It is the domain shader's responsibility to take these coefficients, along with the sampling points generated by the fixed-function tessellator, and evaluate the new geometry that makes up the final curved surface:

$$
\begin{aligned}
P_{uvw} = {} & w^3 p_{300} + u^3 p_{030} + v^3 p_{003} \\
& + 3w^2 u p_{210} + 3u^2 w p_{120} \\
& + 3w^2 u p_{201} + 3v^2 w p_{102} \\
& + 3u^2 v p_{021} + 3v^2 u p_{012} \\
& + 6uvw p_{111};
\end{aligned}
\tag{9.11}
$$

$$
n_{uvw} = w^2 n_{200} + u^2 n_{020} + v^2 n_{002} + uw n_{110} + uv n_{011} + vw n_{101}.
\tag{9.12}
$$

Equation (9.11) shows the cubic function that needs to be evaluated for each output position from the domain shader stage. Similarly, Equation (9.12) shows the quadratic function for computing each normal vector. Listing 9.21 shows an HLSL implementation of these two functions.

```
cbuffer Transforms
{
    matrix mWorld;
    matrix mViewProj;
    matrix mInvTposeWorld;
};

cbuffer TessellationParameters
{
    float4 EdgeFactors;
};
cbuffer RenderingParameters
{
    float3 cameraPosition;
    float3 cameraLookAt;
};
struct DS_OUTPUT
{
    float4 Position       : SV_Position;
    float3 Colour         : COLOUR;
};

[domain("tri")]
DS_OUTPUT dsMain
```

```
    (
        const OutputPatch<HS_OUTPUT, 13> TrianglePatch
        , float3 BarycentricCoordinates : SV_DomainLocation
        , HS_CONSTANT_DATA_OUTPUT input
    )
    {
        DS_OUTPUT output;

        float u = BarycentricCoordinates.x;
        float v = BarycentricCoordinates.y;
        float w = BarycentricCoordinates.z;

        // Original Vertices
        float3 p300 = TrianglePatch[0].position;
        float3 p030 = TrianglePatch[1].position;
        float3 p003 = TrianglePatch[2].position;

        // Edge between v0 and v1
        float3 p210 = TrianglePatch[3].position;
        float3 p120 = TrianglePatch[4].position;
            // Edge between v1 and v2
        float3 p021 = TrianglePatch[5].position;
        float3 p012 = TrianglePatch[6].position;

        // Edge between v2 and v0
        float3 p102 = TrianglePatch[7].position;
        float3 p201 = TrianglePatch[8].position;

        // Middle of triangle
        float3 p111 = TrianglePatch[9].position;

        // Calculate this sample point
        float3 p = (p300 * pow(w,3)) + (p030 * pow(u,3)) + (p003 * pow(v,3))
                + (p210 * 3.0f * pow(w,2) * u)
                + (p120 * 3.0f * w * pow(u,2))
                + (p201 * 3.0f * pow(w,2) * v)
                + (p021 * 3.0f * pow(u,2) * v)
                + (p102 * 3.0f * w * pow(v,2))
                + (p012 * 3.0f * u * pow(v,2))
                + (p111 * 6.0f * w * u * v);
        // Transform world position with view-projection matrix
        output.Position = mul( float4(p, 1.0), mViewProj );

        // Compute the normal - QUADRATIC
        float3 n200 = TrianglePatch[0].normal;
        float3 n020 = TrianglePatch[1].normal;
        float3 n002 = TrianglePatch[2].normal;

        float3 n110 = TrianglePatch[10].normal;
        float3 n011 = TrianglePatch[11].normal;
        float3 n101 = TrianglePatch[12].normal;

        float3 vWorldNorm = (pow(w,2) * n200) + (pow(u,2) * n020) + (pow(v,2) * n002)
                    + (w * u * n110) + (u * v * n011) + (w * v * n101);

        vWorldNorm = normalize( vWorldNorm );
```

```
        // Perform a simple shading calc
    float3 toCamera = normalize( cameraPosition.xyz );
    output.Colour = saturate( dot(vWorldNorm, toCamera) ) * float3( 0.4, 0.4, 1.0 );

    return output;
}
```

Listing 9.21. Domain shader for curved point normal triangles.

Despite the importance of the domain shader stage, the code and mathematics involved are relatively simple and straightforward.

Example Output

When the previously described shader programs are plugged in with traditional triangle geometry, results such as those shown in Figure 9.41 can be generated. Considering that the

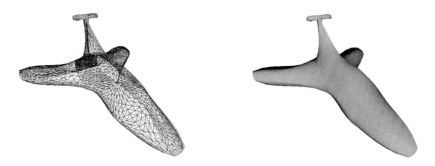

Figure 9.41. Example output with Curved Point Normal Triangles algorithm applied.

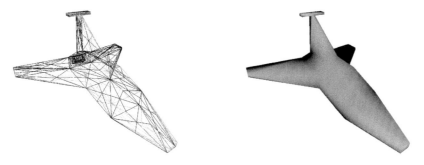

Figure 9.42. Original geometry without Curved Point Normal Triangles algorithm applied.

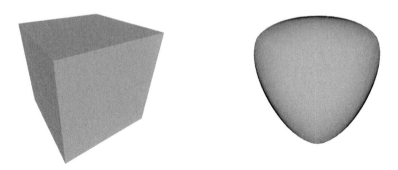

Figure 9.43. Excessive smoothed geometry loses expected sharp edges.

starting point of Figure 9.42 is an obviously low-polygon model, the improvement added by the Curved Point Normal Triangles algorithm is clear.

One characteristic to pay attention to in the wireframe representation in Figure 9.41 is that the tessellation is uniform across the entire mesh. This doesn't pose a problem with regard to the final visual results; but it is inefficient in some cases.

Despite generally good results, there are cases where this algorithm doesn't work very well. In Figure 9.43, notice that the result is probably too rounded (right hand side) and has actually lost the sharp edges that define the intended object (left side).

Encoding Sharp Edges

In this context, a *sharp edge* is an edge where the positional geometry between neighboring triangles matches, but the normal vectors do not. Figure 9.44 shows a simple example of this, in which both sides have a shared position in the centre, but the left side has differing normal vectors, and the right side has both a shared position and a shared normal.

Figure 9.44. A simple example of a sharp edge.

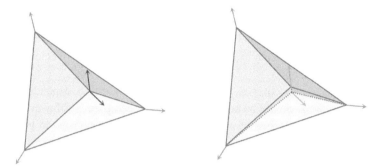

Figure 9.45. Splitting a shared vertex.

The original research paper for this algorithm makes a startling claim (with supporting proof) that simply splitting a vertex will often generate cracks in the surface of the object. Chapter 4 introduced the concept of tessellated surfaces being *watertight*, and this property of Curved Point Normal Triangles can shatter that!

Since the article proves that with only local information available, this problem cannot be avoided purely by using the GPU (remember, a key part of this algorithm was accepting unmodified triangle data without adjacency), a software pre-processing step becomes necessary.

Given that we're interested in edges, there are two possible variations—either one, or both, of the endpoints has differing normals. The software algorithm in use needs to examine all the vertex data in a given mesh and identify vertices where position information is the same, but normal vectors are not. Once these are identified, the simplest approach is to split the offending vertex in two by moving one endpoint very slightly away from the other. Figure 9.45 shows a simple case of this.

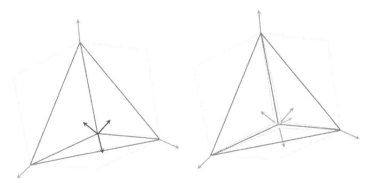

Figure 9.46. Splitting corner geometry.

Unfortunately, the general solution suffers because it can cause many triangles to share many different normal vectors. Figure 9.46 shows an example of corner geometry that first appeared in Figure 9.43. This has three triangles sharing the same position information at the corner, but each triangle requires a different normal vector to accurately represent the intended shape.

There are increasingly complex variations upon this algorithm, depending entirely on how strictly a sharp edge should be honored. At its most complex, a software implementation of the Curved Point Normal triangles implementation can be used to generate intermediary triangles that closely match the surface and control points that would otherwise be generated by the GPU.

While this does solve the problem, it is not an ideal solution, partly because it is nontrivial to implement, but mainly because it breaks the original goal of not having to modify the original geometry.

Back-Face Culling in the Hull Shader

Culling of primitives facing away from the viewer is a classic staple of computer graphics. Modern hardware pipelines will still perform this step (subject to pipeline configuration by the application), and it is largely transparent and oft ignored.

This is interesting with regard to tessellation, due to the ability to amplify geometry (geometry shaders are also interesting for the same reason). With the pipeline configuration used thus far in this chapter, back-face culling will occur before rasterization, exactly as expected, but it is crucial that this is after tessellation has occurred.

Simply put, the hardware may have generated hundreds, if not thousands, of new triangles, only to have them be rejected without making any contribution to the final rendered image. In a high-performance scenario, this wasteful processing is obviously undesirable. It would be far better if it could be avoided entirely.

Direct3D 11 hull shaders do allow this, and before any amplification or complex processing, it is possible to reject an entire patch and cease further processing. Implementing the check to determine if a patch is back facing is the responsibility of the hull shader author, and while this check for a triangle is trivial, it may not be so simple for all tessellation algorithms.

The developer needs to consider the final tessellated output and the possibility that the eventual surface may actually contain elements that aren't entirely back facing. In the context of Curved Point Normal Triangles, the final surface remains relatively close to the underlying control mesh, making this optimization possible.

```
HS_CONSTANT_DATA_OUTPUT hsConstantFunc( InputPatch<VS_OUTPUT, 3> ip, uint
PatchID : SV_PrimitiveID )
{
    HS_CONSTANT_DATA_OUTPUT output;
```

```
    float3 faceNormal
        = normalize
        (
            cross
            (
                ip[2].position - ip[0].position
                , ip[1].position - ip[0].position
            )
        );
    float3 viewDirection = normalize(cameraLookAt - cameraPosition);

    float backFace = sign(0.2 + dot(faceNormal,viewDirection));

    output.Edges[0] = EdgeFactors.x * backFace;
    output.Edges[1] = EdgeFactors.y * backFace;
    output.Edges[2] = EdgeFactors.z * backFace;

    output.Inside = EdgeFactors.w * backFace;

    return output;
}
```

Listing 9.22. Modified hull shader constant function.

Hull shader constant functions that output zero or negative edge tessellation factors will be culled by the pipeline. In Listing 9.22, this is determined by a simple check on the face normal for the incoming triangle. If the inner product of this normal and the current view direction is positive (both point in the same direction), the patch is considered back facing. There are two details to pay attention to with this code. First, the `sign` function which will return either -1.0 or +1.0 and can thus be used to multiply with the edge factors and avoid unnecessary conditionals/branching. Second, the test for being a back face has a "magic" 0.2 factor included, which acts as a simple threshold to stop premature culling of the patch. Due to the curvature of patches, having them culled on the precise turning point can lead to noticeable artifacts in the final image. This factor can be tweaked as necessary, closer to 0.0 will remove more patches, but with increased chance of visual artifacts.

It is important to note that the control point phase of hull shader execution occurs before evaluation of the per-patch constant function. Consequently, the hull shader may perform significant work before it is culled, and further downstream processing can be avoided. Figure 9.47 shows the results of this optimization (right side).

Note that by definition, the difference is hard to detect, but the wireframe rendering appears to be less dense, indicative of the fewer patches being rendered. In this example, the number of domain shader invocations drops from 19,488 to 9,570, and the number of triangles rasterized falls from 16,800 to 8,250—a 51% reduction of both of the unoptimized totals.

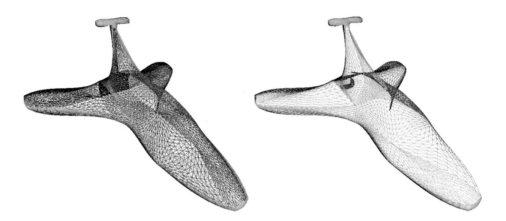

Figure 9.47. Results of using back-face culling.

Given that this optimization has no impact on the final image (unless transparency or similar complex blending operations are being used) it results in a non-trivial reduction in the work that the GPU has to perform.

9.2.2 Conclusion

As previously demonstrated, it is straightforward to implement the original Direct3D-8-era algorithm using Direct3D 11. Given that tessellation is a basic requirement of the specification, this also eliminates the original problem with ATI's TruForm implementation, which only worked on a limited set of hardware. Now, all Direct3D 11 generation hardware supports this algorithm.

The added expressiveness of the Direct3D 11 pipeline allows further extensions to the original algorithm. The main extension to be considered is to break the uniform tessellation across the entire surface of the mesh. As shown in previous images, triangles are generated and distributed evenly across the surface, even where there are only low-frequency variations.

By utilizing mathematics similar to that shown in the back-face culling section, one can dynamically alter the tessellation factors output by the hull shader and weight them to be higher near the edge of the mesh, and in areas of high-frequency variation. However, this requires modification of the input data, since adjacency becomes necessary to avoid cracks between neighboring tiles.

Without any of these changes, the Curved Point Normal Triangles algorithm is a good demonstration of Direct3D 11's new functionality, as well as being a convenient technique to enhance existing triangle-based meshes.

10

Image Processing

The addition of the compute shader to Direct3D 11 is arguably one of its most interesting new features. The compute shader provides a distinctly different technique for using the GPU hardware, and there are many new algorithms that can be implemented with these new abilities. One area in particular that can benefit from the compute shader is the image processing domain. Since the GPU's roots are in image generation and manipulation, it is a natural extension to include general image processing capabilities in the realm of available algorithms. Figure 10.1 demonstrates the results of a blurring filter applied to half of an image.

Figure 10.1. An example image with a blurring filter applied to its right side.

In fact, the compute shader threading model is easily mapped onto an image domain by using the x and y global thread addresses to index the image's pixels (the threading model itself is covered in detail in Chapter 5). This makes it simple to develop an implementation that can use the massively parallel processing capabilities of modern GPUs, and that can provide a fairly significant performance improvement over traditional CPU implementations. In addition, the compute shader provides HLSL atomic operations, group shared memory, device memory resources, and synchronization intrinsic functions to facilitate communication between threads. If these additional tools are used in an appropriate algorithm, even further performance increases can be achieved.

This chapter provides several sample image processing algorithms that attempt to take advantage of the compute shader stage's capabilities. The first algorithm that we inspect is the well-known *Gaussian blur* operation. Due to its wide range of uses and its desirable mathematical properties, it provides a good introduction to image processing with the compute shader. Next, a second filter that has become popular recently is examined—the *bilateral filter*. This is an edge preserving blur filter that can be rather expensive to compute on the CPU, which makes a GPU implementation a very desirable target.

10.1 Image Processing Primer

Before we begin the discussion of implementing image processing algorithms with the compute shader, it would be beneficial to provide a brief introduction to the topic itself. This section will provide some basic entry-level image processing concepts to frame the discussion of the algorithms in this chapter. However, this is by no means an extensive or complete overview of the subject. Many very good image processing books have already been written and are available if the reader would like to investigate the details of the topic further, such as (Gonzalez, 2008).

Image processing is a general term for signal processing operations performed on an image. In our context, the image is likely to be a 2D texture resource that has either been generated in real time or loaded from a file. The input to an image processing algorithm operation is always an image, and the result can be either another image or some information extracted from the input image. In this chapter we will investigate a subset of these algorithms that are used to perform image filtering, but the concepts that we see in this chapter can also be applied to other domains of image processing as well.

10.1.1 Images as Functions

To manipulate the contents of an image, it is helpful to have a more precise definition of what the input image represents. For our purposes, an image is a 2D digital representation of a signal, which is sampled at uniformly spaced sample locations. Each sample in the image is referred to as a *pixel*, and can contain up to four components to represent the sampled signal value at the location of a given pixel. All pixels within an image share the same value format. This description goes along with the familiar 2D grid that is commonly used to show an image, as seen in Figure 10.2, which shows the pixel addressing scheme used by Direct3D 11.

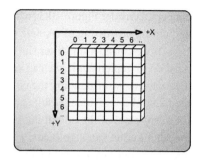

Figure 10.2. A 2D image demonstrating the pixel numbering scheme of Direct3D 11.

It is also noteworthy that there is no restriction to an image being two dimensional. There are many image processing algorithms that can be performed on 1D, 2D, 3D, or even higher dimensional signals. However, since the focus of this book is on real-time rendering, we will restrict the discussion to 2D image processing algorithms.

10.1.2 Image Convolution

With this basic definition of an image, we can look a little deeper into the nature of how image processing algorithms function. Many filtering algorithms are implemented as a convolution between the input image itself and another function, referred to as a *filtering kernel*. For discrete domains like our images, the convolution operation is defined as shown in Equation (10.1):

$$w(x,y) * f(x,y) = \sum_{s=-a}^{a} \sum_{t=-b}^{b} w(s,t)f(x-s,y-t). \qquad (10.1)$$

This mathematical definition may seem somewhat complex at first, but if we take a simplified view of it, we can more clearly see how this operation is applied to images. The convolution is an operation which takes two functions as input and produces an output function. In our image-processing domain, this means that each pixel in the output image is calculated as the summation of the product of the two input images' pixels at various shifted locations. The shifted locations make the two images conceptually move with respect to one another as each pixel is being processed. To visualize this

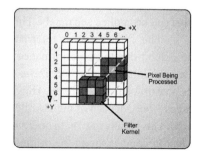

Figure 10.3. The conceptual view of an input image and the filter kernel during processing of a single pixel.

concept, consider Figure 10.3, where the original image is shown, as well as the filter kernel, during processing of a single pixel location.

If we consider this filter shifting operation in the context of what our shader program must do to implement it, we can see that for every pixel in the output image, we must load the image data surrounding the current pixel, apply the filter image to the loaded data, and then sum the individual results and store the value in the output image. This is repeated for every pixel of the output image. With this in mind, we can clearly see how the size of the filter kernel can significantly impact the computational cost of performing a filtering operation, since each pixel requires a number of calculations proportional to the filter size.

This interpretation is relatively simple to visualize, and provides a sufficient view of image processing for the implementation discussions later in the chapter. However, this description is only valid for a subset of image processing filters. Many other filters perform different types of calculations over a pixel's surrounding neighborhood. Fortunately, the same basic concept is still used in the other algorithms as well, in which a small neighborhood of pixels surrounding the current pixel is loaded, some calculations are performed on them, and the results of those calculations are used to determine the value that will be stored in the output image. We will see this pattern in each of the algorithms that we implement in this chapter.

Of course, there is a significant amount of additional mathematical theory and analysis that will be useful to a developer of image filtering algorithms, including frequency domain representations, the various properties of filters in both spatial and frequency domains, and alternative filter implementations and their trade-offs. As mentioned earlier, the reader is invited to further explore these and other details in a complete image processing text.

10.1.3 Separable Filters

There is one final point to discuss before moving on to our sample algorithm implementations. Some filter kernels can be decomposed from a single 2D kernel into two 1D kernels that can be applied separately. This is depicted graphically in Figure 10.4.

This filter kernel is referred to as *separable*, a very important property for a filter to exhibit due to its performance implications. In general, for a 2D filter of size $p \times q$ to be applied to an image of size $m \times n$, the number of operations that need to be performed is proportional to $p*q*m*n$. However, if a filter is separable, it can be performed in two steps—one for each of the 1D filter kernels that the original 2D filter kernel is decomposed into. This results in $p*m*n$ operations in the first step, and $q*m*n$ operations in the second step. In comparison to the original 2D filter kernel, a separable filter reduces the number of operations from $p*q*m*n$ to $(p+q)*m*n$. This is a significant reduction in calculations, especially when the values of p and q are relatively large. We will utilize this separability property in our sample algorithms in this chapter.

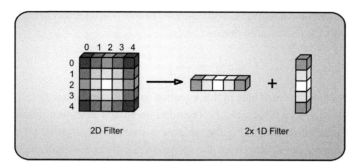

Figure 10.4. A 2D filter kernel decomposed into two 1D filters.

10.2 Gaussian Filter

The first image processing algorithm that we will investigate is the *Gaussian filter*. This filter is one of the most widely known, due to the wide range of applications that can use it, and its relatively low computational cost. The filter itself is named after Carl Friedrich Gauss, a brilliant German mathematician (1777-1885) who contributed to many different areas. The term *Gaussian distribution* is often used to refer to a normal distribution in statistics, which is where the name has become associated with an image processing filter.

The Gaussian filter produces a blurred version of the input image, and is hence referred to as a *low-pass filter* in signal processing terms. This means that the lower-frequency content of the image (the portions that don't change rapidly) is mostly preserved, while the higher-frequency content (sharp edges or transitions) is attenuated.

10.2.1 Theory

The filter receives its name from the use of the Gaussian function for creating the filter kernel weights. The Gaussian function is shown in Equation (10.2) for two dimensional inputs, where x and y are the input variables, and σ is a constant factor determined by the developer(which will be clarified shortly). This equation has two main portions—the exponential, and a constant factor applied to it in the beginning of the equation. The exponent in the exponential term calculates a measure of the distance that an input sample location is from the origin by squaring both input parameters. Since these factors will always result in positive values, the final sign of the exponent will always be negative. This results in the exponential term always producing a value between 1 and 0, with 1 being produced when x and y are both zero, and the value falling off to 0 at some characteristic determined by the σ parameter. The constant term in the front of the equation simply scales the magnitude of the exponential term to maintain several desirable properties, such as having the integral of the function over the entire input domain be 1:

$$g(x, y) = \frac{1}{2\pi\sigma^2} e^{\frac{-(x^2+y^2)}{2\sigma^2}}.$$ (10.2)

The only changing variables in this equation are the x and y locations of the sample. So, the output filter weightings can be regarded as spatial in nature, since the only input into the equation for generating the weights is based on the spatial orientation of the samples. In general, the farther from the center of the filter a sample is, the smaller its resulting weighting on the overall result of the filtered pixel will be. The shape of the filtering kernel can be modified with the constant sigma parameter. Depending on the chosen value of sigma, the amount of blurring can be increased or decreased. The larger the sigma value, the greater the blur that is applied to an image. In Figure 10.5, we can see how an increasingly large sigma value will produce a progressively greater amount of blurring by examining the shape of the filter kernel.

Equation (10.2) can be used to produce the desired set of filter weights for a filter kernel of more or less any size. Since we can't use infinitely sized kernels, we would typically choose an appropriate filter size for the operation being implemented, to produce the appropriate image quality. With a truncated filter kernel size, we can eliminate the constant term from Equation (10.2), since it assumes an infinite input domain. Instead, we will calculate the filter weights at each filter kernel location and then renormalize their total weight by dividing each filter weight by the sum of all of the weights. This maintains the integral-of-1 property with a more computationally friendly filter kernel size. This is required to ensure that the total energy contained within the image signal remains the same before and after the blurring is performed.[1]

The Gaussian filter is used in many areas. Typical uses include performing image blur effects such as blooming, and performing up- and down-sampling operations. In addition, many non-rendering applications also use the Gaussian filter, such as a blur operation being performed prior to a dilation operation in a character recognition system.

10.2.2 Implementation Design

The basic implementation of the Gaussian filter is relatively simple. The desired filter kernel is filled in with values generated by the Gaussian function with a sigma value chosen for the desired level of blurring. This is typically done at design time, so that the filtering kernel is not recalculated for every pixel being processed; but if a dynamic filter size is needed,

---

[1] It is possible to simultaneously achieve other effects by changing this total filter weight as well. For example, if the kernel weights sum to a value less than or greater than 1.0, the resulting image will be either darkened or brightened, respectively.

it is not mandatory to pre-calculate the filter weights. Once the filtering kernel is available, we can easily use the compute shader to invoke a single thread for each pixel to be processed on the input image. We will first consider a naïve, brute-force approach, before moving on to a more elegant solution.

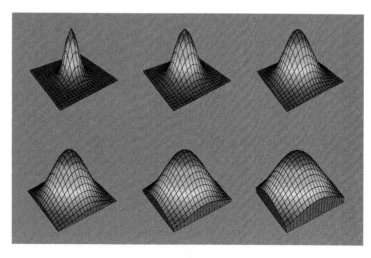

Figure 10.5. Visualizations of various filter kernels with increasing values of sigma, which will produce increasingly blurred output images.

One other consideration must be made for when implementing filter weights. When it is near an edge of the input image, the filter will sample image locations that are actually outside of the image. This must be considered and handled appropriately for a given situation, and is typically compensated for in one of several ways. The exterior samples can be clamped to the edge of the image, which will essentially make the edge pixels a little bit sharper than the interior pixels after the blurring process. Another way to handle this would be to eliminate samples that fall outside of the image. To eliminate a sample, the algorithm must not only skip the addition of the weight filter sample, but must also reduce the overall sample amount. Since we choose the filter weights to add up to 1, we are performing an implicitly neutral operation on the image in the regular, interior pixel cases. If we eliminate some samples, the overall filter weight will add up to less than 1, and thus the resulting value must be renormalized once again by dividing the output value by the new total filter weights. This adds some complexity to the calculation process, and must be implemented with care to ensure that the performance of the algorithm is not significantly reduced. The implementations demonstrated in this chapter simply neglect this effect, since the focus of the chapter is to use the computer shader in an efficient manner.

Brute Force Approach

To invoke the compute shader on each thread, we must select a thread group size that is within the upper limits of the thread count (which is less than 1024 total threads), but that can be invoked in a dispatch size that can cover the entire input image. For example, if we choose the thread groups to be of size [32,32,1], we are within the thread count limit, and our dispatch call can choose the appropriate number of thread groups to request based on

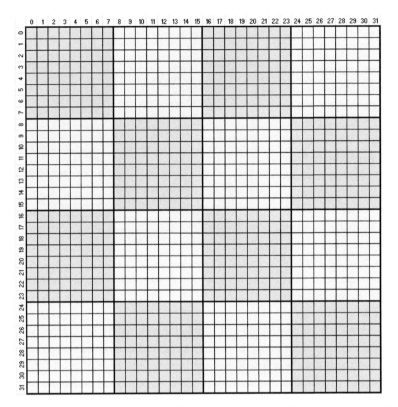

Figure 10.6. Processing an image with a square tile sized thread group, with an example size of 8×8 tiles.

the size of the image to be filtered. If we are processing a 640×480 image, the dispatch size would be [20,15,1]. This tile-based thread group size is demonstrated in Figure 10.6. In general, choosing the thread group size to be a common factor of the available render target sizes is a good initial choice, which will allow the filter shader program to be reused for multiple image sizes.

For the purposes of our first sample, we will assume a 7×7 filter kernel size. The filter is provided as static constants in the shader file, which can be accessed as a two dimensional array of values. Within the shader program, we will use the dispatch thread ID to allow each thread to choose which pixel it will be processing. This is trivial to perform, since the dispatch thread ID will span a range that exactly matches the size of image. Indeed, this is done by design to ensure that we can use this thread addressing scheme! Once each thread is aware of which pixel it should be processing, we can read the input image data for each of the pixels that are needed in our filter kernel. Each pixel value is multiplied by the corresponding filter kernel weight, and all of these products are summed to produce the resultant pixel value for the output image. The output value is stored in the output texture at the pixel location indicated by the dispatch thread ID in much the same way that was used to read the appropriate data from the input image. This initial implementation is shown in Listing 10.1 and visualized for a single pixel in Figure 10.7.

```
// Declare the input and output resources
Texture2D<float4>   InputMap : register( t0 );
RWTexture2D<float4> OutputMap : register( u0 );

// Group size
#define size_x 32
#define size_y 32

// Declare the filter kernel coefficients
static const float filter[7][7] = {
    0.000904706, 0.003157733, 0.00668492, 0.008583607, 0.00668492,
0.003157733, 0.000904706,
    0.003157733, 0.01102157, 0.023332663, 0.029959733, 0.023332663,
0.01102157, 0.003157733,
    0.00668492, 0.023332663, 0.049395249, 0.063424755, 0.049395249,
0.023332663, 0.00668492,
    0.008583607, 0.029959733, 0.063424755, 0.081438997, 0.063424755,
0.029959733, 0.008583607,
    0.00668492, 0.023332663, 0.049395249, 0.063424755, 0.049395249,
0.023332663, 0.00668492,
    0.003157733, 0.01102157, 0.023332663, 0.029959733, 0.023332663,
0.01102157, 0.003157733,
    0.000904706, 0.003157733, 0.00668492, 0.008583607, 0.00668492,
0.003157733, 0.000904706
};

// Declare one thread for each texel of the current block size.
[numthreads(size_x, size_y, 1)]

void CSMAIN( uint3 DispatchThreadID : SV_DispatchThreadID )
{
    // Offset the texture location to the first sample location
    int3 texturelocation = DispatchThreadID - int3( 3, 3, 0 );

    // Initialize the output value to zero, then loop through the
    // filter samples, apply them to the image samples, and sum
    // the results.
    float4 Color = float4( 0.0, 0.0, 0.0, 0.0 );

    for ( int x = 0; x < 7; x++ )
        for ( int y = 0; y < 7; y++ )
            Color += InputMap.Load( texturelocation + int3(x,y,0) ) * filter[x][y];

    // Write the output to the output resource
    OutputMap[DispatchThreadID.xy] = Color;
}
```

Listing 10.1. The brute force approach to implementing the Gaussian filter.

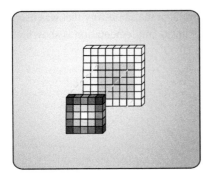

Figure 10.7. A visualization of applying the filter kernel to a single pixel.

This implementation is perfectly adequate for producing a blurred output image in the mathematically correct way. With the highly parallel nature of modern GPUs, this operation is likely to be performed quite quickly. However, there is almost always a need to perform more operations in every rendered frame, to allow either for better image quality or for higher throughput of an operation if it used in a GPGPU application. Therefore, we will attempt to improve the performance of our algorithm, while still producing the same output image.

Separable Gaussian Filter

In our brief image processing primer, we saw that some filter kernels are separable. The Gaussian filter is in fact a separable filter, which we can exploit to gain a significant performance advantage. When a 2D filter is decomposed into two 1D filters, this requires us to execute two processing passes on our image —one for each of the two 1D filters. In the case of the Gaussian filter, both of the 1D filters are simply the cross section of the 2D filter through its center. In the 7×7 filter from our example above, the decomposition into two 1D filters will produce a 1×7 filter and a 7×1 filter. This can be thought of as producing filters in both the x- and y-directions, due to the shape of the resulting filters. This is shown graphically in Figure 10.8.

The first processing pass will read the input image, apply the first filter, and produce an intermediate image result which must be stored in a new texture resource. The second processing pass will read the intermediate result, apply the second filter, and produce the final filtered image. This means that we will require an additional texture resource, which means that additional memory is used by this modified algorithm. However, the potential performance improvement is typically a worthy tradeoff for the extra memory consumption.

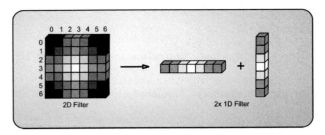

Figure 10.8. Decomposing our 7×7 filter into two 1D filters.

Since we will be modifying the filter kernel that we will be executing, it makes sense to reevaluate the threading setup that we have used in the previous implementation. Instead of using a square thread group size, it makes more sense to flatten out our thread groups to match the shape of the processing kernel being used in each pass. Thus, for the first pass we will use a thread group size of [640,1,1], and the second pass will use a thread group size of [1,480,1]. This will allow us to use the group shared memory to even further reduce the required device memory bandwidth.

In our previous implementation, each thread read all of the input texture values it needed to calculate its own output value. For a 7×7 filter, this amounts to 49 individual values to read from the input image per thread. In a separable implementation, we can expect each thread to only need to read the data for either a 1×7 or a 7×1 filter, for a total of $7 + 7 = 14$. This is already a significant reduction of explicit memory reads, although the effective reduction may be somewhat smaller due to the GPU's texture data caching helping the naïve implementation. In any case, we can further reduce the number of reads from the input texture resource by using the GSM. If each thread reads its own input pixel data and then stores it in the group shared memory for all the threads in the thread group to use, the effective number of device memory reads per thread is reduced from $7 + 7$ to $1 + 1$! Of course we are adding some overhead to this implementation by writing to and reading from the group shared memory, but in general, this should be a performance improvement over performing many more read operations from device memory. After all of the threads have loaded their data into the GSM, we perform a group memory barrier with thread sync to ensure that all of the needed data has been written to the GSM before moving on with the filtering operations. The updated compute shader program for the horizontal filter is shown in Listing 10.2. The vertical version is omitted, since it is identical to the shown version except that it samples in a vertical pattern and declares a different amount of group shared memory.

```
// Declare the input and output resources
Texture2D<float4>   InputMap : register( t0 );
RWTexture2D<float4> OutputMap : register( u0 );

// Image sizes
#define size_x 640
#define size_y 480

// Declare the filter kernel coefficients
static const float filter[7] = {
    0.030078323, 0.104983664, 0.222250419, 0.285375187, 0.222250419,
0.104983664, 0.030078323
};

// Declare the group shared memory to hold the loaded data
groupshared float4 horizontalpoints[size_x];
```

```
// For the horizontal pass, use only a single row of threads
[numthreads(size_x, 1, 1)]

void CSMAINX( uint3 DispatchThreadID : SV_DispatchThreadID )
{
    // Load the current data from input texture
    float4 data = InputMap.Load( DispatchThreadID );

    // Stor the data into the GSM for the current thread
    horizontalpoints[DispatchThreadID.x] = data;

    // Synchronize all threads
    GroupMemoryBarrierWithGroupSync();

    // Offset the texture location to the first sample location
    int3 texturelocation = DispatchThreadID - int3( 3, 0, 0 );

    // Initialize the output value to zero, then loop through the
    // filter samples, apply them to the image samples, and sum
    // the results.
    float4 Color = float4( 0.0, 0.0, 0.0, 0.0 );

    for ( int x = 0; x < 7; x++ )
        Color += horizontalpoints[texturelocation.x + x] * filter[x];

    // Write the output to the output resource
    OutputMap[DispatchThreadID.xy] = Color;
}
```

Listing 10.2. The horizontal pass compute shader listing for our separable implementation.

One important point to note in this listing is that the thread group size has been modified to reflect our desired 1D shape. Since the limit on thread group size is currently 1024, it is possible that one horizontal line will not fit in a single thread group if the input image is sufficiently large. However, this can be overcome by simply modifying the dispatch size to perform each pass with two thread groups for each row or column to be processed. Also, note that the group shared memory is declared such that it can hold the appropriate number of pixels for the current thread group size. Finally, notice that we use the GroupMemoryBarrierWithGroupSync() function to ensure that all threads have executed to this point and that the group memory accesses have been completed before continuing with the filtering process.

Data Sharing with GSM

In the separable Gaussian implementation, we used the group shared memory to significantly reduce the number of device memory accesses needed to perform the desired filtering operation. By using the GSM as a memory cache, we can trade device memory

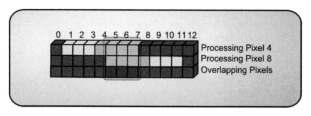

Figure 10.9. A visualization of overlapping calculations in neighboring pixels.

accesses for increasing the number of GSM accesses, plus a memory barrier. However, the group shared memory can be used to share more than cached resource data. We can further modify our current implementation to allow threads to share the results of some calculations to reduce the overall computational burden on the GPU. This may become beneficial after optimizing an algorithm to reduce memory accesses—if the computational portion of the algorithm becomes proportionally large enough, it can be helpful to attempt to minimize even further the number of mathematic operations.

In the case of the Gaussian filter, we can take advantage of the fact that the two individual 1D processing passes use filter weights that are symmetric. This, coupled with our row-and-column based threading scheme, allows us to share intermediate calculations between two different pixels. To aid in explaining this possibility, Figure 10.9 shows the calculations needed for two pixels that are near each another. The key to this concept is that any two pixels equidistant from a pixel between them will both perform the same calculation on that center pixel.

To avoid having duplicate calculations, we can have each thread precalculate the weighted versions of its own pixel value and store it in the GSM before the memory barrier is performed. The precalculated values can then be read by whichever thread needs to use them, which should effectively reduce the number of multiplications needed by a factor of roughly 2. This requires the use of additional group shared memory, as well as additional read and writes to this additional memory. Figure 10.10 shows the layout that will be used to store the additional precalculated values; the modified compute shader program is shown in Listing 10.3.

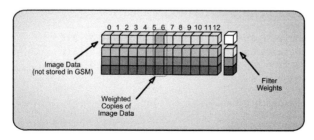

Figure 10.10. The group shared memory layout for caching shared values.

```
// Declare the input and output resources
Texture2D<float4>   InputMap : register( t0 );
RWTexture2D<float4> OutputMap : register( u0 );

// Image sizes
#define size_x 640
#define size_y 480

// Declare the filter kernel coefficients
static const float filter[7] = {
    0.030078323, 0.104983664, 0.222250419, 0.285375187, 0.222250419,
    0.104983664, 0.030078323
};

// Declare the group shared memory to hold the loaded and calculated data
groupshared float4 horizontalpoints[size_x][3];

// For the horizontal pass, use only a single row of threads
[numthreads(size_x, 1, 1)]

void CSMAINX( uint3 DispatchThreadID : SV_DispatchThreadID )
{
    // Load the current data from input texture
    float4 data = InputMap.Load( DispatchThreadID );

    // Stor the data into the GSM for the current thread
    horizontalpoints[DispatchThreadID.x][0] = data * filter[0];
    horizontalpoints[DispatchThreadID.x][1] = data * filter[1];
    horizontalpoints[DispatchThreadID.x][2] = data * filter[2];

    // Synchronize all threads
    GroupMemoryBarrierWithGroupSync();

    // Offset the texture location to the first sample location
    int3 texturelocation = DispatchThreadID - int3( 3, 0, 0 );

    // Initialize the output value to zero, then loop through the
    // filter samples, apply them to the image samples, and sum
    // the results.
    float4 Color = float4( 0.0, 0.0, 0.0, 0.0 );

    Color += horizontalpoints[texturelocation.x + 0][0];
    Color += horizontalpoints[texturelocation.x + 1][1];
    Color += horizontalpoints[texturelocation.x + 2][2];
    Color += data * filter[3];
    Color += horizontalpoints[texturelocation.x + 4][2];
    Color += horizontalpoints[texturelocation.x + 5][1];
    Color += horizontalpoints[texturelocation.x + 6][0];

    // Write the output to the output resource
    OutputMap[DispatchThreadID.xy] = Color;
}
```

Listing 10.3. The cached Gaussian implementation.

While this change to the algorithm does indeed reduce the number of arithmetic operations performed by each thread, it also increases the number of memory access operations as well. The group shared memory is supposed to be a very fast memory area, but there is still some performance penalty for accessing it. In addition, the arithmetic operations we are attempting to minimize in this case are simple multiplications, which modern GPUs can perform quite quickly. This may mean that any potential performance gains will be relatively small. This type of calculation would be more beneficial if we were replacing longer computational strings or more expensive instructions. In the end, the final results will likely vary by the GPU being used and its individual performance characteristics for arithmetic and memory access operations.

10.2.3 Conclusion

The Gaussian filter is a very widely used filter, which provides an adjustable amount of blurring and can be used with a variety of different filter sizes. This allows the filter's performance to be adjusted to the given situation. By using the separable nature of the Gaussian filter, we greatly reduced the number of memory access and arithmetic operations, and then further reduced the number of memory accesses by using the group shared memory as a customized memory cache. We even further reduced the number of arithmetic operations by precalculating some filter elements that are shared by more than one pixel. This type of optimization trades arithmetic operations for memory access operations, and can be beneficial in some cases, when sufficiently heavy arithmetic operations are being replaced. However, it may not always produce a net performance improvement. This means that the algorithm should be carefully designed to use the GSM only when it is beneficial to do so.

10.3 Bilateral Filter

The second filter algorithm we will investigate is the *bilateral filter*. This filter was originally proposed in (Tomasi, 1998). It is similar in nature to the Gaussian filter and provides another form of a blurring filter. However, the bilateral filter blurs an image while still preserving the sharp edges and object boundaries in the image content. This is in contrast to the Gaussian filter, which blurs the complete image regardless of the content. If a Gaussian filter is used to blur the image several times in succession, the image can become washed out, and it can become difficult to discern between objects. Figure 10.11 demonstrates the difference between the two filter outputs after performing several passes of the filter.

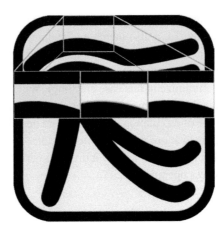

Figure 10.11. A comparison of the original image, Gaussian filtered (center) and the bilateral filtered (right).

Due to its edge-preserving properties, the bilateral filter can be used to perform a different class of functions than the Gaussian filter. For example, algorithms that use stochastic methods to sample the properties of a complex object often result in a noisy output, due to the randomized nature of the sampling technique. The bilateral filter can be used to reduce or minimize the noise in the signal, while still retaining the important boundaries that define various objects in the image. A common example of this is using a bilateral filter to smooth the results of screen space ambient occlusion calculations.

10.3.1 Theory

The bilateral filter shares some properties with the Gaussian filter. With the bilateral filter, weights are based on the spatial distances from the center of the filter, as with the Gaussian filter, but an additional factor is added to the weight calculation. The filter weights are also scaled by the relative difference in color values between the current pixel being processed and the sample pixel. If the two colors are somewhat similar, the scaling of the weight leaves the weight value mostly unchanged. However, if there is a large difference between the two color values, the weight is scaled to a smaller value, which has less influence on the overall result of the pixel. Thus, the bilateral filter modifies its weights based not only on the spatial arrangement of samples, but also on the intensity values contained within the image. The equation for calculating the weights of each individual sample is shown in Equation (10.3):

$$BF[I_p] = \frac{1}{W_p} \sum_{q \in S} G_s(\| p - q \|) G_r(I_p - I_q) I_q. \tag{10.3}$$

As seen in Equation (10.2) (Paris), the bilateral filter weights are a product of a Gaussian function based on the spatial distance (indicated by G_s), and a Gaussian function based on the color delta value (indicated by G_r). This dependence on the intensity values is the key to performing a filtering operation that preserves edges in the image. However, since every pixel in the image has a potentially different value, the weights must be calculated dynamically for every sample that is used to produce an output pixel value. This is a relatively large difference from the Gaussian filter, where we could easily precalculate the desired filter weight based on the desired sigma value. In addition, since the weights change from pixel to pixel, the bilateral filter is a nonlinear filter, which also means that it is generally not separable. This can have profound performance implications when sufficiently large filter sizes are used.

Once all of the weights have been calculated, the resulting output pixel value is determined in the same way as in as the Gaussian filter. The appropriate neighboring pixels are sampled, and then scaled by the current pixel's corresponding weighting value. All of these weighted samples are then summed to produce an output pixel value. Another very important point to consider when calculating the bilateral filter weights is that the content of the image may produce filter weights that add up to significantly less than 1. This can

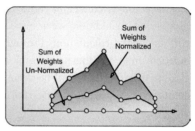

Figure 10.12. Renormalizing of the bilateral filter weights.

produce an abnormally darkened area in an area that is otherwise relatively bright, which would appear out of place. This is in contrast to Gaussian filter weights, which are chosen to always add up to a total of 1. To compensate for this fluctuation in total weight, the total contributions from all of the pixels must be renormalized with one another. This is done by trivially dividing the resulting pixel value by the sum of the weights themselves. This is depicted graphically in Figure 10.12.

Another complication to using this filter is the fact that there are three independent channels in a normal color texture. This means that the intensity parameter that is used for the Gaussian filter weighting must either be performed for all three channels independently, or some mechanism must be used to convert from a three-dimensional color to a one-dimensional equivalent value prior to calculating the weight values. Both possibilities will increase the number of calculations needed to determine the weight values, and the latter option will reduce the image quality of the filter somewhat, by approximating the color space with a single scalar value. In principle, since the GPU is capable of performing vector style calculations, it should be possible to calculate each of the color channels independently with the same number of instructions.[2]

---

[2] In the past, all GPU architectures were vector-register based. However, this is not the case any longer, with recent NVIDIA architectures using a scalar execution pipeline (while recent AMD architectures remain vector based). Thus, the comments presented here about vector operations may or may not apply to a particular GPU.

10.3.2 Implementation Design

The basic implementation setup for the bilateral filter is very similar to that of the Gaussian filter. It will execute in the compute shader, with the input image bound with a shader resource view, and the resulting image will be bound with an unordered access view so the shader program can write the appropriate data to it. We will consider how a simple brute force implementation looks, and then later develop a more efficient approach.

Brute Force Approach

The straightforward technique is quite similar to the Gaussian method, with the exception that the color-based weight value must be calculated and applied to each sample used in the summation. In addition, we assume that the input color image will use all four channels and will use float4 calculations throughout the shader to allow each channel to be processed independently. For this naïve implementation, each pixel will process a single output pixel, and the thread groups will be arranged in the 32×32 sized groups once again. The implementation is shown in Listing 10.4.

```
// Declare the input and output resources
Texture2D<float4>   InputMap : register( t0 );
RWTexture2D<float4> OutputMap : register( u0 );

// Group size
#define size_x 32
#define size_y 32

// Declare the filter kernel coefficients
static const float filter[7][7] = {
    0.000904706, 0.003157733, 0.00668492, 0.008583607, 0.00668492,
0.003157733, 0.000904706,
    0.003157733, 0.01102157, 0.023332663, 0.029959733, 0.023332663,
0.01102157, 0.003157733,
    0.00668492, 0.023332663, 0.049395249, 0.063424755, 0.049395249,
0.023332663, 0.00668492,
    0.008583607, 0.029959733, 0.063424755, 0.081438997, 0.063424755,
0.029959733, 0.008583607,
    0.00668492, 0.023332663, 0.049395249, 0.063424755, 0.049395249,
0.023332663, 0.00668492,
    0.003157733, 0.01102157, 0.023332663, 0.029959733, 0.023332663,
0.01102157, 0.003157733,
    0.000904706, 0.003157733, 0.00668492, 0.008583607, 0.00668492,
0.003157733, 0.000904706
};

// Declare one thread for each texel of the current block size.
[numthreads(size_x, size_y, 1)]
```

```
void CSMAIN( uint3 GroupID : SV_GroupID, uint3 DispatchThreadID :
SV_DispatchThreadID, uint3 GroupThreadID : SV_GroupThreadID, uint GroupIndex
: SV_GroupIndex )
{
    // Offset the texture location to the first sample location
    int3 texturelocation = DispatchThreadID - int3( 3, 3, 0 );

    // Each thread will load its own depth/occlusion values
    float4 CenterColor = InputMap.Load( DispatchThreadID );

    // Range sigma value
    const float rsigma = 0.051f;

    float4 Color = 0.0f;
    float4 Weight = 0.0f;

    for ( int x = 0; x < 7; x++ )
    {
        for ( int y = 0; y < 7; y++ )
        {
            // Get the current sample
            float4 SampleColor = InputMap.Load( texturelocation + int3( x, y, 0 ) );

            // Find the delta, and use that to calculate the range weighting
            float4 Delta = CenterColor - SampleColor;
            float4 Range = exp( ( -1.0f * Delta * Delta )
                                / ( 2.0f * rsigma * rsigma ) );

            // Sum both the color result and the total weighting used
            Color += SampleColor * Range * filter[x][y];
            Weight += Range * filter[x][y];
        }
    }

    // Store the renormalized result to the output resource
    OutputMap[DispatchThreadID.xy] = Color / Weight;
}
```

Listing 10.4. The brute force bilateral filter implementation.

We can see from Listing 10.4 that the color-weighting factor is based on the delta value between the center pixel and the sample pixel, which is then used as the input parameter to a Gaussian function. The spatial weights are precalculated in the same way we saw in the Gaussian filter implementation, but they could just as easily be converted to be dynamically calculated, or even provided in a constant buffer by the application. All of the samples are looped through to build the overall weighted combination, and the total combination of the weights that are applied to the samples is summed in the Weight variable. This is then used to renormalize the summed output combination of the samples, producing a final result that is written to the output resource through the unordered access view.

Like to the Gaussian brute force approach, this implementation performs a number of operations roughly proportional to the number of samples used in the filter kernel. For our 7×7 filter example, this means that the primary portion of our algorithm is executed 49 times, with each loop performing a device memory resource read, followed by some computations on the data; and finally, the result is written to the device memory resource. Clearly, this is not an optimal solution, since memory accesses can introduce significant time delays while the algorithm waits for the requested data to be fetched by the GPU memory system.

Separable Bilateral Filter

In our Gaussian filter implementation, we used the filter's separable nature to reduce the amount of work required to process a pixel. This allows us to perform the algorithm in two steps, which not only reduces the number of operations, but also lets us use the group shared memory to reduce the number of device memory accesses even further. Overall, this provides a significant performance improvement over the naïve implementation. Ideally, we would like to use a similar technique to reduce the number of calculations and memory accesses needed for the bilateral filter as well.

However, as we mentioned earlier, the bilateral filter is nonlinear and is generally not separable. Strictly speaking, this means that we would not be able to perform the same style of optimizations with the bilateral filter. With this in mind, in many cases, it is still possible to use a separable implementation, even though it is not mathematically correct. The resulting output image will not be identical to the basic implementation, but since this filter performs a blurring operation, it is less noticeable if the results are not precisely the same as those of the true algorithm. The performance benefits of using a separable filter generally outweigh the imperfect results, and we will make this tradeoff in the next implementation.

After deciding to use a separable version of the filter, we will set up the algorithm in much the same way as the separable Gaussian filter, with the exception of the new per-sample calculations that are required for the bilateral filter. We can also use the group shared memory to cache the required color values, just as we have seen in the Gaussian implementation. The remainder of the filter remains the same, with the exception that we must execute two passes to perform the algorithm now, instead of a single pass, as before. The updated filter implementation is shown in Listing 10.5.

```
// Declare the input and output resources
Texture2D<float4>   InputMap : register( t0 );
RWTexture2D<float4> OutputMap : register( u0 );

// Image sizes
#define size_x 640
#define size_y 480
```

```
// Declare the filter kernel coefficients
static const float filter[7] = {
    0.030078323, 0.104983664, 0.222250419, 0.285375187, 0.222250419,
    0.104983664, 0.030078323
};

// Declare the group shared memory to hold the loaded data
groupshared float4 horizontalpoints[size_x];

// For the horizontal pass, use only a single row of threads
[numthreads(size_x, 1, 1)]

void CSMAINX( uint3 DispatchThreadID : SV_DispatchThreadID )
{
    // Load the current data from input texture
    float4 CenterColor = InputMap.Load( DispatchThreadID );

    // Stor the data into the GSM for the current thread
    horizontalpoints[DispatchThreadID.x] = CenterColor;

    // Synchronize all threads
    GroupMemoryBarrierWithGroupSync();

    // Offset the texture location to the first sample location
    int3 texturelocation = DispatchThreadID - int3( 3, 0, 0 );

    // Range sigma value
    const float rsigma = 0.051f;

    float4 Color = 0.0f;
    float4 Weight = 0.0f;

    for ( int x = 0; x < 7; x++ )
    {
        // Get the current sample
        float4 SampleColor = horizontalpoints[texturelocation.x + x];

        // Find the delta, and use that to calculate the range weighting
        float4 Delta = CenterColor - SampleColor;
        float4 Range = exp( ( -1.0f * Delta * Delta )
                            / ( 2.0f * rsigma * rsigma ) );

        // Sum both the color result and the total weighting used
        Color += SampleColor * Range * filter[x];
        Weight += Range * filter[x];
    }

    // Store the renormalized result to the output resource
    OutputMap[DispatchThreadID.xy] = Color / Weight;
}
```

Listing 10.5. A separable implementation of the bilateral filter.

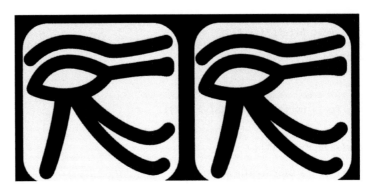

Figure 10.13. A comparison of the standard implementation and the separable implementation of the bilateral filter.

The results from this version of the algorithm are significantly faster than the brute force method. In this case, the resulting image quality is similar to what it would have been and does not produce objectionable artifacts in the output image. Figure 10.13 shows a sample image with both the result of the original filter and the approximated result from the separable implementation in Figure 10.13.

10.3.3 Conclusion

The bilateral filter is a very important tool to have in any image processing library. Its ability to perform a blurring operation that mostly preserves strong edges makes it quite versatile. It has applications in post-processing, with a particularly strong case for being used to smooth the results of stochastic, or randomized, sampling techniques. While the brute force implementation produces a slightly higher image quality result, by use the separable filter techniques, we can gain a big performance advantage and perhaps even use larger filter sizes, because of the computational savings.

11

Deferred Rendering

11.1 Overview

As graphics hardware has become more and more generic and programmable, applications employing real-time 3D graphics have begun to explore alternatives to traditional rendering pipelines, in order to avoid their disadvantages. One of the most popular techniques currently in use is known as *deferred rendering*. This technique is primarily geared toward supporting large numbers of dynamic lights without a complex set of shader programs (see Figure 11.1). It has been successfully integrated into engines made by Crytek (Mittring, 2009), Naughty Dog (Balestra & Engstad, 2008) (Hable, 2010), and Guerrilla Games (Valient, 2007).

This chapter provides a brief introduction of deferred rendering that covers its advantages and disadvantages, while the following sections explore a basic Direct3D 11 implementation of both traditional deferred rendering and *light pre-pass deferred rendering*. The final three sections take things a step further and explore how Direct3D 11 features can be exploited to improve both the quality and performance of a deferred renderer.

11.1.1 Problems with Forward Rendering

One of the most important aspects in designing a renderer is determining how to handle lighting. Lighting is extremely important, since it deals with calculating the intensity and color of light reflecting off surface geometry at a particular point, which becomes the most

Figure 11.1. The final output and intermediate g-buffer textures of a deferred renderer. Model courtesy of Radioactive Software.

significant factor in determining the final color of a pixel. Typically, it consists of the following steps:

1. Determine which lights need to be applied to a particular pixel, based on the light type and attenuation properties

2. Evaluate a *bidirectional reflection distribution function (BRDF)*[1] for each pixel, using material properties (*albedo*), surface properties (*normal* and *position*), and the properties of each light that affects the pixel.

3. Apply the results of a light visibility calculation, or in other words, determine whether or not a pixel is "in shadow."

---

[1] A *BRDF* is an equation used to evaluate the amount of light that reflects off a surface. Typically, a BRDF will use several input parameters, including the properties of the light, the material properties of the surface, and the position of the eye or camera. (Zink, Hoxley, Engel, Kornmann, & Suni)

The traditional way to handle this process is known as *forward rendering*. With forward rendering, geometry is rendered with various surface properties stored in the vertex attribute data, in textures, and in constant buffers. Then, for each pixel that is rasterized for the input geometry, the material data is used to apply the lighting equation for one or more lights. The results of evaluating the lighting equation are then output for each pixel, and possibly summed with the previous contents of the render target. This approach is straightforward and intuitive, and it was the overwhelmingly dominant approach for real-time 3D graphics before the advent of programmable graphics hardware. These fixed-function GPUs (and the earlier Direct3D APIs for using them) supported three different light types, as follows:

1. *Point lights*—these lights have an equal contribution in all directions and are attenuated based on the distance from the light source to the surface being lit. As a result, they have a spherical area of effect.

2. *Spot lights*—these have a direction associated with them, and the light contribution is attenuated based on the angle between the surface and the light direction. Consequently, spot lights have a conical area of effect.

3. *Directional lights*—unlike the other two light sources, this type is considered a "global" light source. This is because the position of the surface is not taken into account when determining the light contribution. In fact, directional light sources have no position, and are instead defined by only a direction. The light is treated as if it were a point light infinitely far away from the surface, in the direction specified for the light.

As real-time rendering engines have scaled to accommodate programmable hardware and more complex scenes, these basic light types have mostly remained, due to their flexibility and ubiquity. However, the means in which they are applied have begun to show to show their age. This is because with modern hardware and APIs, forward rendering exhibits several key weaknesses when it comes supporting large numbers of dynamic light sources.

The first disadvantage is that the application of light sources is tied to the granularity at which scene geometry is drawn. In other words, when we enable a light source, we must apply it to all the geometry that is rasterized during any particular draw call. At first it may not be obvious why this is a bad thing, especially for simple scenes with few lights. However, as the number of lights scales up, it becomes important to apply a light source selectively, to reduce the number of lighting calculations performed in the pixel shader. But since changing which lights are applied can only be done in between draw calls, we're limited in how much we can cull lights that aren't needed. So for cases where a light only affects a small portion of the geometry rasterized by a draw call, we must still apply that light to all geometry, since we can't selectively apply the light per-pixel. We could split the meshes in the scene into smaller parts to improve granularity, but this increases the number of draw calls in a frame (and thus increases the CPU overhead). It would also increase the number of intersection tests required to determine if a light affects a mesh, which also increases the amount of work performed by the CPU. Another related consequence is that our ability to render multiple instances of geometry in a single batch is reduced, since the set of lights used has to be the same for all of the geometry instances in a batch. Since instances may appear in different areas of a scene, they typically will be affected by different combinations of lights.

The other major disadvantage of forward rendering is shader program complexity. Being able to handle various numbers of lights and light types with various material variations can require an explosion in the number of required shader permutations.[2] A high number of shader permutations is undesirable, since it increases memory usage, as well as overhead from switching between shader programs. Depending on how the permutations are compiled, they can also significantly increase compilation times. An alternative to using many shader programs is to use dynamic flow control, which has various GPU performance implications. Another alternative is to only render one light at a time and additively blend the result into the render target, which is known as *multipass rendering*. However, this approach requires paying the cost for transforming and rasterizing geometry multiple times. Even ignoring the issues associated with using many permutations, the resulting pixel shader programs can become very expensive and complex on their own. This is because of the need to evaluate material properties *and* perform necessary lighting and shadowing calculations for all active light sources. This makes the shader programs difficult to author, maintain, and optimize. Their performance is also tied to the overdraw of the scene, since overdraw results in shading pixels that aren't even visible. A z-only prepass can significantly reduce overdraw, but its efficiency is limited by the implementation of the Hi-Z unit in the hardware. Shader executions will also be wasted, since pixel shaders must run in 2×2 quads at a minimum, which can be particularly bad for highly-tessellated scenes with many small triangles.

These disadvantages collectively make it difficult to scale the number of dynamic lights in a scene while maintaining adequate performance for real-time applications. However if you look at the descriptions very carefully, you'll notice that all of these disadvantages stem from one primary problem that is inherent to forward rendering: lighting is tightly coupled with the rasterization of scene geometry. This means that if we were to decouple those two steps, we could potentially limit or completely bypass some of those disadvantages. But how can we do this? If we look at step 2 in the lighting process, we see that to calculate lighting, we need both the material properties and the geometric surface properties for our scene geometry. This means that if we have a step where we rendered out all

---

[2] The term *shader permutations* refers to compiling many variants of shader program using common HLSL source code as the base. Typically, conditional compilation and macros are used to enable or disable certain portions of the shader program, and the permutations are compiled with various combinations of those features

of these properties to a buffer (which we would call a *geometry buffer*, or *g-buffer* (Saito & Takahashi, 1990)), we could have a second step where we iterate through the lights in the scene and calculate the lighting values for each pixel. This is exactly the premise of deferred rendering, and we will explore many of the properties of this rendering technique throughout the remainder of this chapter.

11.1.2 The Deferred Rendering Pipeline

The first step of the deferred rendering pipeline (outlined in Figure 11.2) is to render all scene geometry into a g-buffer, which typically consists of several render target textures. The individual channels of these textures are used to store geometry surface information per-pixel, such as surface normal or material properties. The second step is the lighting phase. In this step, geometry representing the light's area of influence on the screen is rendered, and for each resulting fragment that is shaded, the g-buffer information is sampled from the render target textures. The g-buffer information is then combined with the light properties to determine the resulting lighting contribution to that pixel. This contribution is then summed with the contribution of all other light sources affecting that pixel, to determine the final, lit color of the surface.

Since deferred rendering avoids the primary drawback of forward rendering (lighting too tightly coupled with geometry), it has the following advantages:

- The number of shader permutations is drastically reduced, since lighting and shadowing calculations can be moved into their own separate shader programs.

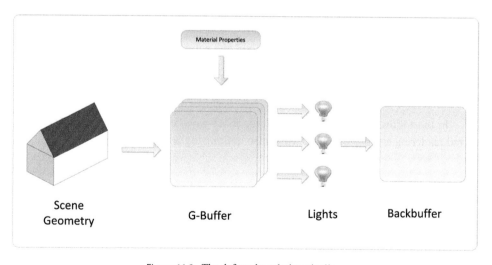

Figure 11.2. The deferred rendering pipeline.

- Mesh instances can be batched more frequently, since lighting parameters are no longer needed when rendering scene geometry. Consequently, the active light sources do not need to be the same for all instances of a mesh.

- Scene geometry only needs to be rendered once, since there's no need to resort to multipass rendering for lighting.

- It's no longer necessary to perform CPU work to determine which lights affect different portions of the screen. Instead, bounding volumes or screen-space quads can be rasterized over the portion of the screen that a light affects. Depth and stencil testing can also be used to further reduce the number of shader executions.

- The overall required shader and rendering framework architecture can be simplified, because lighting and geometry have been decoupled.

These advantages have caused deferred rendering to become extremely popular among modern real-time rendering engines. Unfortunately, the approach also comes with several drawbacks of its own:

- A significant amount of memory and bandwidth must be dedicated to the generation and sampling of the g-buffer, since it needs to store any information required to calculate lighting for that pixel.

- Using hardware MSAA is nontrivial, and generally requires that the lighting calculations be performed for each subsample. This can introduce a significantly increased computational burden.

- Transparent geometry can't be handled in the same manner as opaque geometry, since it can't be rendered into the g-buffer. This is because the g-buffer can only hold properties for a single surface, and rendering transparent geometry requires calculating the lighting for multiple overlapping per-pixel and combining the resulting color.

- Having different BRDFs for different materials is no longer straightforward, since the evaluation of a pixel's final color has been moved to the lighting pass.

In the following sections, we will talk about ways in which the advantages of deferred rendering can be exploited, while also minimizing it drawbacks.

11.2 Classic Deferred Rendering

In this section, we will work through an extremely basic deferred rendering implementation in D3D11. To maintain simplicity in this introduction to the topic, we will assume that all materials use a *Blinn-Phong BRDF* (Zink, Hoxley, Engel, Kornmann, & Suni). We will also omit any performance optimizations for now, as these will be covered in the following section.

11.2.1 G-Buffer Layout

To start off, we must first design the layout and format of the g-buffer. As mentioned in the previous section, the g-buffer contains the scene geometry's surface properties that are needed to evaluate the BRDF. This means that we must examine the equations for our BRDF to determine which properties will be needed. For reference, the equations for the variant of Blinn-Phong that we will use are listed in Equation (11.1):

$$\text{DiffuseColor} = \text{Albedo}_{\text{diffuse}} \times \text{LightColor}_{\text{diffuse}} \times (N \cdot L),$$

$$\text{SpecularColor} = \text{Albedo}_{\text{specular}} \times \text{LightColor}_{\text{specular}} \times (N \cdot H)^P, \qquad (11.1)$$

$$H = \frac{L + V}{|| L + V ||}.$$

In the above equations N is the surface normal direction vector, L is the direction vector from the surface to the light source, V is the direction vector from the surface to the camera, and P is the specular power (which is a measure of shininess/roughness in this lighting model). The *Albedo* terms refer to the reflectiveness of the material being shaded, and the specular and diffuse albedos are modulated with the specular and diffuse lighting contributions, respectively. To provide all of the parameters needed to complete the above equations, we will need to store diffuse albedo, specular albedo, the specular power, the surface normal vector, and the surface position vector in our g-buffer. To have our g-buffer contain all of the values required to evaluate the above equations, we will need to store 13 distinct floating-point values in our g-buffer render targets. Since we can store at most 4 values per texel in a texture,[3] we will need to use 4 render targets in our g-buffer. Figure 11.3 shows the formats of the render targets that we will use, along with the layout of the data within those render targets.

With these render target formats, the resulting footprint for each pixel is 64 bytes. This means that at 1280×720 resolution, the total size of the g-buffer is 56.25 megabytes, or 126.56 megabytes for 1920×1080. Keep in mind that this is *in addition* to the memory footprint of the back buffer and depth-stencil buffer, which is the minimum number of

---

[3] This is because the available texture element formats supply at most four components.

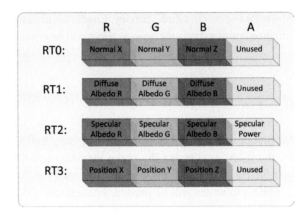

Figure 11.3. G-buffer layout.

resources required for a forward rendering setup. While this is obviously a very large footprint, various optimizations can be applied to greatly reduce the memory usage. See the section entitled "G-Buffer Attribute Packing" for more details.

11.2.2 Rendering the G-Buffer

G-Buffer Rendering without Normal Mapping

Direct3D 11 supports rendering to 8 simultaneous render targets, which means that we can fill our g-buffer by rendering a single pass for each mesh in our scene.[4] Since this pass only has to output material and surface properties, without calculating any lighting, we only require simple vertex and pixel shader programs. The vertex shader transforms vertex positions to clip space (for use by the rasterizer stage) and world space (for eventual storage in the g-buffer), and also transforms the vertex normal vectors to world space (also for eventual storage in the g-buffer). The pixel shader simply outputs the vertex position, vertex normal, diffuse albedo, specular albedo, and specular power. The code for these shader programs is provided in Listing 11.1.

---

[4] While a render target array could also hold the g-buffer data, it would require using the *SV_RenderTargetArrayIndex* system value semantic, and the geometry would be rasterized once for each of the render targets in the array. Since we aren't changing any of the viewing or perspective parameters for each render target, it doesn't make sense to perform multiple rasterizations. Thus, in this case, it is a better choice to use multiple render targets, instead of render target arrays.

```
// Constants
cbuffer Transforms
{
      matrix WorldMatrix;
      matrix WorldViewMatrix;
      matrix WorldViewProjMatrix;
};

cbuffer MatProperties
{
      float3 SpecularAlbedo;
      float SpecularPower;
};

// Textures/Samplers
Texture2D DiffuseMap          : register( t0 );
SamplerState AnisoSampler     : register( s0 );

// Input/Output structures
struct VSInput
{
      float4 Position         : POSITION;
      float2 TexCoord         : TEXCOORDS0;
      float3 Normal           : NORMAL;
};

struct VSOutput
{
      float4 PositionCS       : SV_Position;
      float2 TexCoord         : TEXCOORD;
      float3 NormalWS         : NORMALWS;
      float3 PositionWS       : POSITIONWS;
};

struct PSInput
{
      float4 PositionSS       : SV_Position;
      float2 TexCoord         : TEXCOORD;
      float3 NormalWS         : NORMALWS;
      float3 PositionWS       : POSITIONWS;
};

struct PSOutput
{
      float4 Normal           : SV_Target0;
      float4 DiffuseAlbedo    : SV_Target1;
      float4 SpecularAlbedo   : SV_Target2;
      float4 Position         : SV_Target3;
};

// G-Buffer vertex shader
VSOutput VSMain( in VSInput input )
{
      VSOutput output;
```

```
        // Convert position and normals to world space
        output.PositionWS = mul( input.Position, WorldMatrix ).xyz;
        output.NormalWS = normalize( mul( input.Normal, (float3x3)WorldMatrix ) );

        // Calculate the clip-space position
        output.PositionCS = mul( input.Position, WorldViewProjMatrix );

        // Pass along the texture coordinate
        output.TexCoord = input.TexCoord;

        return output;
}

// G-Buffer pixel shader
PSOutput PSMain( in PSInput input )
{
        PSOutput output;

        // Sample the diffuse map
        float3 diffuseAlbedo = DiffuseMap.Sample( AnisoSampler, input.TexCoord ).rgb;

        // Normalize the normal after interpolation
        float3 normalWS = normalize( input.NormalWS );

        // Output our G-Buffer values
        output.Normal = float4( normalWS, 1.0f );
        output.DiffuseAlbedo = float4( diffuseAlbedo, 1.0f );
        output.SpecularAlbedo = float4( SpecularAlbedo, SpecularPower );
        output.Position = float4( input.PositionWS, 1.0f );

        return output;
}
```

Listing 11.1. G-buffer generation shader code.

It should also be mentioned that the tessellation pipeline can also be used when rendering the g-buffer. G-buffer generation is the only pass where the scene geometry is being used, so if tessellation is desired, it should be used in this phase so that the g-buffer information reflects the finely tessellated meshes.

G-Buffer Rendering with Normal Mapping

Since the final surface normal vector is required for the lighting pass, any normal mapping[5] must be applied during the g-buffer pass. A common approach for forward rendering is to transform light positions and vectors to tangent space (Lengyel, 2001) in the vertex shader, and then apply lighting in that space to facilitate the use of normal maps. However, this

---

[5] *Normal mapping* is a technique that simulates more complex geometry by storing perturbed surface normal vectors in a texture, and then looks up these normal vectors in the pixel shader for use in lighting calculations.

approach doesn't work for deferred rendering, since the tangent frame isn't available in the lighting pass.[6] Instead, the normal map value must be transformed to view space or world space, so that it can be used with the light position and direction. In our implementation we perform the lighting calculations in world space, so we will transform the normal map value to world space, as well. The first step in doing this is to transform the vertex tangent frame to world space in the vertex shader. Then the per-vertex tangent frame is interpolated for each invocation of the pixel shader, which then samples the normal map value and uses the tangent frame to transform it to world space. The updated vertex and pixel shader programs are shown in Listing 11.2.

```
// Constants
cbuffer Transforms
{
     matrix WorldMatrix;
     matrix WorldViewMatrix;
     matrix WorldViewProjMatrix;
};

cbuffer MatProperties
{
     float3 SpecularAlbedo;
     float  SpecularPower;
};

// Textures/Samplers
Texture2D DiffuseMap          : register( t0 );
Texture2D NormalMap           : register( t1 );
SamplerState AnisoSampler     : register( s0 );

// Input/Output structures
struct VSInput
{
     float4 Position  : POSITION;
     float2 TexCoord  : TEXCOORDS0;
     float3 Normal    : NORMAL;
     float4 Tangent   : TANGENT;
};

struct VSOutput
{
     float4 PositionCS        : SV_Position;
     float2 TexCoord          : TEXCOORD;
     float3 NormalWS          : NORMALWS;
     float3 PositionWS        : POSITIONWS;
     float3 TangentWS         : TANGENTWS;
     float3 BitangentWS       : BITANGENTWS;
};
```

[6] These tangent space basis vectors could be added to the g-buffer, but this would require as many as nine floating point values to describe the space correctly.

```
struct PSInput
{
    float4 PositionSS        : SV_Position;
    float2 TexCoord          : TEXCOORD;
    float3 NormalWS          : NORMALWS;
    float3 PositionWS        : POSITIONWS;
    float3 TangentWS         : TANGENTWS;
    float3 BitangentWS       : BITANGENTWS;
};

struct PSOutput
{
    float4 Normal            : SV_Target0;
    float4 DiffuseAlbedo     : SV_Target1;
    float4 SpecularAlbedo    : SV_Target2;
    float4 Position          : SV_Target3;
};

// G-Buffer vertex shader, with normal mapping
VSOutput VSMain( in VSInput input )
{
    VSOutput output;

    // Convert position and normals to world space
    output.PositionWS = mul( input.Position, WorldMatrix ).xyz;
    float3 normalWS = normalize( mul( input.Normal, (float3x3)WorldMatrix ) );
    output.NormalWS = normalWS;

    // Reconstruct the rest of the tangent frame
    float3 tangentWS = normalize( mul( input.Tangent.xyz,
                                       (float3x3)WorldMatrix ) );
    float3 bitangentWS = normalize( cross( normalWS, tangentWS ) )
                                       * input.Tangent.w;

    output.TangentWS = tangentWS;
    output.BitangentWS = bitangentWS;

    // Calculate the clip-space position
    output.PositionCS = mul( input.Position, WorldViewProjMatrix );

    // Pass along the texture coordinate
    output.TexCoord = input.TexCoord;

    return output;
}

// G-Buffer pixel shader, with normal mapping
PSOutput PSMain( in PSInput input )
{
    PSOutput output;

    // Sample the diffuse map
    float3 diffuseAlbedo = DiffuseMap.Sample( AnisoSampler,
                                       input.TexCoord ).rgb;
```

```
// Normalize the tangent frame after interpolation
float3x3 tangentFrameWS = float3x3( normalize( input.TangentWS ),
                                    normalize( input.BitangentWS ),
                                    normalize( input.NormalWS ) );

// Sample the tangent-space normal map and decompress
float3 normalTS = NormalMap.Sample( AnisoSampler, input.TexCoord ).rgb;
normalTS = normalize( normalTS * 2.0f - 1.0f );

// Convert to world space
float3 normalWS = mul( normalTS, tangentFrameWS );

// Output our G-Buffer values
output.Normal = float4( normalWS, 1.0f );
output.DiffuseAlbedo = float4( diffuseAlbedo, 1.0f );
output.SpecularAlbedo = float4( SpecularAlbedo, SpecularPower );
output.Position = float4( input.PositionWS, 1.0f );

return output;
}
```

Listing 11.2. G-buffer generation shader code with normal mapping.

11.2.3 Rendering the Lights

After filling the g-buffer with the complete scene geometry, we perform a second rendering pass, where all lights in the scene are rendered. For each light, we will render a quad (comprised of two triangles) that covers the entire render target. The vertex shader used is extremely simple, as it merely passes the vertex position to the next stage. The pixel shader is more complex, and starts off by calculating a texture coordinate based on the screen-space pixel position (which is obtained using the SV_Position system value semantic). This texture coordinate is then used to sample the g-buffer textures, so that the material and surface parameters can be obtained. The shader then uses these parameters to evaluate the Blinn-Phong equations from Equation (11.1) to calculate the diffuse lighting contribution and the specular lighting contribution. These values are then summed, and an attenuation factor is applied, with the attenuation factor being based on the type of light source. For a point light, a simple linear distance attenuation model is used. For a spot light, the distance attenuation model is used, as well as an angular attenuation factor. For a directional light, no attenuation factor is applied, since directional lights are considered "global" light sources. Once the attenuation factor is applied, the resulting color value is output from the pixel shader with additive blending enabled. Listing 11.3 contains the shader code for performing these calculations

```
// Textures
Texture2D NormalTexture                 : register( t0 );
Texture2D DiffuseAlbedoTexture          : register( t1 );
Texture2D SpecularAlbedoTexture         : register( t2 );
Texture2D PositionTexture               : register( t3 );

// Constants
cbuffer LightParams
{
      float3 LightPos;
      float3 LightColor;
      float3 LightDirection;
      float2 SpotlightAngles;
      float4 LightRange;
};

cbuffer CameraParams
{
      float3 CameraPos;
};

// Helper function for extracting G-Buffer attributes
void GetGBufferAttributes( in float2 screenPos, out float3 normal,
                           out float3 position,
                           out float3 diffuseAlbedo, out float3 specularAlbedo,
                           out float specularPower )
{
    // Determine our indices for sampling the texture based on the current
    // screen position
    int3 sampleIndices = int3( screenPos.xy, 0 );

    normal = NormalTexture.Load( sampleIndices ).xyz;
    position = PositionTexture.Load( sampleIndices ).xyz;
    diffuseAlbedo = DiffuseAlbedoTexture.Load( sampleIndices ).xyz;
    float4 spec = SpecularAlbedoTexture.Load( sampleIndices );

    specularAlbedo = spec.xyz;
    specularPower = spec.w;
}

// Calculates the lighting term for a single G-Buffer texel
float3 CalcLighting( in float3 normal,
                     in float3 position,
                     in float3 diffuseAlbedo,
                     in float3 specularAlbedo,
                     in float specularPower )
{
    // Calculate the diffuse term
    float3 L = 0;
    float attenuation = 1.0f;
    #if POINTLIGHT || SPOTLIGHT
            // Base the the light vector on the light position
            L = LightPos - position;
```

```
                // Calculate attenuation based on distance from the light source
                float dist = length( L );
                attenuation = max( 0, 1.0f - ( dist / LightRange.x ) );

                L /= dist;
        #elif DIRECTIONALLIGHT
                // Light direction is explicit for directional lights
                L = -LightDirection;
        #endif

        #if SPOTLIGHT
                // Also add in the spotlight attenuation factor
                float3 L2 = LightDirection;
                float rho = dot( -L, L2 );
                attenuation *= saturate( ( rho - SpotlightAngles.y )
                                        / ( SpotlightAngles.x -
                                            SpotlightAngles.y ) );
        #endif

        float nDotL = saturate( dot( normal, L ) );
        float3 diffuse = nDotL * LightColor * diffuseAlbedo;

        // Calculate the specular term
        float3 V = CameraPos - position;
        float3 H = normalize( L + V );
        float3 specular = pow( saturate( dot( normal, H ) ), specularPower )
                                        * LightColor * specularAlbedo.xyz * nDotL;

        // Final value is the sum of the albedo and diffuse with attenuation applied
        return ( diffuse + specular ) * attenuation;
}

// Lighting pixel shader
float4 PSMain( in float4 screenPos : SV_Position ) : SV_Target0
{
        float3 normal;
        float3 position;
        float3 diffuseAlbedo;
        float3 specularAlbedo;
        float specularPower;

        // Sample the G-Buffer properties from the textures
        GetGBufferAttributes( screenPos.xy, normal, position, diffuseAlbedo,
                                specularAlbedo, specularPower );

        float3 lighting = CalcLighting( normal, position, diffuseAlbedo,
                                        specularAlbedo, specularPower );

        return float4( lighting, 1.0f );
}
```

Listing 11.3. Lighting shader code.

Since the blend state used has additive blending enabled, the output is summed with the previous contents of the render target. Thus, the output from each light is summed to create the final color value for each pixel.

11.3 Light Pre-Pass Deferred Rendering

Deferred rendering has several advantages, but it also suffers in performance and flexibility due to the need to have all of the surface and material properties contained in the g-buffer. Storing many attributes in the g-buffer consumes a great deal of memory and bandwidth (particularly when MSAA is used), and also makes it difficult to implement different types of materials and lighting models. light pre-pass deferred rendering(Engel, 2009) is a technique that attempts to alleviate these problems somewhat by further splitting the lighting pipeline into three steps, as opposed to two. The end result is that the g-buffer can be made much smaller, and the lighting pass is less expensive, but all deferred scene geometry must be rendered twice, instead of once. The basic steps are as follows:

1. Render a minimal g-buffer with only the surface properties required to evaluate the core portions of the lighting equation.

2. Evaluate the core portion of the lighting equation for all lights in the scene by rendering geometry covering the affected pixels, and sum the results into a light buffer.

3. Render the scene geometry a second time, in which the light buffer value is sampled and the material albedos are applied to determine the final surface color.

This section will walk through a basic implementation of a light pre-pass deferred renderer without optimizations, similar to the outline in the previous section. It should be noted that light pre-pass deferred rendering is often referred to as *deferred lighting* to distinguish it from *deferred rendering* or *deferred shading*. This emphasizes the fact that the basic lighting calculations are done in a deferred step, but the final shading step is done while rasterizing the scene geometry.

11.3.1 Light Pre-Pass G-Buffer Layout

As mentioned in the overview, with light pre-pass deferred rendering, we can render a minimal g-buffer. Essentially, we only want the surface properties required to evaluate the core diffuse and specular components of the Blinn-Phong BRDF, without the albedos.

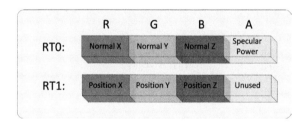

Figure 11.4. G-buffer layout for light prepass deferred rendering.

Thus, we only need surface position, surface normal vector, and specular power in our g-buffer. Figure 11.4 shows the resulting layout in our render targets:

For our basic implementation, this leaves us with only 2 render targets, instead of the 4 required for the classic deferred renderer. For our unoptimized implementation, it also cuts our g-buffer memory footprint in half, leaving it at 32 bytes instead of 64. It should be noted that it is possible to have only 1 render target if position is reconstructed from the depth buffer instead of being explicitly stored in the g-buffer (see the section entitled "G-Buffer Attribute Packing" for details). This can provide an advantage on hardware on which using multiple render targets has a significant performance penalty.

11.3.2 Rendering the G-Buffer

The approach for rendering the g-buffer is very similar to the one used for the classic deferred rendering implementation. The only difference is that we no longer need to write the diffuse albedo or the specular albedo. The shader program code for doing this is listed in Listing 11.4:

```
// Textures
Texture2D       NormalMap        : register( t0 );
SamplerState    AnisoSampler     : register( s0 );

// Constants
cbuffer Transforms
{
    matrix WorldMatrix;
    matrix WorldViewMatrix;
    matrix WorldViewProjMatrix;
};

cbuffer MaterialProperties
{
    float SpecularPower;
};
```

```
// Input/output structures
struct VSInput
{
    float4 Position          : POSITION;
    float2 TexCoord          : TEXCOORDS0;
    float3 Normal            : NORMAL;
    float4 Tangent           : TANGENT;
};

struct VSOutput
{
    float4 PositionCS        : SV_Position;
    float2 TexCoord          : TEXCOORD;
    float3 NormalWS          : NORMALWS;
    float3 TangentWS         : TANGENTWS;
    float3 BitangentWS       : BITANGENTWS;
    float3 PositionWS        : POSITIONWS;
};

struct PSInput
{
    float4 PositionSS        : SV_Position;
    float2 TexCoord          : TEXCOORD;
    float3 NormalWS          : NORMALWS;
    float3 TangentWS         : TANGENTWS;
    float3 BitangentWS       : BITANGENTWS;
    float3 PositionWS        : POSITIONWS;
};

struct PSOutput
{
    float4 Normal            : SV_Target0;
    float4 Position          : SV_Target1;
};

// G-Buffer vertex shader for light prepass deferred rendering
VSOutput VSMain( in VSInput input )
{
    VSOutput output;

    // Convert normals to world space
    float3 normalWS = normalize( mul( input.Normal, (float3x3)WorldMatrix ) );
    output.NormalWS = normalWS;

    // Reconstruct the rest of the tangent frame
    float3 tangentWS = normalize( mul( input.Tangent.xyz,
                                       (float3x3)WorldMatrix ) );
    float3 bitangentWS = normalize( cross( normalWS, tangentWS ) )
                                        * input.Tangent.w;

    output.TangentWS = tangentWS;
    output.BitangentWS = bitangentWS;

    // Calculate the world-space position
```

```
        output.PositionWS = mul( input.Position, WorldMatrix ).xyz;

        // Calculate the clip-space position
        output.PositionCS = mul( input.Position, WorldViewProjMatrix );

        // Pass along the texture coordinate
        output.TexCoord = input.TexCoord;

        return output;
}

// G-Buffer pixel shader for light prepass deferred rendering
PSOutput PSMain( in PSInput input )
{
        // Normalize the tangent frame after interpolation
        float3x3 tangentFrameWS = float3x3( normalize( input.TangentWS ),
                                            normalize( input.BitangentWS ),
                                            normalize( input.NormalWS ) );

        // Sample the tangent-space normal map and decompress
        float3 normalTS = NormalMap.Sample( AnisoSampler, input.TexCoord ).rgb;
        normalTS = normalize( normalTS * 2.0f - 1.0f );

        // Convert to world space
        float3 normalWS = mul( normalTS, tangentFrameWS );

        // Output our G-Buffer values
        PSOutput output;
        output.Normal = float4( normalWS, SpecularPower );
        output.Position = float4( input.PositionWS, 1.0f );
        return output;
}
```

Listing 11.4. G-buffer generation shader code for light prepass deferred rendering.

11.3.3 Rendering the Light Buffer

For the lighting pass, we will only evaluate a portion of the lighting equation, and will the result to the render target. The simplified equations that we will use are listed below in Equation (11.2):

$$\text{DiffuseLight} = \text{LightColor}_{\text{diffuse}} \times (N \circ L),$$

$$\text{SpecularLight} = \text{LightColor}_{\text{specular}} \times (N \circ H)^P, \qquad (11.2)$$

$$H = \frac{L + V}{\| L + V \|}.$$

Since *DiffuseLight* and *SpecularLight* each have 3 components, storing the full values in our light buffer would require storing 6 individual floating-point values. This would require writing to at least two render targets simultaneously, which is often not optimal from a performance and memory-consumption point of view. Thus, a common optimization is to store only the specular intensity rather than the color, which allows the light buffer to consist of a single 4-component texture. This approach is taken in the lighting shader code in Listing 11.5.

```
// Textures
Texture2D NormalTexture        : register( t0 );
Texture2D PositionTexture      : register( t1 );

// Constants
cbuffer LightParams
{
        float3 LightPos;
        float3 LightColor;
        float3 LightDirection;
        float2 SpotlightAngles;
        float4 LightRange;
};

cbuffer CameraParams
{
        float3 CameraPos;
}

// Vertex shader for lighting pass of light prepass deferred rendering
float4 VSMain( in float4 Position : POSITION ) : SV_Position
{
        // Just pass along the position of the full-screen quad
        return Position;
}

// Helper function for extracting G-Buffer attributes
void GetGBufferAttributes( in float2 screenPos, out float3 normal,
                           out float3 position, out float specularPower )
{
        // Determine our indices for sampling the texture based on the current
        // screen position
        int3 sampleIndices = int3( screenPos.xy, 0 );

        float4 normalTex = NormalTexture.Load( sampleIndices );
        normal = normalTex.xyz;
        specularPower = normalTex.w;
        position = PositionTexture.Load( sampleIndices ).xyz;
}

// Calculates the lighting term for a single G-Buffer texel
float4 CalcLighting( in float3 normal, in float3 position, in float specularPower )
```

```
{
        // Calculate the diffuse term
        float3 L = 0;
        float attenuation = 1.0f;
    #if POINTLIGHT || SPOTLIGHT
                // Base the the light vector on the light position
                L = LightPos - position;

                // Calculate attenuation based on distance from the light source
                float dist = length( L );
                attenuation = max( 0, 1.0f - ( dist / LightRange.x ) );

                L /= dist;
    #elif DIRECTIONALLIGHT
                // Light direction is explicit for directional lights
                L = -LightDirection;
    #endif

    #if SPOTLIGHT
                // Also add in the spotlight attenuation factor
                float3 L2 = LightDirection;
                float rho = dot( -L, L2 );
                attenuation *= saturate( ( rho - SpotlightAngles.y )
                                    / ( SpotlightAngles.x - SpotlightAngles.y ) );
    #endif

        float nDotL = saturate( dot( normal, L ) );
        float3 diffuse = nDotL * LightColor * attenuation;

        // Calculate the specular term
        float3 V = CameraPos - position;
        float3 H = normalize( L + V );
        float specular = pow( saturate( dot( normal, H ) ), specularPower )
                                    * attenuation * nDotL;

        // Final value is diffuse RGB + mono specular
        return float4( diffuse, specular );
}

// Pixel shader for lighting pass of light prepass deferred rendering
float4 PSMain( in float4 screenPos : SV_Position ) : SV_Target0
{
        float3 normal;
        float3 position;
        float specularPower;

        // Get the G-Buffer values
        GetGBufferAttributes( screenPos.xy, normal, position, specularPower );

        return CalcLighting( normal, position, specularPower );
}
```

Listing 11.5. Light buffer generation pixel shader code.

Once the light buffer is rendered, it is possible to apply screen-space operations to the contents of the light buffer to simulate more complex light interactions. For example, a bilateral blur can be used to simulate subsurface scattering for skin and other translucent materials (Mikkelsen, 2010).

11.3.4 Rendering the Final Pass

In the final pass, opaque scene geometry is rendered for a second time. The vertex shader is quite simple, as it only needs to pass along the mesh's diffuse map texture coordinate to the pixel shader. The pixel shader samples the light buffer based on the screen-space position of the pixel being shaded, and the values in the light buffer are combined with material albedos to determine the final color of the surface. The shader in Listing 11.6 demonstrates this concept.

```
// Textures
Texture2D      DiffuseMap      : register( t0 );
Texture2D      LightTexture    : register( t1 );
SamplerState   AnisoSampler    : register( s0 );

// Constants
cbuffer Transforms
{
      matrix WorldMatrix;
      matrix WorldViewMatrix;
      matrix WorldViewProjMatrix;
};

cbuffer MaterialProperties
{
      float3 SpecularAlbedo;
}

// Input/Output structures
struct VSInput
{
      float4 Position  : POSITION;
      float2 TexCoord  : TEXCOORDS0;
};

struct VSOutput
{
      float4 PositionCS : SV_Position;
      float2 TexCoord   : TEXCOORD;
};

struct PSInput
```

```
{
    float4 ScreenPos  : SV_Position;
    float2 TexCoord   : TEXCOORD;
};

// Vertex shader for final pass of light prepass deferred rendering
VSOutput VSMain( in VSInput input )
{
    VSOutput output;

    // Calculate the clip-space position
    output.PositionCS = mul( input.Position, WorldViewProjMatrix );

    // Pass along the texture coordinate
    output.TexCoord = input.TexCoord;

    return output;
}

// Pixel shader for final pass of light prepass deferred rendering
float4 PSMain( in PSInput input ) : SV_Target0
{
    // Sample the diffuse map
    float3 diffuseAlbedo = DiffuseMap.Sample( AnisoSampler, input.TexCoord ).rgb;

    // Determine our indices for sampling the texture based on the current
    // screen position
    int3 sampleIndices = int3( input.ScreenPos.xy, 0 );

    // Sample the light target
    float4 lighting = LightTexture.Load( sampleIndices );

    // Apply the diffuse and specular albedo to the lighting value
    float3 diffuse = lighting.xyz * diffuseAlbedo;
    float3 specular = lighting.w * SpecularAlbedo;

    // Final output is the sum of diffuse + specular
    return float4( diffuse + specular, 1.0f );
}
```

Listing 11.6. Shader code for final pass.

It should be noted that other terms that contribute to the final pixel color can also be added in, such as emissive lighting (glow), Fresnel/rim lighting, or reflection maps. This makes light pre-pass deferred rendering more natural for implementing these sorts of material variations.

11.4 Optimizations

The deferred rendering implementations that we have seen so far would present subopti-
mal performance characteristics in many situations, since their aim was to demonstrate the
core concepts of deferred rendering. To make the approach practical for use in real-time
applications, it is often necessary to leverage the flexibility and programmability of the
Direct3D 11 pipeline to improve the performance. The following sections will focus on
various methods for reducing the required memory and bandwidth usage associated with
rendering and sampling the g-buffer, as well as minimizing the number of pixel shader
executions that are needed in the lighting pass.

11.4.1 G-Buffer Attribute Packing

One of the primary disadvantages of using a deferred approach to rendering is that surface
and material properties must be rendered out to one or more render targets. A significant
amount of bandwidth must be dedicated to writing the values during the g-buffer generation pass, as well as when sampling the textures during the lighting pass. As a result, reducing the memory footprint of the g-buffer is a simple and effective method for optimizing
the performance of a deferred renderer. This is generally true even when some extra math is
performed in the shader programs to pack or unpack the values from a compressed format,
since almost all modern GPUs have a large number of ALUs relative to their bandwidth
and texturing units. With this in mind, we will look at the various types of data typically
stored in a g-buffer and determine a compressed format for storage, based on the typical
range of values, the precision required to avoid unacceptable artifacts, and the expense of
packing and unpacking the data.

Normal Vectors

For compressing normal vectors, we can take advantage of the fact that a normal vector
is a unit-length direction vector. This greatly restricts the domain of the normal values,
which can allow us to use a signed 8-bit integer format (DXGI_FORMAT_R8G8B8A8_SNORM)
for storage, instead of a floating point format. It can also allow us to store the direction in a
representation that requires only 2 values instead of 3, thus dropping a component entirely.

Perhaps the most obvious approach is to convert the normal vector to Spherical
Coordinates and omit ρ, since it is implicit that $\rho = 1$ for a normalized direction vector.
The equations for converting a vector from a Cartesian coordinate system to a spherical
coordinate system are well-known and are trivial to implement in HLSL. The conversion can also be algebraically simplified for the case where $\rho = 1$, reducing the number

of shader instructions required for compression and decompression (Wilson, 2008). This simplification is demonstrated by the shader code in Listing 11.7.

```
float2 CartesianToSpherical(float3 cartesian)
{
    float2 spherical;

    spherical.x = atan2(cartesian.y, cartesian.x) / 3.14159f;
    spherical.y = cartesian.z;

    return spherical;
}

float3 SphericalToCartesian(float2 spherical)
{
    float2 sinCosTheta, sinCosPhi;

    sincos(spherical.x * 3.14159f, sinCosTheta.x, sinCosTheta.y);
    sinCosPhi = float2(sqrt(1.0 - spherical.y * spherical.y), spherical.y);

    return float3(sinCosTheta.y * sinCosPhi.x,
                            sinCosTheta.x * sinCosPhi.x,
                  sinCosPhi.y);
}
```

Listing 11.7. Optimized spherical coordinate conversion for normals.

The main downside of using spherical coordinates is that they require use of the trigonometric intrinsics *sin()*, *cos()*, and *atan2()*. These operations are among a special group of arithmetic operations known as *transcendentals*, and they often use hardware resources that are somewhat scarce on modern GPUs. As a result, it can often be desirable from a performance point of view to avoid their usage altogether when an alternative method exists.

Another approach is to simply store the *X*- and *Y*- components of the Cartesian normal vector and drop the *Z* component (Valient, 2007), since the *Z* component of a unit vector can be reconstructed as long as *X*, *Y*, and the sign of *Z* are known. Equation (11.3) shows the equation used for this reconstruction:

$$z = \pm\sqrt{1 - (x^2 + y^2)}. \tag{11.3}$$

It is possible to pack the sign bit into the stored *X* or *Y* values, making this technique suitable for normal vectors in any coordinate space. However some implementations have stored view space normal vectors, and then assumed that the sign of *Z* is always negative (or positive, if a right-handed coordinate system is used). At first this seems like a plausible assumption to make, since an object must be facing the viewer if it is visible. However, it

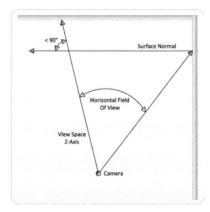

Figure 11.5. Negative view-space normals.

is important to note while it is *generally* true that the sign of Z will be negative for a view-space normal, with a perspective projection it is possible for surfaces to be rasterized that face in the positive Z direction (Lee, 2009). This is illustrated in Figure 11.5.

In addition, normal maps completely override the vertex normal, making it possible for the normal to face in any direction. This will cause visual artifacts due to incorrect lighting results, which may or may not be unacceptable, depending on the error introduced by having an incorrect sign.

A third approach to normal vector compression is to use a *sphere map transformation* (Mittring, 2009).[7] This transform was originally developed for mapping a reflection vector to the [0, 1] range, but it also works for a normal vector. It essentially works by storing a 2D location on a map, where each location corresponds to the normal vector on a sphere. The shader code for performing the packing and unpacking is shown in Listing 11.8.

```
float2 SpheremapEncode( float3 normal )
{
    return normalize( normal.xy ) * ( sqrt( -normal.z * 0.5f + 0.5f ) );
}

float3 SpheremapDecode( float2 encoded )
{
    float4 nn = float4( encoded, 1, -1 );
    float l = dot( nn.xyz, -nn.xyw );
    nn.z = l;
    nn.xy *= sqrt( l );
    return nn.xyz * 2 + float3( 0, 0, -1 );
}
```

Listing 11.8. Sphere map transformation shader code.

---

[7] A discussion of sphere mapping and its properties are available in (Zink, Hoxley, Engel, Kornmann, & Suni).

Diffuse Albedo

Diffuse albedo is a simple case, since the source data is typically a color value of the range [0,1]. This means that we can use an unsigned, normalized, 8-bit integer format such as `DXGI_FORMAT_R8G8B8A8_UNORM`. It may also be desirable to store the values in sRGB color space,[8] since this is typically the storage format for diffuse textures. This requires using a format such as `DXGI_FORMAT_R8G8B8A8_UNORM_SRGB`, which causes the hardware to perform the sRGB conversion on the value output from the pixel shader. Alternatively, a 10-bit format such as `DXGI_FORMAT_R10G10B10A2_UNORM` can provide additional precision if the fourth component isn't needed for other data.

Specular Albedo and Power

Specular albedo is a similar case to diffuse albedo, in that it doesn't require a great deal of precision. The 255 discrete values provided by an 8-bit integer are generally adequate for both the albedo and the power. Additionally, it is common to store only a monochrome specular value, rather than RGB components. This allows us to store only 2 values for specular albedo, as opposed to 4.

Position

Position is a value that typically requires high precision, since it can have a potentially large domain and is used for high-frequency shadow computations. For storing the *XYZ* components of a world or view space position, even a 16-bit floating point format is generally inadequate for avoiding artifacts. Fortunately, there is an alternative to storing the full *XYZ* value. When executing a pixel shader, the screen space *XY* position is implicit and available to shader through the `SV_Position` semantic. By using the view and projection matrices used when rendering the scene geometry, it is possible to reconstruct the view space or world space position of a pixel from only a single value.

To begin with, we will discuss the basics of a perspective projection. For any pixel on the screen, there is an associated direction vector that begins at the camera position and points towards the pixel's position on the far clipping plane. You can calculate this direction vector by using the screen-space *XY* position of the pixel to linearly interpolate between the positions of the corners on the view frustum's far clipping plane, subtracting the camera position, and then normalizing (you don't have to subtract the camera position if you're doing this in view space, since the camera position is located at the origin of view space). If geometry is rasterized at that pixel position, this means that the surface at that pixel lies somewhere along the vector from the origin to that pixel. The distance along that

---

[8] The *sRGB* color space is a standard RGB color space often used by image files and display devices. Using sRGB causes more precision to be used for darker color values, which matches the human eye's natural sensitivity to those color regions.

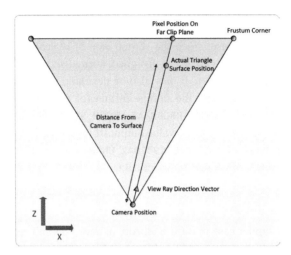

Figure 11.6. Reconstructing surface position using view ray and camera distance.

vector will vary, depending on how far that geometry is from the camera, but the direction is always the same. This is very similar to the method used for casting primary rays in a ray tracer: for a pixel on the near or far clip plane, get the direction from the camera to that pixel, and check for intersections. What this ultimately means is that if we have the screen space pixel position and the camera position, we can figure out the position of the triangle surface if we have the distance from the camera to the surface. Refer to Figure 11.6 for an illustration of this concept.

We're now ready to work out a simple implementation of this concept in a deferred renderer. The first step will be in our g-buffer pass, where we'll store the distance from the camera to the triangle surface in a single floating-point value. The simplest and cheapest way to do this is to transform the vertex position to view space in the vertex shader (since doing this implicitly makes the position relative to the camera position, as described above) and then compute the magnitude of the resulting position vector in the pixel shader. Listing 11.9 provides the g-buffer shader code for calculating camera distance.

```
// -- G-Buffer vertex shader --
// Calculate view space position of the vertex and pass it to the pixel
// shader
output.PositionVS = mul(input.PositionOS, WorldViewMatrix).xyz;

// -- G-Buffer pixel shader --
// Calculate the length of the view space position to get the distance from
// camera->surface
output.Distance.x = length(input.PositionVS);
```

Listing 11.9. G-buffer shader code for camera distance calculation.

The next step is to calculate the view ray vector during our lighting pass. We'll do this by first determining the vector from the camera to the vertex position in the vertex shader. Then in the pixel shader, we'll normalize this vector to get the final view ray vector. This vector is then multiplied with the distance value sampled from the g-buffer and added to the camera position to reconstruct the original surface position. Note that the vertex position, camera position, and resulting camera ray can be in any coordinate space, as long as the same space is used for all three of them. This allows reconstructing either a view space or a world space position. The code in Listing 11.10 uses world space for simplicity of presentation.

```
// -- Light vertex shader --
#if VOLUMES
        // Calculate the world space position for a light volume
        float3 positionWS = mul(input.PositionOS, WorldMatrix);
#elif QUADS
        // Calculate the world space position for a full-screen quad (assume input
        // vertex coordinates are in [-1,1] post-projection space)
        float3 positionWS = mul(input.PositionOS, InvViewProjMatrix);
#endif

// Calculate the view ray
output.ViewRay = positionWS - CameraPositionWS;

// -- Light Pixel shader --
// Normalize the view ray, and apply the distance to reconstruct position
float3 viewRay = normalize(input.ViewRay);
float viewDistance = DistanceTexture.Sample(PointSampler, texCoord);
float3 positionWS = CameraPositionWS + viewRay * viewDistance;
```

Listing 11.10. Light shader code for view ray calculation and position reconstruction.

This provides us with a flexible and fairly efficient method for reconstructing position from only a single high-precision value stored in the g-buffer. However, we can still make further optimizations if we restrict ourselves to view space, since in view space the view frustum is aligned with the Z-axis.

When reconstructing a view space position, we can extrapolate the view ray until it intersects with the far clipping plane. Computing this intersection point is trivial, since it's the point where the Z component is equal to the far clip distance. When rasterizing a quad for a directional light, it's even simpler, since the frustum corner position can be linearly interpolated. With the Z component of the view ray set to a known value, it is no longer necessary to normalize the view ray vector in the pixel shader. Instead, we can multiply it by a value that scales along the camera's z-axis to get the final reconstructed position. In the case where Z is the far clip distance, we want to scale by a ratio of the original surface depth relative to the far clip plane. In other words, the surface's view space Z divided by the far

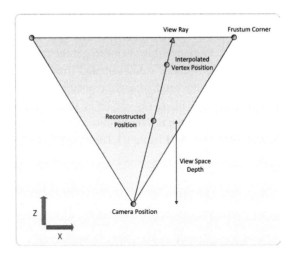

Figure 11.7. Optimized view-space position reconstruction.

clip distance. This scaling value can be calculated in the g-buffer pass and stored in the g-buffer, so that it can be used in the lighting pass. This technique is illustrated in Figure 11.7.

```
// -- G-Buffer vertex shader --
// Calculate view space position of the vertex and pass it to the pixel
shader
output.PositionVS = mul(input.PositionOS, WorldViewMatrix).xyz;

// -- G-Buffer pixel shader --
// Divide view space Z by the far clip distance
output.Depth.x = input.PositionVS.z / FarClipDistance;

// -- Light vertex shader --
#if VOLUMES
    // Calculate the view space vertex position
    output.PositionVS = mul(input.PositionOS, WorldViewMatrix);
#elif QUADS
    // Calculate the view space vertex position (you can also just directly
    // map the vertex to a frustum corner to avoid the transform)
    output.PositionVS = mul(input.PositionOS, InvProjMatrix);
#endif

// Light Pixel shader
#if VOLUMES
    // Extrapolate the view space position to the far clip plane
    float3 viewRay = float3(input.PositionVS.xy *
                                    (FarClipDistance / input.PositionVS.z),
                    FarClipDistance);
#elif QUADS
```

```
// For a directional light, the vertices were already on the far clip plane so
// we don't need to extrapolate
        float3 viewRay = input.PositionVS.xyz;
#endif

// Sample the depth and scale the view ray to reconstruct view space position
float normalizedDepth = DepthTexture.Sample(PointSampler, texCoord).x;
float3 positionVS = viewRay * normalizedDepth;
```

Listing 11.11. Shader code for optimized view space position reconstruction.

Avoiding normalization of the view ray in the pixel shader reduces the number of math operations, particularly in the directional light case where the view ray is already extrapolated. Another point to be aware of is that the depth value stored in the g-buffer is always of the range $[0, 1]$. This means you can store it in a normalized integer format (such as DXGI_FORMAT_R16_UNORM) without having to do any rescaling after sampling it in the pixel shader.

While the above techniques work well when a depth or distance value is explicitly written to the g-buffer, it is also possible to avoid storing a value altogether. This can be done by sampling the depth stencil buffer used when rendering the g-buffer pass. A depth buffer will store the post-projection Z value divided by the post-projection W value, where W is equal to the view-space Z component of the surface position. This makes the value initially unsuitable for our needs, but fortunately it's possible to recover the view-space Z component using the parameters of the perspective projection(Baker). Once we do that, we can convert it to a normalized depth value and proceed with the same approach outlined above. However, this is unnecessary. Instead of extrapolating the view ray to the far clip plane, if we instead clamp it to the plane at $Z = 1$ we can scale it by the view-space Z without having to manipulate it first. Essentially, we scale the view ray in the vertex shader, rather than scaling the Z value in the pixel shader, which saves us an extra pixel shader math operation.

```
// Light vertex shader
#if VOLUMES
        // Calculate the view space vertex position
        output.PositionVS = mul(input.PositionOS, WorldMatrix);
#elif QUADS
        // For a directional light we can clamp in the vertex shader, since we
        // only interpolate in the XY direction
        float3 positionVS = mul(input.PositionOS, InvProjMatrix);
        output.ViewRay = float3(positionVS.xy / positionVS.z, 1.0f);
#endif
```

```
// Light Pixel shader
#if VOLUMES
        // Clamp the view space position to the plane at Z = 1
        float3 viewRay = float3(input.PositionVS.xy / input.PositionVS.z, 1.0f);
#elif QUADS
        // For a directional light we already clamped in the vertex shader
        float3 viewRay = input.ViewRay.xyz;
#endif

// Calculate our projection constants (this can be done in the application code
// and passed in a constant buffer)
ProjectionA = FarClipDistance   / (FarClipDistance - NearClipDistance);
ProjectionB = (-FarClipDistance * NearClipDistance)
                                / (FarClipDistance - NearClipDistance);

// Sample the depth and convert to linear view space Z
float depth = DepthTexture.Sample(PointSampler, texCoord).x;
float linearDepth = ProjectionB / (depth - ProjectionA);
float3 positionVS = viewRay * linearDepth;
```

Listing 11.12. Shader code for view space position reconstruction using a depth-stencil buffer.

If it's necessary to work in an arbitrary coordinate space, the linear depth value can be used in conjunction with the first technique to reconstruct position. This can be done by projecting the view ray onto the camera's local *Z* axis and using the result to figure out a proper scaling value.

```
// -- Light Pixel Shader --
// Normalize the view ray
float3 viewRay = normalize(input.ViewRay);

// Sample the depth buffer and convert it to linear depth
float depth = DepthTexture.Sample(PointSampler, texCoord).x;
float linearDepth = ProjectionB / (depth - ProjectionA);

// Project the view ray onto the camera's z-axis
float viewZProj = dot(EyeZAxis, viewRay);

// Scale the view ray by the ratio of the linear z value to the projected
// view / ray
float3 positionWS = CameraPositionWS + viewRay * (linearDepth / viewZProj);
```

Listing 11.13. Shader code for position reconstruction from a depth-stencil buffer using an arbitrary coordinate space.

When choosing a method for recovering position data, careful attention should be paid to precision and error resulting from the storage format used. The *Z/W* value stored

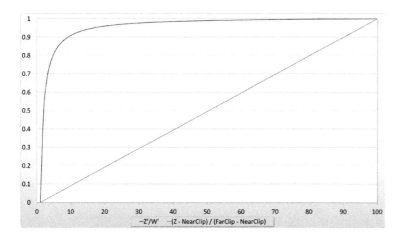

Figure 11.8. Post-perspective z/w vs. view space z.

in a depth buffer is non-linear with respect to the distance from the camera, and increases exponentially. Figure 11.8 shows a graph of Z/W for a range of Z values between the near clipping plane and far clipping plane (1 and 100 in this case).

Because of this non-linearity, more precision is used for depth ranges closer to the near clipping plane than to the far clipping plane, as demonstrated by the fast rise in the Z/W plot in Figure 11.8. This can cause error by causing reconstructed positions to become high for surfaces that are close to the far clipping plane. It should be noted that D3D11 has 32-bit floating point depth stencil formats available. However floating point numbers also have an inherent non-linear distribution of precision that causes error to increase as values move away from zero(Goldberg, 1991), which can actually compound the errors. To counteract this, it is possible to flip the distances for the near and far plane when creating a perspective projection(Persson, A couple of notes about Z, 2009), (Kemen, 2009). With such a projection Z/W will begin at 1.0 at the near clipping plane and decrease to 0.0, causing the two non-linear precision distributions to mostly balance out. It should be noted that if this technique is used, the direction of the depth test will also need to be reversed.

11.4.2 Light Shading Optimizations

As mentioned in the previous sections, applying lighting in a deferred renderer requires sampling the g-buffer textures, performing lighting calculations in a pixel shader, and summing the results with the contents of the render target. Consequently, the performance of a particular scene is often directly tied to the number of pixels shaded during the lighting pass. This

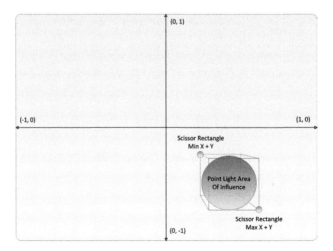

Figure 11.9. Scissor rectangle calculation for a point light.

section will focus on optimizations that can reduce the number of pixels shaded, to minimize the amount of bandwidth and/or shader ALU resources consumed by the lighting pass.

Scissor Test

The rasterizer stage of the pipeline can perform what is known as the *scissor test*.[9] This test checks whether the screen space position of each rasterized pixel intersects with a "scissor" rectangle and avoids shading any pixels that fail the test. This can be leveraged by a deferred renderer as a coarse-grained mechanism for culling pixels that won't be affected by a particular light. However, it should be noted that since this technique is based on the screen-space coverage of a light source, it only works for local light sources (point lights and spot lights), and not for global light sources (directional lights).

To effectively use the scissor test, we must calculate a screen-space bounding rectangle for the light source. Before we can do this however, we must determine the area of the screen that will be affected by a light source, which can be calculated based on the light's parameters, the camera position and orientation, and the projection parameters. This process is simplest for a point light source. Since a point light contributes equally in all directions, we can assume that the light's area of effect is perfectly spherical in both world space and view space. Thus, we can trivially fit a bounding sphere in view space to the light's area of effect. Once this bounding sphere is determined, an axis-aligned bounding box can be fitted to encapsulate the sphere. The corners of the screen-space bounding rectangle can then be determined by transforming the eight corners of the box by the projection matrix,

---

[9] The use of the scissor test is described in the rasterizer section of Chapter 3.

and finding the min and max X and Y values of the eight resulting coordinates. Figure 11.9 illustrates this process.

This process can be optimized if it is assumed that the projection is perspective, and that it is symmetrical with respect to the camera's X and Y axes. In this case, we can determine if the minimal or maximal X or Y value will be on the face of the box closest to the camera, or on the face of the box farthest from the camera, based on the view-space X and Y coordinates of the sphere. The code in Listing 11.14 uses this optimization to determine the final screen-space rectangle that will be passed to the `ID3D11DeviceContext::RSSet ScissorRects` method.

```cpp
D3D11_RECT ViewLights::CalcScissorRect( const Vector3f& lightPos,
                                        float lightRange )
{
    // Create a bounding sphere for the light, based on the position
    // and range
    Vector4f centerWS = Vector4f( lightPos, 1.0f );
    float radius = lightRange;

    // Transform the sphere center to view space
    Vector4f centerVS = ViewMatrix * centerWS;

    // Figure out the four points at the top, bottom, left, and
    // right of the sphere
    Vector4f topVS = centerVS + Vector4f( 0.0f, radius, 0.0f, 0.0f );
    Vector4f bottomVS = centerVS - Vector4f( 0.0f, radius, 0.0f, 0.0f );
    Vector4f leftVS = centerVS - Vector4f( radius, 0.0f, 0.0f, 0.0f );
    Vector4f rightVS = centerVS + Vector4f( radius, 0.0f, 0.0f, 0.0f );

    // Figure out whether we want to use the top and right from quad
    // tangent to the front of the sphere, or the back of the sphere
    leftVS.z = leftVS.x < 0.0f ? leftVS.z - radius : leftVS.z + radius;
    rightVS.z = rightVS.x < 0.0f ? rightVS.z + radius : rightVS.z - radius;
    topVS.z = topVS.y < 0.0f ? topVS.z + radius : topVS.z - radius;
    bottomVS.z = bottomVS.y < 0.0f ? bottomVS.z - radius
                                   : bottomVS.z + radius;

    // Clamp the z coordinate to the clip planes
    leftVS.z = Clamp( leftVS.z, m_fNearClip, m_fFarClip );
    rightVS.z = Clamp( rightVS.z, m_fNearClip, m_fFarClip );
    topVS.z = Clamp( topVS.z, m_fNearClip, m_fFarClip );
    bottomVS.z = Clamp( bottomVS.z, m_fNearClip, m_fFarClip );

    // Figure out the rectangle in clip-space by applying the
    // perspective transform. We assume that the perspective
    // transform is symmetrical with respect to X and Y.
    float rectLeftCS = leftVS.x * ProjMatrix( 0, 0 ) / leftVS.z;
    float rectRightCS = rightVS.x * ProjMatrix( 0, 0 ) / rightVS.z;
    float rectTopCS = topVS.y * ProjMatrix( 1, 1 ) / topVS.z;
    float rectBottomCS = bottomVS.y * ProjMatrix( 1, 1 ) / bottomVS.z;
```

```cpp
// Clamp the rectangle to the screen extents
rectTopCS = Clamp( rectTopCS, -1.0f, 1.0f );
rectBottomCS = Clamp( rectBottomCS, -1.0f, 1.0f );
rectLeftCS = Clamp( rectLeftCS, -1.0f, 1.0f );
rectRightCS = Clamp( rectRightCS, -1.0f, 1.0f );

// Now we convert to screen coordinates by applying the
// viewport transform
float rectTopSS = rectTopCS * 0.5f + 0.5f;
float rectBottomSS = rectBottomCS * 0.5f + 0.5f;
float rectLeftSS = rectLeftCS * 0.5f + 0.5f;
float rectRightSS = rectRightCS * 0.5f + 0.5f;

rectTopSS = 1.0f - rectTopSS;
rectBottomSS = 1.0f - rectBottomSS;

rectTopSS *= m_uVPHeight;
rectBottomSS *= m_uVPHeight;
rectLeftSS *= m_uVPWidth;
rectRightSS *= m_uVPWidth;

// Final step is to convert to integers and fill out the
// D3D11_RECT structure
D3D11_RECT rect;
rect.left = static_cast<LONG>( rectLeftSS );
rect.right = static_cast<LONG>( rectRightSS );
rect.top = static_cast<LONG>( rectTopSS );
rect.bottom = static_cast<LONG>( rectBottomSS );

// Clamp to the viewport size
rect.left = max( rect.left, 0 );
rect.top = max( rect.top, 0 );
rect.right = min( rect.right, static_cast<LONG>( m_uVPWidth ) );
rect.bottom = min( rect.bottom, static_cast<LONG>( m_uVPHeight ) );

return rect;
}
```

Listing 11.14. C++ code for fitting scissor rectangle to a point light.

For a spot light, we can use the same process of fitting a bounding sphere and de-termining the screen-space extents. The code in Listing 11.14 will work for a spot light without any modifications, since the resulting bounding sphere will completely encompass the conical area of effect, regardless of the light's orientation and angular attenuation pa-rameters. If necessary, the angular attenuation factors can be used to fit a bounding cone to the light, to which a tighter bounding sphere can be fit.

An alternative to this approach is to approximate the light's bounding cone using a mesh of vertices, and then calculate the bounding box, based on the vertices' projected positions in screen space. This requires more calculations at runtime, but it can potentially result in a tighter fit than using a bounding sphere.

GPU-Generated Quads

The algorithm used for generating the extents of a scissor rectangle can be trivially adapted for use in a vertex shader or a geometry shader. This effectively offloads the computation to the GPU, and removes the need for a state change before drawing each light source. The removal of the state change allows multiple light sources to be batched together into a single draw call, which can dramatically reduce CPU overhead for large numbers of lights. See the light prepass sample for a demonstration of this approach.

Bounding Geometry

As mentioned in the previous section, it's possible to determine bounding geometry for point and spot lights, based on their position and attenuation properties, for use with the scissor test. It's possible to take this concept a step further, and create a triangle mesh representing a light's bounding volume and then render that, instead of using a full screen quad. This frees us from the limitations of screen-space rectangles and also makes natural use of the GPU's vertex processing and rasterization capabilities.

To generate the geometry, we do not need to take into account an individual light's properties. Instead, we can generate a single sphere and a single cone to be used with all lights, and then apply a world transform by calculating translation, rotation, and scale components based on the light properties. For a point light, it is simplest to create the sphere mesh with a radius of 1 and centered at the origin. Then our translation can be set to the light's position, and the X, Y, and Z scales can all be set to the light's effective range. For a spot light, a cone aligned with the Z axis (with the tip at the origin) and an angle of 45 degrees should be used. Then the translation and orientation can be set to the light's position and orientation, and the Z scale can be set to the effective range. The X and Y scales can then be set by determining the radius of the cone, which is done by calculating the tangent of the spotlight angle and multiplying with the range.

When the bounding geometry is rendered, we must also choose whether the pixels should be rendered with front-face or back-face culling. We can't simply disable culling, because pixels would get lit twice. This decision must be carefully made, because if the bounding geometry intersects with the near clipping plane or the far clipping plane of the camera, some of the front-facing or back-facing triangles will not be rasterized. Thus, if a bounding volume intersects with the near clipping plane, back-face culling should be used. If it intersects with the far-clipping plane, back-face culling should be used. If it intersects with both, this means the light is very large and should be rendered with a quad or some other method. As an alternative to switching culling modes, the vertices can also be clamped to the near and far clipping planes in the vertex shader. This can be useful if lights need to be batched together into a single draw call.

Aside from also limiting pixel shading to the bounding volume's screen space projection, using bounding geometry also allows us to make use of hardware depth testing to cull additional pixels. The idea is that we only want to light pixels where the scene geometry intersects with our light bounding meshes. This means that we want to reject pixels where the light is occluded by scene geometry, as well as pixels where the light is "floating in air." When rendering back-face geometry, we can reject the first case by setting the depth test to D3D11_COMPARISON_LESS_EQUAL as our depth comparison mode. For rendering front-face geometry, we can reject the second case by using D3D11_COMPARISON_GREATER_EQUAL. Since we must use either front-face or back-face culling, we can't reject both cases simultaneously in a single pass. As such, it can be advantageous to use a heuristic to estimate which culling/depth testing mode would result in more pixels being culled.

Stencil Test

The stencil buffer lets us store a unique 8-bit integer value per pixel, which we can use in simple logical tests that determine whether or not a pixel should be culled. With a deferred renderer, we can use the stencil buffer to store a mask that indicates which geometry needs to be lit, and thus avoid shading pixels that don't require lighting. This can result in large savings if a significant portion of the screen contains a *skybox*[10] or background geometry, or if there is a lot of scene geometry with materials that don't require dynamic lighting.

To implement this, we simply set the stencil comparison function to D3D11_ COMPARISON_ALWAYS for lit geometry while filling the g-buffer, which causes the reference value to be written to the stencil buffer for each pixel rasterized. For the reference value, we use some known value that is different than what the stencil buffer was cleared to. Then when rendering the lights, we use the same reference value and set the test to D3D11_ COMPARISON_EQUAL. This causes only pixels with the correct stencil value to be written, and any others to be culled. Since most modern graphics hardware features Hi-*Z* units that can perform *Z*-culling and stencil culling before the pixel shader is executed, the stencil test can effectively prevent pixels from being shaded at all.

This concept can also be extended to implement *light masking*. Light masking allows certain lights to only affect certain scene geometry. For example, a set of lights could be used to only light a player character, and not the level geometry. Or if a light is in a room, it can be made to only affect the walls of that room, without relying on shadowing or attenuation to keep it from "bleeding through" into adjacent rooms. Since a stencil buffer has eight bits that can be tested independently, the logical way to implement light masks is to have each bit represent a *light group* that a light or object in the scene can belong to. Essentially, a light will only affect geometry that belongs to the same group. So in the case of adjacent rooms, each room would have a unique light group that the lights and geometry

---

[10] A *skybox* is a cube of geometry rendered with the camera at the center of the cube. A cube texture is typically applied to it, giving the appearance of geometry very far from the viewer. Since it represents objects that are very far away, it would not receive lighting of this sort and can hence be eliminated.

would belong to. This way the light source doesn't "bleed through" the wall, and only affects the room it is in.

When rendering the scene geometry, we can use the same premise as above, except that we set the reference value to an 8-bit unsigned integer created by bitwise OR'ing all of the light groups to which it belongs. Then during the lighting pass, we also determine the same 8-bit value using the light groups that a light belongs to. However, instead of using that value as the reference, we set it as the stencil read mask. This mask is bitwise AND'ed with the value in the stencil buffer, and the reference value is compared with the stencil buffer value, using the specified test function. Thus, if we use a reference value of 0 and D3D11_COMPARISON_LESS as the test function, the stencil test will pass if the geometry at that pixel and the light both belong to any of the same light groups.

Screen Space Tile Sorting

This technique is another exploitation of the spatial coherence of local light sources. The basic concept involves splitting the screen into tiles of a fixed size and determining which light sources need to be rendered for each tile (Balestra & Engstad, 2008), (Lauritzen, 2010). Then the lights are all rendered in one batch or a series of batches, where the pixel shader evaluates and sums the lighting contribution for all lights in the batch. The diagram in Figure 11.10 illustrates the concept.

At first glance, it may seem that this technique is inferior to the other techniques mentioned earlier, since pixels can only be culled at the level of granularity used for the tiles. This can be especially inefficient when a small light intersects with multiple tiles. However, this technique has three key advantages. The first is that the g-buffer only needs to be sampled once for each batch, which potentially reduces the bandwidth required for rendering each light, as long as the batches contain multiple lights. This also means that the g-buffer's attributes only have to be unpacked or reconstructed once per batch, which saves shader ALU cycles. The second is that the lighting contribution only has to be written to the render target (and possibly blended) once per batch, which can further reduce the bandwidth usage of the lighting pass. The third is that it makes the lighting computations suitable for evaluation on general-purpose computation resources such as CPUs or compute shaders (Lauritzen, 2010), since no rasterization is required and the memory access is extremely coherent.

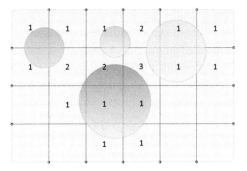

Figure 11.10. Tiled light binning.

If this technique is performed using the standard GPU rendering pipeline, it can be combined with some of the other optimizations mentioned earlier in this section. For instance, depth testing can still be used by rendering a quad and setting the view-space Z of each vertex to the maximum or minimum depth of all lights in the batch. In addition, the scissor test can also still be used to cull pixels for smaller lights.

11.5 Anti-Aliasing

Deferred rendering has many advantages, which result from the way it reorganizes the rendering pipeline. However at the same time, this reorganization makes it incompatible with certain techniques and enhancements that have previously worked with traditional forward rendering. One of the biggest examples is multisample anti-aliasing, or MSAA.[11] MSAA works by altering the rasterization process so that triangle coverage is computed at a subpixel level, so that the output of each pixel can be written to none, some, or all of the subsamples in the render target. The resulting subsamples are then blended to produce the final output. For this to work, we must be rasterizing the scene geometry while rendering the final pixels. Also, for those pixels where geometry intersects, the subsamples must contain the proper color values that would result from fully lighting and shading those triangles. With classic deferred rendering, we satisfy neither of those conditions, while with light prepass rendering, we only satisfy one. In this chapter we will discuss alternatives and workarounds for providing anti-aliasing in a deferred renderer.

11.5.1 Supersampled Anti-Aliasing

Supersampled anti-aliasing, commonly abbreviated *SSAA*, is the simplest alternative type of anti-aliasing to implement, and was mentioned briefly in Chapter 3. With SSAA, we simply render the scene at a sample rate (resolution) that is higher than the output, and then filter (downsample) the resulting image before presenting it to the display. Some GPU hardware features SSAA modes that can be toggled in the driver settings, but it can be easily implemented in software as well. In a deferred renderer, we simply make the g-buffer and light buffer be render targets larger than the back buffer, and then downsample the result of the lighting pass (or the second geometry pass, in the case of light prepass). Or alternatively, the g-buffer can simply be rendered at a higher resolution. Then, during the lighting pass multiple, g-buffer texels can be sampled and evaluated, so that the result can be filtered.

Naturally, the main problem with SSAA is performance. Increasing the resolution by a factor of 2 or 4 puts additional pressure on nearly every part of the GPU pipeline, and also greatly increases the amount of GPU memory (and corresponding bandwidth) consumed

---

[11] For more details on MSAA, see Chapter 3, "The Rendering Pipeline."

by render targets. However, SSAA is certainly effective, and it may be worthwhile to provide as an option if no other type of AA is implemented.

11.5.2 Multisampled Anti-Aliasing

As mentioned previously, breaking the rendering pipeline into 2 or 3 passes makes it incompatible with MSAA. However, we can still use hardware support for MSAA to produce an antialiased output, without having to resort to supersampling.

Let's start by analyzing the g-buffer pass. If we make our g-buffer render targets multisampled and enable MSAA while rendering them, this gives us a multisampled description of the surface and material properties. This means that for pixels where a triangle doesn't have complete coverage, the g-buffer properties for all triangles that were rasterized will be contained in the individual subsamples. Thus, if we sample and light all of these subsamples individually and then filter or average the results, we would produce an antialiased image. A sample function for performing this is provided in Listing 11.15.

```
// Textures
Texture2DMS<float4> NormalTexture            : register( t0 );
Texture2DMS<float4> DiffuseAlbedoTexture     : register( t1 );
Texture2DMS<float4> SpecularAlbedoTexture    : register( t2 );
Texture2DMS<float4> PositionTexture          : register( t3 );

// Helper function for extracting G-Buffer attributes
void GetGBufferAttributes( in float2 screenPos,
                           in int ssIndex,
                           out float3 normal,
                           out float3 position,
                           out float3 diffuseAlbedo,
                           out float3 specularAlbedo,
                           out float specularPower )
{
    // Determine our indices for sampling the texture based
    // on the current screen position
    int2 samplePos = int2( screenPos.xy );

    normal = NormalTexture.Load( samplePos, ssIndex).xyz;
    position = PositionTexture.Load( samplePos, ssIndex).xyz;
    diffuseAlbedo = DiffuseAlbedoTexture.Load(samplePos, ssIndex).xyz;
    float4 spec = SpecularAlbedoTexture.Load(samplePos, ssIndex);

    specularAlbedo = spec.xyz;
    specularPower = spec.w;
}

// Lighting pixel shader that performs MSAA resolve
float4 PSMain( in float4 screenPos : SV_Position ) : SV_Target0
```

```
{
        PSOutput output;
        float3 lighting = 0;

        for ( int i = 0; i < NUMSUBSAMPLES; ++i )
        {
                float3 normal;
                float3 position;
                float3 diffuseAlbedo;
                float3 specularAlbedo;
                float specularPower;

                GetGBufferAttributes( screenPos.xy, i, normal, position,
                                      diffuseAlbedo, specularAlbedo,
                                      specularPower );
                lighting += CalcLighting( normal, position, diffuseAlbedo,
                                          specularAlbedo, specularPower );
        }

        lighting /= NUMSUBSAMPLES;

        return float4( lighting, 1.0f );
}
```

Listing 11.15. Pixel shader code for lighting MSAA subsamples.

One issue we must be aware of with this approach is that it is not always desirable to sample all subsamples from the g-buffer and compute lighting for them. This is the case when the geometry rendered for the lighting doesn't fully cover a pixel, or when the depth/stencil tests don't pass for all subsamples. To ensure that only the appropriate samples are used, we can take SV_Coverage as an input to the pixel shader. This semantic provides a uint value, in which each bit corresponds to an MSAA sample. A bit with a value of 1 indicates that the triangle coverage test passed for the corresponding MSAA sample point, while a value of 0 indicates that it failed. If we perform a bitwise AND of this mask with a bit value for the current sample in the loop, we can determine whether we should skip the subsample. The code in Listing 11.16 demonstrates how the code from Listing 11.15 can be modified to follow this approach.

```
float4 PSMain( in float4 screenPos : SV_Position,
in uint coverageMask : SV_Coverage ) : SV_Target0
{
        PSOutput output;
        float3 lighting = 0;
        float numSamplesApplied = 0.0f;
        for ( int i = 0; i < NUMSUBSAMPLES; ++i )
        {
                if ( coverageMask & ( 1 << i ) )
```

```
                   {
                            float3 normal;
                            float3 position;
                            float3 diffuseAlbedo;
                            float3 specularAlbedo;
                            float specularPower;

                            GetGBufferAttributes( screenPos.xy, i, normal,
                                                  position, diffuseAlbedo,
                                                  specularAlbedo, specularPower );
                            lighting += CalcLighting( normal, position,
                                                      diffuseAlbedo, specularAlbedo,
                                                      specularPower );

                            ++numSamplesApplied;
                   }
         }

         lighting /= numSamplesApplied;

         return float4( lighting, 1.0f );
}
```

Listing 11.16. Pixel shader code for lighting MSAA subsamples with coverage test.

Another way to ensure proper g-buffer sampling is to run the pixel shader at per-sample frequency instead of at per-pixel frequency. In this case, the pixel shader only runs if the triangle coverage test and the depth/stencil test both passed for a given sample, so there is no need to check the input coverage mask.

Using MSAA gives us better performance than supersampling, because we don't have to render the g-buffer at a higher resolution. However, we can do even better. When rendering with MSAA, the majority of the pixels on the screen will be fully covered by a triangle from the scene geometry. Consequently, all of the subsamples of a given pixel within the g-buffer will be the same. We can take advantage of this to optimize the approach outlined above by only lighting one subsample for pixels where all subsamples are identical. One possible approach is to read all subsamples of one of the g-buffer textures in the lighting pass, compare for equivalence, and then dynamically branch based on that result. While this can work in some cases, it's possible that the subsamples will be identical (or within the threshold used) for one g-buffer texture, while one or more of the other textures are different. In addition, performance will vary depending on the granularity of dynamic branching in the GPU. If one pixel takes the slow branch, where it evaluates all subsamples, this can cause all adjacent pixels within a certain tile size to take both sides of the branch.

To avoid this issue, a stencil mask can be used, rather than branching in the pixel shader. The mask can be generated with a full-screen pass that samples the g-buffer and performs the comparison, and then uses the `clip()` or `discard` intrinsics to prevent that pixel from being output. However this can still suffer the same problems inherent with using an equality comparison on the g-buffer contents. Once the stencil mask is generated, the lighting must be rendered with two passes: one where the subsamples are lit individually and one where only one sample is used (Persson, Deferred shading 2, 2008).

Ideally, we would want to mark "edge" pixels (pixels where subsamples aren't all identical), based on the actual subpixel triangle coverage, rather than performing a comparison after the fact. In Direct3D 11 the pixel shader can get this information directly through the `SV_Coverage` semantic, making it trivial to determine if a pixel is fully covered, and then output an appropriate value to the g-buffer. The code in Listing 11.17 demonstrates how this can be implemented.

```
// G-Buffer Pixel shader that generates an MSAA mask using SV_Coverage
float4 PSMain( in PSInput input, in uint coverageMask : SV_Coverage ) : SV_Target0
{
    // Compare the input mask with a "full" mask to determine if all
    // samples passed the coverage test
    const uint FullMask = ( 1 << NUMSUBSAMPLES ) - 1;
    float edgePixel = coverageMask != FullMask ? 1.0f : 0.0f;


    ...
}
```

Listing 11.17. Shader code for edge mask generation using SV_Coverage.

MSAA with Light Pre-Pass Deferred Rendering

With light pre-pass deferred rendering, our rendering pipeline is split into three steps, instead of two. Thus, if the MSAA subsamples are resolved during the lighting pass and the final pass is rendered without MSAA, the albedo from one surface can incorrectly be applied to a lighting value that is the combined result of two different surfaces. Or alternatively, if MSAA is used for the final pass, the two separate albedo values will be used, but the lighting values will already have been combined. This will produce artifacts, since the lighting equation is nonlinear.

Ideally, the lighting pass should calculate lighting for each individual subsample in the g-buffer and then store the results to the corresponding subsample of an MSAA render target. While Direct3D 11 allows loading the subsamples of a texture, it doesn't have a mechanism for specifying separate values for the individual subsamples of a render target when a pixel shader runs at a per-pixel frequency. To work around this, we can have the lighting pixel shader run at the per-sample frequency, so that the light values for each subsample are preserved. The shader code in Listing 11.18 demonstrates this approach.

```
// Light pixel shader that executes once per MSAA subsample
float4 PSMainPerSample( in float4 screenPos : SV_Position,
in uint subSampleIndex : SV_SampleIndex )   : SV_Target0
{
        float3 normal;
        float3 position;
        float specularPower;

        // Get the G-Buffer values for the current sub-sample, and calculate
        // the lighting
        GetGBufferAttributes( screenPos.xy, ViewRay, subSampleIndex,
                                    normal, position, specularPower );

        return CalcLighting( normal, position, specularPower );
}
```

Listing 11.18. Per-sample light buffer pixel shader code.

Naturally, running pixel shaders at per-sample frequency incurs a higher overhead than running them per-pixel. To mitigate this effect, the stencil mask technique mentioned earlier can be used to mask off pixels that don't need to be shaded per-sample.

For the final pass, we no longer need to output separate values per-sample. Instead, we want to sample the light buffer for all relevant subsamples, filter the results, and finally apply the albedo terms. Thus, running the pixel shader at per-pixel frequency and relying on the coverage and depth/stencil tests is perfectly adequate. However, we must be careful not to inadvertently apply lighting from subsamples not covered by the triangle currently being rasterized, or artifacts will occur. To avoid these artifacts, we can make use of the input coverage mask, much like the lighting pass of a classic deferred renderer. The code in Listing 11.19 demonstrates how to do this.

```
float4 PSMain( in PSInput input, in uint coverageMask : SV_Coverage ) : SV_Target0
{
        // Determine our coordinate for sampling the texture based on
        // the current screen position
        int2 sampleCoord = int2( input.ScreenPos.xy );

        float3 diffuse = 0;
        float3 specular = 0;
        float numSamplesApplied = 0.0f;
        // Loop through the MSAA samples and modulate diffuse and specular
        // lighting by the albedo
        for ( uint i = 0; i < NUMSUBSAMPLES; ++i )
        {
                // We only want to apply a lighting sample if the geometry we've
                // rasterized passed the coverage test for the corresponding
                // MSAA sample point
                if ( coverageMask & ( 1 << i ) )
```

```
        {
                // Sample the light target for the current sub-sample
                float4 lighting = LightTexture.Load( sampleCoord, i );

                // Seperate into diffuse and specular components
                diffuse += lighting.xyz;
                specular += lighting.www;

                ++numSamplesApplied;
        }
    }

    // Apply the diffuse normalization factor
    const float DiffuseNormalizationFactor = 1.0f / 3.14159265f;
    diffuse *= DiffuseNormalizationFactor;

    // Apply the albedos
    float3 diffuseAlbedo = DiffuseMap.Sample( AnisoSampler, input.TexCoord ).rgb;
    float3 specularAlbedo = float3( 0.7f, 0.7f, 0.7f );
    diffuse *= diffuseAlbedo;
    specular *= specularAlbedo;

    // Final output is the sum of diffuse + specular divided by number of
    // samples covered
    float3 output = ( diffuse + specular ) / numSamplesApplied;
    return float4(output , 1.0f );
}
```

Listing 11.19. Final pass pixel shader code for light prepass deferred rendering with MSAA.

As with the lighting pass, it's possible to use a stencil mask or dynamic branching to prevent sampling and applying multiple subsamples from the light buffer. However we should keep in mind that the work done per-subsample is very little in the final pass, and thus the savings in terms of pixel shading performance are much smaller than in the lighting pass. Also, if a stencil mask is used, this means rendering all of the scene geometry twice, which makes it very likely that performance could actually be worse than if we simply looped through all of the subsamples. Either way, profiling should be done on the target hardware to determine which approach produces the best performance.

11.5.3 Screen-Space Anti-Aliasing

In this section, we'll describe techniques that apply an anti-aliasing effect in screen space after the scene has been rendered and shaded. The main advantage of such techniques is that they are mostly orthogonal to the techniques used to render the geometry, which makes them suitable not just for forward rendering, but also for deferred rendering, or any non-traditional rendering pipelines. They can also be quite a bit cheaper than MSAA, since MSAA can greatly increase the bandwidth usage and memory footprint (especially when deferred rendering is used). This also helps screen-space techniques scale better with higher resolutions. One added benefit from these techniques is that they can be applied after HDR tone mapping. When an MSAA resolve happens before tone mapping, the quality often suffers in high-contrast areas, because tone mapping uses a non-linear operator. This manifests as triangle edges that appear to have no anti-aliasing applied to them (Persson, Post-tonemapping resolve for high quality HDR anti-aliasing in D3D10, 2008). Moving the resolve to after the tone mapping step can greatly reduce the artifacts that occur in high contrast areas.

The primary disadvantage of these techniques is that they typically don't have any sort of subpixel information available to them, which makes it difficult or impossible to reconstruct signals for small or thin geometry. It also means that geometry edges can't snap to subpixel positions, as they can with MSAA, which makes it difficult to alleviate temporal aliasing artifacts.

Edge Blur

One of the oldest and simplest implementations of screen-space AA involves detecting "edges" in the scene and applying a blur to the pixels (Policarpo & Fonseca, 2005). Typically, an edge-detection operator such as a Sobel filter is used to detect edges in screen space, with the detection applied to scene depth and normal vectors. Using normal vectors and depth, rather than color or luminosity, restricts edge detection to triangle edges, preventing triangle interiors from inadvertently being blurred. Only depth is required to detect edges at mesh silhouettes, while use of normal vectors is required to detect edges where triangles intersect. However, even with normal vectors, this approach will still fail for coplanar triangles that have different material properties.

The obvious downside to this technique is that quality is generally rather poor. Applying a box filter or Gaussian filter to pixel colors can reduce aliasing artifacts somewhat, but it can also destroy detail in the process.

Morphological Anti-Aliasing

Morphological anti-aliasing (Reshetov, 2009), commonly abbreviated as *MLAA*, is another screen-space technique that uses sophisticated pattern recognition techniques to identify triangle edges. Rather than using a simple edge detection operator like Sobel that operates on a single pixel at a time, the algorithm works by splitting the screen into tiles and detecting all edge patterns within a tile. These patterns are then used to estimate the actual triangle coverage for a pixel, which is then used to blend the colors on both sides of the edges. Since the algorithm reconstructs the actual triangle edges and doesn't just rely on a blur, the results can be excellent for still images. However moving images can still suffer from temporal artifacts (due to the fundamental limitation of working in screen space with no subpixel information), although those are generally reduced compared to not having any anti-aliasing at all. The algorithm will also fail frequently on thin, wire-like geometry, where the resulting aliasing patterns are very low frequency.

The original reference implementation provided by Intel runs entirely on the CPU, using streaming SIMD instructions. Using the CPU can reduce the GPU load and can be an effective use of otherwise idle CPU cores; however, the cost of transferring render target data to and from the GPU can significantly reduce overall performance. Instead, it is likely to be more desirable to keep the implementation entirely on the GPU, which can be done using only pixel shaders (Biri, Herubel, & Deverly, 2010). A compute shader implementation is also possible and could potentially benefit from the added flexibility and shared memory resources.

11.6 Transparency

The concept of deferred rendering works extremely well for opaque geometry, since we can make the assumption that for each pixel (or subsample, in the case of MSAA), only one surface is visible. This allows us to use a g-buffer to store the surface information, since we have one g-buffer texel available for each shaded pixel. But transparent geometry causes a problem with this approach, since it breaks our assumption of 1 surface per pixel. Instead we have 0 or 1 opaque surfaces visible, with an unbounded number of transparent surfaces layered on top of it. Because of this, we have to deal with transparency as a special case when using deferred rendering.

11.6.1 Forward Rendering

The simplest way to handle transparent geometry is to fall back to forward rendering. The opaque geometry can be rendered and fully shaded using deferred techniques, and afterwards, the transparent geometry can be forward rendered and blended right on top of it. For simple transparent materials that don't require dynamic lighting or shadowing, this approach works well, without any major problems. However, if dynamic lighting is required, it essentially means that a full forward rendering pipeline has to be implemented side by side with the deferred pipeline. This negates one of the biggest advantages of deferred rendering, which is simplifying the rendering implementation. In fact, with this case, we end up with a renderer that is *more* complex than a traditional forward renderer, since two separate pipelines are needed. The transparent geometry will also suffer from all of the usual forward-rendering performance drawbacks that we discussed in the beginning of this chapter, such as poor efficiency for small triangles, and reduced batching.

Due to the downsides associated with using dynamic lighting with transparent materials, it's desirable to use simplified materials as much as possible. For many glass-like materials, dynamic lighting and shadowing don't provide the most important visual cues required for realism. Typically it's more important to provide reflections, which can be implemented with static reflection maps. The overall ambient or diffuse lighting term can also be precomputed and stored in vertices or *lightmaps*,[12] which allows the transparent geometry to blend in with the opaque geometry.

11.6.2 Multiple G-Buffer Layers

A natural extension of deferred rendering to support transparent geometry is to use multiple g-buffers to store surface information. Having N g-buffers allows up to N surfaces to be independently lit and blended using standard deferred lighting techniques, an arrangement suitable for use with overlapping transparent geometry. The major downside is that additional memory and bandwidth are required to store and write out the multiple g-buffer layers.

Coarse Binning

To fill the g-buffer layers, a simple approach is to do a *coarse binning* of transparent geometry on the CPU. With this case, the first layer is reserved for opaque surfaces, and the transparent objects are binned into the rest of the layers. Which layer an object is sent to can be determined by reserving depth ranges for certain layers, or by attempting to determine where objects overlap by using simple bounding volumes. With both approaches, it is impossible to properly handle the case of intersecting triangles, which is a fundamental limitation of order-dependent transparency methods. Then, during the lighting pass, each light can sample all of the layers and light them individually, and afterward, each layer can be blended, based on the opacity of the surface stored in the g-buffer.

Unfortunately, in this case, depth-buffer-based optimizations for light volumes won't work, since there is no longer one depth value that can be used. However if the depth is sampled in the shader for each layer, dynamic branching can be used to skip the lighting calculations if the depth is outside of the light bounds. The stencil buffer can also be used to mark pixels where a layer doesn't need to be lit, but this requires multiple passes per light. It is also possible to maintain a separate light buffer for each layer, which allows hardware depth optimizations to be applied, at the expense of more memory and an additional pass.

---

[12] A *lightmap* is a 2D texture that contains precomputed lighting values for triangle surfaces.

Depth Peeling

Another approach to handling the layers is to use *depth peeling* (Everitt, 2001). With depth peeling, all geometry is rendered for each layer. When rendering, the depth from the previous layer is sampled and compared with the depth of the surface being rendered. A given pixel is written only if its depth is greater than the same pixel in the previous layer. The result is that the overlapping geometry is "automatically" sorted, based on depth at a per-pixel level. Then, in the lighting pass, the layers are once again lit in order and blended together. The main advantage of this technique is that it is not dependent on the order of submission for geometry, which means that no CPU-side sorting is required. It also properly handles cases where transparent triangles overlap, since depth is sorted per-pixel. The obvious downside is that all scene geometry has to be rendered N times, where N is the number of depth layers in the scene. The added pixel shader cost can be mitigated somewhat if a depth-only prepass is performed first, which allows depth testing to be used to ensure that only one pixel per layer is shaded when writing out the g-buffer (Persson, Deep deferred shading, 2007).

11.6.3 Inferred Lighting

Inferred lighting (Kircher & Lawrance, 2009) is a technique that was initially developed to allow lighting to be run at a resolution lower than the final output resolution. It is a variation of light pre-pass deferred rendering, where the results of the lighting passes are bilaterally upsampled while rendering the scene geometry a second time. The bilateral upsample is performed by taking four samples from the g-buffer and comparing depth, normal, and object ID values. Samples that aren't within a threshold are rejected, which ensures that lighting performed for a different surface isn't applied to the geometry being rendered.

While the aim of the technique was primarily to enable lower-resolution shading as a performance optimization, it also enabled a novel method for handling transparent geometry. Since the bilateral filter automatically rejects pixels from the light buffer that aren't from the same mesh, it is possible to interleave transparent geometry into the g-buffer, using a stipple pattern. If the g-buffer is divided into a grid of 2×2 quads, a pattern can be used, where one texel is reserved for opaques while the other three provide three layers of transparent geometry. Thus, the opaque geometry would be rendered to the top-left texel, the first transparent layer would be in the top-right area, and so on. The diagram in Figure 11.11 illustrates this concept.

To assign transparent geometry to a particular layer, a coarse sorting pass can be used, as described in the "Multiple G-Buffer Layers" section. In this case, it is desirable to keep the sorting on the CPU, so that the shader has advance knowledge of which stipple pattern must be used. During the lighting pass, the lights can sample the g-buffer normally, without knowing or caring whether a texel belongs to opaque or transparent geometry. Finally, all

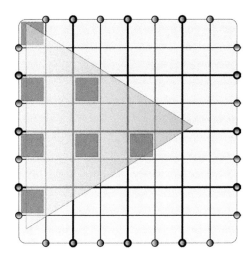

Figure 11.11. Stipple pattern for rendering transparent geometry with inferred lighting.

of the scene geometry is rendered again. The lighting buffer is sampled again, and samples from other meshes are discarded.

The result is that for quads where there is no transparent geometry, the opaque geometry is shaded at the resolution used for the g-buffer. However, for each layer of overlapping transparency, the sampling rate decreases by 25%. So in the worst case, where there are 3 layers of overlapping transparency, the opaque geometry is shaded at 1/4 rate. Transparent geometry is always shaded at 1/4 rate, due to the stipple pattern. This can cause noticeable artifacts when high frequency normal maps are used, or when shadow mapping is used. However the diffuse albedo map is still sampled at the full rate of the output resolution, preserving any high-frequency details contained in them. Aside from the sampling rate issue, there is an added performance cost from the bilateral upsample. This cost can be compensated by reducing the resolution of the g-buffer and lighting buffer; however this compounds the sampling frequency issues mentioned above.

12

Simulations

As seen in Chapter 10, "Image Processing," the compute shader can be used to implement new types of algorithms on the GPU in a very intuitive way. However, these new algorithms are certainly not limited to the image processing domain. Many different algorithms lend themselves to being executed on a massively parallel architecture, and this chapter aims to introduce several such applications. In general, as the GPU has continued to become more powerful, there has been a general trend to move more of the workload off of the CPU and onto the GPU. With this trend comes the guideline to minimize the interaction needed between the CPU and GPU during any given frame. These sample programs were designed with this concept in mind—minimal input is required by the CPU to both execute a simulation and to render it.

The first sample algorithm implements a fluid surface simulation technique. In particular, water simulation has become a very important topic in real-time rendering. Providing a fluid simulation that reacts to its environment in a realistic manner is becoming more and more desirable. This sample demonstrates how to use the compute shader to dynamically update the state of a 2D grid of fluid columns. This approximation provides convincing results, while still maintaining a data structure that can take advantage of the new features available in the compute shader.

The second sample algorithm demonstrates one of the oldest effects in computer graphics—a particle system. Even though the concept has been around for a long time, there are some new tricks to be applied to the old problem. This particle system is implemented entirely on the GPU and only interacts with the CPU through parameters provided to the system in constant buffers. This sample simulates how particles would behave around a black hole. While the physics are not calculated to scale, they provide a convincing effect, and demonstrate the general concept of a particle system that can be used as the basis for other types of simulations as well.

These two algorithms demonstrate techniques that use the compute shader to perform calculations that would have been performed on the CPU in older generations of GPU hardware. By leveraging the GPU to perform these simulations, we can directly use the results to render its current state. This reduces the required CPU work, in addition to allowing the massively parallel GPU be used to its fullest potential.

12.1 Water Simulation

Providing believable water rendering in a scene adds a significant sense of realism to its overall appearance. Using a realistic water rendering requires performing a simulation to determine the current state of the system before it can be visualized with its geometry and various optical properties. Water, or any other liquid for that matter, exhibits properties that are very complex, and that are the result of billions of individual molecules interacting with one another and their surrounding environment. Even with the processing power of modern GPUs, performing a true physical simulation of this scale is still out of reach for currently available hardware. However, to add a fluid rendering into a scene does not actually require having access to the state of every molecule in the simulation. Instead, we are only interested in the primary visual properties of the fluid. More specifically, we are interested in the *surface* of the water, since that is what we will end up rendering. Because of this, we can consider the macro-behavior of the fluid, instead of the micro-behavior of the individual elements that make up the simulation body.

This allows us to make a number of simplifications to a potential simulation that can still realistically predict the behavior of the fluid surface, but that can reduce the required computational complexity of the simulation. In the algorithm presented here, we replace the massive number of individual particles with virtual columns of fluid, arranged in a grid-like fashion. Our virtual fluid can move from column to column, but only between physically touching neighbor columns. When considered all together, the height of each column defines the height of the fluid surface at that location in the grid. This property can then be extracted from the simulation and used to generate a number of vertices that will represent the fluid surface. A grid of vertices is precisely the desired result, since it can be used directly in the subsequent rendering of the fluid surface.

By making this model simplification, we can increase or decrease the size of the simulation to allow for either a larger fluid body, or a higher resolution simulation over a smaller physical area. However, the simplification is not without its costs as well. Because of the representation we are choosing to use, it is not possible to include advanced features of a fluid body, such as splashes or breaking waves. However, the resulting behavior of the simulated fluid is still realistic in appearance, and is sufficient for a real-time rendering usage.

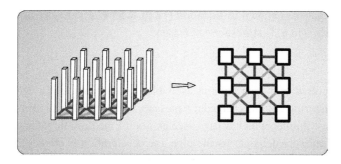

Figure 12.1. The virtual water column and pipe connections between each of them.

12.1.1 Theory

The concept presented in this article is based on a Web chat by Yann Lombard(Lombard), which references a paper entitled "Dynamic Simulation of Splashing Fluids" by James F. O'Brien and Jessica K. Hodgins (O'Brien, 1995). The paper outlines a method of simulating water interactions based on "columns" of water that are connected together by pipes. The columns are organized in a regular grid arrangement, and each column is connected to its immediate neighbors by a virtual pipe. One pipe is used in each of the eight directions outward from each column: up, down, right, left, upper left, upper right, lower right, and lower left. The algorithm is used to approximate the amount of fluid that would be transferred through each of the pipes over a given fixed time step, based on the delta in height values between each column. This water transfer is then used to modify the virtual volumes contained in the columns. This then modifies the height values in each column once again, to be used in the next simulation step. The setup of the columns and pipes is shown in Figure 12.1.

Using this model lets us provide physical parameters for both the fluid being simulated and the physical characteristics of the columns and grids. Each water column is represented as a column with a provided constant cross-sectional area. By using a constant cross-sectional area, we ensure that the volume of fluid contained within the column is directly proportional to the height of the water in the column, as shown in Equation (12.1):

$$h_{ij} = \frac{V_{ij}}{A_{ij}}.$$ (12.1)

Once the height of the fluid column is available, we can determine the maximum pressure within the column based on gravity, the density of the fluid, and the atmospheric pressure above the column. The greatest pressure in the column will occur at the bottom of the column, which is where we will assume the pipes are connected at. This maximum pressure is calculated as shown in Equation (12.2), where ρ is the density of the fluid, g represents gravity, and p_0 is atmospheric pressure:

$$H_{ij} = h_{ij}\rho g + p_0. \tag{12.2}$$

With the pressure in each column available, we can calculate a delta pressure between two neighboring columns. The delta pressure is then applied across the virtual pipe between the two columns, which causes acceleration of the fluid flow through the pipe. If there is a large pressure delta across the pipe, the acceleration is greater than that from a small delta. This relationship is shown in Equation (12.3), where c is the cross section of the virtual pipe connecting columns and m is the mass of the fluid in the column, which can be calculated from the volume of the fluid and its density:

$$a_{ij \to kl} = \frac{c(H_{ij} - H_{kl})}{m}. \tag{12.3}$$

If we assume that the flow through the pipe is constant over a time interval, the flow through the virtual pipe can be calculated as shown in Equation (12.4), where Q is the flow through the pipe:

$$Q_{ij \to kl}^{t+\Delta t} = Q_{ij \to kl}^{t} + \Delta t(ca_{ij \to kl}). \tag{12.4}$$

This can be used to calculate the total flow through every virtual pipe, which essentially specifies the total change in volume in a fluid column for a given period of time when all pipes for a fluid column are considered together. This is demonstrated in Equation (12.5), which uses an average of the previous and current flow rates over each virtual pipe:

$$\Delta V_{ij} = \Delta t \sum_{kl \in n_{ij}} \left(\frac{Q_{ij \to kl}^{t} + Q_{ij \to kl}^{t+\Delta t}}{2} \right). \tag{12.5}$$

With a change in volume available to us, we can use Equation (12.1) to determine the change in height for a water column. This is ultimately what drives the entire simulation— a difference in fluid heights between neighboring columns causes a fluid flow between them, which ultimately causes a fluid flow between columns, which subsequently alters the volume of the fluid in the columns and their corresponding heights. The process is repeated over and over again to deliver a time varying simulation that provides realistic changes in the surface of the fluid. A sample pair of neighboring columns is provided in Figure 12.2, which visually shows the parameters of our equations.

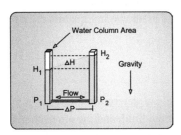

Figure 12.2. A graphical depiction of the important parameters in our simulation.

To accommodate this simulation, we need to maintain two pieces of information for each column of fluid. We need the height of each column, as well as the current flow in each virtual pipe. The height value is the property that we will be using to render the results of our simulation, and the flow values are needed to maintain the inertia of the system from time step to time step. This allows for disturbances in the fluid level to be propagated from one side of the simulation field to another and then be reflected back again.

12.1.2 Implementation Design

With a clear understanding of the theory behind this algorithm, we can explore how to use the tools that are available Direct3D 11 to implement the simulation and render the results. The update phase of the simulation is performed first, and will be executed in the compute shader. This will allow the use of some of the special capabilities of the compute shader, including data sharing and thread synchronization.

Resource Selection

As we have seen in the section on theory, we are interested in maintaining the status of the simulation as a series of height values. Since the simulation is composed of a grid of fluid columns, this height data can naturally be mapped onto a 2D texture resource with a single floating-point component. It is also necessary to maintain a record of the current flow values into and out of each of the water columns. Since there eight different neighbors that can interact with a fluid column, it would seem that we would require two four-component

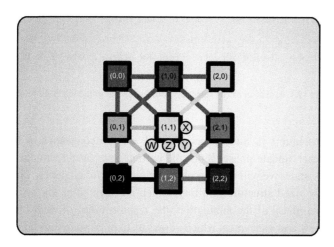

Figure 12.3. The bidirectional flow values managed for each fluid column.

floating-point textures to manage all of the needed data. However, we can take advantage of the fact that the flow into one virtual pipe has an equal and opposite flow, depending on which of the two water columns is using it. This means that we can store only four flow values for every fluid column, and that the remaining four flow values can be taken from the neighboring fluid columns. This concept is depicted in Figure 12.3.

This means that together with height data of the simulation, we need a total of five floating-point variables for each fluid column. Since the maximum number of components per-texel in a texture is four, it would be necessary to use two texture resources together to hold all of the needed data. Instead of doing this, we have chosen to use a structured buffer resource that can provide all five variables together in a single buffer element. This allows us to reference the complete state of a fluid column in a single location, and we can used a simple array syntax to access the data. Unfortunately, structured buffers are always 1D in nature, so we will have to manually convert from our 2D grid coordinates to the appropriate 1D buffer index. In practice this is fairly trivial, since we will know in advance what the total size of the simulation is, so the compromise does not introduce a large amount of additional work. Listing 12.1 demonstrates how the structured buffer is created.

```
// Here size is the number of elements, and structsize is
// the size of the structure to be used.
m_State.ByteWidth = size * structsize;
m_State.BindFlags = D3D11_BIND_SHADER_RESOURCE | D3D11_BIND_UNORDERED_ACCESS;
m_State.MiscFlags = D3D11_RESOURCE_MISC_BUFFER_STRUCTURED;
m_State.StructureByteStride = structsize;
m_State.Usage = D3D11_USAGE_DEFAULT;
m_State.CPUAccessFlags = 0;

// here pData is assumed to contain the system memory pointer to
// the initial buffer data.
ID3D11Buffer* pBuffer = 0;
HRESULT hr = m_pDevice->CreateBuffer( &m_State, pData, &pBuffer );
```

Listing 12.1. Creation of the structured buffer for use in the water simulation.

With the resource type and format selected, we can consider how the state will be updated during each update phase. Since the current state of the simulation is needed to calculate the new state, we will need to maintain two individual states. These will be kept in two identically sized structured buffers, which are created as shown in Listing 12.1. After each update phase of the simulation, the responsibility of each buffer is reversed from current to new and vice versa, which allows us to ping-pong the states back and forth between the two buffers.

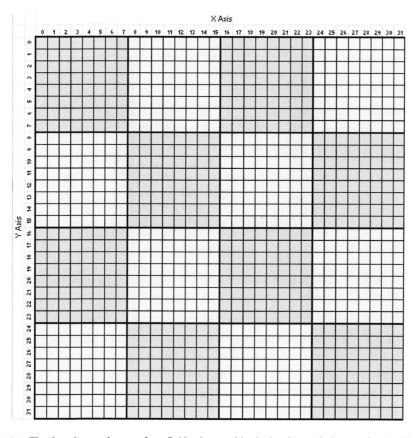

Figure 12.4. The thread group layout of our fluid column grids, depicted as an 8×8 group for visualization.

Threading Scheme

Next, we must determine what type of threading system will be used during the simulation update phase. If we assume that we will use a single thread for each fluid column to be updated, we must consider what types of data accesses it will need to make so that we can choose an appropriate layout scheme. Each fluid column will require data from all of its immediate neighbors to determine how much fluid will be flowing into and out of it. This would indicate that a square-shaped thread group would be appropriate, since it would allow adjacent fluid columns to use the group shared memory for device memory accesses, as well as for intermediate calculations. Thus, the rectangular grid of fluid columns would be split into thread groups as shown in Figure 12.4. We must choose the dispatch size to ensure that enough thread groups are created to update the entire simulation grid. In our sample application, we will be processing a 16×16 area of the grid with each thread group, so the total size of the simulation must fit to multiples of this size in each dimension.

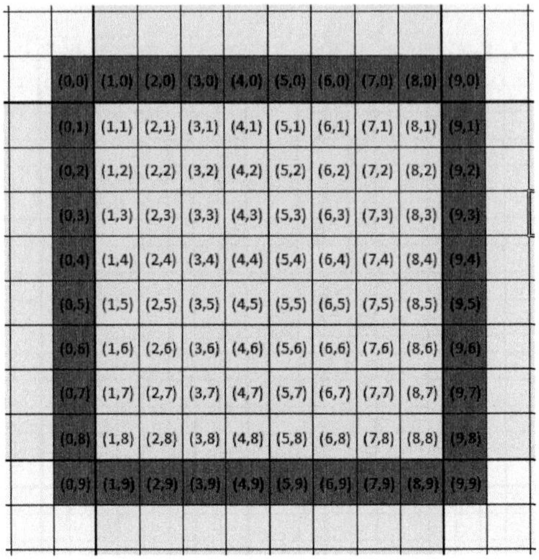

Figure 12.5. The additional threads added around the thread group, depicted with an 8×8 thread group for visualization.

With this orientation in mind, we must also consider how to handle the updating of fluid columns that fall on the outer edges of a thread group. Since each fluid column requires flow data from all of its neighbors, and the data that is required must be calculated in the current simulation pass, it would seem that we would need to write all of the updated flow calculations back to the device resources before updating the height values. This is highly undesirable, since we already have all of the flow and height data loaded to calculate the new flow values; writing them to device memory and then reloading all of that data in a subsequent shader pass essentially doubles the required device memory bandwidth. Fortunately, we can avoid this additional memory accesses by adding some additional threads to the thread group, which can calculate the additional flow values for the fluid columns on the perimeter of the thread group. In this case, we will add an additional row of threads to the complete outer perimeter of the thread group, as shown in Figure 12.5. With the additional threads, each thread group will now use a thread group size of [18,18,1], even though we will only be updating a 16×16 area of the simulation with each thread group.

Each of the threads indicated in Figure 12.5 will behave exactly like the other threads in the group. This means that they will load their height values and calculate the new flow values into and out of themselves. This data can then be stored in the group shared memory, for all threads in the original thread group to use in their final height updates. These perimeter threads will simply not calculate their updated height values, and they will also not update the output resource. The fluid columns represented by these perimeter threads will have their state updated by the adjacent thread group—these flow calculations

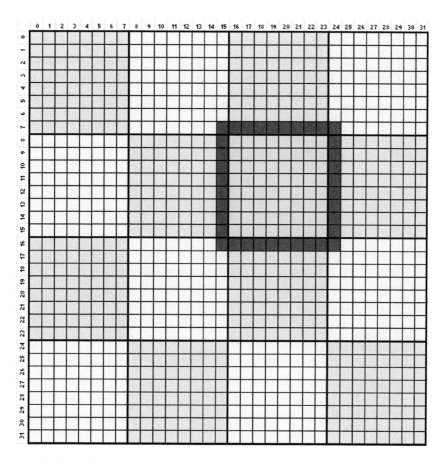

Figure 12.6. Perimeter threads overlapping with adjacent thread groups, depicted with an 8×8 thread group for visualization.

are used strictly to update the current grid of fluid columns, as we identified in Figure 12.4. The concept of using updated thread groups, with the perimeter threads shown in adjacent thread groups, is shown in Figure 12.6.

Fluid Simulation Update Phase

With a general threading model in hand, we can move on to determining the actual work that each individual thread will perform. Each thread will perform four actions. Depending on where a thread is located in the thread group and the overall simulation grid, some of these steps may be reduced or skipped. We will cover these details as they arise in the following discussion.

Accessing the current simulation state. To begin the update process, we must have access to the current state of the system. We would like to load the required information into the group shared memory to minimize device memory reads, in much the same manner as we saw in Chapter 10. To do this, we must declare an appropriately sized array of group shared memory that can hold the relevant simulation state data. This declaration is shown in Listing 12.2.

```
// Declare the structure that represents one fluid column's state
struct GridPoint
{
    float  Height;
    float4 Flow;
};

// Declare the input and output resources
RWStructuredBuffer<GridPoint> NewWaterState     : register( u0 );
StructuredBuffer<GridPoint>   CurrentWaterState : register( t0 );

// Declare the constant buffer parameters
cbuffer TimeParameters
{
    float4 TimeFactors;
    float4 DispatchSize;
};

// Simulation Group Size
#define size_x 16
#define size_y 16

// Group size with the extra perimeter
#define padded_x (1 + size_x + 1)
#define padded_y (1 + size_y + 1)

// Declare enough shared memory for the padded group size
groupshared GridPoint loadedpoints[padded_x * padded_y];

// Declare one thread for each texel of the simulation group. This includes
// one extra texel around the entire perimeter for padding.
[numthreads(padded_x, padded_y, 1)]
```

Listing 12.2. Thread group and group shared memory size declarations.

In this listing, we can see the declaration of the `GridPoint` structure, whose contents comprise the state required to represent one fluid column. We can also see that the two structured buffer resources are declared with the `GridPoint` structure as the argument to their template types. This allows us to use the structure members directly on each element of the structured buffer, making the shader code a bit easier to read. Next, we declare the constant buffer that we will use in this simulation update. These parameters will be discussed individually in the following sections as needed.

With the resources declared, we move on to the declaration of the thread group size and the group shared memory. The size of the area that is being updated with this thread group is defined with the size_x and size_y parameters. The corresponding padded size is then declared with an extra 1 on both sides of it, to represent the additional perimeter of threads around the thread group. We use this padded size to declare both the number of threads to use in this thread group and the group shared memory array of GridPoint structures. The ability to declare a structure as the array element type for the group shared memory is a nice feature, and it demonstrates the flexibility of GSM usage to represent the various data types that are needed. With this layout, we can access the group shared memory with the same array-like syntax that we use with the structured buffers, which further simplifies the shader program.

With all of the needed memory and resources allocated, the algorithm can begin. As mentioned above, the current state of the simulation must be loaded into the group shared memory. Each thread in the thread group, including the extra perimeter of threads, will load the state of the fluid column that it represents. The individual states that correspond to a thread can be visualized in Figure 12.6. The calculation of the proper index to load must take into account the fact that the dispatch thread ID will no longer be a one-to-one mapping with the simulation grid, because of the additional perimeter threads. This is shown in Listing 12.3 with the calculation of the location variable, which provides the 2D location in the simulation grid for the current thread based on the sizes provided in the DispatchSize variable. This variable is provided by the application through a constant buffer and provides the dispatch size as well as the overall simulation grid dimensions. The location is then flattened into the 1D buffer index with the textureindex variable. The data at the textureindex offset is loaded by each thread, then stored directly into the group shared memory for use later in the shader. Once the data is loaded, we perform a memory barrier with a group synchronization, to ensure that the entire thread group has loaded its data into the GSM before proceeding.

```
// Grid size - this matches your dispatch call size!
int gridsize_x = DispatchSize.x;
int gridsize_y = DispatchSize.y;

// The total texture size
int totalsize_x = DispatchSize.z;
int totalsize_y = DispatchSize.w;

// Perform all texture accesses here and load the group shared memory with
// last frame's  height and flow. These parameters are initialized to zero to
// account for  'out of bounds'texels.
```

```
loadedpoints[GroupIndex].Height = 0.0f;
loadedpoints[GroupIndex].Flow = float4( 0.0f, 0.0f, 0.0f, 0.0f );

// Given your GroupThreadID and the GroupID, calculate the thread's location
// in the buffer.
int3 location = int3( 0, 0, 0 );
location.x = GroupID.x * size_x + ( GroupThreadID.x - 1 );
location.y = GroupID.y * size_y + ( GroupThreadID.y - 1 );
int textureindex = location.x + location.y * totalsize_x;

// Load the data into the GSM
loadedpoints[GroupIndex] = CurrentWaterState[textureindex];

// Synchronize all threads before moving on to ensure everyone has loaded
// their state into the group shared memory.
GroupMemoryBarrierWithGroupSync();
```

Listing 12.3. Loading the current simulation state into the group shared memory.

Calculate the updated flow values. After the current simulation state is available in the GSM, we can start calculating the new state of the simulation. This requires us to determine how much flow is being induced in each of the virtual pipes between the current thread's neighbors. As mentioned earlier, we only need to keep track of four flow values for each thread, and the remaining four flow values can be read from the neighbor fluid column's flow variables. We will choose the convention that each of the four components of the **float4** flow variable will indicate the flow to the right, to the lower right, to the bottom, and to the lower left neighbors of the fluid column. This is demonstrated in Figure 12.7, which also shows how the neighboring fluid columns can be used to read the corresponding flow values.

With this orientation in mind, we must calculate the delta in height values between the current fluid column and its neighbors. To do this, we initialize a NewFlow variable to zero flow. Then we calculate the appropriate delta value for each component of the variable, but only if the current thread isn't at the edge of the simulation. It is very important to ensure that we don't calculate a delta value with a non-existent fluid column, because we will either access an incorrect fluid column (due to the 1D buffer storage) or access an out-of-range index. The former will probably introduce erroneous delta values at strange locations in the simulation, and the latter will cause a delta value to be created against a zero

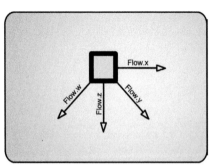

Figure 12.7. The flow values, and which virtual pipes they represent.

value that is returned in out-of-range accesses. In either case, this is undesirable and should
be avoided. The process for calculating these delta values is shown in Listing 12.4.

```
// Initialize the flow variable to the last frames flow values
float4 NewFlow = float4( 0.0f, 0.0f, 0.0f, 0.0f );

// Check for 'not' right edge
if ( ( GroupThreadID.x < padded_x - 1 ) && ( location.x < totalsize_x - 1 ) )
{
    NewFlow.x = ( loadedpoints[GroupIndex+1].Height
                - loadedpoints[GroupIndex].Height );

    // Check for 'not' bottom edge
    if ( ( GroupThreadID.y < padded_y - 1 ) && ( location.y < totalsize_y - 1 ) )
    {
        NewFlow.y = ( loadedpoints[(GroupIndex+1) + padded_x].Height
                    - loadedpoints[GroupIndex].Height );
    }
}

// Check for 'not' bottom edge
if ( ( GroupThreadID.y < padded_y - 1 ) && ( location.y < totalsize_y - 1 ) )
{
    NewFlow.z = loadedpoints[GroupIndex+padded_x].Height
              - loadedpoints[GroupIndex].Height;

    // Check for 'not' left edge
    if ( ( GroupThreadID.x > 0 ) && ( location.x > 0 ) )
    {
        NewFlow.w = ( loadedpoints[GroupIndex + padded_x - 1].Height
                    - loadedpoints[GroupIndex].Height );
    }
}
```

Listing 12.4. Calculating the height delta between fluid columns.

The delta height values can then be combined, using a number of constants to represent the physical properties of the simulation. These constants are declared as literal
constants in the shader, but they could just as easily be loaded into the shader through a
constant buffer if they will change over the lifetime of the application. The calculation of
the new flow values is shown in Listing 12.5, where the NewFlow variable is updated to
hold the new flow values, instead of the delta heights. The new flow value is a combination
of the existing flow in the virtual pipe from the previous simulation step and the additional
flow induced by the delta in height values, and is also based on the size of the time step
that has passed since the last update. The application provides the elapsed frame time in
the TimeFactors.x parameter, and a minimum selection is made between the actual time

step and a maximum step size constant. This allows the simulation to be applied over reasonable time steps, even if the reference device is being used to perform the calculations. Finally, a damping factor is used to gradually reduce the amount of flowing inertia in the system. This will eventually allow the system to return to an equilibrium state after an appropriate period of time.

The recalculated flow values are then written back to the group shared memory, so they may be used by all of the threads in the simulation to update their current height values. An important consideration here is that the perimeter threads still calculate their updated flow values, but only write them into the group shared memory. These flow values will eventually be discarded, but they will be needed in the next phase of the implementation to produce the correct height values for the fluid columns that they are in contact with. After the group shared memory is written to, we perform another memory barrier to synchronize the thread group's execution.

```
const float TIME_STEP = 0.05f;
const float PIPE_AREA = 0.0001f;
const float GRAVITATION = 10.0f;
const float PIPE_LENGTH = 0.2f;
const float FLUID_DENSITY = 1.0f;
const float COLUMN_AREA = 0.05f;
const float DAMPING_FACTOR = 0.9995f;

float fAccelFactor = ( min( TimeFactors.x, TIME_STEP ) * PIPE_AREA * GRAVITATION )
                     / ( PIPE_LENGTH * COLUMN_AREA );

// Calculate the new flow, and add in the previous flow value as well. The
// damping factor degrades the amount of inertia in the system.
NewFlow = ( NewFlow * fAccelFactor + loadedpoints[GroupIndex].Flow )
          * DAMPING_FACTOR;

// Store the updated flow value in the group shared memory for other threads
// to access
loadedpoints[GroupIndex].Flow = NewFlow;

// Synchronize all threads before moving on
GroupMemoryBarrierWithGroupSync();
```

Listing 12.5. Calculating the new flow values based on the delta heights in the current simulation step, as well as the previous step's flow values.

This is an appropriate time to consider the memory accesses that were required to perform the new flow calculations. All of the height values and the existing flow values are read from the GSM, meaning that this phase of the calculation does not touch the device memory at all. In addition, since we synchronized all of the threads in the thread group after filling the GSM, we effectively eliminated the need for further low-level synchronization during the calculation of the flow values. This makes an efficient use of the GSM and balances out memory barrier synchronization with a fair amount of arithmetic, making it worthwhile to use it.

Calculate the updated height values. With all of the flow values available in the GSM, we can trivially read the current height values and apply the flows out of the current fluid column. We can also read the flow values from the neighboring fluid columns and negate their value to apply the proper magnitude flow into those columns. Since we are only reading from data that was already synchronized in the GSM in the previous step, we can read all of these data values without any further synchronization. This process is shown in Listing 12.6. In contrast with the thread masking that was required earlier, there is no need to perform the conditional instructions in this case. This is because any flow values to out-of-bounds fluid columns will have a zero value. This is in fact why we masked the threads off earlier in the process.

```
// Calculate the new height for each column, then store the height and
modified flow values.
// The updated height values are now stored in the loadedheights shared memory.

// Out of the current column...
loadedpoints[GroupIndex].Height = loadedpoints[GroupIndex].Height + NewFlow.x
                                + NewFlow.y + NewFlow.z + NewFlow.w;

// From left columns
loadedpoints[GroupIndex].Height = loadedpoints[GroupIndex].Height
                                - loadedpoints[GroupIndex-1].Flow.x;

// From upper left columns
loadedpoints[GroupIndex].Height = loadedpoints[GroupIndex].Height
                                - loadedpoints[GroupIndex-padded_x-1].Flow.y;

// From top columns
loadedpoints[GroupIndex].Height = loadedpoints[GroupIndex].Height
                                - loadedpoints[GroupIndex-padded_x].Flow.z;

    // From top right columns
loadedpoints[GroupIndex].Height = loadedpoints[GroupIndex].Height
                                - loadedpoints[GroupIndex-padded_x+1].Flow.w;
```

Listing 12.6. Calculating the new height values for this simulation step.

Storing the simulation state. The final process in updating the simulation state is to update the output structured buffer resource with the newly calculated height and flow values, so they can be used in the next simulation step. This is done through an unordered access view, which is the only means of output for the compute shader. The writing of the values

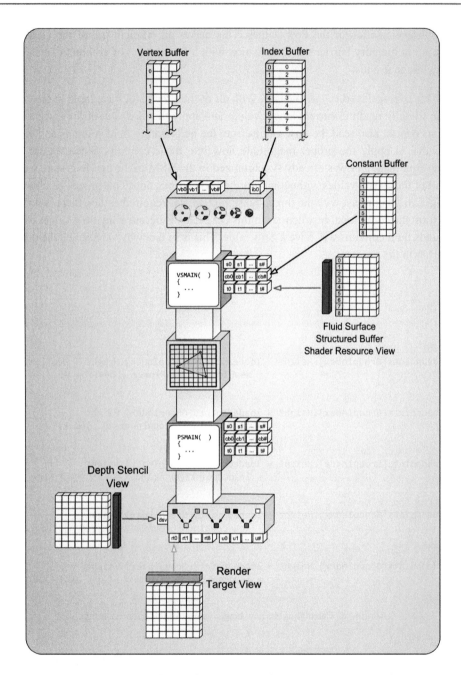

Figure 12.8. The rendering pipeline layout used for visualizing the fluid surface.

to the output resource is shown in Listing 12.7. This final step must mask off the perimeter threads to ensure that they don't modify the contents of the output resource. In theory, all of these values would be overwritten by the other thread groups being executed, but depending on the order that the thread groups are processed in it may lead to invalid data being written structured buffer. The writing of the output data is the first and only access that is performed through the unordered access view. This is a good property to replicate in other algorithms, as it indicates that the device memory has a minimum number of accesses.

```
// Finally store the updated height and flow values only for the threads
// within the actual group (i.e. excluding the perimeter). Otherwise there would
// be a double calculation for each perimeter texel, making the simulation
// inaccurate!
if ( ( GroupThreadID.x > 0 ) && ( GroupThreadID.x < padded_x - 1 )
     && ( GroupThreadID.y > 0 ) && ( GroupThreadID.y < padded_y - 1 ) )
{
    NewWaterState[textureindex] = loadedpoints[GroupIndex];
}
```

Listing 12.7. Storing the results of the simulation to the output structured buffer.

Rendering the Fluid Simulation

After the simulation state is updated, the surface of the fluid can be rendered. However, by this point all of the hard work has already been completed and we simply need to read the height values in the structured buffer that contains the current state of the simulation. The pipeline layout is shown in Figure 12.8.

We perform this height lookup in the vertex shader using the same calculation that was used in the compute shader to find the location within the structured buffer to load. The difference in this case is that we use input vertex attributes to determine which point in the 2D simulation grid each vertex should represent. To demonstrate the boundaries of each thread group that is used in the simulation, we color the vertices with one of

Figure 12.9. The final result of the rendered fluid surface.

two colors, which are alternated for each successive thread group. This produces a checkerboard pattern that can be seen in the final rendered output, as shown in Figure 12.9.

12.1.3 Conclusion

This fluid simulation provides an efficient method to represent realistic, time-varying fluid surfaces. The simulation is run completely on the GPU, freeing up the CPU to perform other tasks. Since the simulation grid size is not fixed, it can be expanded to any desired size, as long as it fits within a structured buffer resource. This allows for potentially huge simulation spaces, or extremely high resolution simulations of smaller spaces. In addition, since it leverages a highly parallel algorithm, the performance of the technique will scale well with future hardware performance improvements.

12.2 Particle Systems

Fluid rendering is a very useful and visually appealing simulation to add to a scene. However, its use is restricted to a fairly specific domain of situations due to the fact that it represents a fluid. There is a whole class of additional natural and man-made phenomena that are desirable to add to a scene, and that are fundamentally different than fluids. These effects include smoke, fire, sparks, or debris from an explosion, just to name a few. As with fluid simulation, the visual appearance of these phenomena is the result of many millions or billions of individual molecules continually interacting with one another in many different ways, over a period of time. This is simply not possible to directly simulate in a real-time rendering context; hence, we must find a more efficient method to produce a rendered image sequence that can approximately produce a similar appearance.

One potential technique for implementing these effects is the use of particle systems. A particle system is a construct composed of many individual elements, referred to as particles. Each particle has a unique set of variables associated with it to define its current status; when considered at the same time, the particles form a particle system. The simulation portion of this concept requires us to define an algorithm for creating new particles, destroying old particles, and incrementally updating the state of the particles between creation and destruction events for a small increment of time. Each particle is created, then is updated once in each simulation step, and finally destroyed after it is no longer needed. The main constraint in a particle system is that all particles must use the same set of variables, which allows their update method to operate in the same way on each particle.

This technique of using many unique elements and updating them with the same set of rules is used to derive complexity out of many simple components. For example, to simulate the smoke rising from a fire, each particle represents one puff of smoke. Each particle maintains a position, a velocity, and perhaps a rotation and scale, all of which are updated every time step of the simulation. Here, the update method generally makes each particle drift in some direction, with a semi-perturbed pathway. After each simulation step, the particle system is rendered with each particle being represented as a small quad with a smoke-puff texture applied to it. With all of the particles moving independently, but with the same rules applied to them, we can model a more complex system with a simple one. The same concept applies to the other examples mentioned above, except that perhaps the particle properties, the rendering attributes, and the update method would be changed to the appropriate version for that type of particle system.

Particle systems have been used in computer graphics for a very long time. In the past, they were implemented on the CPU and then simply rendered after each update. However, with the parallel nature of the GPU and the very parallel nature of the updating mechanism, the GPU is an ideal processor to perform the simulation. In addition, since the simulation results are already residing in video memory, the rendering process doesn't need to transfer it out of system memory, which results in faster rendering. The sample program discussed in this section implements such a GPU-based particle system, while taking advantage of some of the new features in Direct3D 11. This particle system provides a specific implementation of a particular type of system, but it can easily be adapted to support other particle systems as well.

12.2.1 Theory

The particle system we will build in this section represents a particle emitter and consumer, where the emitter creates particles and the consumer destroys them when they get too close to it. The system will be governed by a simple gravity system based around the consumer. This could be thought of as a simplified black hole, which exhibits a large gravitational pull on each particle and will swallow them up once they pass the event horizon. Before moving to the implementation of the particle system, we will examine the basic laws of gravitation in order to implement our particle update method in a physically plausible manner.

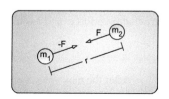

Figure 12.10. Two point masses attracting one another, based on Newton's law of universal gravitation.

Newton's law of universal gravitation states that there is an attractive gravitational force between two point masses, which is proportional to the product of the two masses, and inversely proportional to the square of the distance between them. This relationship is depicted in Figure 12.10, and is defined in Equation (12.6), where F is the gravitational force, G is the gravitational constant, m_1 and m_2 are the masses of the two objects, and r is the distance between their center of mass:

$$F = G\frac{m_1 m_2}{r^2}.$$ (12.6)

According to this relationship, we can see that the closer two objects are to each other, the greater the attractive gravitational force between them becomes. In addition, the more massive the objects are the greater this force is as well. This makes sense when considering a black hole, which represents a singularity of infinitely large mass (this is a simplification, but it will serve as an adequate explanation for our purposes). If an object approaches a black hole, at some point it will be too close to the black hole and will be dragged into it. The acceleration of the particle caused by the gravitational force is calculated with Newton's second law, which is shown in Equation (12.7). The acceleration can be found by dividing the gravitational force by the mass of the object that it is acting on:

$$F = ma.$$ (12.7)

In the context of our simulation, the black hole will represent a very large mass instead of an infinite mass, due to the obvious calculation issues with using infinite numbers. Each particle will have a fixed mass and will be subjected to the gravitational pull of the black hole. In addition to the gravitational effects on the particles, they will be created with a randomized initial velocity as they are emitted from the particle emitter. This will let the user see where the particle emissions are coming from, in addition to where they are being attracted to. To calculate the particle velocity after each simulation step, we will use Equation (12.8) where v_0 is the initial velocity at the beginning of the time step, a is the acceleration caused by the gravitational force, and t is the amount of time that has passed in this time step:

$$v = v_0 + at.$$ (12.8)

After the new velocity of the particle has been determined, we can determine the modified position of the particle over the current time step. This is performed as shown in Equation (12.9). With these basic physical interactions clarified, we can continue to the implementation design that will use the GPU to efficiently simulate how these bodies will interact:

$$p = p_0 + vt.$$ (12.9)

12.2.2 Implementation Design

The concept of a particle system is well known, but implementing one with the help of the new features of Direct3D 11 is not. In this section we will explore one possible implementation and discuss the features that are used. Our particle system will be simulated in the compute shader, and then rendered using the rendering pipeline in a similar fashion as we have seen in the fluid simulation sample. We will begin with a discussion of what type of resources to use, followed by a description of how we will add particles into the system. We will then clarify how the particle update phase is performed, and finally will discuss the rendering technique used to present the results of the simulation.

Resource Selection

To select an appropriate resource type, we once again must consider the type of data we will be storing in it. We would like to store a list of all of the particles that exist within the particle system. Since the particles can be stored in a list, this implies a 1D storage type. In addition, the data that will be stored for each particle will consist of at least a position and a velocity, plus whatever additional specialized data is needed for a particular type of particle system. Since there are more than four individual data items to store (the position and velocity are both vectors), it is not possible to use a single 1D texture to hold the data. Instead, we will use a structured buffer to hold our particles. This will allow the creation of a customized structure for our particle data and will provide a very easy method of accessing the data.

The nature of a particle system allows for the use of one of the special variants of a structured buffer. We have seen in Chapter 2 that a specially made unordered access view allows a structured buffer to be used as an `AppendStructuredBuffer` or a `ConsumeStructuredBuffer` from within HLSL. The append and consume buffers allow for a simple storage mechanism, in which the order of the elements is not required to be preserved, which in turn allows for some optimizations to be implemented in the GPU when reading and writing to and from the resource. Our particle system is comprised of a varying number of particles, and there is no reason to be concerned with the order that they appear within the buffer since they are all treated in exactly the same

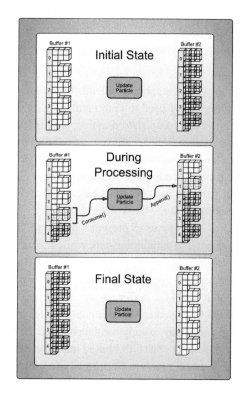

Figure 12.11. Two structured buffer resources used as append and consume buffers to manage particle data.

manner. In other words, they all are updated in the same fashion, so there is no need to know precisely which particle is being processed by a particular thread. This is a fundamental difference between the fluid simulation and the particle system—the fluid simulation columns all rely on information from their neighbors, which requires that each column know its exact location within the simulation. The unordered nature of the particle system makes it particularly well suited to using append and consume buffers to hold them.

With this in mind, we will use two structured buffer resources to contain our particle system. One will hold the current particle data, and the other will receive the updated particle data after an update sequence is performed. Figure 12.11 depicts how these resources are used to provide and then receive the particle data. After an update pass is performed, the buffer that received the updated data will then become the current state of the simulation, and the process can be repeated in the next simulation step to refill the other buffer. Before a simulation update is performed, the current state buffer holds all of the particle data while the other buffer is empty. During processing, the elements are consumed from the current state buffer, and then appended to the new state buffer. After processing, the roles are reversed, and the new state buffer holds all of the particle data.

Threading Scheme

Now that we have chosen our resource model, we must consider how we will invoke an appropriate number of threads to process these data sets. As we described above, the particles do not need to communicate with one another at all, which means there is no need to use the group shared memory. This lets us choose a threading orientation based solely on ensuring that an appropriate number of threads are instantiated to process all of the particles that are currently active. The upper limit to the number of particles in the system is dependent on the size of the buffer resources that are created to hold their data. The number of particles can also vary downward to a smaller number if particles have died off or are destroyed in the black hole.

This seems to present a problem. The number of particles can change from simulation step to simulation step, while the number of active particles is available in the buffer resource. However, the CPU is in control of specifying how many threads to instantiate through its dispatch methods. There is a device context method for reading the element count out of the buffer resource, but this count is copied into another buffer resource. If the CPU tried to map the secondary buffer into system memory, it would require quite a significant delay to copy the data back to the CPU for reading. That would negate the benefits of having the fast GPU-based implementation, since we would be synchronizing data to the CPU in every frame. So instead of trying to read back the number of particles, we can take a more conservative approach with our thread invocation. We can estimate the number of particles in the system, based on the properties of the creation and destruction mechanisms that are used, and then round up to the next higher number of threads specified by our thread group size. For example, if we choose the maximum size thread group of [1024,1,1], we would round our estimated number of particles up to the next multiple of 1024.

But what happens when the additional threads from rounding up try to read from the consume buffer and no particle data is available? Trying to consume more data than is present produces an undefined behavior, so we must ensure that this does not happen. We can manage this by copying the element count to a constant buffer with the CopyStructureCount method, and then masking off the particle processing code with a conditional statement to ensure that the appropriate number of elements is processed. The threading setup is shown in Listing 12.8, along with the required resource declarations.

```
struct Particle
{
    float3 position;
    float3 velocity;
    float  time;
};

AppendStructuredBuffer<Particle>  NewSimulationState : register( u0 );
ConsumeStructuredBuffer<Particle> CurrentSimulationState   : register( u1 );

cbuffer SimulationParameters
{
    float4 TimeFactors;
    float4 EmitterLocation;
    float4 ConsumerLocation;
};

cbuffer ParticleCount
{
    uint4 NumParticles;
};

[numthreads( 512, 1, 1)]

void CSMAIN( uint3 DispatchThreadID : SV_DispatchThreadID )
{
    // Check for if this thread should run or not.
    uint myID = DispatchThreadID.x
             + DispatchThreadID.y * 512
             + DispatchThreadID.z * 512 * 512;

    if ( myID < NumParticles.x )
    {
        // Perform update process here...
    }
}
```

Listing 12.8. The threading setup used to update all of the particles in the system.

In this listing, we see the particle structure defined as a position, velocity, and a scalar time value. The particle structure is then used to declare the append and consume buffers. Next is the declaration of the constant buffers that are used, which consist of the simulation parameters and the number of particles will be stored in the `ParticleCount` buffer. After these resources are specified, we declare the thread group size of [512,1,1], which means that our granularity for instantiating threads is in 512-thread increments. In the body of the shader, we determine the global dispatch thread ID of the current thread by flattening the 3D dispatch thread ID. It is then compared to the number of particles in the consume buffer, which was copied into the constant buffer with the `CopyStructureCount` method.

This arrangement has some additional benefits besides the unordered access optimizations. Since the append and consume buffers manage the counting of the total number of elements that are present in the buffers, we can customize our processing to consider only those buffer elements that are currently active. This decreases the application's sensitivity to knowing how many particles are in existence, as long as the maximum number of elements is not exceeded. With this type of processing, we effectively decouple the CPU from the GPU for this processing action and allow the GPU to perform as quickly as it is capable of, without requiring synchronization with the CPU.

Particle Creation

Now that we know how we will store our particles, and how we will approach them from a threading perspective, we can begin looking into how to add particles into the simulation. There are two different methods used in this sample. The first possibility is that we can initialize the system with a particular number of particles at startup. The second possibility is to add particles into the simulation during runtime. We will look deeper into these two options in the following sections.

System initialization. When we begin the simulation, we have the option to begin with a given number of particles from the very first simulation step. This can be accomplished in two steps. First we must initialize the contents of the structured buffers to hold our desired data. As we have seen in Chapter 2, we can provide an initial set of data for a resource in its creation function. For use in a particle system, this would entail filling in the initial data with an array of particles. If we don't want to initialize the entire contents of the particle system, we must still provide a complete set of initial data to initialize the resource. The second step to providing the initial set of particles is to tell the unordered access view how many particles should be considered to exist within the structured buffer. This is done through setting the initial counts parameter in the **ID3D11DeviceContext::CSSetUnordered AccessViews()** method.

Specification of how many elements exist in the buffer is only needed during the first use of the buffers. Once the first update phase has been completed, all of the active particles will have been updated and appended to the output buffer through the compute shader program. Since the writing of the data uses the append method, the count is automatically incremented for each particle added to the output buffer. Likewise, the buffer that provides the initial state of the simulation will have its count automatically decremented for each consume operation performed on it, and should end up with a count of zero after the first update phase is performed. Thus, the buffer that contains the initial particle data should get an initial count that indicates the initial number of particles in the system, and the second structured buffer will get an initial count of zero only for the first binding sequence. For each update phase after the first one, both buffers should be bound with an initial count of -1 to indicate that the internal count value should be used.

Dynamic injection of particles. Since we want to have a dynamic particle system, it is necessary to be able to add new particles into the simulation during runtime. However, after the initial resource creation, the CPU cannot directly modify the contents of the buffer (since it uses a default usage). Even so, the CPU is still able to perform particle injection indirectly, by running a separate compute shader program in which each thread that is invoked will create a new particle and add it to the output buffer resource. In this way, we can allow the CPU to control the emission of new particles, and can still use the parallel nature of the GPU to efficiently implement the desired number of new particles in batches.

For our sample program, we want to produce particles that emanate from an emitter location, and to have them initialized with a randomized velocity direction. To do this, we will create eight particles at a time for each thread group that is dispatched. We will use a static array of direction vectors which point to the eight corners of a unit sized cube, and then reflect the vectors about a random vector provided by the application. Then, each of the eight threads in the thread group will read one of the static vectors, reflect it, and use the resulting vector to specify the initial velocity vector for our system. The intended use of this compute shader program is to dispatch a single thread group at a time, with a single randomized vector supplied in a constant buffer to aid in the initialization of the particles. If more thread groups are needed at the same time, a randomized vector could be read out of a resource for each thread group by using the SV_GroupID system value semantic. This compute shader to insert particles is shown in Listing 12.9.

```
struct Particle
{
    float3 position;
    float3 direction;
    float  time;
};

AppendStructuredBuffer<Particle> NewSimulationState : register( u0 );

cbuffer ParticleParameters
{
    float4 EmitterLocation;
```

```
    float4 RandomVector;
};

static const float3 direction[8] =
{
    normalize( float3(  1.0f,  1.0f,  1.0f ) ),
    normalize( float3( -1.0f,  1.0f,  1.0f ) ),
    normalize( float3( -1.0f, -1.0f,  1.0f ) ),
    normalize( float3(  1.0f, -1.0f,  1.0f ) ),
    normalize( float3(  1.0f,  1.0f, -1.0f ) ),
    normalize( float3( -1.0f,  1.0f, -1.0f ) ),
    normalize( float3( -1.0f, -1.0f, -1.0f ) ),
    normalize( float3(  1.0f, -1.0f, -1.0f ) )
};


[numthreads( 8, 1, 1)]

void CSMAIN( uint3 GroupThreadID : SV_GroupThreadID )
{
    Particle p;

    // Initialize position to the current emitter location
    p.position = EmitterLocation.xyz;

    // Initialize direction to a randomly reflected vector
    p.direction = reflect( direction[GroupThreadID.x], RandomVector.xyz ) * 5.0f;

    // Initialize the lifetime of the particle in seconds
    p.time = 0.0f;

    // Append the new particle to the output buffer
    NewSimulationState.Append( p );
}
```

Listing 12.9. The compute shader for dynamically adding particles into the system.

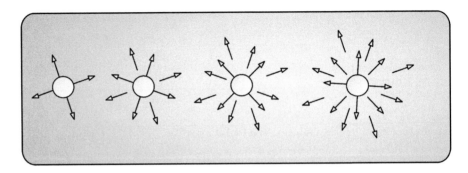

Figure 12.12. Particles emitted in randomized directions from an emission point.

The result of executing this compute shader is to insert eight particles into the system at uniformly distributed, but randomized directions. If this is performed several times in succession, the result is a halo of particles emitted from the emission point in all directions. This is depicted in Figure 12.12, which is a screen shot where the particles are allowed to continue travelling in their initial velocity directions, as determined by the randomized vectors.

Particle State Update Phase

Once all of the particles have been added into the particle system, we can finally execute the actual simulation calculations to update the state of the particle system. This is performed in a compute shader that follows the threading model described in the "Threading Scheme" section. Essentially we dispatch an appropriate number of thread groups to ensure that we have enough threads to process all of the particles in the simulation. In our sample, we can simply retain a count of the number of particles that have been created, and assume that the particles remain in existence for their entire possible lifetime. This estimate is then used to round up to the next higher multiple of 512 to determine how many thread groups to dispatch.

The particle system update shader is set up as shown in Listing 12.8. The update method that is used within the inner loop to actually update the particle is directly based on the physically based gravitation equations described in the "Theory" section. The portion of the shader for executing this update method is shown in Listing 12.10.

```
static const float G = 10.0f;
static const float m1 = 1.0f;
static const float m2 = 1000.0f;
static const float m1m2 = m1 * m2;
static const float eventHorizon = 5.0f;

[numthreads( 512, 1, 1)]

void CSMAIN( uint3 DispatchThreadID : SV_DispatchThreadID )
{
    // Check for if this thread should run or not.
    uint myID = DispatchThreadID.x
            + DispatchThreadID.y * 512
            + DispatchThreadID.z * 512 * 512;

    if ( myID < NumParticles.x )
    {
        // Get the current particle
        Particle p = CurrentSimulationState.Consume();

        // Calculate the current gravitational force applied to it
        float3 d = ConsumerLocation.xyz - p.position;
        float r = length( d );
```

```
float3 Force = ( G * m1m2 / (r*r) ) * normalize( d );

// Calculate the new velocity, accounting for the acceleration from
// the gravitational force over the current time step.
p.velocity = p.velocity + ( Force / m1 ) * TimeFactors.x;
// Calculate the new position, accounting for the new velocity value
// over the current time step.
p.position += p.velocity * TimeFactors.x;

// Update the life time left for the particle.
p.time = p.time + TimeFactors.x;

// Test to see how close the particle is to the black hole, and
// don't pass it to the output list if it is too close.
if ( r > eventHorizon )
{
    if ( p.time < 10.0f )
    {
        NewSimulationState.Append( p );
    }
}
}
}
```

Listing 12.10. Updating the state of each particle.

If the current thread corresponds to a live particle, it consumes one particle from the consume buffer. It then calculates the gravitational force from the black hole, the acceleration caused by this force, the new velocity caused by the acceleration, and finally the new position is calculated from the updated velocity. Once the particle state has been updated, we finally test to see if we are within an eventHorizon radius of the black hole. If we are, then the particle is not appended to the output append buffer. This effectively eliminates the particle from the system, since it isn't propagated to the next simulation step. Similarly, if a particle has been alive longer than a threshold amount of time, it is not added to the output buffer.

Rendering the Particle System

After performing a simulation update phase, we have a structured buffer that contains the current state of the particle system. The next step is to render the results of our simulation in an efficient manner. The pipeline configuration is shown in Figure 12.13 for reference during the description. As discussed earlier, if at all possible, we are trying to keep the CPU from having to read back any information about the particle system. For rendering the results of the particle system, we want to do the same thing and avoid having to read the particle data back to system memory. Unfortunately, we cannot directly use the structured buffer as a vertex buffer, since vertex buffers are not allowed to have other bind

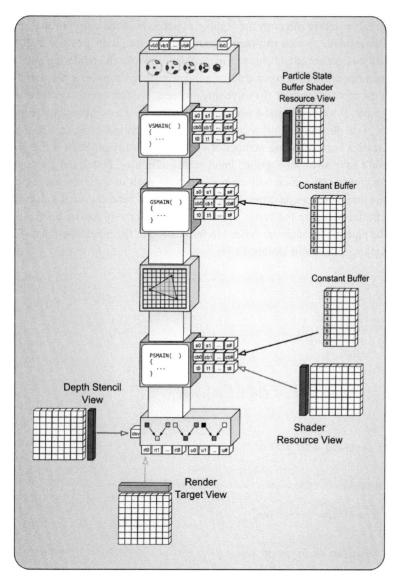

Figure 12.13. The rendering pipeline configuration for rendering the particle system.

flags, which are required for using the append/consume functionality with an unordered access view.

We could copy the contents of the particle system to a third buffer resource that would serve as a vertex buffer. However, depending on how large the particle system is, this could consume a large amount of bandwidth. Instead of doing this, we can use indirect drawing. In this case, we can copy the number of particles in the buffer to an indirect arguments buffer resource, which can then execute the pipeline with a call to `ID3D11Device Context::DrawInstancedIndirect`. This will let us invoke the rendering pipeline an appropriate number of times, but we still have the problem of supplying the particle data as vertices to the input assembler. To solve this problem, we can bypass the input assembler stage altogether. If we don't bind a vertex buffer to the input assembler, but we still invoke the pipeline with our indirect `draw` call using a point-primitive type, the input assembler will still create a corresponding number of vertices. The only problem is that these input vertices won't have any user-supplied input vertex attributes. Fortunately, we can still declare the `SV_VertexID`, which will uniquely identify each of the vertices that are passed into the pipeline. Then we can use the vertex shader to read the vertex data out of the structured buffer based on the vertex identifier. This effectively lets us funnel the particle data into the pipeline through the vertex shader, instead of the input assembler. This vertex shader program is shown in Listing 12.11.

```
struct Particle
{
    float3 position;
    float3 direction;
    float  time;
};

StructuredBuffer<Particle> SimulationState;

struct VS_INPUT
{
    uint vertexid : SV_VertexID;
};

struct GS_INPUT
{
    float3 position : Position;
};

GS_INPUT VSMAIN( in VS_INPUT input )
{
    GS_INPUT output;

    output.position.xyz = SimulationState[input.vertexid].position;

    return output;
}
```

Listing 12.11. Loading the particle data into empty input vertices using the SV_VertexID system value semantic.

With the particle data introduced into the pipeline, we can now render the individual point primitives in the locations of each particle. However, this would be a fairly unsuitable technique for rendering the particle system, since each particle could only occupy a single pixel of the output render target. Instead, we will use the geometry shader to expand our individual point primitives into two triangles that form a quad. The shader program for this operation is shown in Listing 12.12.

```
cbuffer ParticleRenderParameters
{
    float4 EmitterLocation;
    float4 ConsumerLocation;
};

static const float4 g_positions[4] =
{
    float4( -1, 1, 0, 0 ),
    float4( 1, 1, 0, 0 ),
    float4( -1, -1, 0, 0 ),
    float4( 1, -1, 0, 0 ),
};

static const float2 g_texcoords[4] =
{
    float2( 0, 1 ),
    float2( 1, 1 ),
    float2( 0, 0 ),
    float2( 1, 0 ),
};

[maxvertexcount(4)]
void GSMAIN( point GS_INPUT input[1], inout TriangleStream<PS_INPUT>
SpriteStream )
{
    PS_INPUT output;

    float dist = saturate( length( input[0].position - ConsumerLocation.xyz )
                                    / 100.0f );
    float4 color = float4( 0.2f, 0.2f, 1.0f, 0.0f ) * ( dist )
                + float4( 1.0f, 0.2f, 0.2f, 0.0f ) * ( 1.0f - dist );

    // Transform to view space
    float4 viewposition = mul( float4( input[0].position, 1.0f ),
                            WorldViewMatrix );

    // Emit two new triangles
    for ( int i = 0; i < 4; i++ )
    {
        // Transform to clip space
        output.position = mul( viewposition + g_positions[i], ProjMatrix );
        output.texcoords = g_texcoords[i];
        output.color = color;
```

```
        SpriteStream.Append(output);
    }

    SpriteStream.RestartStrip();
}
```

Listing 12.12. Expanding points into quads with the geometry shader.

As seen in Listing 12.12, we expand the points into quads in view space which lets us use a static set of offsets for the new vertex locations. These offset positions are then transformed into clip space for use in the rasterizer stage. Similarly, we use a static array of texture coordinates to provide a full texture over the created quad. We also generate a color value that is based on the distance from the black hole, and then pass the color to the output vertices as well. These output vertices are then rasterized and passed to the pixel shader, where it samples a particle texture and writes the output to the output merger. This is shown in Listing 12.13.

```
Texture2D    ParticleTexture : register( t0 );
SamplerState LinearSampler : register( s0 );

float4 PSMAIN( in PS_INPUT input ) : SV_Target
{
    float4 color = ParticleTexture.Sample( LinearSampler, input.texcoords );
    color = color * input.color;

    return( color );
}
```

Listing 12.13. Applying a texture to the output pixels.

After the pixel shader generates the color that is destined for the render target, we must also configure the output merger to accommodate our desired rendering style. Our particle system is going to use what is called *additive blending*, which essentially means that we will modify the blend state so that each pixel produced by the pixel shader will be added to the contents of the render target. This will composite particles onto each other if they overlap, and will create a glowing effect if many particles are occupying the same location simultaneously. In addition, since we are additively blending it is important to disable depth-buffer writing in the depth stencil state, while still using the depth buffer for depth testing. Disabling depth writing requires that any objects in the scene that partially occlude the particle system must be rendered before it, to allow the resulting rendering to be properly sorted according to depth. The resulting rendering can be seen in Figure 12.14.

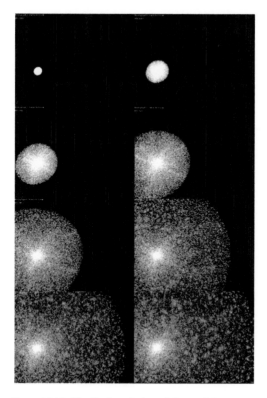

Figure 12.14. The final rendering of the particle system.

12.2.3 Conclusion

The ability of the GPU to efficiently process many parallel data items simultaneously allows the processing of many particles within a particle system. By using some new features of Direct3D 11, we can maintain the particle system completely on the GPU and simply control the update and rendering of the simulation with the CPU. This minimizes any synchronization needed between the GPU and CPU, which allows the GPU to run as fast as possible. In addition, we have seen how to render the resulting particles as quads, using an additive blending technique.

13

Multithreaded Paraboloid Mapping

We have seen in Chapter 7 that Direct3D 11 provides a very thorough and consistent API for handling multithreaded use cases. Two different types of multithreaded actions are available for use by an application. The first is multithreaded resource creation, and the second is the ability to perform multithreaded draw submission sequences. While multithreaded resource creation can allow for a simpler implementation for many operations, the ability to submit drawing operations in parallel has the potential to significantly reduce the amount of overhead on the CPU required to render a particular frame. The sample application presented in this chapter seeks to exercise this multithreaded submission capability and investigate how it impacts the overall rendering performance of a particular test scene.

We have also seen in Chapter 7 that there are several different ways that multithreaded drawing can be used. Depending on the granularity of the rendering task that a command list is generated for, an application can theoretically achieve better performance from the multithreaded rendering system. We will focus on the largest granularity that was discussed in Chapter 7, capturing a complete "view" of a scene as a single rendering pass. This granularity level essentially incorporates all of the actions that take place from when a render target is cleared until it has been completely filled and can be unbound from the pipeline. Therefore, a test scene that takes advantage of this mechanism requires more than one of these "view"-level rendering passes, which can be recorded in command lists in parallel, and then can be executed by a single thread to submit the command lists to the GPU in the appropriate order.

The scene we will use to test this capability will employ a technique called *dual-paraboloid environment mapping* (Heidrich, 1998), (Brabec, 2002). This is an environment mapping technique that simulates reflections around an object by rendering the object's surroundings into a pair of special render targets, using a specialized transformation based on paraboloids. With only two render targets, the entire surroundings of an object can be captured and then used to perform a per-pixel lookup to find approximately what would be visible at each pixel on a reflective object. The technique requires these special render targets to be filled for each reflective object in a scene, so we can use these rendering passes as a vehicle to test our multithreaded submission system's performance by using multiple reflective objects that are near one another. By varying the number of reflective objects, we should be able to determine roughly what impact the multithreading features have on the overall performance of the application.

This chapter will explore the details of the dual paraboloid environment mapping technique, develop an implementation for it, and then execute the algorithm with and without multithreading, to attempt to characterize the benefits of performing multithreading draw submission. The chapter closes with a discussion of how this ability can be used, and when it should be beneficial.

13.1 Theory

The basic concept behind paraboloid mapping can be approached by considering what would be visible if you viewed a perfectly reflecting paraboloid surface. In many ways,

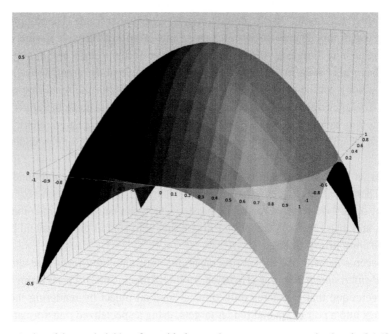

Figure 13.1. A plot of the paraboloid surface with the x and y parameters constrained to the [-1,1] domain.

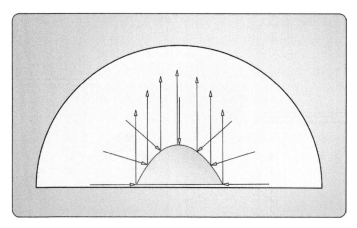

Figure 13.2. The region of space that would be visible through a reflective paraboloid.

this is similar to viewing your surroundings through a Christmas tree bulb—you will see a distorted view of the objects that are in the room with you. However, a paraboloid is shaped differently than a spherical bulb, and hence will produce a different view of the objects around it. To better characterize what would be visible in this surface, we will examine some of the basic properties of the paraboloid surface. We begin with the mathematical definition of the paraboloid surface itself, which is presented in Equation (13.1):

$$f(x,y) = \frac{1}{2} - \frac{1}{2}(x^2 + y^2).$$

(13.1)

The surface that results from evaluating this equation over an x- and y-domain of [-1,1] is depicted in Figure 13.1. In this figure, we see that the surface has a maximum when x and y are zero, and the value falls off as x and y move away from zero. We will be interested in the region of this surface where the result of this equation is positive, meaning that the sum of the square of x and y must be less than or equal to 1.

Returning to our initial consideration of the reflective surface, we can see that if we view the paraboloid from above over the domain [-1,1], the reflections will vary over the hemisphere above its zero plane. This region is depicted in Figure 13.2 by drawing arrows from the viewing direction and seeing the direction that they would be reflected toward.

This ability to "see" an entire hemisphere is a very important property, since it means that we can use a single paraboloid projected onto a plane to capture precisely one half of the surroundings of a particular point in space. With our viewing direction from above the paraboloid, we can visualize this projection of the paraboloid onto a viewing plane between us and the paraboloid. This is roughly depicted in Figure 13.3.

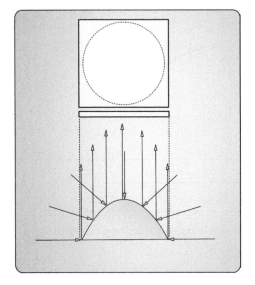

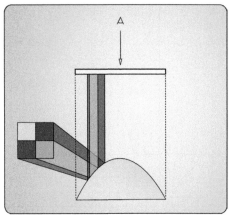

Figure 13.4. A profile view of the virtual viewer, the paraboloid, and an object being rendered into the paraboloid map.

Figure 13.3. The projection of a paraboloid onto a plane.

The concept shown in Figure 13.3 is the key to the entire paraboloid environment mapping technique. This ability to project the paraboloid contents onto a plane can be mapped to a render target, where the region of the render target represents the domain we have already described—where x and y are constrained to the [-1,1] region. We can dynamically generate the contents of this render target by performing a paraboloid projection on the scene contents, and can effectively produce a form of an environment map for half of the scene. If we use two "paraboloid maps" that are oriented 180 degrees from one another, we can capture the entire contents (both halves) of a scene surrounding the origin of the coordinate space. The challenge is to define precisely how we will perform the paraboloid projection when generating the paraboloid maps, and inversely, how to properly look up a particular point in the paraboloid maps when they are used to query the scene while rendering a reflective object.

13.1.1 Paraboloid Map Generation

When generating a paraboloid map, the desired effect is to place each vertex of a model into the location of the render target that corresponds to where that vertex would be seen in the reflective paraboloid. Thus we need to know where we would consider the viewer to be looking at the paraboloid, where the paraboloid is positioned, and where the vertices are that are being transformed. If we assume that the scene surroundings are sufficiently far

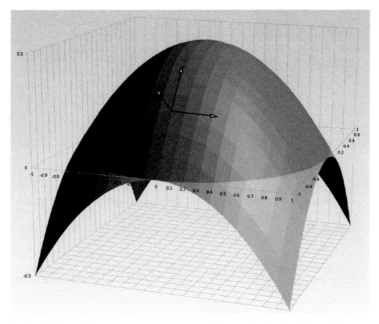

Figure 13.5. A visualization of the two tangent vectors, and the result of calculating their cross product.

away from the object,1 the viewing vectors can be assumed to be parallel over the entire region of the paraboloid that we are interested in. This scenario is depicted in Figure 13.4.

With this in mind, we would like to calculate the paraboloid surface normal vector that is needed to reflect the view vector into the direction of the object being rendered. This normal vector will provide information that uniquely identifies where on the paraboloid the object will be visible. We will use it to specify where to place each vertex within the paraboloid map, as well as to know where to look in the paraboloid map when sampling it. We will use two different approaches to define our normal vector. The first approach will find the normal vector on the paraboloid by finding two vectors that are perpendicular to its surface, and then taking the cross product of these two vectors. This is done trivially by taking the partial derivative with respect to x and y, respectively, to find our two vectors. This process is shown in Equations (13.2)–(13.5), and is demonstrated in Figure 13.5:

$$P = (x, y, f(x, y));\qquad(13.2)$$

---

[1] This is an assumption that isn't necessarily physically correct, since a viewer at a single point would cast non-parallel viewing rays onto the paraboloid, in much the same manner that we have seen with perspective projections. However, this is a simple approximation that makes generating and sampling the paraboloid map much less expensive to compute with minimal negative effects. The same effect can be found by assuming that the paraboloid itself is physically infinitesimally small within the scene.

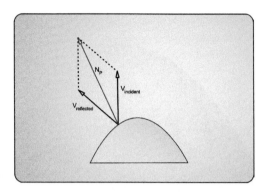

Figure 13.6. A visualization of adding the incident and reflected vectors.

$$T_x = \frac{\delta P}{\delta x} = \left(1, 0, \frac{\delta f(x,y)}{\delta x}\right) = (1, 0, -x); \tag{13.3}$$

$$T_x = \frac{\delta P}{\delta x} = \left(1, 0, \frac{\delta f(x,y)}{\delta x}\right) = (1, 0, -x); \tag{13.4}$$

$$N_P = T_x \times T_y = (x, y, 1). \tag{13.5}$$

This series of equations will allow us to calculate the normal vector when given a location on the paraboloid to look up. However, in general we won't have a prior knowledge of where to look in the paraboloid when generating and sampling its contents. Instead, we will have the incident vector (taken from the viewing direction) and the reflected vector (calculated from the paraboloid origin to the vertex being projected) and must determine the normal vector that will produce the reflected vector from the incident vector. This is also a trivial operation if we assume that both of these vectors originate at the paraboloid surface. When this is the case, we can add the two vectors and the resulting vector will point in the direction of the normal vector, although it will not have a normalized length. This is shown in Equation (13.6) and Figure 13.6:

$$N_P = V_{inc} + V_{refl}. \tag{13.6}$$

By combining the results of Equation (13.5) and Equation (13.6), we can produce a simple technique to find the normal vector with the information that we will have readily available to us. We can add the incident and reflected vectors to find our normal vector, and then divide all three components by the z-component to produce the form shown in Equation (13.5). This equality is shown in Equation (13.7):

$$V_{inc} + V_{refl} = V_{sum} = (x_{sum}, y_{sum}, z_{sum}) = \frac{1}{z_{sum}} (x, y, 1). \qquad (13.7)$$

With this equation, we have the desired x and y coordinates that will define where in the paraboloid map the object will appear. We will use this equality to perform our paraboloid projection while generating the paraboloid maps later in this chapter. By placing each vertex of an object into the appropriate location in the paraboloid map, we can use the rasterizer stage to fill in its contents. Since each paraboloid can represent one half-space around itself, we will use two paraboloid maps to represent the entire area around the point that we are interested in.

13.1.2 Paraboloid Map Sampling

With the ability to generate a pair of paraboloid maps around a desired point in our scene, we also need to be able to know where to sample our paraboloid maps to find the appropriate reflection contents. Typically, when rendering the reflective object that the paraboloid maps have been generated for, we will have that object's surface normal vector, as well as the viewing vector from the camera to that point in space. In this situation, the viewing vector and surface normal can be used to calculate the direction that the view vector is reflected into. It is this vector that we want to find in our paraboloid maps. The relationship between these vectors is depicted in Figure 13.7.

Fortunately, the lookup process is quite similar to what we did when we generated the paraboloid map. We will know what viewing direction was used to generate the paraboloid

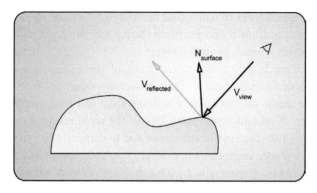

Figure 13.7. The vectors associated with sampling a paraboloid map.

map, and we will also know the reflected vector that we are looking up. These are precisely the incident and reflected vectors that we used while performing the paraboloid projection. Thus, we can use Equation (13.7) and add these two vectors, and divide the result by its z-component. The resulting x and y coordinates will then indicate the location in the paraboloid map to sample in order to recreate the reflection. This completes the lookup process, and allows us to use our paraboloid maps to approximate what would be seen in the reflected scene.

13.2 Implementation Design

With the theory of paraboloid mapping clearly defined, we can now consider how best to implement this technique in the Direct3D 11 rendering pipeline. We will consider two different operations: the generation of the paraboloid maps, and the subsequent use of the paraboloid maps during a rendering pass. In addition, we will also consider how to perform both operations at the same time, such as when rendering one reflective object into another reflective object's paraboloid maps.

13.2.1 Resource Selection

Before filling and using our paraboloid maps, we must first decide what would be the most appropriate resource format to use. The paraboloid maps themselves are 2D projections of the scene around them, and thus a 2D texture is the natural choice. However, we still have some freedom to choose how our two paraboloid maps are associated with one another. At first thought, we could use one 2D texture resource for each of the two paraboloid maps that are needed. Unfortunately, we can't use MRT rendering to fill both render targets simultaneously, since each object will appear in a different location in each of the paraboloid maps. Furthermore, to allow for proper depth sorting with the depth buffer, we need to have unique depth buffers for each paraboloid map.

However, there is still a way to use some of the newer features of Direct3D 11 to fill both paraboloid maps with a single rendering sequence. If we create a two-element 2D texture array, we can contain both the front and back paraboloid maps within a single resource. With both maps accessible in the same resource, we can use the SV_ RenderTargetArrayIndex system value semantic to dynamically determine in the geometry shader which render target element to send a batch of geometry to. This will let us pass different versions of the geometry to each paraboloid map from within a single draw call.

This is a good enough reason to choose the 2D texture array resource type. However, this rendering technique also allows the use of another feature of Direct3D 11 to further simplify the process of rendering the geometry into our paraboloid maps. Since we are using the geometry shader to pass the geometry to the appropriate render target, we will need to create one copy of the geometry for each paraboloid map. This could either be statically performed by copying the paraboloid projection code twice, or it could also be implemented with a simple loop. Both of these methods are less than optimal, and either clutter the code (in the case of repeating the code) or introduce additional work to implement (in the case of looping). Instead, we can simply use geometry shader instancing to produce multiple copies of each primitive that the geometry shader receives. Then the SV_GSInstanceID system value semantic can be used to determine which paraboloid map each invocation of the geometry shader should be applied to. This results in very clean shader code that can efficiently fill multiple paraboloid maps simultaneously.[2]

13.2.2 Rendering Paraboloid Maps

The first step in using our paraboloid maps is to render the scene into one for a reflective object. For the sake of discussion, we will begin by assuming that we have a single reflective object and a single textured object in our scene. Thus, we need a technique to render the textured object into a paraboloid map. The majority of this work will be split between the vertex shader and geometry shader stages. Listing 13.1 provides the vertex shader for this operation, and Listing 13.2 provides the geometry shader.

As seen in Listing 13.1, the vertex shader receives vertices with just a position and a texture coordinate. We assume that the matrix passed in to the shader in the WorldViewMatrix is the concatenation of a world matrix for the object being rendered, and a view matrix that defines the basis of the paraboloid being filled. In reality, this is just a view matrix taken from the location and orientation of the reflective object where the orientation will define the forward and backward directions of the two paraboloids. The first step in the vertex shader is to transform the incoming vertex position into this "paraboloid basis."

Next, the distance from the origin of the paraboloid basis is calculated and used to generate a normalized length vector pointing to the vertex. The normalized z-component of this vector is passed to the next stage in the OUT.z_value variable, and a scaled version of the distance to the vertex is also passed to the next shader stage for eventual use in the depth buffer. Finally, the w-coordinate is simply set to 1, since we haven't performed any perspective projections, and the input texture coordinates are simply passed through to the next stage.

---

[2] In fact, we don't need to limit ourselves to only filling the two paraboloid maps of a single object in each rendering pass. If there are four reflective objects in a scene, we can fill all eight of the corresponding paraboloid maps in the same rendering pass. For simplicity, we will use only two geometry shader instances at a time for the remainder of this chapter.

```
cbuffer ParaboloidTransforms
{
    matrix WorldViewMatrix;
};

Texture2D      ColorTexture : register( t0 );
SamplerState   LinearSampler : register( s0 );

struct VS_INPUT
{
    float3 position : POSITION;
    float2 tex      : TEXCOORDS0;
};

struct GS_INPUT
{
    float4 position : SV_Position;
    float2 tex      : TEXCOORD0;
    float  z_value  : ZVALUE;
};

struct PS_INPUT
{
    float4 position : SV_Position;
    float2 tex      : TEXCOORD0;
    float  z_value  : ZVALUE;
    uint   rtindex  : SV_RenderTargetArrayIndex;
};

GS_INPUT VSMAIN( VS_INPUT IN )
{
    GS_INPUT OUT;

    // Transform the vertex to be relative to the paraboloid's basis.
    OUT.position = mul( float4( IN.position, 1 ), WorldViewMatrix );

    // Determine the distance between the paraboloid origin and the vertex.
    float L = length( OUT.position.xyz );

    // Normalize the vector to the vertex
    OUT.position = OUT.position / L;

    // Save the z-component of the normalized vector
    OUT.z_value = OUT.position.z;

    // Store the distance to the vertex for use in the depth buffer.
    OUT.position.z = L / 500;

    // Set w to 1 since we aren't doing any perspective distortion.
    OUT.position.w = 1;

    // Pass through texture coordinates
    OUT.tex = IN.tex;

    return OUT;
}
```

Listing 13.1. The vertex shader for rendering a textured object into a paraboloid map.

In the geometry shader, we first declare that the maximum number of vertices will be 3, since we are only generating one triangle primitive with each invocation. In addition, we declare that we want 2 instances of each primitive to be created with the `instance` function attribute. To process the primitive being sent to the geometry shader, we begin by initializing the order and direction variables that will be used to determine if we are processing the front or back paraboloid. This is determined by the `id` input attribute, which is the `SV_GSInstanceID` system value semantic. We choose the paradigm such that when the `id` is 0, we are working on the front paraboloid, while a value of 1 indicates that we are working on the back paraboloid.[3] If we are working on the back paraboloid, we simply switch the ordering of the vertices by swizzling its components, and we then set the direction variable to -1. Then the input primitive vertices are only accessed using the `order` variable, which allows for a simple way to choose the right winding order.[4]

The next step is to loop through each of the vertices of the primitive and perform the paraboloid projection on them. This is done by calculating a `projFactor`, which is essentially the z-component of the vector to the vertex (which was generated in the vertex shader) multiplied by the direction variable, and added to 1. The multiplication by the direction variable will convert the negative half-space into the positive half-space when we are processing the back paraboloid. Then, the addition of 1 is actually performing the addition of the view vector from Equation (13.7). Since we are working in the paraboloid basis, the incident vector (or view vector) is (0,0,1), so adding it to the vertex direction vector is the same as adding 1 to its z-component. We then use the resulting `projFactor` to divide the x and y components, which produces the output location within the paraboloid map. The available range of the results of these components provides values that vary between (-1,1) for both the x- and the y-coordinates, which fits nicely into the clip space positions that must be provided for the rasterizer stage.

Another useful point to consider is that any vertices that exist in the other half-space (when the positive z-valued vertices are being processed in the negative half-space, and vice-versa), they will end up with a negative depth value. We can see this by considering the two possible cases. When geometry exists in the positive half-space around the paraboloid basis and we are processing the positive half-space paraboloid map, then the vertices' z-values will be positive and will be rasterized, while any geometry in the negative half space will be culled. In the opposite case, we are processing the negative half space. The *direction* variable is used to change any negative z-values to positive values, and the vertex

---

[3] The terms *front* and *back* here only refer to the orientation of the paraboloids, which is determined by the view matrix that is passed into the vertex shader. The paraboloid maps themselves can be viewed from any angle, regardless of their orientation, so the choice of basis can be made in whatever way is convenient. Many implementations choose to use a world-space-aligned orientation to simplify the implementation, but we include the transformation in the vertex shader for flexibility in choosing the basis.

[4] Since the built-in vector data types can be accessed with array-like syntax, we can swizzle the components of the vector with the member syntax (xyz → xzy) and then use the vector variable like an array (order[0], order[1], and so forth).

winding order is flipped. This effectively reverses the prior situation, and clips or culls the positive half-space geometry. In this way, we only apply the appropriate sets of geometry to each of the paraboloid maps.

```
[maxvertexcount(3)]
[instance(2)]
void GSMAIN( triangle GS_INPUT input[3],
             uint id : SV_GSInstanceID,
             inout TriangleStream<PS_INPUT> OutputStream )
{
    PS_INPUT output;

    // Initialize the vertex order and the direction of the paraboloid.
    uint3 order = uint3( 0, 1, 2 );
    float direction = 1.0f;

    // Check to see which copy of the primitive this is. If it is 0, then it
    // is considered the front facing paraboloid. If it is 1, then it is
    // considered the back facing paraboloid. For back facing, we reverse
    // the output vertex winding order.
    if ( id == 1 )
    {
        order.xyz = order.xzy;
        direction = -1.0f;
    }

    // Emit three vertices for one complete triangle.
    for ( int i = 0; i < 3; i++ )
    {
        // Create a projection factor, which determines which half space
        // will be considered positive and also adds the viewing vector
        // which is (0,0,1) and hence can only be added to the z-component.
        float projFactor = input[order[i]].z_value * direction + 1.0f;
        output.position.x = input[order[i]].position.x / projFactor;
        output.position.y = input[order[i]].position.y / projFactor;
        output.position.z = input[order[i]].position.z;
        output.position.w = 1.0f;

        // Simply use the geometry shader instance as the render target
        // index for this primitive.
        output.rtindex = id;

        // Pass through the texture coordinates to the pixel shader
        output.tex = input[order[i]].tex;

        output.z_value = input[order[i]].z_value * direction;

        // Write the vertex to the output stream.
        OutputStream.Append(output);
    }

    OutputStream.RestartStrip();
}
```

Listing 13.2. The geometry shader for rendering a textured object into dual paraboloid maps.

Once the output primitives have been rasterized to their appropriate render targets, the pixel shader is invoked for each fragment. In this case, we simply sample the object's texture using the passed-through texture coordinates. This allows the object being rendered to appear properly across each primitive being rasterized. The pixel shader is shown in Listing 13.3.

```
float4 PSMAIN( in PS_INPUT IN ) : SV_Target
{
    float4 OUT;

    clip( IN.z_value + 0.05f );

    // Sample the texture to find the appropriate color value
    OUT = ColorTexture.Sample( LinearSampler, IN.tex );

    return OUT;
}
```

Listing 13.3. The pixel shader for rendering a textured object into a paraboloid map.

A sample pair of front and back paraboloid maps for a sample scene is provided in Figure 13.8. The warped view produced by the paraboloid projection can clearly be

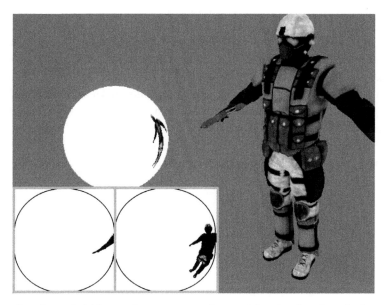

Figure 13.8. Sample paraboloid maps for a test scene, demonstrating the view from the front and back paraboloids. Model courtesy of Radioactive Software.

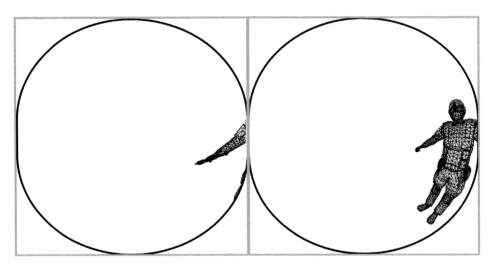

Figure 13.9. The same sample scene as shown in Figure 13.8, except with the geometry rasterized in wireframe. Model courtesy of Radioactive Software.

seen in the figures. In addition, Figure 13.9 shows the same scene rendered in wireframe rasterization mode.

13.2.3 Sampling Paraboloid Maps

Now that we have generated the paraboloid maps, we must properly sample them when rendering a reflective object into the final output render target. This process uses only the vertex and pixel shader stages, and it follows a more traditional rendering technique. The only required input vertex data are the object space position and normal vectors. The vertex shader begins by transforming the vertex position to world space. This world-space position is then used to calculate the world-space vector from the camera to the vertex. In addition, the vertex shader calculates the world-space normal vector, and of course produces the clip space position for passing on to the rasterizer stage.

```
cbuffer ParaboloidLookupParams
{
    matrix WorldViewProjMatrix;
    matrix WorldMatrix;
    matrix ParaboloidBasis;
    float4 ViewPosition;
}

Texture2DArray  ParaboloidTexture : register( t0 );
SamplerState    ParaboloidSampler : register( s0 );
```

```
struct VS_INPUT
{
    float3 position : POSITION;
    float3 normal   : NORMAL;
};

struct PS_INPUT
{
    float4 position : SV_Position;
    float3 normal   : TEXCOORD0;
    float3 eye      : TEXCOORD1;
};

struct pixel
{
    float4 color    : COLOR;
};

PS_INPUT VSMAIN( VS_INPUT IN )
{
    PS_INPUT OUT;

    // Find the world space position of the vertex.
    float4 WorldPos = mul( float4( IN.position, 1 ), WorldMatrix );

    // Output the clip space position for the rasterizer.
    OUT.position    = mul( float4( IN.position, 1 ), WorldViewProjMatrix );

    // Find world space normal and eye vectors.
    OUT.normal      = normalize( mul( IN.normal, (float3x3)WorldMatrix ) );
    OUT.eye         = normalize( WorldPos.xyz - ViewPosition.xyz );

    return OUT;
}
```

Listing 13.4. The vertex shader for rendering a reflective object that will sample a paraboloid map.

After the reflective object's geometry has been rasterized, the pixel shader is invoked to determine what color should appear at each fragment location. This is done by first normalizing the world-space surface normal vector and the view-direction vector. Next, we calculate the reflection of the eye vector about the surface normal. This essentially indicates where in world space will be visible from that particular fragment for the given viewing angle. We then convert this reflected vector to be relative to the paraboloid's basis, so that it will be in the same coordinate frame as our paraboloid maps.

This new reflected vector is used to determine where in the paraboloid maps to look up. In essence, we perform the exact same process that we have seen before. The vector's z component is added to 1 to find the z-coordinate of the summed vector in Equation (13.7). This is then used to project the x and y reflection vector components by dividing by this summed vector *z*-component. After this division, the resulting *x* and *y* coordinates indicate the paraboloid location to look up, but are scaled to the [-1,1] range. Since the paraboloid is stored in a texture, we must convert these coordinates to the appropriate texture coordinates. This is done with a scaling factor and a biasing factor to remap the coordinates. Finally, we choose which paraboloid map to sample based on the sign of the reflection vector's *z*-component.[5] A small scaling is also applied to the resulting sampled value to serve as an attenuation factor. This will help to discern the difference between multiple reflections later on.

```
float4 PSMAIN( PS_INPUT IN ) : SV_Target
{
    float4 OUT;

    // Normalize the input normal and eye vectors
    float3 N = normalize( IN.normal );
    float3 E = normalize( IN.eye );

    // Calculate the world space reflection vector, and then transform it to
    // the paraboloid basis.
    float3 R = reflect( E, N );
    R = mul( R, (float3x3)ParaboloidBasis );

    // Calculate the forward paraboloid map texture coordinates, with z
    // determining which paraboloid map to sample (front or back).
    float3 front;
    front.x = (R.x / (2*(1 + R.z))) + 0.5;
    front.y = 1-((R.y / (2*(1 + R.z))) + 0.5);
    front.z = 0.0f;

    // Calculate the backward paraboloid map texture coordinates, with z
    // determining which paraboloid map to sample (front or back).
    float3 back;
    back.x = (R.x / (2*(1 - R.z))) + 0.5;
    back.y = 1-((R.y / (2*(1 - R.z))) + 0.5);
    back.z = 1.0f;

    // Sample the appropriate paraboloid map based on which direction
    // the reflection vector is pointing.
    if ( R.z > 0 )
        OUT = ParaboloidTexture.Sample( ParaboloidSampler, front );
    else
        OUT = ParaboloidTexture.Sample( ParaboloidSampler, back );

    OUT *= 0.8f;

    return OUT;
}
```

Listing 13.5. The pixel shader for rendering a reflective object that will sample a paraboloid map.

---

[5] The 2D texture array element is selected by setting the *z*-component of the texture coordinates being used for sampling. In our example, we use either a 0 or a 1 to select the front and back paraboloid maps, respectively.

Figure 13.10. The results of sampling a paraboloid map. Model courtesy of Radioactive Software.

The sampled paraboloid map value is then passed to the output merger to be written to the final render target. An example rendering using this lookup technique is shown in Figure 13.10.

13.2.4 Sampling and Rendering Paraboloid Maps

In our testing scenario, we are interested in gaining an understanding of how multithreaded draw submission performs when we are rendering several rendering passes over a single frame. In this case, we would be interested in having multiple reflective objects in the same scene. Unless we restricted ourselves to situations where these reflective objects were occluded from one another, we would need to be able to handle rendering one reflective object's representation into another's paraboloid maps. In fact, as we allow more and more reflective objects to be added to the scene, there may be many reflective objects rendered in a particular set of paraboloid maps. Thus, we must prepare a rendering pipeline configuration that can be used for this situation.

We will use a combination of the two methods that have been described above. Since each reflective object will have its own set of paraboloid maps, the process of applying one reflective object to another requires us to perform the paraboloid projection on its geometry, followed by sampling its own paraboloid maps to indicate what is visible on its surface. In this way, reflections can bounce back and forth from one reflective object to another. In the next section, we will discuss the implications of this effect, as well as how to handle this from a scene level. For now, we will consider the shader programs used to perform this rendering operation.

This configuration utilizes a vertex shader, a geometry shader, and a pixel shader, shown in Listings 13.6, 13.7, and 13.8, respectively. The vertex shader is a direct combination of the previous two techniques, performing the paraboloid projection in the first portion of the shader and then calculating the needed world-space vectors for its own paraboloid lookup in the latter portion of the shader.

```
struct VS_INPUT
{
    float3 position : POSITION;
    float3 normal   : NORMAL;
};

struct GS_INPUT
{
    float4 position : SV_Position;
    float  z_value  : ZVALUE;
    float3 normal   : TEXCOORD0;
    float3 eye      : TEXCOORD1;
};

struct PS_INPUT
{
    float4 position : SV_Position;
    float  z_value  : ZVALUE;
    float3 normal   : TEXCOORD0;
    float3 eye      : TEXCOORD1;
    uint   rtindex  : SV_RenderTargetArrayIndex;
};

GS_INPUT VSMAIN( VS_INPUT IN )
{
    GS_INPUT OUT;

    // Transform the vertex to be relative to the paraboloid's basis.
    OUT.position = mul( float4( IN.position, 1 ), WorldViewMatrix );

    // Determine the distance between the paraboloid origin and the vertex.
    float L = length( OUT.position.xyz );

    // Normalize the vector to the vertex
    OUT.position = OUT.position / L;

    // Save the z-component of the normalized vector
```

```
    OUT.z_value = OUT.position.z;

    // Store the distance to the vertex for use in the depth buffer.
    OUT.position.z = L / 500;

    // Set w to 1 since we aren't doing any perspective distortion.
    OUT.position.w = 1;

    // Find the world space position of the vertex.
    float4 WorldPos = mul( float4( IN.position, 1 ), WorldMatrix );

    // Find world space normal and eye vectors.
    OUT.normal      = normalize( mul( IN.normal, (float3x3)WorldMatrix ) );
    OUT.eye         = normalize( WorldPos.xyz - ViewPosition.xyz );

    return OUT;
}
```

Listing 13.6. The vertex shader for rendering a reflective object into another object's paraboloid map.

The geometry shader is also a combination of the previous methods. The paraboloid projection calculations are performed (including the render target selection and vertex winding modification), then the world-space vectors are passed to the pixel shader. This geometry shader uses the same instancing technique we have seen previously to fill both paraboloid maps.

```
[maxvertexcount(3)]
[instance(2)]
void GSMAIN( triangle GS_INPUT input[3],
            uint id : SV_GSInstanceID,
            inout TriangleStream<PS_INPUT> OutputStream )
{
    PS_INPUT output;

    // Initialize the vertex order and the direction of the paraboloid.
    uint3 order = uint3( 0, 1, 2 );
    float direction = 1.0f;

    // Check to see which copy of the primitive this is. If it is 0, then it
    // is considered the front facing paraboloid. If it is 1, then it is
    // considered the back facing paraboloid. For back facing, we reverse
    // the output vertex winding order.
    if ( id == 1 )
    {
        order.xyz = order.xzy;
        direction = -1.0f;
    }

    // Emit three vertices for one complete triangle.
```

```
for ( int i = 0; i < 3; i++ )
{
    // Create a projection factor, which determines which half space
    // will be considered positive and also adds the viewing vector
    // which is (0,0,1) and hence can only be added to the z-component.
    float projFactor = input[order[i]].z_value * direction + 1.0f;
    output.position.x = input[order[i]].position.x / projFactor;
    output.position.y = input[order[i]].position.y / projFactor;
    output.position.z = input[order[i]].position.z;
    output.position.w = 1.0f;

    // Simply use the geometry shader instance as the render target
    // index for this primitive.
    output.rtindex = id;

    // Propagate the normal and eye vectors.
    output.normal = input[order[i]].normal;
    output.eye = input[order[i]].eye;

    output.z_value = input[order[i]].z_value * direction;

    // Write the vertex to the output stream.
    OutputStream.Append(output);
}

OutputStream.RestartStrip();
}
```

Listing 13.7. The geometry shader for rendering a reflective object into another object's paraboloid map.

The final step this rendering configuration is to look up the paraboloid map of the object being rendered. This is done in the pixel shader, which is in fact the same pixel shader that we used when rendering the reflective object into a standard render target.[6]

```
float4 PSMAIN( PS_INPUT IN ) : SV_Target
{
    float4 OUT;

    clip( IN.z_value + 0.05f );

    // Normalize the input normal and eye vectors
    float3 N = normalize( IN.normal );
    float3 E = normalize( IN.eye );
```

---

[6] At this point in the pipeline, the pixel shader isn't concerned with whether we are rendering into another paraboloid map, or if we are rendering into a standard perspective-projection-based render target. This is one of the advantages of designing the rendering pipeline in stages, which promotes the reuse of individual shader programs.

```
    // Calculate the world space reflection vector, and then transform it to
    // the paraboloid basis.
    float3 R = reflect( E, N );
    R = mul( R, (float3x3)ParaboloidBasis );

    // Calculate the forward paraboloid map texture coordinates, with z
    // determining which paraboloid map to sample (front or back).
    float3 front;
    front.x = (R.x / (2*(1 + R.z))) + 0.5;
    front.y = 1-((R.y / (2*(1 + R.z))) + 0.5);
    front.z = 0.0f;

    // Calculate the backward paraboloid map texture coordinates, with z
    // determining which paraboloid map to sample (front or back).
    float3 back;
    back.x = (R.x / (2*(1 - R.z))) + 0.5;
    back.y = 1-((R.y / (2*(1 - R.z))) + 0.5);
    back.z = 1.0f;

    // Sample the appropriate paraboloid map based on which direction
    // the reflection vector is pointing.
    if ( R.z > 0 )
        OUT = ParaboloidTexture.Sample( ParaboloidSampler, front );
    else
        OUT = ParaboloidTexture.Sample( ParaboloidSampler, back );

    OUT *= 0.8f;

    return OUT;
}
```

Listing 13.8. The pixel shader for rendering a reflective object into another object's paraboloid map.

Figure 13.11. The result of rendering reflective objects into another reflective object's reflection. Model courtesy of Radioactive Software.

With this rendering configuration, we can now render reflective objects in the reflections of other reflective objects. This can lead to multiple reflections being performed between objects, producing a "hall of mirrors" effect. The results of this rendering technique are shown in Figure 13.11.

13.3 Multithreading Scenario

While rendering reflections based on an object's surroundings is an interesting and effective technique, we developed these effects in order to stress the multithreading capabilities of Direct3D 11, and to gain an insight into how they function. To perform some performance analysis, we will first define our test scene and some of the characteristics that it will exhibit. After we have a defined scene to test with, we will describe the variables that we will use in our testing setup, and why they are chosen, including the number of threads being used and the number of rendering passes being performed (which is roughly equivalent to the number of reflective objects in the scene).

13.3.1 Scene Definition

Our test scene will include a number of different reflective objects, each of which will require a separate rendering pass to generate the needed paraboloid maps. In addition, the final rendering sequence that uses all of the various paraboloid maps to render the reflective objects in the scene will also be considered as another rendering pass. All together, we will need one scene rendering for each reflective object, plus the final rendering for presenting the results of the scene rendering. Figure 13.12 provides a basic layout of the scene and depicts which objects will be reflective and which will use traditional rendering techniques.

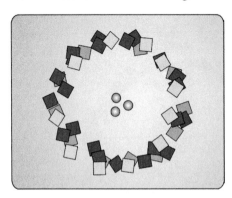

Figure 13.12. The general layout of our test scene, with objects in the center of the scene that will use a reflective material.

For each reflective object that will be rendered, we must render the entire scene from its current location to generate the paraboloid environment maps. Once this has been performed for all of the reflective objects, the final rendering pass can be performed, which will also render all of the objects within the scene one last time. This more or less means that for every rendering pass, we will be rendering all of the objects in the scene. By producing a large number of individual objects to render, we are effectively increasing the number of API calls that need to be performed by the CPU, since each object requires the pipeline to be configured before its geometry is rendered with one of the draw calls. With this in mind, we can think of our scene as a series of rendering operations that must be performed for every object, each of which must be processed for each of our rendering passes. This is roughly depicted in Figure 13.13.

When each of the reflective objects is rendered (in both the paraboloid map generation phase and the final rendering pass) it will sample its own paraboloid maps and use them to determine what color will be visible at each pixel of its geometric surfaces. With multiple highly reflective objects located near each other, we will see a "hall of mirrors" effect, where each mirror reflects its neighboring mirror surface back and forth multiple times. To ensure that there are no recursion hazards caused by this semi-infinite environment bouncing, we simply only allow each set of paraboloid maps to update its contents once per frame. If a set of maps is requested to generate its environment maps more than once in a frame, we will simply use the previous frame's maps, instead of regenerating them. This will cause some small amount of lag in some of the reflections, as well as missing reflections in the first few rendered frames; but ultimately, this limitation won't cause noticeable artifacts. Since a second or

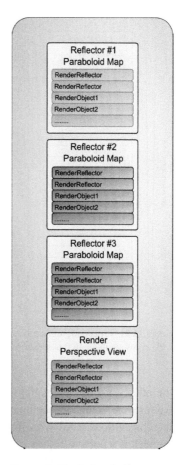

Figure 13.13. A "rendering operation" view of how the test scene will be processed.

third bounce of the environment will be difficult for the user to accurately recognize, there will likely be little or no visual impact on the output scene rendering. We are primarily interested in testing the multithreaded performance, so we will accept these small artifacts.

13.3.2 Multithreading Solution

If this scene rendering were performed by a single thread, we would essentially need to serially process each paraboloid map generation pass one at a time. This would require each sequence of the rendering operations shown in Figure 13.13 to be executed one after

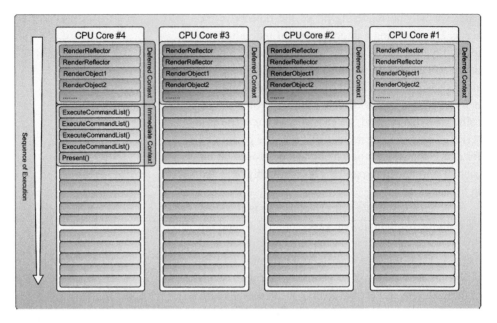

Figure 13.14. Processing our rendering operations in parallel. This depiction assumes that there are four CPU cores available in this scenario.

the other, effectively creating one long chain of operations. After all of these rendering passes were complete, the final rendering pass could be performed. By performing all of these operations serially, we need to execute $(n+1)*m$ rendering operations, where n is the number of reflective objects in the scene, and m is the number of objects in the scene (both reflective and non-reflective).

By having multiple threads to process all of the various rendering passes, we can take one complete rendering pass and process it in a separate thread to produce a command list. The command list can then be executed on the main rendering thread in an efficient manner. This effectively reduces the CPU costs of the draw submission portion of the scene by a factor that is proportional to the number of CPU cores on the user's computer.[7] Of course, there will be some overhead associated with creating and processing the command list objects, but with a sufficiently large scene (a sufficiently large number of rendering operations), we would expect to see a performance improvement. This parallel processing of our scene is depicted in Figure 13.14.

---

[7] Of course, many factors influence the performance of a particular computer system, especially when multi-threading is involved. The memory configuration is quite important, as well as the amount of other work that is being done by other threads. However, if all other things are equal, we will assume that processing the drawing operations in parallel will reduce their overall cost by a factor proportional to the number of CPU cores.

13.3.3 Empirical Results

With this concept in mind, we will turn our attention to some empirical tests taken on two different development computers. One of the configurations uses a dual-core processor and the other uses a quad core processor, so this should provide some indication of the relative performance difference when the same application is executed over more processor cores. The relevant specifications of the two test machines are listed below.

	Processor	System Memory	Video Card	Operating System
PC 1	AMD Phenom II X4 @ 3.2 GHz	4 GB	AMD 5700	Windows 7 Ultimate 64 bit
PC 2	Intel Core 2 Duo @ 2.4 GHz	4 GB	AMD 5830	Windows 7 Ultimate 64 bit

The tests were performed on each configuration with three reflective objects in the scene, and a varying number of non-reflective objects. By changing the number of non-reflective objects, we can control the number of draw submission API calls (which includes the pipeline configuration, as well as draw calls) that must be performed in each frame. This will let us characterize how multiple threads operating on multiple cores process these API calls in parallel. The results of these tests are shown in Figure 13.15. The graph

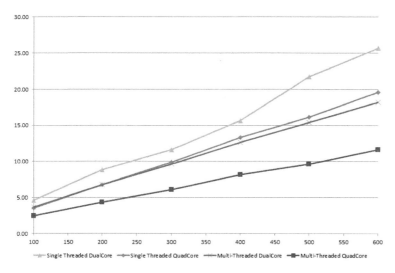

Figure 13.15. A graph showing the time required to generate a frame under varying configurations and scene conditions.

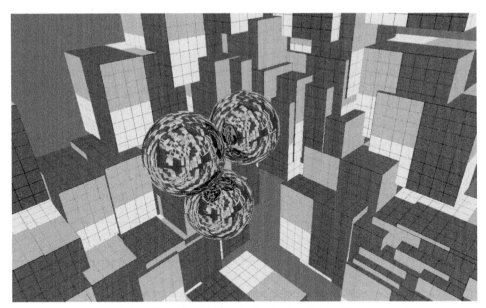

Figure 13.16. The final result of our test scene rendering.

presents the time in milliseconds required to render a frame for a particular test. Thus, a smaller number along the *y*-axis indicates better performance, while the scene complexity increases to the right of the graph. Separate data plots are provided for each computer. The test was run in both a single-threaded mode and a multi-threaded mode.[8]

From this graph, we see a very clear performance improvement when using a multithreaded configuration, for both the dual-core and quad-core configurations. In addition, there is a very linear change in frame time with the increasing number of objects in the scene. This indicates that the application itself is CPU-bound—that is, rendering speed is limited by how quickly the CPU can submit the needed API calls. By spreading the cost of performing these API calls over multiple processor cores, we can effectively speed up our overall frame rendering time. Figure 13.16 shows the final output of our test scene.

13.4 Conclusion

In this chapter we have developed an efficient dual-paraboloid-based technique to provide dynamic environment mapping to objects within our scene. By using some of the new

---

[8] The single-threaded mode renders directly with the immediate context, while the multi-threaded mode renders one rendering pass into a command list, on a deferred context. The command lists are then submitted to the immediate context for submission to the GPU.

features in Direct3D 11, we can generate a complete environment map in a single rendering pass. This technique can be used to create a sample scene that stresses the ability of the host computer to submit API calls quickly enough by rendering multiple reflective objects surrounded by a large number of non-reflective objects.

With a large CPU workload, we have explored how the multithreading capabilities of Direct3D 11 can be used to reduce the time required to submit all of the draw submission calls to the API. There is significant improvement in overall rendering speed when using multiple threads on a computer that uses a multi-core CPU. The increase in performance can be attributed to the multithreaded rendering approach, in which the scene rendering speed increases with the number of threads/cores used.

Appendix A:
Source Code

Throughout this book, there have been numerous references to source code, but notably in short "snippet" forms. The intention behind this was to focus only on the code specifically related to the topic being discussed, and in particular, to avoid long blocks of code that distract from the main text. This appendix describes the full source code provided as an open-source project hosted as part of the CodePlex project. This freely available code contains complete, working sample code for each of the sample algorithm chapters in this book.

This model of providing the source code online as an open source project was chosen for several reasons. By not providing a CD with the book, we can continue to maintain, update, and even extend the samples long after the book has been published. This means that in general, the code will continue to improve over time, ultimately providing the user with a better reference. In addition, since all of the code is built from the same library, readers can easily use the library for their own needs. In this way, the sample programs serve as examples of how to use the library.

Our hope is that this model will ultimately allow the sample applications to provide a maximum benefit to the reader in more than one way.

The Hieroglyph 3 Engine

The library that the samples are built upon is named *Hieroglyph 3*. It was initially developed by Jason Zink as a tool to learn about both Direct3D and software engineering. As such, its development began sometime in the year 2000, starting with Direct3D 8, and has evolved along with the API into its current form, which uses Direct3D 11. It consists of a single static library that an application can link to, and attempts to provide a flexible system that can be used to explore what is possible with Direct3D. Hieroglyph 3 was made an open source project in February, 2010. This was done in part to provide an open way to demonstrate the use of various parts of Direct3D, as well as to give beginners a quick and easy way to see how an application works. Jack Hoxley and Matt Pettineo have both contributed to the library and sample programs throughout the development of this book as well.

In addition, computer graphics books often provide their sample code in a customized library that is delivered with the book. The same library is rarely used across more than one book, and thus users must always acquaint themselves with whatever library is being used by the book they are currently reading. While this is not a terrible situation (after all, having at least some source code to work from is better than none...) it would be better to have continuity across multiple projects with the same library. Since Hieroglyph 3 is an open project, it can be reused over and over and will be available as the basis for future books as well.

Again, this has been done in the hope that it will provide a better learning experience for the reader, which is the ultimate motivation for writing this book in the first place!

Key Features

The library is intended to provide a simple and efficient way to use the various features of Direct3D 11 to try out new rendering techniques. As mentioned above, the Hieroglyph 3 library is based on Direct3D 11. While there are many individual components and features that make up its feature set, this section will briefly describe the most important and visible features.

In general, the library provides two major sets of functionality. The first is support for application development, including an extensible base application class, along with many of the individual features that an application needs to use such as an event system, a mechanism for receiving Win32 messages, high precision timing classes, math and geometry classes, among many others. In general, the concept is to make the application writing process less work-intensive, so that programmers can spend more time developing rendering algorithms.

The second portion of the library is entirely concerned with defining a scene and rendering it in a customizable way. This includes the definition of the objects that make up a 3D scene and how to combine them into a scene-graph-based representation. Each renderable object can be configured to have customizable geometry, as well as a configurable material. These are then used to represent the object in any rendering of the scene, and are hence the primary inputs into the rendering system.

The rendering library is essentially built around the concept of using the geometry and material classes to provide input to the rendering pipeline (as geometry) and to the pipeline configuration (as a material). There are several higher-level mechanisms for defining how a particular material will render an object in different situations, which allows for very flexible rendering setups. The same general mechanism is used to implement a wide variety of algorithms, including all of the techniques described in this book.

Where to Get It

The library can be downloaded from CodePlex at the following URL: http://hieroglyph3 .codeplex.com. It is distributed under the MIT license; additional details about the license can be found on the project's CodePlex page. To download the library, you can either obtain a snapshot zip file containing the library and all of its sample applications, or you can connect directly to the source code repository and easily update to the most recent version as it becomes available.

The library currently uses Subversion as its source code revision control tool. A common and popular Subversion client ,TortoiseSVN, is available at the following URL: http:// tortoisesvn.tigris.org. However, this is not the only SVN client program out there, and as mentioned above, you do not even need to use a SVN client program at all to inspect and use the library and its samples. The directions for doing both are available on the CodePlex project's source code page.

After obtaining a copy of the library, there is a single solution file that can be opened. All of the contents of the library are defined with relative paths, which should be accessible from wherever the library is located on your hard disk. The individual project files are configured such that the Hieroglyph 3 library itself will be compiled first, and then each of the sample applications will be built in turn (since they depend on the library, they will be built after it). After the successful building of the library and its applications, they can be individually executed, either through the IDE or directly through the running the newly produced EXE file.

13.4.1 Dependencies and Tools

Hieroglyph 3 and its associated samples were created using Microsoft's Visual C++ 2008 IDE and compiler. Any version of this and later versions should be compatible, but they may require minor edits (such as upgrading the project file).

Additionally, the following libraries are required:

- Microsoft DirectX SDK (June 2010)

- Boost C++ Libraries (1.44)

The DirectX SDK (or DXSDK for short) is periodically updated by Microsoft, so newer versions may be available after this book has been published. As noted above, the book samples were created using the June, 2010 SDK, so it has been tested with this iteration. Later versions should in general be compatible, but the release notes for each SDK must be consulted for additional details. The DXSDK (both the June, 2010 and future versions) can be downloaded from the Microsoft DirectX Developer Center at the following URL: http://msdn.microsoft.com/directx.

The Boost C++ library is also required for building the Hieroglyph 3 library. It is a very well-known and thoroughly tested C++ library that provides solutions to many common problems encountered in C++ development. The Boost library can be downloaded from the following URL: http://www.boost.org. The reader's Visual C++ IDE must be configured to make the boost include directories available to the compiler. To do this, select **Tools > Options...** from the VC++ menu, which will open the general options dialog window. Then select the **Projects and Solutions** item in the left side of the dialog, and then the **VC++ Directories** subitem. Then, on the right side of the dialog, the user can select **Include files** from the drop down-list and add an entry for your boost directory location.

Contributing to Hieroglyph 3

Since Hieroglyph 3 is an open source project, it can also benefit from the open source development model. If you have found or identified a bug, have thought of a new feature or sample algorithm, or have already implemented new content, please consider submitting it to the CodePlex project page. There are discussion areas and an issue tracker area that can be used to communicate with the developers about desired features or contributions.

In addition, the authors can frequently be found on the www.gamedev.net forums, so posting questions there can also be a good way to communicate with them, and with a large group of DirectX users.

Creating a New Direct3D 11 Project with Hieroglyph 3

All of the sample applications in Hieroglyph 3 are in the Hieroglyph3/Applications folder. In this folder, there is a subfolder for each of the sample applications. Each project is defined with a Visual C++ project file, in addition to the source code files it is comprised of. To create a new application, simply create a new folder in this directory and create a project within it. The simplest way to create a new project is to copy the project file from one of the samples (a project file has the extension *.vcproj) and then open it with a text editor. Replace the old name and path information in the project file with the new project name, as needed. In addition, the App.h and App.cpp files can also be copied to the new directory as a starting point for the application.

After you create the project in this manner, it is nearly ready to be compiled and run. First, the project must be added to the Hieroglyph 3 solution. To do this, right-click on the solution from within the IDE, select **Add -> Existing Project...**, and then select the new project file. The final step is to set the solution properties to make the new project depend on the Hieroglyph 3 project, which is done by right clicking the solution and choosing **Properties**. From the resulting dialog, choose **Common Properties > Project Dependencies** and select the new project from the drop-down list. Then check the Hieroglyph 3 project as a dependency, and the new project will correctly be built only after the Hieroglyph 3 project is.

In addition to the project folders, there is a **Data** folder that houses the file based data for the samples, including shader programs, textures, Lua script files, and geometric models. The paths in the project files are set up to be relative, so the files in this data directory are available to all of the projects. Any new data files should be added to these directories in the appropriate place. Because of this data folder layout, you must provide the files in a similar manner when running them outside of the Visual C++ IDE. If you want to directly run the ***.EXE**, you should copy it to the **Application/Bin** directory, which will provide the same reference paths.

Compilation and Deployment

Compiling a Hieroglyph 3 application can be done easily through Visual Studio. Simply open the Hieroglyph 3 solution, then select Build > Build Solution from the window menu. This will cause the Hieroglyph 3 project and all application projects to be compiled and linked. For debugging, projects should be built with the Debug solution configuration, which disables optimizations and produces full debugging symbols. This configuration can be selected from the left-most drop-down box in the toolbar beneath the window menu. To enable full optimizations, select the Release configuration instead. To debug an application, right-click on the project for the desired application and select Set as StartUp Project. Then select Debug >Start Debugging from the window menu to start the application in the debugger.

If a Hieroglyph 3 application will be deployed to a non-development computer, it is important to distribute not only the compiled Release-mode executable and shader/texture/model content, but also to ensure that DLL dependencies are installed on the host PC. By default, a Hieroglyph 3 application will have two such dependencies: the DirectX End-User Runtime, and the Visual C++ 2008 Runtime.

There are two options for distributing the DirectX runtimes: the Web installer, and the offline installer. The Web installer is available from the DirectX Developer Center, and will automatically download and install the latest DirectX components on the host PC. The offline installer is also available from the DirectX Developer Center, or can be found in the **Redist** folder of the DirectX SDK install folder. The offline installer works a bit differently, as it selectively installs components based on the .cab files present when running the installer. Consequently, a developer can choose to only include the files for the components actually used by an application, and thus minimize the size of an installer package or **.zip** file. If compiled using the June, 2010 DirectX SDK, by default, a Hieroglyph 3 application will only require the following files to be included:

- dsetup.dll

- dsetup32.dll

- dxdllreg_x86.cab

- dxsetup.exe

- dxupdate.cab

- Jun2010_d3dx10_43_x86.cab

- Jun2010_d3dx11_43_x86.cab

- Jun2010_D3DCompiler_43_x86.cab

Note that if an application is compiled as 64-bit rather than 32-bit file (and linked against the corresponding 64-bit versions of the DirectX libraries) the **x64** .cab files should be included instead. Also, if any additional DirectX components such as XInput or XAudio are used, the corresponding **.cab** files should also be included. Also note that while it is possible to manually redistribute the appropriate DLLs to the host PC, doing so is against the End User License Agreement provided with the DirectX SDK. Thus, the provided redistributable installers should be used instead. Additional information about redistributing DirectX components can be found at the following URL: http://msdn.microsoft.com/en-us/library/ee418267%28v=vs.85%29.aspx#DirectX_Redistribution.

For the Visual C++ 2008 Runtime, the corresponding DLLs can be deployed manually or by using the provided redistributable package. To deploy manually, copy the files found in the **VC\redist\x86\Microsoft.VC90.CRT** folder of the Visual Studio 2008 install directory to the folder on the host PC where the executable will be located. The redistributable package is a standard installer that automatically installs and registers the CRT DLLs, and can be found on the Microsoft Download Center at the following URL: http://www.microsoft.com/downloads/en/details.aspx?FamilyID=9b2da534-3e03-4391-8a4d-074b9f2bc1bf&displaylang=en. Note that the redistributable only installs the Release version of the runtime, so if there is a need to deploy the Debug version of an application (for example, to initiate a remote debugging session), the manual deployment method must be used. However, if only the Release version is required, it is recommended that the redistributable be used instead, to avoid potential complications. Also, note that if Service Pack 1 for Visual Studio is installed, there is a newer version of the Visual C++ redistributable that can be found here: http://www.microsoft.com/downloads/en/details.aspx?familyid=ba9257ca-337f-4b40-8c14-157cfdffee4e&displaylang=en. In addition, if an application is built as 64-bit instead of 32-bit, the corresponding x64 redistributable must be used, instead. For manual deployment, use the files found in **VC\redist\amd64\Microsoft.VC90. CRT**. For more details on Visual C++ deployment, consult the MSDN documentation: http://msdn.microsoft.com/en-us/library/ms235299%28v=VS.90%29.aspx.

Source Code for This Book

As described above, the sample programs in this book have been added to the Hieroglyph 3 project as sample applications. To minimize the amount of discussion on the applications themselves, we have focused more on the relevant code at hand, instead of directly referencing which sample application is being used to demonstrate an algorithm. However, as an aid to the reader, we provide the following reference information to link the source code to the appropriate portion/section of the book.

Sample Code Reference

Each chapter in the book included references to the sample code covered, but for completeness, the following table contains a cross reference for finding the appropriate portion of the library for each chapter.

Chapter	Title	Section	Sample Project Name
1	Overview of Direct3D 11	NA	NA
2	Direct3D 11 Resources	all sections	(built into Hieroglyph 3)
3	The Rendering Pipeline	all sections	(built into Hieroglyph 3)
4	The Tessellation Pipeline	Parameters	TessellationParams

5	The Computation Pipeline	all sections	(built into Hieroglyph 3)
6	High Level Shading Language	NA	NA
7	Multithreaded Rendering	all sections	(built into Hieroglyph 3)
8	Mesh Rendering	all sections	SkinAndBones
9	Dynamic Tessellation	"Interlocking Terrain Tiles"	InterlockingTerrainTiles
		"Curved Point Normal Triangles"	CurvedPointNormalTriangles
10	Image Processing	all sections	ImageProcessor
11	Deferred Rendering		DeferredRendering LightPrePass
12	Simulations	"Water Simulation Particle Systems"	WaterSimulationI ParticleStorm
13	Multithreaded Paraboloid Rendering	all sections	MirrorMirror

Table A.1. Sample Code Reference.

B

Appendix B:
Direct3D 11 Queries

Throughout this book, we have explored in great detail the operations that Direct3D 11 makes available to a developer. This includes the ability to manipulate and control the individual actions carried out by modern high-performance GPUs. Since the GPU has evolved into such a sophisticated and specialized processor, there are many instances where the biggest bottleneck in an application is introduced when the CPU requires data that forces the GPU to stop what it is doing to respond to CPU's request. For example, if the CPU requests that a buffer be mapped into system memory for reading, that buffer cannot be used by the GPU until the CPU releases it. This can lead to situations where the CPU is waiting for a response from the GPU, or vice versa. In either case, one of the two processing units is considered to be "stalled," since it is not performing any useful operations while it is waiting on the other.

Direct3D 11 provides a mechanism with which some of these stalling situations can be avoided, or at least minimized. There is an interface that provides asynchronous access to some pipeline information and operations that the CPU can use without the GPU having to halt its operation and respond. This interface is the ID3D11Query interface. We will take a closer look at this interface while investigating how it operates, and finally, will consider a few sample usages for this functionality.

Using Query Interfaces

As mentioned above, the ID3D11Query interface provides the ability to acquire data asynchronously from the GPU. This means that the CPU can request information from the GPU about the operations performed between a beginning and ending point in a sequence of API calls, and the driver will respond as soon as the information is available. By allowing the driver to respond at a later point, the CPU can continue and check the query result later on, after the GPU has finished the tasks that the query is concerned with. This provides a communication mechanism between the two processing units that can be used in a number of different ways, which we will explore in the following sections.

As mentioned above, various different types of queries can be performed. However, only a single interface is used to carry out the queries—ID3D11Query. The individual types of queries are selected during the creation of the query. As with all objects in Direct3D 11, the query objects are created with a method from the device—the ID3D11Device::CreateQuery() method. This method takes a pointer to a description structure, plus a pointer to the returned query interface. In the case of the query description, the structure contains only two variables. The description structure and its subelements are shown in Listing B.1.

```
struct D3D11_QUERY_DESC {
  D3D11_QUERY Query;
  UINT        MiscFlags;
};
enum D3D11_QUERY {
  D3D11_QUERY_EVENT,
  D3D11_QUERY_OCCLUSION,
  D3D11_QUERY_TIMESTAMP,
  D3D11_QUERY_TIMESTAMP_DISJOINT,
  D3D11_QUERY_PIPELINE_STATISTICS,
  D3D11_QUERY_OCCLUSION_PREDICATE,
  D3D11_QUERY_SO_STATISTICS,
  D3D11_QUERY_SO_OVERFLOW_PREDICATE,
  D3D11_QUERY_SO_STATISTICS_STREAM0,
  D3D11_QUERY_SO_OVERFLOW_PREDICATE_STREAM0,
  D3D11_QUERY_SO_STATISTICS_STREAM1,
  D3D11_QUERY_SO_OVERFLOW_PREDICATE_STREAM1,
  D3D11_QUERY_SO_STATISTICS_STREAM2,
  D3D11_QUERY_SO_OVERFLOW_PREDICATE_STREAM2,
  D3D11_QUERY_SO_STATISTICS_STREAM3,
  D3D11_QUERY_SO_OVERFLOW_PREDICATE_STREAM3
}
enum D3D11_QUERY_MISC_FLAG {
  D3D11_QUERY_MISC_PREDICATEHINT
}
```

Listing B.1. The D3D11_QUERY_DESC structure and its components.

From Listing B.1, we can see that the primary purpose of this description structure is to select the type of the query, with the D3D11_QUERY enumeration providing the available choices. There are queries for acquiring pipeline event and timing information, for getting statistics about the pipeline operation over the selected sequence of API calls, and for a number of queries related to each of the stream output streams (making a total of four different streams to query). In addition to the query type, the D3D11_QUERY_DESC structure also provides a predicate hint option in the D3D11_QUERY_MISC_FLAG enumeration, which will be discussed briefly in the predicated rendering section.

We will provide an example of using the pipeline statistics query, since this is generally a useful tool for developers. The sequence that must be followed to query the pipeline statistics (as one might expect) is as follows:

- Create the query object.

- Begin the query.

- End the query.

- Retrieve the results.

As mentioned above, the query is created through the ID3D11Device::CreateQuery() method. Since we want to query the pipeline statistics, we would fill in the description structure accordingly and pass it to the create query method. After the query has successfully been created, we will use the ID3D11DeviceContext interface to begin and end the query. The methods ID3D11DeviceContext::Begin() and ID3D11DeviceContext::End() will mark the beginning and the end of a query, where all of the corresponding API calls between these two methods will affect the outcome of the query. It is important to note that the semantics for when to call the begin and end methods depend on the type of query being performed, and that some queries don't even require the call to begin.

After the query sequence has been indicated, the application checks on the status of the query by calling the ID3D11DeviceContext::GetData() method and passing in the desired query. This method may return S_OK or S_FALSE, depending on if the query has completed executing. If it returns S_OK, the data returned in the user-supplied pointers to the GetData method will contain the results of the query. Otherwise, the query is still being executed, and the CPU must check back at a later time for the results. The actual data that is returned will vary by query type. In the case of the pipeline statistics, the result will be a D3D11_QUERY_DATA_PIPELINE_STATISTICS structure, which is shown in Listing B.2.

```
struct D3D11_QUERY_DATA_PIPELINE_STATISTICS {
    UINT64 IAVertices;
    UINT64 IAPrimitives;
    UINT64 VSInvocations;
    UINT64 GSInvocations;
    UINT64 GSPrimitives;
    UINT64 CInvocations;
    UINT64 CPrimitives;
```

```
    UINT64 PSInvocations;
    UINT64 HSInvocations;
    UINT64 DSInvocations;
    UINT64 CSInvocations;
}
```

Listing B.2. The D3D11_QUERY_DATA_PIPELINE_STATISTICS structure.

Each of the members of this structure provides information on the general statistics
for each stage of the pipeline. This can be used to understand how the pipeline is performing, including what effect LOD changes have, as well as debugging the pipeline when the
results of a rendering operation are not as expected. The other query types will provide
different result types, so the developer must check which data type to supply in order to
receive the appropriate results. After the query has been used, it can be released in the typical COM manner.

Predicated Rendering

Some of the query types allow the application to determine if a particular draw call will
affect the current render target, if it were to be executed. In particular, the D3D11_QUERY_
OCCLUSION query type returns the number of pixels that pass the depth and stencil tests for
a particular rendering sequence. This is typically used to determine if a draw call will produce a significant amount of visual difference before it is rendered. Similarly, the D3D11_
QUERY_OCCLUSION_PREDICATE query type performs the same test, except that it returns a
Boolean answer to the query, instead of a number of fragments.

The occlusion predicate actually uses a subclass of the ID3D11Query interface—
the ID3D11Predicate interface. This special query type must be created with the
ID3D11Device::CreatePredicate() method, in a fashion similar to the generic query
technique shown above.

This predicated version of the query allows for a special method of rendering, called
predicated rendering. Essentially, the query is started with the call to Begin, and then
any objects being rendered will count toward the occlusion query until the End method is
called. Once this predicate query has been started, the application can bind it to the pipeline
as an indication of whether or not subsequent API calls should take effect or not. This is
done through the ID3D11DeviceContext::SetPredication() method, which also takes
a Boolean parameter to allow selection of how the predicate is interpreted. If no pixels pass
the depth and stencil tests, the predicate will return false. If true is passed in this parameter, the subsequent rendering calls will not be executed when the predicate returns false.

The predication of the subsequent rendering calls can be disabled again by setting a NULL value with the SetPredication() method.

This allows for the application to create a predicate query for an object, and then only render that object if it will affect the output render target. This is an optimization technique that reduces the amount of unneeded processing in a very complex scene.

Appendix C: Tessellation Summary

Chapter 4 gave an overview of the parameters used to control fixed-function tessellation, as well as the entry points where you can provide wholly custom HLSL code. Chapter 9 took this introduction to tessellation and demonstrates how to apply this powerful Direct3D 11 feature to real-world problems.

For convenience, this appendix is included to summarize the key parameters and considerations for using this new pipeline feature.

Summary of Tessellation Parameters

When using tessellation, consider the following:

1. There are three basic primitive types that can be tessellated:

    a. Line

    b. Triangle

    c. Quad

2. There are two independent axes of tessellation:

    a. Edge

    b. Inside

3. The fixed function tessellator determines the final triangles that are rasterized. Its partitioning method can be set to control how it interprets the parameters you provide:

    a. `integer`

    b. `pow2`

    c. `fractional_odd`

    d. `fractional_even`

Some key considerations when implementing a tessellation algorithm:

1. Water-tightness is important—when tessellating edges, always consider how neighboring geometry will be tessellated as well.

2. Temporal changes can look ugly—take care to avoid popping, and consider using morphing techniques for smooth transitions.

3. Varying the tessellation factors is your key to scaling quality versus performance.

The main steps in implementing tessellation are:

1. Provide a control mesh: the vertex and index buffers provided by your application through the API now define control points, and not individual vertices specifying each individual triangle.

2. Write a hull shader: create a custom HLSL shader that processes the control mesh for two purposes: first to provide parameters to the fixed-function tessellator, and second, to provide appropriate inputs to the domain shader.

3. Write a domain shader: another custom HLSL shader that uses individual sample locations generated by the fixed-function tessellator, along with algorithm-specific outputs from the hull shader to construct the final triangles to be sent to a geometry shader (optional) and rasterized as part of the final image.

Bibliography

Akenine-Moeller, T. a. (2002). Real-Time Rendering 2nd Ed. Natick, MA: A K Peters, Ltd.

AMD & Microsoft. (n.d.). *PNTriangles11*. Retrieved from Microsoft Developer Network (MSDN): http://msdn.microsoft.com/en-us/library/ee416573(VS.85).aspx

ATI. (2001). *TruForm Resources*. Retrieved from AMD Developer Central: http://developer.amd.com/archive/gpu/trueform/Pages/default.aspx

Baker, S. (n.d.). Learning to Love your Z-buffer. http://www.sjbaker.org/steve/omniv/love_your_z_buffer.html.

Balestra, C., & Engstad, P.-K. (2008). The Technology of Uncharted: Drake's Fortune. *Game Developers Conference 2008*.

Biri, V., Herubel, A., & Deverly, S. (2010). Practical morphological antialiasing on the GPU. *SIGGRAPH 2010: Games & Real Time*. ACM.

Brabec, S. A. (2002). Shadow mapping for hemispherical and omnidirectional light sources. *Proceedings of Computer Graphics International*.

Catmull, E., & Clark, J. (1978). Recursively generated B-spline surfaces on arbitrary topological meshes. *Computer-Aided Design*, 350–355.

Eberly, D. H. (2007). *3D Game Engine Design, Second Edition*. San Francisco: Morgan Kaufmann Publishers.

Engel, W. (2009). *ShaderX 7: Advanced Rendering Techniques*. Boston, MA: Course Technology PTR.

Engel, W. (2009). The Light Pre-Pass Renderer: Renderer Design for Efficient Support of Multiple Lights. *SIGGRAPH 2009: Advances in Real-Time Rendering in 3D Graphics and Games*. ACM.

Everitt, C. (2001). *Interactive Order-Independent Transparency*. Nvidia Corporation.

Fatahalian, K. (n.d.). *From Shader Code to a Teraflop: How GPU Shader Cores Work.* Retrieved January 16, 2011, from http://bps10.idav.ucdavis.edu/talks/03-fatahalian_gpuArchTeraflop_ BPS_SIGGRAPH2010.pdf

Goldberg, D. (1991). What Every Computer Scientist Should Know About Floating-Point Arithmetic. *ACM Computing Surveys Volume 23 Issue 1.*

Gonzalez, R. C. (2008). *Digital Image Processing, 3rd Edition.* Upper Saddle River, NJ: Pearson Prentice Hall.

Hable, J. (2010). Uncharted 2: HDR Lighting. *Game Developers Conference 2010.*

Heidrich, W. a.-H. (1998). View independent environment maps. *Proceedings of the ACM SIGGRAPH/EUROGRAPHICS workshop on Graphics hardware*, 39-ff.

Hoxley, J. (n.d.). *Programming Vertex, Geometry, and Pixel Shaders.* Retrieved January 16, 2011, from GameDev.net: http://wiki.gamedev.net/index.php/D3DBook:Table_of_Contents#Lighting

Kaplanyan, A. (2010). Cascaded Light Propagation Volumes for Real-time Indirect Illumination. *ACM SIGGRAPH Symposium on Interactive 3D Graphics and Games*, 99–107.

Kemen, B. (2009). Floating Point Depth Buffers. *Gamasutra*, http://www.gamasutra.com/blogs/ BranoKemen/20091231/3972/Floating_Point_Depth_Buffers.php.

Kircher, S., & Lawrance, A. (2009). Inferred Lighting: Fast dynamic lighting and shadows for opaque and translucent objects. *SIGGRAPH 2009: Game Paper Proceedings.* ACM.

Lauritzen, A. (2010). Deferred Rendering for Current and Future Rendering Pipelines. *SIGGRAPH 2010: Beyond Programmable Shading.* ACM.

Lee, M. (2009). Pre-lighting in Resistance 2. *Game Developers Conference 2009.*

Lengyel, E. (2001). Computing Tangent Space Basis Vectors for an Arbitrary Mesh. *Terathon Software 3D Graphics Library*, http://www.terathon.com/code/tangent.html.

Lombard, Y. (n.d.). Retrieved January 16, 2011, from www.andyc.org: http://www.andyc.org/lecture/ viewlog.php?log=Realistic%20Water%20Rendering,%20by%20Yann%20L

Lorensen, W. E. (1987). Marching Cubes: A high resolution 3D surface reconstruction algorithm. *SIGGRAPH '87 Proceedings of the 14th annual conference on Computer graphics and interactive techniques*, 163–169.

Microsoft Corporation. (n.d.). *Process and Thread Functions.* Retrieved January 16, 2011, from MSDN: http://msdn.microsoft.com/en-us/library/ms684847(v=vs.85).aspx

Mikkelsen, M. S. (2010). *Skin Rendering by Pseudo-Separable Cross Bilateral Filtering.* Naughty Dog, Inc.

Mittring, M. (2009). A bit more Deferred. *Triangle Game Conference.* Crytek.

Nguyen, H. (2008). *GPU Gems 3.* Upper Saddle River, NJ: Addison-Wesley.

O'Brien, J. a. (1995). Dynamic Simulation of Splashing Fluids. *Proceedings of Computer Animation Conference*, 198–205, 220 .

OpenMP Architecture Review Board. (n.d.). *OpenMP*. Retrieved January 16, 2011, from http://www .OpenMP.org: http://openmp.org/wp/

Paris, S. K. (n.d.). *A Gentle Introduction to Bilateral Filtering and its Applications*. Retrieved February 20, 2011, from http://people.csail.mit.edu/sparis/bf_course/

Persson, E. (2007). Deep deferred shading. *Humus*, http://www.humus.name/index .php?page=3D&ID=75.

Persson, E. (2008). Deferred shading 2. *Humus*, http://www.humus.name/index.php?page=3D&ID=81.

Persson, E. (2008). Post-tonemapping resolve for high quality HDR antialiasing in D3D10. In W. Engel, *ShaderX6: Advanced Rendering Techniques*. Charles River Media.

Persson, E. (2009). A couple of notes about Z. *Humus*, http://www.humus.name/index .php?page=News&ID=255.

Policarpo, F., & Fonseca, F. (2005). *Deferred Shading Tutorial*. CheckMate Games.

Reshetov, A. (2009). Morphological Antialiasing. *Proceedings of the 2009 ACM Symposium on High Performance Graphics*. ACM.

Saito, T., & Takahashi, T. (1990). Comprehensible rendering of 3D shapes. *SIGGRAPH '90 Proceedings of the 17th annual conference on Computer graphics and interactive techniques* (pp. 197–206). New York, NY: ACM.

Snook, G. (2001). Simplified Terrain Using Interlocking Tiles. In M. DeLoura, *Game Programming Gems 2*. Charles River Media.

Spherical Coordinates. (n.d.). *Wolfram MathWorld*, http://mathworld.wolfram.com/ SphericalCoordinates.html.

Sutter, H. (n.d.). *The Free Lunch Is Over: A Fundamental Turn Toward Concurrency in Software*. Retrieved January 16, 2011, from www.GotW.ca: http://www.gotw.ca/publications/ concurrency-ddj.htm

Tomasi, C. a. (1998). Bilateral Filtering for Gray and Color Images. *Sixth International Conference on Computer Vision*, 839–846.

Valient, M. (2007). Deferred Rendering in Killzone 2. *Develop Conference*. Brighton, UK.

Valve Software. (n.d.). *Source Engine Features*. Retrieved from The Valve Developer Community: http://developer.valvesoftware.com/wiki/Source_Engine_Features

Vlachos, A. e. (2001). *Curved PN Triangles*.

Vlachos, A., Peters, J., Boyd, C., & Mitchell, J. L. (2001). *Curved PN Triangles*.

Weisstein, E. W. (n.d.). *NURBS Surface*. Retrieved from Wolfram MathWorld: http://mathworld .wolfram.com/NURBSSurface.html

Williams, L. (1978). Casting curved shadows on curved surfaces. *SIGGRAPH '78 Proceedings of the 5th annual conference on Computer graphics and interactive techniques*, 270–274.

Wilson, P. (2008). G-Buffer Normals and Trig Lookup Textures. *Torque Games Blog*, http://www
.garagegames.com/community/blogs/view/15340.

Zink, J. (n.d.). *Programming Vertex, Geometry, and Pixel Shaders*. Retrieved January 16, 2011,
from GameDev.net: http://wiki.gamedev.net/index.php/D3DBook:Single_Pass_Environment_
Mapping

Zink, J., Hoxley, J., Engel, W., Kornmann, R., & Suni, N. (n.d.). *Programming Vertex, Geometry,
and Pixel Shaders*. Retrieved January 16, 2011, from GameDev.net: http://wiki.gamedev.net/
index.php/D3DBook

Index

Printed and bound by CPI Group (UK) Ltd, Croydon, CR0 4YY

24/10/2024

01778294-0001